Often Used Convers

	To convert from	To	Multiply by
Length	feet	meters	3.048×10^{-1}
	inches	centimeters	2.54
	miles (statute)	kilometers	1.609
Mass	pounds (avdp)	kilograms	4.536×10^{-1}
	tons (US short)	kilograms	9.072×10^{2}
	tons (US long)	kilograms	1.106×10^{3}
	tonnes (metric)	kilograms	1.000×10^{3}
Time	years (sidereal)	seconds	3.156×10^{7}
Area	hectares	square meters	1.000×10^{4}
	hectares	square kilometers	1.000×10^{-2}
	acres	square meters	4.047×10^{3}
	square miles	square kilometers	2.590
Volume	liters	cubic meters	1.000×10^{-3}
	liters	cubic decimeters	1.000
	gallons (US liquid)	cubic meters	3.785×10^{-3}
	barrels (US liquid)	cubic meters	1.192×10^{-1}
	barrels (US dry)	cubic meters	1.156×10^{-1}
	barrels (US petroleum)	cubic meters	1.590×10^{-1}
Force	dynes	newtons	1.000×10^{-5}
Pressure	atmospheres	pascals	1.013×10^{5}
	mm Hg (torr)	pascals	1.333×10^{2}
	pounds per square inch (psi)	pascals	6.893×10^{3}
Viscosity	poises	pascal seconds	1.000×10^{-1}
Energy	British thermal units (Btu)	joules	1.055×10^{3}
	ergs	joules	1.000×10^{-7}
	calories	joules	4.184
	Calories (nutritional)	joules	4.184×10^{3}
	kilowatt hours	joules	3.600×10^{6}
	Q (10^{18} Btu)	terajoules	1.055×10^{9}
Power	horsepower	watts	7.457×10^{2}

* The heat content of various fuels is given on the back endpaper.

Environmental Chemistry

Environmental Chemistry

John W. Moore
Eastern Michigan University

Elizabeth A. Moore

ACADEMIC PRESS New York San Francisco London

A Subsidiary of Harcourt Brace Jovanovich, Publishers

ACADEMIC PRESS, INC.
111 Fifth Avenue, New York, New York 10003

United Kingdom Edition published by
ACADEMIC PRESS, INC. (LONDON) LTD.
24/28 Oval Road, London NW1

Library of Congress Cataloging in Publication Data

Moore, John W
    Environmental chemistry.

    Bibliography: p.
    Includes index.
    1.   Environmental chemistry.   I.   Moore, Eliza-
beth A., joint author.   II.   Title.
QD31.2.M63   540   75-26348
ISBN 0−12−505050−X

Contents

v

Preface

Precisely what is meant by "environmental chemistry" is not especially easy to define, and there is really no consensus as to what should be included in this kind of textbook. Specific environmental problems such as photochemical smog, the degradation of stratospheric ozone, or mercury pollution of natural waters require the expertise of a variety of disciplines to determine appropriate approaches to their alleviation, but there is no doubt that chemistry and chemists must play a central role if successful approaches are to be found.

This book is predicated upon a rather broad interpretation of what is meant by environmental chemistry. Topics such as the origin and evolution of the environment, the energy crisis, mineral resources, solid wastes, recycling, and the effects of foreign substances on living systems have been added to the more conventional environmental chemistry relating to air and water pollution. An attempt has been made to apply chemical principles to all of these problems and to show the interrelationships among them. Often this has been done by choosing as examples specific problems of the type already mentioned and deriving from them general characteristics and principles.

One consequence of this approach is the inclusion of a large number of cross-references from one section of the text to another. Another is the numerous problems at the end of each chapter, many of which attempt to test the student's understanding of the applicability and interrelationships of the underlying general principles. Solutions to the problems have not been included, but when necessary information cannot be found within the text itself references have usually been given to outside sources for helpful facts or analysis. It is hoped that the interested reader will make extensive use of these references since other relevant information is also available in many of them.

The broad approach to environmental chemistry employed has evolved from the syllabus of a course first taught at Eastern Michigan University during the summer of 1972. The minimum prerequisite was one year of general chemistry, and students were strongly advised to complete quantitative analysis and/or a semester of organic chemistry prior to enrollment. Since no textbooks were available in this "intermediate" environmental chemistry category at the time the syllabus was developed, handouts of lecture notes were provided, and considerable reliance was placed on assigned readings in advanced treatises.

Because of its popularity among chemistry, biology, and other science majors, ranging from sophomore to graduate levels, the course has been taught at least once a year since 1972. Each such offering has provided further opportunities to refine the presentation of the material and to obtain feedback from students regarding the usefulness to them of including a broad range of topics. The requirement of a general chemistry prerequisite and recommendation of analytical and organic courses has allowed the subject matter to be treated in sufficient depth to bridge the gap between the often superficial texts intended for nonscience majors and the highly specialized treatises found on the bookshelves of professional chemists who are attacking environmental problems. Despite the breadth of coverage, many seniors and graduate students were challenged by the course because it included aspects of applied chemistry that they had not become aware of through the conventional undergraduate curriculum. A very important aspect of this text is its attempt to provide such students with a broad overview of the applications of chemical facts and principles. It is hoped that this will prove useful when choices relating to specialized areas for research and/or employment opportunities must be made.

Two other important threads which bind the book together are evolution (chemical, geological, and biological) and an attempt to develop what economist Kenneth Boulding has called "generalized eyes and ears." To this latter might be added "pens" since the discoverer of new knowledge has some obligation to disseminate it in ways that are accessible to others who might use it. Many environmental problems have had their genesis in the failure of decision-makers to consider all aspects of a problem and the failure of specialists with relevant information to make it available or understandable. The need for improved communication and broadened perspectives are obvious when environmental problems are studied. These aspects of the course have been discussed in more detail elsewhere [see John W. Moore, *J. Coll. Sci. Teaching* **4**(5), 319–321 (May 1975)].

The Suggested Reading lists at the ends of Parts I through VII should prove extremely useful to both students and teachers. The literature of environmental chemistry is diverse, and much important information is to be found in government documents and research reports which are less accessible than the journals familiar to most chemical researchers. We have attempted to collect the most useful references from all sources and to indicate their level and contents as accurately and as briefly as possible. For those who wish to delve further we suggest a study of the Appendix which surveys sources available in the environmental literature.

The International System (SI) of units, as described by Martin A. Paul, *J. Chem. Doc.* **11**, 3–8 (1971), has been adopted throughout the text. A summary of SI units and conversion factors appears on the front endpapers. This uniform and consistent set of units is especially helpful in the discussions of energy (Part II). The nomenclature *dioxygen* has been used throughout to distinguish gaseous O_2 from atomic or other forms of the element, and the prefix di has been attached in the case of other diatomic molecules as well. In many cases this aids in avoiding ambiguity, and we hope it will not appear unusual to our readers.

We firmly believe that many of the problems, techniques, and approaches to solutions discussed are of profound importance to both chemists and other scientists, as well as to the public at large. We hope that our readers will agree, and that they will give serious consideration to what contributions they can make toward the solution or abatement of particular environmental problems.

John W. Moore
Elizabeth A. Moore

Acknowledgments

A number of persons have made contributions to this book. Ronald W. Collins, Clark G. Spike, and K. Rengan, as well as other colleagues in the Department of Chemistry at Eastern Michigan University have provided valuable support and information. Mrs. Margaret Schultz cheerfully typed much of the manuscript on stencils and ran off copies for student use. Professors Henry A. Bent, John Hayes, Henry Foth, Margaret F. Fels, Roger Minear, and Dr. George L. Waldbott read and provided useful comments on portions of the manuscript. Professors Amos Turk and Barry Huebert each provided a more extensive review. Mr. Mark Riddle read and commented on an early version of the manuscript, and many of his suggestions were incorporated in its final form. Students who have used the materials in Eastern Michigan classrooms since 1972 have corrected errors and helped us clarify our prose.

One of us (JWM) is grateful for a fellowship in the Eastern Michigan University Center for the Study of Contemporary Issues during 1974–1975 which allowed additional time for writing and permitted numerous contacts with others in the environmental field. We gratefully acknowledge the library sources of Eastern Michigan University and the University of Michigan upon which we heavily relied for references and reports.

Special thanks are due Professor Theodor Benfey, editor of *Chemistry,* for permission to reproduce in its entirety the poem "Necessary Ethic" by Henry A. Bent which appears at the beginning of Part VII. Parts of Section 17.6 in expanded form have also appeared in *Chemistry* [**48**(6), June 1975] under the title "The Vinyl Chloride Story." We are also grateful for the permissions granted by other publishers to reproduce several of the figures and tables. Each of these is indicated in the text.

Finally, we wish to thank the staff of Academic Press for making our first attempt at textbook authorship proceed so smoothly.

By acknowledging contributions of others we do not intend to shift responsibility for any errors or omissions which remain. We are aware that in such a rapidly developing field no textbook can hope to be completely up-to-date. We hope that readers who detect errors or misstatements will be good enough to inform us of them so that they can be corrected in future printings.

I
Origins

What's past is prologue.

Shakespeare

The Present is the living sum total of the whole Past.

Thomas Carlyle

THE THREE CHAPTERS THAT CONSTITUTE PART I of this book are concerned with the application of a number of chemical subdivisions—nuclear chemistry, spectroscopy, acid-base theory, electrochemistry, thermodynamics, and biochemistry—to the understanding of the truly ancient history of the chemical elements, the earth, and the biosphere. At first glance such material may seem irrelevant, since environmental chemists are primarily concerned with today's problems and their effects on the future. On closer examination, however, many of the facts and principles contained in these three chapters are intimately related to more specific environmental problems which will be dealt with later on.

Chemistry can be applied fruitfully to many more than just those carefully defined problems encountered in the laboratory, and indeed it must be so applied if environmental problems are to be attacked successfully. The reader of this part of the book should constantly bear in mind that the techniques and philosophy of those chemists who have studied the origins of the earth and of life are highly useful in the evaluation of environmental threats. In both cases a broad knowledge is required to devise plausible scenarios or hypotheses which can be tested by application of scientific facts and logic.

Finally, it is the purpose of these three chapters to evoke an appreciation of the slow evolutionary development of the biosphere and its myriad, interconnected components. Only recently has any species achieved the levels of population and technology which today enable mankind to alter this evolutionary process significantly, rapidly, and on a global scale.

1

Origin of the Elements and Interstellar Molecules

*We are brothers to the wind, the sun, the stars—
and perhaps to more.*

St.-John Perse

There is considerable (but far from conclusive[1]) evidence that the universe originated in a giant fireball which emerged from an infinitely dense collection of neutrons between ten and twenty billion years ago. This so-called "big bang" resulted in the formation of hydrogen and some helium nuclei from which galaxies and stars eventually evolved. It also is assumed to be responsible for the observation that the universe seems continually to be expanding and for the presence of background microwave radiation in outer space. The big bang began a process of evolution which continues to this day and seems capable of going on many billions of years into the future. The origin of the earth and its biosphere represent only a minor part of this universal evolution, but one which it is useful for us to examine in some detail.

1.1 NUCLEOSYNTHESIS OF THE ELEMENTS

Ninety chemical elements (those up through $_{92}U$ excepting $_{43}Tc$ and $_{61}Pm$) have been discovered in the substances that make up the crust and

---

[1] D. W. Sciama, "Modern Cosmology." Cambridge Univ. Press, London and New York, 1971.

3

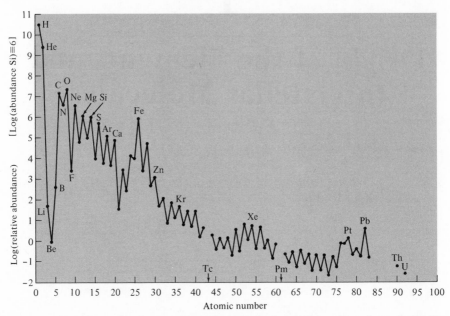

Figure 1.1 Relative abundances of the elements in the universe. Data from A. G. W. Cameron, A critical discussion of abundance of nuclei. *In* "Explosive Nucleosynthesis" (D. N. Schramm and W. D. Arnett, eds.), p. 3. Univ. of Texas Press, Austin, 1973.

atmosphere of the earth. Fifteen more have been synthesized through man's ingenuity or discovered elsewhere in the universe. From these fundamental elements all of the compounds and mixtures that make up man's environment, and indeed man himself, can be obtained by appropriate combinations. Today most astrophysicists accept the theory that elements heavier than hydrogen and helium have been synthesized from these remnants of the big bang by nuclear reactions in stars. At least two lines of evidence support such a view. First, hydrogen and helium are the most abundant elements in the universe (see Figure 1.1 and Table 1.1). There is nearly ten times as much hydrogen as helium in most stars, and the abundance of oxygen (which is third on the list) is less than a hundredth that of helium. Second, our knowledge of nuclear reactions is extensive enough so that reaction sequences or mechanisms can be devised which account for the production of the elements in the relative abundances shown in Figure 1.1.

According to astrophysical theory, the life history of a star might go something like this: In some region of the universe the densities of hydrogen, helium, and other atoms or molecules are increased by random processes to the point where gravitational attraction causes a slow contraction to begin. As gravitational potential energy is converted into kinetic energy the velocities of the atoms increase and the temperature of the region rises along with the density of matter. When the temperature

TABLE 1.1   Elemental Composition, by Percent of Total Atoms, of Selected Systems[a]

Earth crust		Meteorite		Human body		Universe[b]	
O	46.6	O	32.3	H	63	H	93.3
Si	27.2	Fe	28.8	O	25.5	He	6.49
Al	8.13	Si	16.3	C	9.5	O	0.063
Fe	5.00	Mg	12.3	N	1.4	C	0.035
Ca	3.63	S	2.12	Ca	0.31	N	0.011
Na	2.83	Ni	1.57	P	0.22	Ne	0.010
K	2.59	Al	1.38	Cl	0.03	Mg	0.003
Mg	2.09	Ca	1.33	K	0.06	Si	0.003
Ti > H > C				S > Na > Mg		Fe	0.0001

[a] Source: J. Selbin, The origin of the chemical elements. *J. Chem. Educ.* **50**(5), 306 (1973).
[b] Data from Figure 1.1.

reaches 10^7 to 10^8 K, the activation barrier to fusion of hydrogen nuclei is overcome and a nuclear reaction called *hydrogen burning* begins to occur.

$$^1_1H + ^1_1H \longrightarrow ^2_1H + \beta^+ \text{ (positron)} + \nu \text{ (neutrino)}$$
$$^2_1H + ^1_1H \longrightarrow ^3_2He + \gamma \text{ (γ ray)}$$
$$^3_2He + ^3_2He \longrightarrow ^4_2He + 2\,^1_1H$$
$$\overline{4\,^1_1H \longrightarrow ^4_2He + 2\beta^+ + 2\gamma + 2\nu} \qquad (1.1)$$

The activation energy for nuclear reactions such as hydrogen burning is extremely high because two positively charged nuclei must be brought very close together before nuclear forces of attraction (which are large only when two particles are separated by less than 10^{-14} m) become strong enough to fuse the particles together. According to Coulomb's law, the electrostatic force between two particles increases as each of their charges increase and as the square of the distance between them decreases. Two hydrogen nuclei must have very high kinetic energies in order to overcome their mutual repulsion at very short distances. As the charges on the particles increase, the force of repulsion also increases, accounting for the larger activation energies for fusion of nuclei containing larger numbers of protons. Since the nuclear force depends on the total number of nuclear particles (nucleons), the presence of neutrons can often reduce the activation barrier to fusion.

If the star is large enough and the gas from which it condensed contained a small amount of $^{12}_6C$, a catalytic nuclear reaction called the *carbon-nitrogen* (or *Bethe*) *cycle* may be the major pathway for hydrogen burning.

$$^{12}_6C + ^1_1H \longrightarrow \gamma + ^{13}_7N \longrightarrow \beta^+ + \nu + ^{13}_6C \qquad (1.2a)$$

$$^{13}_6C + ^1_1H \longrightarrow ^{14}_7N + \gamma \qquad (1.2b)$$

$$^{14}_7N + ^1_1H \longrightarrow \gamma + ^{15}_8O \longrightarrow \beta^+ + \nu + ^{15}_7N \qquad (1.2c)$$

$$^{15}_7N + ^1_1H \longrightarrow ^4_2He + ^{12}_6C \qquad (1.2d)$$

$$\overline{4\,^1_1H \longrightarrow ^4_2He + 3\gamma + 2\beta^+ + 2\nu} \qquad (1.2)$$

In this sequence of reactions unstable nuclides such as $^{13}_{7}\text{N}$ and $^{15}_{8}\text{O}$ decompose rapidly, but stable ones such as $^{13}_{6}\text{C}$, $^{14}_{7}\text{N}$, and $^{15}_{7}\text{N}$ have long enough lifetimes to be hit by a proton. Note that although $^{12}_{6}\text{C}$ is a reactant in the first step of the carbon-nitrogen cycle, it is regenerated in step 1.2d and thus serves as a catalyst.[2] While the hydrogen-burning reactions are occurring the energy released is sufficient to increase the velocities of the atoms to the point where further gravitational collapse is balanced by increased pressure. Usually the star remains stable for billions of years. As the denser, more massive helium nuclei are produced they tend to concentrate toward the center of the star and an inner shell (Figure 1.2) forms which is distinct from the outer, hydrogen-burning region. When a temperature of 10^8 K and a density of about 10^5 g/cm^3 are attained in this inner shell, the activation barrier to *helium burning* is exceeded and helium begins to be converted to carbon and oxygen (see equations 1.3).

$$
\begin{aligned}
^3_2\text{He} + {}^4_2\text{He} &\longrightarrow {}^7_4\text{Be} + \gamma \\
^7_4\text{Be} + {}^1_1\text{H} &\longrightarrow {}^8_5\text{B} + \gamma \\
^8_5\text{B} &\longrightarrow {}^8_4\text{Be} + \beta^+ + \nu \\
^8_4\text{Be} + {}^4_2\text{He} &\longrightarrow {}^{12}_6\text{C} \\
^{12}_6\text{C} + {}^4_2\text{He} &\longrightarrow {}^{16}_8\text{O}
\end{aligned} \tag{1.3}
$$

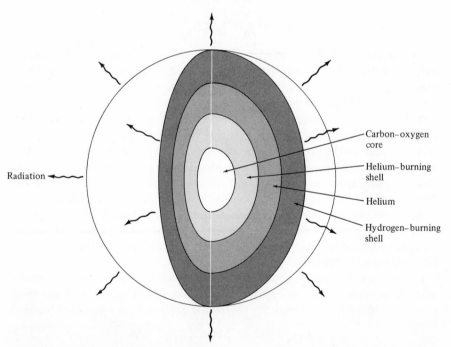

Figure 1.2 Schematic diagram of the structure of a star whose mass is several times that of the sun in the double shell Red Giant phase.

---

[2] Reaction 1.2d produces $\gamma + {}^{16}_{8}\text{O}$ about once in every 2000 times; for further details, see J. Selbin, *J. Chem. Educ.* **50**(6), 380 (1973).

If the star is large enough (about 3.5 or more times the mass of the sun), sufficient gravitational energy will be available to increase the temperature of the inner core of the star and the densities (concentrations) of carbon and oxygen to the point where these heavier nuclei begin to react:

$$
{}^{12}_{6}C + {}^{12}_{6}C \nearrow \begin{matrix} {}^{20}_{10}Ne + {}^{4}_{2}He \\ \text{or} \\ {}^{23}_{11}Na + {}^{1}_{1}H \end{matrix} \qquad (1.4)
$$

$$
{}^{16}_{8}O + {}^{16}_{8}O \longrightarrow \begin{matrix} {}^{28}_{14}Si + {}^{4}_{2}He \\ \text{or} \\ {}^{31}_{15}P + {}^{1}_{1}H \\ \text{or} \\ {}^{31}_{16}S + {}^{1}_{0}n \end{matrix} \qquad (1.5)
$$

Should the temperature reach $\sim 3 \times 10^9$ K, the nuclei would have sufficient kinetic energy to overcome the activation barriers to all nuclear reactions. Under these circumstances an equilibrium should be reached among protons, neutrons, and nuclei. In addition, large amounts of energy would be released.

What nuclei would be produced in the *equilibrium process* described in the preceding paragraph? Obviously those that are most stable. The stability of an atomic nucleus, like the stability of a molecule, depends on the amount of energy required to separate the subatomic particles from which it is made. In the case of a nucleus this quantity is called the *binding energy*. For instance, in the case of one mole of the isotope of iron having mass number 56 there is an increase in mass of $26(1.00783) + 30(1.00867) - 55.93493 = 0.52875$ g when all of the nucleons are separated.

$$
{}^{56}_{26}Fe \longrightarrow 26\ {}^{1}_{1}p + 30\ {}^{1}_{0}n \qquad (1.6)
$$

Using Einstein's equation $E = mc^2$ to convert this to energy units, 0.52875×10^{-3} kg $\times (2.99792 \times 10^8$ m/s$)^2 = 4.75215 \times 10^{13}$ J. The complete dissociation of one mole of iron nuclei requires a very large quantity of energy since a considerable increase in mass occurs. The stability of ${}^{54}_{26}Fe$ relative to other nuclei may be confirmed by working Exercise 1.

Since different nuclei contain different numbers of nucleons, comparisons of stabilities must determine how tightly each nucleon is held. Thus, binding energies are usually normalized by dividing by the number of nucleons. In Fig. 1.3 the binding energy per nucleon is plotted versus mass number for the most stable isotope of each element. The zero point on the energy axis has been taken as the energy of completely separated protons and neutrons. The minimum in the graph at ${}^{56}_{26}Fe$ represents the maximum binding energy (or dissociation energy), corresponding to the most stable nucleus. In an equilibrium process, where all nuclear reactions can occur at appreciable rates, the energy of the total system would be reduced by

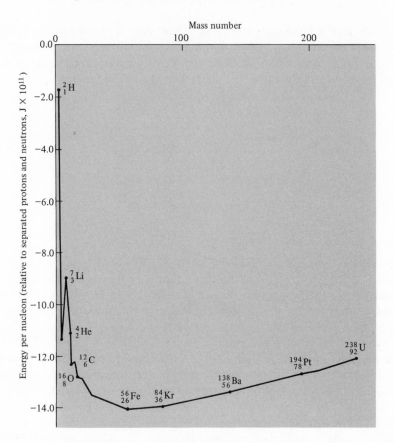

Figure 1.3   Stabilities of atomic nuclei.

moving from either end of the binding energy curve toward the middle. Therefore the nuclei in the vicinity of iron (mass numbers between 46 and 66) should be produced from the heavier as well as the lighter elements. This conclusion raises two important questions: First, why isn't iron the most abundant element in the universe instead of hydrogen; and, second, how are the elements of mass number greater than 66 produced?

Looking back at Fig. 1.1, it is easy to see that there is a hump in the abundance curve at atomic number 26. Iron is the ninth most abundant element. The fact that it is not more abundant is one more argument in favor of the evolutionary hypothesis that the heavier elements are in the process of being built up from hydrogen, and that the process is not yet complete. The previously described equilibrium among nuclear reactions has apparently only been attained in a small fraction of stars. In the centers of such stars large amounts of iron and its neighbors do build up (see Figure 1.4). No additional nuclear fuels are available since nearly all nuclei have been converted to the most stable ones. When this has taken place a final

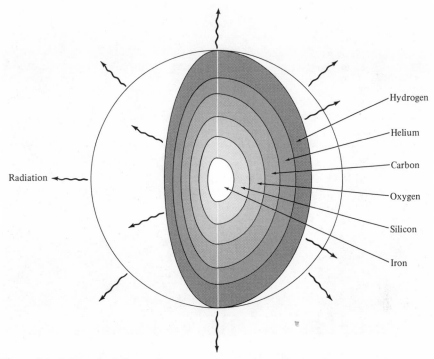

Figure 1.4 Schematic diagram of structure of a star immediately prior to supernova explosion.

catastrophic gravitational collapse suddenly draws in large quantities of the hydrogen, helium, and other light elements from the outer layers of the star, fuses them together, and causes a supernova explosion. One such explosion was recorded by Chinese astronomers in the year 1054, in the same region of the sky as what is today called the Crab Nebula (Figure 1.5.) Since 1054 this vast cloud of gas has been expanding at the rate of 1600 km/s, about 50 times as fast as the earth's orbital velocity around the sun. Such supernova explosions scatter the newly synthesized chemical elements throughout the universe.

Eventually, of course, the scattered atoms become trapped by a gravitational field and condense into a second- or third-generation star. This time, however, the quantities of the heavier elements should be larger since their abundance in the interstellar gas has been increased by the above-mentioned reactions in the first-generation supernova. Studies of stars which have independently been identified as being of more recent origin (by estimates of the rate at which their hydrogen is being converted to helium, for instance) do indicate greater abundances of iron relative to hydrogen. This further supports the hypothesis that all of the elements have been synthesized from primordial hydrogen.

Returning to the second of the questions posed a little earlier, nuclei

Figure 1.5 Photograph of Crab Nebula (courtesy of Hale Observatories).

having mass numbers greater than 66 apparently were built up by two types of neutron capture and beta particle (electron) emission processes.[3] The *s-process* occurs under the usual conditions within a star where the number of neutrons is relatively small. When a supernova occurs it is augmented by the *r-process* which makes use of the far larger concentration of neutrons available during that short period. The latter process seems to be the only one capable of accounting for the abundances of neutron-rich isotopes and elements heavier than bismuth. The nuclei produced by these two mechanisms are dispersed rapidly enough by the supernova explosion that they cannot reach equilibrium.

1.2 INTERSTELLAR ATOMS AND MOLECULES

While studying the relative abundances of the chemical elements reported in Figure 1.1, you may have wondered how such numbers are obtained since only a few persons have ventured the (astronomically) short distance from the earth to the moon, and no space flights have even attempted to study the regions outside the solar system. Obviously, one does not simply take a sample of outer space into the laboratory and analyze its elemental composition. Since the earth is as large (relative to man) as it is,

[3] J. Selbin, *J. Chem. Educ.* **50**(5), 306 (1973).

it is not even possible to determine its overall composition without using considerable indirect (and necessarily controversial) evidence. The information in Figure 1.1 is an intelligently weighted average of data from three distinctly different sources: (1) quantitative chemical analyses of the earth's crust and of meteorite fragments; (2) spectroscopic analyses of the sun, other stars, and the interstellar regions; and (3) theoretical calculations using the mechanisms of nucleosynthesis briefly outlined in Section 1.1.

Direct chemical analysis, although it is the oldest and simplest method of determining the relative abundances of the elements, suffers from several drawbacks when it is extrapolated to the universe, the solar system, or even to the earth as a whole. By comparison with the entire sphere of the earth, which has a radius of about 6400 km, man's deepest penetration of less than 10 km seems rather trivial. Extrapolation of analyses based on such limited sampling are not likely to give unbiased results regarding composition of the earth or solar system. Better evidence can be obtained from analysis of meteorite fragments because it appears that meteorite fragments might represent an undifferentiated sample of the initial material from which the earth condensed.

There is considerable evidence that the earth and the meteorites which strike it had a common origin. The ages of meteorites can be determined from the relative amounts of different elements within them and are similar to the estimated age of the earth. For instance, the half-life for the nuclear reaction

$$^{40}_{19}\text{K} \longrightarrow {}^{40}_{18}\text{Ar} + \beta^+ \qquad t_{1/2} = 1.4 \times 10^9 \text{ years} \qquad (1.7)$$

is long enough so that all $^{40}_{19}\text{K}$ would not be converted to argon within the lifetime (about 4.5×10^9 years) of the earth. By measuring the ratio of $^{40}_{19}\text{K}$ to $^{40}_{18}\text{Ar}$ it has been established that the ages of most meteorites are in excess of 4×10^9 years. Only a lower limit to the age can be set because some gaseous argon might have been lost when the meteorite was heated to a high temperature while passing through the earth's atmosphere. Other evidence of the common origin of earth and meteorites is the fact that the quantities of different isotopes of the same element are often found to be in the same proportion in both locations, as are quantities of chemically related elements such as the alkali metals potassium and rubidium.

In addition to the possible loss of volatiles through heating, several other problems beset the analysis of meteorite samples. A major one is their relative scarcity and small size, which requires that microanalytical techniques of extremely high sensitivity be employed. Furthermore, great care must be taken that removal of material by weathering has not occurred and that extraneous material has not contaminated the sample in the course of the analysis. For example, many older determinations of lead were too high by a factor of ten as a result of impurities in reagents and material leached from glass containers. Even if accurate chemical analysis can be done, there is a problem related to the relative abundance of the two main types of meteorites: stony and iron. Although the majority of meteorites that are found are in the latter category, this seems to be a result of their distinctive appearance, which makes them more noticeable to collectors, and their chemical durability, which allows them to resist weathering much longer. When samples are collected immediately after a meteor has been seen to fall, more than 90% are found to be stony meteorites. Abundance ratios are usually based primarily on the analysis of this type.

The second technique for determining the abundances of the elements does not require the physical presence of a sample for analysis at all. By using *spectroscopy* as a probe it is possible to determine how much of each kind of element is present in the stars and even the interstellar regions. The first studies of this type involved visible radiation, but with the advent of radar and radiotelescopes much longer wavelengths may be used as well. The latter are useful in detecting molecules as well as atoms, and in the last few years a great many interstellar molecules have been found.

The fundamental concept upon which spectroscopic analysis depends is that of distinct *energy levels* which are available to atoms or molecules. The absorption or emission of electromagnetic radiation is thought to require a quantum jump from one level to another. Since energy must be conserved when a particle jumps from a state whose energy is E_1 to one whose energy is E_2, the quantum of radiation (photon) emitted or absorbed must contain

$$|E_2 - E_1| = E_{\text{photon}} = h\nu \qquad (1.8)$$

units of energy. In the last part of equation 1.8 (first derived by Planck in 1900) ν represents the frequency of the light and h ($= 6.626 \times 10^{-34}$ J s) is Planck's constant. A positive value of $E_2 - E_1$ corresponds to absorption of radiation while a negative value would occur upon emission of a photon.

According to the quantum theory an individual atom or molecule can gain or lose energy by jumping from one discrete energy level to another. The amounts of energy which can be gained or lost (which determine the frequencies of radiation absorbed or emitted) depend on the spacings between the energy levels. These separations depend in turn on the mechanisms by which the energy levels arise. For example, the position of an electron relative to the nucleus of an atom yields a series of *electronic energy levels*. For a hydrogen atom the spacings between these levels may be obtained either by solution of the Schrödinger equation or from the Bohr theory.[4] The separation between electronic energy levels is on the order of 10^{-18} J/atom, corresponding to a frequency $\nu = E/h = (10^{-18}$ J$)/(6.626 \times 10^{-34}$ J s$) = 1.5 \times 10^{15}$ s$^{-1} = 1.5 \times 10^{15}$ Hz. Such frequencies are found in the ultraviolet and visible regions of the electromagnetic spectrum. This is indicated in Figure 1.6, which shows the wavelengths and frequencies corresponding to different types of radiation.

---

[4] See, for example, W. L. Masterton and E. J. Slowinski, "Chemical Principles," 3rd ed., pp. 127ff. Saunders, Philadelphia, Pennsylvania, 1973.

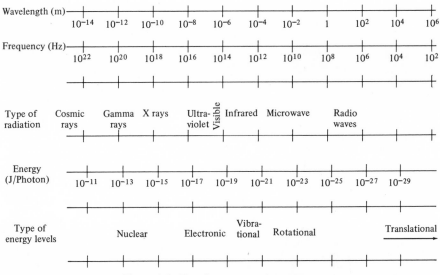

Figure 1.6 The electromagnetic spectrum.

The energies emitted or absorbed by electronic quantum jumps are invariably large and correspond to well-separated lines (specific frequencies). Each of the chemical elements exhibits a characteristic line spectrum when sufficient energy is supplied (by the high temperature of a flame or a star, for example) to excite its atoms above the lowest energy electronic state. The intensities of these lines can be used to determine the quantity of a given element which is present. The first application of electronic line spectra in determining the composition of the universe was the observation by Bavarian optician Joseph von Fraunhofer of black lines superimposed on continuous radiation from the sun. Later in the nineteenth century Kirchoff correctly interpreted von Fraunhofer's lines as resulting from absorption of solar radiation by hydrogen in the sun's outer atmosphere. The element helium (Greek: *helios* = sun) was discovered spectroscopically on the sun 20 years before it was isolated on earth.

In addition to electronic levels atoms exhibit *nuclear energy levels* whose energies depend on the structural arrangements of neutrons and protons. These levels are separated by rather large energies and transitions among them result in emission of γ rays (note that γ rays appear in nuclear reactions 1.1, 1.2, and 1.3). The identification of characteristic γ-ray frequencies when nuclei have been excited by neutron absorption is the basis for *neutron activation analysis,* a sensitive tool which has been very useful in quantitative analysis of trace quantities of pollutants such as the heavy metals.

An individual atom also gains energy if it is accelerated to a greater velocity. Such energy is called *translational energy* because it corresponds

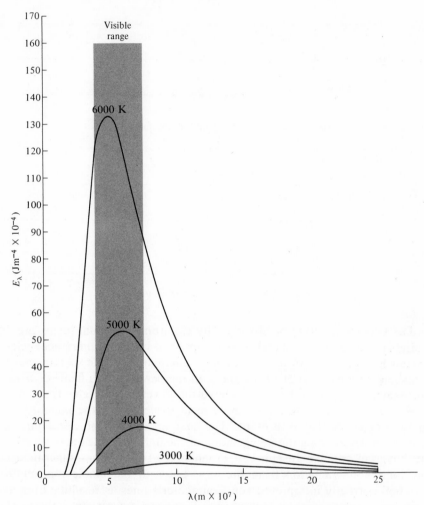

Figure 1.7   Blackbody radiation at various temperatures.

to the atom's translating (literally, in Latin, carrying across) from one place to another. Quanta of translational energy vary in magnitude depending on the mass of an atom and the volume of space throughout which the atom is free to move, but usually contain about 10^{-39} J/atom. Since quanta of translational energy are very small, a large number must be absorbed by an atom before it can be accelerated to the average room temperature velocity. This means that the number of different combinations of initial and final states in equation 1.8 is large, and the distinct frequencies of radiation which may be absorbed or emitted will be large in number and very close together. [The approximate energy difference of 10^{-39} J/atom corresponds to a frequency, $\nu = \Delta E/h = (10^{-39}$ J$)/(6.62 \times 10^{-34}$ J/Hz$) = 1.5 \times 10^{-5}$ Hz.

Thus, adjacent translational frequencies may be separated by as little as 10^{-5} Hz, far too small a difference to be resolved or distinguished by any spectrometer.] Transitions among the translational energy levels of atoms (or molecules) give rise to what is known as a continuous spectrum. The quantity of radiation emitted as a function of wavelength varies slowly, producing a continuous graph as shown in Figure 1.7. No lines or sharp peaks are observed. The theoretical interpretation of such "blackbody" radiation curves by Max Planck was, of course, the earliest application of the quantum hypothesis.

The result of Planck's analysis, the blackbody radiation law, is shown in equation 1.9:

$$E_\lambda = \frac{8\pi ch\lambda^{-5}}{e^{ch/\lambda kT} - 1} \qquad (1.9)$$

where c is the velocity of light, k is Boltzmann's constant, and T is the absolute temperature. The radiant energy (E_λ) emitted within the narrow range of wavelength $\lambda + d\lambda$ depends on both λ^{-5} and $e^{-ch/\lambda kT}$. The former decreases and the latter increases with larger values of λ, producing the maxima observed in Figure 1.6. As T increases the maximum in E_λ shifts to shorter wavelengths, permitting the determination of the temperature of a remote object from its spectrum. This "color temperature" determines the visible color of a star, a smaller value corresponding to a red or orange color and a larger value corresponding to white or blue-white. The color temperature of the sun is about 6000 K.

A number of techniques for detecting *molecules* in outer space have been developed during the past 30 years. In addition to translational, electronic, and nuclear energy levels there are two other ways in which energy may be added to molecules. Because more than one atom is present a molecule can be made to vibrate or rotate. Bonds may periodically lengthen and shorten (stretching vibrations), bond angles may change (bending vibration), or the entire molecule may rotate around some axis. Figure 1.8 diagrams some of these motions using the water molecule as an example.

As was true of translational, electronic, and nuclear energy, only certain vibrational and rotational states are permitted. Quanta of *rotational energy* are of the order of 10^{-23} J/molecule and quanta of *vibrational energy* are of the order of 10^{-20} J/molecule. Referring to equation 1.8 it is found that transitions among vibrational levels are found in the infrared region of the spectrum and rotational transitions are at microwave frequencies. Both types of energy levels yield characteristic spectra which can be used to "fingerprint" molecules, but since infrared radiation is absorbed by CO_2 and H_2O in the earth's atmosphere microwave spectroscopy is the best tool for the study of interstellar species.

Table 1.2 lists the interstellar molecules which have been discovered so far, together with some of their properties and their abundances. Most of

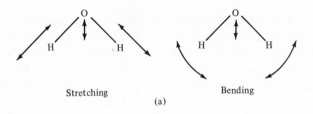

Stretching   Bending

(a)

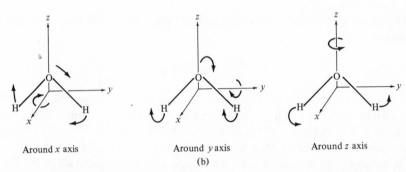

Around *x* axis   Around *y* axis   Around *z* axis

(b)

Figure 1.8 (a) Vibrational and (b) rotational motion of the water molecule. There are two stretching vibrations, symmetric and asymmetric. In the former both bonds lengthen at the same time. In the latter one O—H bond lengthens as the other shortens.

the interstellar regions of space consist of a near-perfect vacuum populated by about one hydrogen atom per cubic centimeter, but there are a number of clouds, such as the one pictured in Figure 1.9, whose densities are much larger. Usually these contain about 10^4 H_2 molecules per cubic centimeter, a concentration millions of times smaller than the best vacuum attainable at the earth's surface. Other molecules occur at the concentrations shown in Table 1.1, and there are quantities of dust (grains of silicates, graphite, and possibly iron having diameters on the order of 0.2 μm) as well. Surprisingly enough, although their densities are low, these clouds contain more matter than any other objects in our galaxy. Some of their masses are as much as 10^5 times the mass of the sun.

Most chemists did not expect that so many and such complex molecules could be found in the inimical environment of outer space, but plausible mechanisms can be devised for their formation. One might expect a correlation between the stability of a molecule (in terms of bond order or bond energy) and its abundance, provided that its constituent atoms are present in large enough concentration. This seems to be true with certain notable exceptions. For example, H_2 is the most common molecule by a factor of ten thousand, and this is probably because of the very large abundance of H atoms (Figure 1.1). The next most abundant molecule, CO, has a bond order of three and the largest bond dissociation energy (1077 kJ/mol) of

TABLE 1.2 Molecules Observed in the Interstellar Medium[a]

Molecule	Year of discovery	Frequency of discovery line (GHz)	Approximate density in molecular clouds (molecules/m³)
		Diatomic molecules	
H—H	1970	Ultraviolet (rocket)	10^{10} (estimated)
C≡O	1970	115.271	10^{6}
O—H	1963	1.665	10^{4}
C≡S	1971	146.969	10^{3}
C≐N	1937	Visible	
	1970	113.492	10^{2}
Si≡O	1971	130.268	10 (approximate)
C—H⁺	1937	Visible	Not observed in H_2 clouds
C—H	1937	Visible	Not observed in H_2 clouds
S=O	1973	—	—
Si≡S	1974	—	—
		Triatomic molecules	
H—C≡N	1970	86.339	10^{3}
H—S—H	1972	168.762	10^{2}
O=C=S	1971	109.463	10^{2} (approximate)
H—O—H	1969	22.235	—
H—O—D	1974	—	—
		Polyatomic molecules	
NH_3	1968	23.694	10^{4}
CH_3OH	1970	0.834	10^{3}
$H_2C=O$	1969	4.830	10^{2}
H—N=C=O	1971	87.925	10^{2} (approximate)
$H_2C=NH$	1973	—	—
H—C≡C—C≡N	1970	9.098	10^{2} (approximate)
H_3C—C≡N	1971	110.383	10^{2} (approximate)
H_3C—C≡C—H	1971	85.457	10^{2} (approximate)
$H_2C=S$	1972	3.139	10
H—C(=O)—NH₂	1971	4.619	>10 (approximate)
H_3C—C(=O)—H	1971	1.065	>10 (approximate)
H—C(=O)—OH	1970	1.639	—
CH_3CH_2OH	1974	85.3, 90.1	—
Interstellar dust	1920–30	visible	1% of total mass

[a] Data from: W. D. Metz, *Science* **182,** 466–468 (1973); P. M. Solomon, *Phys. Today,* March, pp. 32–40 (1973); D. M. Rauk, C. H. Townes, and W. J. Welch, *Science* **174,** 1083–1101 (1971).

Figure 1.9 The Trifid Nebula in constellation Sagittarius (courtesy of Hale Observatories).

any diatomic molecule. Third on the list of diatomics is OH, whose occurrence is favored because H is so common, followed by CS and CN, both of which have large bond orders. Their dissociation energies are 761 and ~ 1000 kJ/mol, respectively.

Several molecules which might be expected to be present have not been discovered. For instance, C_2, N_2, O_2, and NO have bond orders of two or more, but are not listed in Table 1.2. Since C_2, N_2, and O_2 have no permanent dipole moment their microwave emission should be very weak and they may simply not have been detected as yet. Absence of NO is quite surprising, but apparently there is no good mechanism for its formation.[5]

One further aspect of interstellar chemistry is of interest. The fact that molecules as complex as cyanoacetylene or formamide can exist in interstellar space is convincing evidence that molecules of equal or greater complexity could have formed spontaneously during the time immediately following the origin of the earth. Several of these compounds are precursors or intermediates in the reaction sequences which lead to the forma-

---

[5] P. M. Solomon and W. Klemperer [*Astrophys. J.* **178,** 389 (1972)] have used the results of mass spectrometric studies to infer mechanisms by which interstellar molecules may form. Similar techniques have been widely used to simulate photochemical smog and other air pollution problems.

tion of amino acids and carbohydrates in terrestrial experiments on chemical evolution (Section 3.3). Although it has been argued that such earthbound experiments may not imitate correctly the conditions of a newly formed planet, the exceedingly hostile environment of outer space provides a much more convincing test for the evolution of complex organic structures from atoms or very simple molecules.

Although we shall not have time to discuss details of the spectroscopic analytical methods whose basic principles have been enunciated in this section, it should be evident that they constitute extremely powerful tools for discovering and monitoring environmental contaminants. Ozone, for example, is observed in the atmosphere by means of its ultraviolet absorption. We shall refer to these methods again in conjunction with specific pollutants, but the reader is advised to consult a text on instrumental analysis or the literature on accepted standard methods relating to individual substances for experimental details.

EXERCISES

(The first group of exercises is based primarily on material found in this text.)

1. Calculate the binding energy for each of the following nuclei (Nuclear masses may be obtained from the "Handbook of Chemistry and Physics"):

    a. $^{54}_{26}$Fe   b. $^{7}_{3}$Li   c. $^{235}_{92}$U
    d. $^{55}_{25}$Mn   e. $^{52}_{24}$Cr   f. $^{59}_{27}$Co

2. Which of the nuclei in Exercise 1 is most stable? Are any of them more stable than $^{56}_{26}$Fe?

3. Although its nucleus is extremely stable, iron is not the most abundant element in the universe. Hydrogen is far more abundant. Give a reasonable explanation for this fact.

4. Give some of the types of evidence which support the conclusion that the earth and the meteorites which strike its surface had a common origin. What difficulties are encountered in the determination of cosmic abundances of the elements by analysis of meteorite samples?

5. What types of energy levels are available to atoms? To molecules? In what region of the electromagnetic spectrum would radiation absorbed or emitted as a result of quantum jumps involving each of these types of energy levels be found?

6. Of what significance is the discovery of interstellar molecules with regard to the theory of evolution?

(Subsequent exercises may require more extensive research or thought.)

7. The photolytic dissociation of O_2 to form atomic oxygen requires photons whose wavelengths are shorter than 240 nm. Calculate the minimum energy of

such photons and compare it with the bond dissociation energy of O_2. Why does much photochemistry depend on uv radiation?

8. x Rays and γ rays are often referred to as "ionizing radiation." Calculate the energies of photons in the x-ray and γ-ray regions of the spectrum and compare them to the ionization energies of lithium, carbon, and fluorine atoms. Why can ionizing radiation be dangerous to human health?

9. Most general chemistry texts give an equation for calculating the energy levels of the hydrogen atom using the Bohr theory. This theory may be extended to *ions* such as He^+ and Li^{2+} which contain a single electron (see J. E. Huheey, "Inorganic Chemistry," pp. 12–14. Harper, New York, 1971). Explain (giving the wavelengths at which spectral lines would appear) how you could detect the relative amounts of H and Li^{2+} present in a star. Which line in the Li^{2+} spectrum would correspond to the red line in the hydrogen spectrum at about 656 nm?

10. Explain clearly the difference between the r-process and the s-process for neutron capture. Use a table of nuclides to follow the s-process beginning with $^{87}_{38}Sr$. What isotopes of strontium and yttrium could be produced assuming the average nucleus collides with a neutron once every 100 years? Could $^{122}_{50}Sn$ or $^{128}_{52}Te$ be formed by the s-process?

11. The cosmic abundances of lithium, beryllium, and boron are unusually low (Table 1.1). What explanations can be given?

12. D. Weiss, B. Whitten, and D. Leddy [*Science* **178**, 69–70 (1972)] reported that "Antique hair contained a significantly greater amount of lead than did contemporary hair." Many environmentalists have argued against the use of leaded gasoline on the basis of health effects. Discuss the results of Weiss *et al.* with respect to:

    a. Their implications for the "Get the lead out!" campaign.

    b. Possible sources of error in the analyses [see *Science* **180**, 1080 (1973)].

13. Think about the question: "Why is the visible region (400–750 nm) of the electromagnetic spectrum in this particular wavelength range?" Consider both environmental, external factors and the chemistry of visual receptors [see G. Wald, *Science* **162**, 230–239 (1968)]. What problems might be encountered during the evolution of an organism that could "see" in the uv or infrared regions? (Apparently some insects do recognize plants by means of ultraviolet pigmentation in flower petals.)

14. What types of reactions are possible for the formation of molecules in interstellar space? What mechanisms are available for their dissociation?

2

The Development of the Solid Earth

Many an Aeon moulded earth before
her highest, man, was born.
Many an Aeon too may pass when
earth is manless and forlorn.

Tennyson

So far we have concerned ourselves with the chemistry of the farthest regions of space and the origins of atoms and molecules. With this background in mind we can come a little closer to home, in space if not in time, to consider the origin of our earthly environment and its implications with regard to present-day environmental problems.

2.1 FORMATION OF THE EARTH

According to the most widely accepted scientific theory, atoms, molecules and cosmic dust particles from an interstellar cloud became trapped by each other's gravitational fields approximately 6×10^9 years ago and began to fall together into what we now know as the solar system. As the cloud swirled together the center portion became a star, our sun, while the outer portion separated into revolving rings of high density (see Figure 2.1). Eventually these rings also condensed into the precursors of what we now call the planets.

As the molecules, dust particles, and larger aggregates accelerated toward each other to form the sun and planets, gravitational potential energy was released and the temperature rose. This increase was aided by the

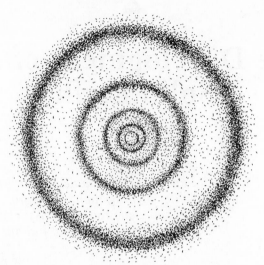

Figure 2.1 Separation of the solar nebula into concentric rings. Source: H. P. Berlage, "The Origin of the Solar System," p. 91. Pergamon, Oxford, 1968.

spontaneous, exothermic decay of radioactive elements such as $^{40}_{19}K$ and $^{235}_{92}U$. Because the sun contained by far the greatest mass, its temperature reached 2×10^7 K, the activation barrier to hydrogen fusion was overcome, and it began to burn as a star. Since the planets were less massive, they did not attain such a high temperature and eventually began to cool. Their components began to condense to liquids and eventually to solidify. Thus the earth, the sun, other planets, and the asteroids and meteorites of the solar system all had the common origin suggested by the evidence already mentioned in Section 1.2.

Although the foregoing scenario is of necessity highly speculative, it does provide a basis for considering the distribution of the chemical elements among the solid, liquid, and gaseous portions of the present-day earth. Regardless of whether the earth and solar system were formed exactly as outlined above or by some other mechanism, they must originally have had much the same elemental composition as an interstellar cloud or the sun. If we assume (as was done in the discussion of the origin of the elements in Section 1.1) that the same laws of chemical and physical behavior which are familiar today applied equally well three to six billion years ago, it is possible to account for many of the changes that must have taken place during the evolution of the earth from its initial state to its current one.

2.2 THE LOSS OF GASEOUS ELEMENTS

It is apparent from the data in Table 1.1 that the composition of the earth's crust varies considerably from that estimated for the universe as a

TABLE 2.1 Deficiency Factors [log (Cosmic Abundance/Terrestrial Abundance)] for Selected Elements[a]

Element	Atomic number	Molecular weight of typical volatile molecule	Deficiency factor
H	1	2.0	6.58
He	2	4.0	14.17
C	6	16.0 (CH_4)	3.22
N	7	28.0	5.25
O	8	32.0	0.79
F	9	38.0	0.96
Ne	10	20.2	10.46
Na	11	—	0.12
Mg	12	—	0.08
Al	13	—	−0.04
Si	14	—	0.00
P	15	34.0 (PH_3)	−0.02
S	16	34.1 (H_2S)	0.70
Cl	17	71.0	0.25
Ar	18	40.0	6.29
Fe	26	—	−0.56
Kr	36	83.8	6.89
Xe	54	131.3	7.03

[a] Source for cosmic abundance data: A. G. W. Cameron, A critical discussion of abundance of nuclei. *In* "Explosive Nucleosynthesis" (D. N. Schramm and W. D. Arnett, eds.). Univ. of Texas Press, Austin, 1973; source for terrestrial abundance data: B. Mason, "Principles of Geochemistry," 3rd ed. Wiley, New York, 1966.

whole. These variations in composition appear more clearly when one calculates *deficiency factors* for the elements. (The deficiency factor is the logarithm of the ratio of cosmic abundance/terrestrial abundance.) This has been done in Table 2.1. A number of processes, referred to collectively as differentiation of the elements, have contributed to these differences between terrestrial and cosmic abundances.

The first of these processes has to do with the loss of volatile, inert elements such as N_2 and the noble gases. It is a well-known consequence of the kinetic theory of gases[1] that the smaller the molecular weight of a gaseous substance the greater the average molecular speed at a given temperature, T:

$$\text{Average molecular speed} = (3\ RT/M)^{1/2} \qquad (2.1)$$

where R is the ideal gas constant and M the molecular weight. Furthermore, it can easily be shown that in order to escape from a gravitational field[2] a particle's vertical velocity must exceed

$$V_{\text{escape}} = (2\ GM_b/r)^{1/2} \qquad (2.2)$$

---

[1] See, for example, W. L. Masterton and E. J. Slowinski, "Chemical Principles," 3rd ed., pp. 113–116. Saunders, Philadelphia, Pennsylvania, 1973.

[2] R. Gymer, "Chemistry: An Ecological Approach," pp. 134–136. Harper, New York, 1973.

where $G(= 6.672 \times 10^{-11} \text{ m}^3 \text{ s}^{-2} \text{ kg}^{-1})$ is the gravitational constant, M_b is the mass of the body whose gravitational field is to be escaped, and r is the distance from the particle to the center of the body. At the surface of the earth $M_b = 6.0 \times 10^{24}$ kg and $r = 6.4 \times 10^6$ m, giving an escape velocity of 1.1×10^4 m/s. Detailed calculations indicate that due to the spread of molecular velocities about the average, only those gases whose average molecular speed is less than one-fifth the escape velocity can be expected to remain on a planet over the billions of years of a geological time scale.

At the current surface temperature of the earth only dihydrogen molecules, hydrogen atoms, and helium atoms attain sufficiently high speeds to escape. However, temperatures probably were higher during the period when earth was forming. Thus molecules such as NH_3, CH_4, CO, N_2, and O_2 might have attained speeds approaching escape velocity. If account is taken of the fact that some elements (oxygen, for example) form quite stable, nonvolatile compounds, many of the deficiency factors in Table 2.1 can be rationalized. Thus He and Ne, which are completely inert, would be expected to be most deficient, and H and N should not be far behind. Carbon, which might be found in CH_4, CO, or CO_2, but also as carbonate salts, ought to be intermediate, and those elements (Mg, Al, Si, Fe) forming stable oxides would be expected in about the same abundance on earth as in the cosmos.

Even when chemical affinities are taken into account, however, a discrepancy arises in the case of Kr and Xe. For Xe atoms to achieve an average speed approaching earth's escape velocity would require a temperature on the order of 10^5 K (see Exercise 10), under which conditions most chemical compounds decompose to individual atoms or very simple molecules. Thus lighter elements such as oxygen should be lost far more readily than xenon.

One way around this problem is to assume that as the earth began to condense a large number of *planetesimals,* bodies ranging in diameter from several to thousands of meters, were formed. Because the mass of the planetesimals was spread around the earth's orbit, escape velocities would have been much smaller than they are today. Nitrogen and the noble gas elements could have been lost from the planetesimals at low enough temperatures so that oxides, silicates, carbonates, etc., could remain as solids. Although it can account for the terrestrial abundances of the noble gases the planetesimal theory is by no means acceptable to all astrophysicists, and numerous other hypotheses are available.[3] Whichever is accepted, however, it is reasonable to suppose that immediately following its formation the earth had little or no atmosphere.

Because they are gases in elemental form and many nitrogen compounds are volatile, the small amounts of He, Ne, Ar, Kr, Xe and N which remain

---

[3] See, for example, Edward Anders, Meteorites and the early solar system. *Annu. Rev. Astron. Astrophys.* **9,** 1–34 (1971).

on earth tend to concentrate in the atmosphere and they are generally referred to as *atmophiles* to distinguish them from the other elements which form nonvolatile compounds.

2.3 THE PRIMARY DIFFERENTIATION OF THE ELEMENTS

As the earth condensed, the release of both gravitational and nuclear energy probably kept the temperature well above that of the modern earth's surface. Most of the oxygen was present as oxides of the more electropositive elements such as magnesium, and sulfur was probably in the form of sulfides. Since silicon forms a highly stable oxide (SiO_2) as well as silicate salts, it almost certainly was in oxidized form. Of the dozen or so metals present in greatest abundance at this stage of the earth's evolution, iron is the least readily oxidized, and, since metals were in excess over oxygen and sulfur, much of the iron probably remained in a molten, metallic phase.

Because of its greater density this liquid iron slowly migrated toward the center of the earth, reducing and alloying a number of even less reactive metals like nickel, gold, and platinum along the way. The less dense silicates, oxides, and sulfides floated to the surface. In this way the three main structural components of the earth—core, mantle, and crust—were formed. These are illustrated in Figure 2.2. The study of the propagation of earth-

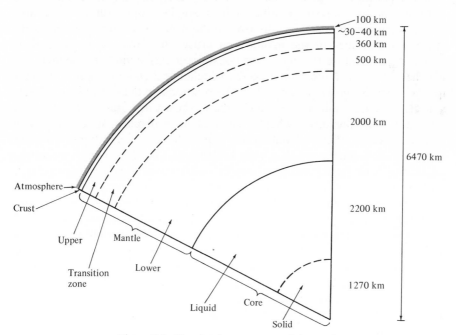

Figure 2.2 Structural components of the earth.

TABLE 2.2   Goldschmidt's Geochemical Classification of the Elements

Term	Characteristic property	Elements included[a]
Siderophile	Association with metallic *iron*	Fe, Co, Ni, Ru, Rh, Pd, Os, Ir, Pt, Mo, W, Re, Au, Ge, Sn, C, P, (Pb, As, S)
Chalcophile	Tendency to form *sulfide minerals*	Cu, Ag, Zn, Cd, Hg, Ga, In, Tl, (Ge, Sn), Pb, As, Sb, Bi, S, Se, Te, (Fe, Mo, Cr)
Lithophile	Tendency to be bound to *oxide* ions	Li, Na, K, Rb, Cs, Be, Mg, Ca, Sr, Ba, B, Al, Sc, Y, Rare Earths, (C), Si, Ti, Zr, Hf, Th, (P), V, Nb, Ta, Cr, (W), U, F, Cl, Br, I, Mn, (H, Tl, Ga, Ge, Fe)
Atmophile	Tendency to occur as a *gaseous* component of the atmosphere	N, He, Ne, Ar, Kr, Xe

[a] Elements in parentheses show the indicated characteristic to a lesser extent. (Data from B. Mason, "Principles of Geochemistry," 3rd ed. Wiley, New York, 1966; L. H. Ahrens, "Distribution of the Elements in Our Planet." McGraw-Hill, New York, 1965.)

quake vibrations has led geophysicists to the conclusion that the core consists of iron, partly molten and partly solid. The mantle is composed of silicates whose crystal structures vary so that they become denser with increasing depth. The crust, which contains only about 1% of the total mass of the earth, is less dense and much more diverse in composition than the core or mantle.

The separation of the iron core during the early history of the earth was very similar to modern smelting of iron ore in a blast furnace (Section 11.3). Some metals tended to dissolve in the iron, some preferred the oxide and silicate lattices of the mantle and crust, and others concentrated in crustal phases whose major anionic constituent was sulfide. These facts led geochemist V. M. Goldschmidt to the geochemical classification of the elements which is shown in Table 2.2. The relationship between the geochemical classification and the periodic table is shown in Figure 2.3. Although there are numerous cases where a given element displays several types of behavior under different conditions — iron behaves as a lithophile in the earth's crust, sulfur as a siderophile in the core, and dioxygen is obviously atmophilic — the geochemical property most often displayed correlates fairly well with the periodic properties of the elements.

The *lithophiles* tend to be found in oxide or silicate lattices. They consist of elements such as sodium, magnesium, chlorine, and bromine which tend to form ions having the same electronic configuration as a noble gas. Also included are transition metals having less than a half-filled d subshell, the lanthanides, boron, aluminum, and silicon. In general, the metallic lithophiles are more readily oxidized (have more negative reduction potentials, see Table 2.3) than iron. Their presence in the oxide phase of the primitive earth is the result of the inability of metallic iron to replace them from their compounds.

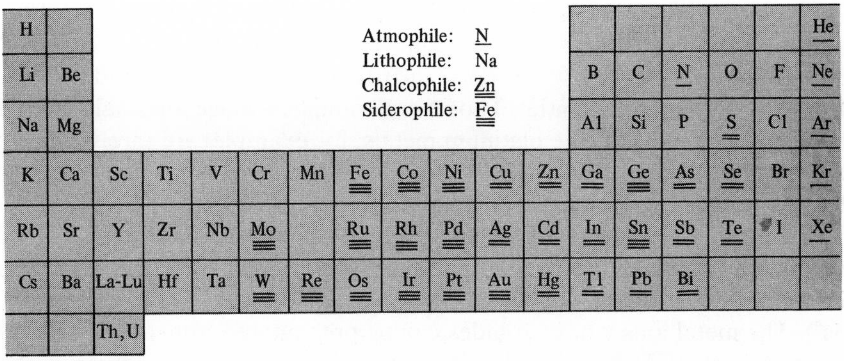

H																	He
Li	Be											B	C	N	O	F	Ne
Na	Mg											Al	Si	P	S	Cl	Ar
K	Ca	Sc	Ti	V	Cr	Mn	Fe	Co	Ni	Cu	Zn	Ga	Ge	As	Se	Br	Kr
Rb	Sr	Y	Zr	Nb	Mo		Ru	Rh	Pd	Ag	Cd	In	Sn	Sb	Te	I	Xe
Cs	Ba	La-Lu	Hf	Ta	W	Re	Os	Ir	Pt	Au	Hg	Tl	Pb	Bi			
	Th,U																

Atmophile: N
Lithophile: Na
Chalcophile: Zn
Siderophile: Fe

Figure 2.3 The geochemical classification of the elements in relation to the periodic system. Source: B. Mason, "Principles of Geochemistry," 3rd ed. Wiley, New York, 1966. Only the most characteristic geochemical property is indicated in the case of elements that display more than one type of behavior.

TABLE 2.3 Electrode Potentials and Solubilities of Selected Elements Related to Geochemical Classification

Element	Reduction potential	Solubility of sulfide in H_2O (M)
	Lithophiles	
Li/Li$^+$	−3.045	Soluble
Ca/Ca^{2+}	−2.866	—
Al/Al^{3+}	−1.662	Hydroxide precipitates more readily than sulfide
Cr/Cr^{3+}	−0.774	Hydroxide precipitates more readily than sulfide
	Chalcophiles	
Zn/Zn^{2+}	−0.7628	1.58×10^{-11}
Ga/Ga^{3+} (borderline lithophile)	−0.529	—
Cu/Cu^{2+}	+0.337	8.94×10^{-19}
Ag/Ag$^+$	+0.7991	1.76×10^{-17}
	Siderophiles	
Fe/Fe^{2+}	−0.429	6.32×10^{-10}
Ni/Ni^{2+}	−0.250	5.48×10^{-11}
Sn/Sn^{2+}	−0.136	1.00×10^{-13}
Rh/Rh^{3+}	+0.80	—
Pd/Pd^{2+}	+0.987	—
Au/Au$^+$	+1.691	—

The *siderophiles* all have reduction potentials more positive than iron and their oxidized forms can be reduced by metallic iron. As the earth's core differentiated they were replaced from their compounds and removed from the crust and mantle. Those siderophiles whose original (cosmic) abundances were low (the platinum metals, for example) are rarely found at the earth's surface. *Chalcophiles* may be distinguished by their great affinity for sulfide and other large anions. This is indicated by the insolubility of their sulfides in aqueous solution as seen from the data in Table 2.3. Often chalcophiles form pseudo-noble gas ions (Cd^{2+}, $4s^2 4p^6 4d^{10}$) or ions having a filled d subshell plus an inert pair of electrons (Bi^{3+}, $5s^2 5p^6 5d^{10} 6s^2$). The metal ions whose sulfides can be precipitated from acidic aqueous solution in the scheme of qualitative inorganic analysis are all chalcophiles.

A few ores have apparently formed in processes very similar to the primary differentiation of the elements. There is only a small amount of sulfur (about one atom of sulfur for every thousand atoms of oxygen) in the crust and mantle. Excess iron traps much of it in the form of pyrite, FeS_2, and most sulfides have relatively low melting and boiling points. Therefore, it is reasonable that the chalcophile elements were among the last to crystallize from the molten mixture. The greatest nickel deposits in the world were apparently created in this way in the vicinity of Sudbury, Ontario. A molten magma was trapped in a bowl-like depression (lopolith) roughly 25 miles in diameter. As the melt cooled the denser nickel and copper sulfides crystallized and settled below the silicates of aluminum, iron, magnesium, and calcium to form a high-grade ore deposit.

2.4 SECONDARY DIFFERENTIATION OF THE ELEMENTS

Since oxygen is much more abundant than sulfur in the earth's mantle and crust, most of the elements found there are lithophiles. Within this geochemical class, however, there are smaller groups of elements which usually occur together as a result of systematic, minor variations in the structures of silicate and oxide minerals. The distribution of the elements as a result of such variations is referred to as secondary differentiation. It is quite important in determining the nature of some ores as well as the structure of clay and soil minerals (Sections 12.1–12.3).

The structures of most silicates and oxides may be thought of as consisting of close-packed (or very nearly so), three-dimensional arrays of oxide ions among which positive ions such as K^+, Mg^{2+}, Fe^{2+}, Fe^{3+}, and Si^{4+} are interspersed. Of course, the bonds between a highly charged ion of a relatively electronegative element (such as silicon) and the oxide ion probably have a good deal of covalent character, and we shall later distinguish structures on the basis of differing linkages of SiO_4 tetrahedra which involve covalent bonds. For the moment, however, let's consider only the ionic type of bonding. Four rules govern the formation of stable ionic lattices:

1. The principle of hard and soft acids and bases[4] applies. Since O^{2-} is a hard base, soft cations such as Cu^{2+} or Ag^+ will be found at very low concentrations. Hard cations such as Ca^{2+} or Rb^+ will be preferred. This effect has already been observed in connection with primary differentiation of chalcophiles from lithophiles.

2. Two cations of similar radius (within about 10–20%) and the same charge can enter a lattice and provide nearly the same stability. However, the smaller ion will form slightly stronger bonds.

3. Of two cations having nearly the same radius, the one having greater charge will form a more stable lattice.

4. When substitutions of ions having different charges (rule 3) are made, electroneutrality must be maintained. The total number of positive charges must equal the total negative charge of the oxide ions.

A close-packed lattice such as that of the oxide ions in silicates contains empty spaces known as triangular, tetrahedral, or octahedral holes. These holes (shown in Figure 2.4) occur because it is impossible to pack spheres together so that they touch at all points.

As a molten silicate crystallizes, oxide ions pack together and positive ions are distributed among the holes in a regular way to make up the crystal lattice. It is because all holes of a given type in a close-packed lattice are identical that substitution of one ion for another is so common, when both have nearly the same size and are equally hard acids (rules 1 and 2). Such *isomorphous substitutions* account for the fact that certain elements are usually found together while others, which might be expected to occur together because of their similar chemical properties, are not. For example, sodium ions ($r = 97$ pm)[5] and calcium ions ($r = 101$ pm) form a series of feldspars (see Table 2.4) which represent a continuous variation (solid solution) between anorthite, $CaAl_2Si_2O_8$, and albite, $NaAlSi_3O_8$. Note that for each Na^+ which replaces a Ca^{2+} in one of the octahedral holes, a Si^{4+} must replace an Al^{3+} in one of the tetrahedral holes (rule 4) in order to maintain electrical neutrality. The density figures in Table 2.4 decrease as Na^+ replaces Ca^{2+}, as do the melting points, indicating that the monovalent ion does not hold the oxide lattice together as tightly as the divalent one (rule 3).

As the primitive earth cooled, this isomorphous substitution probably

---

[4] Hard Lewis acids and bases are small and not polarizable. Soft acids and bases have large radii and are easily polarizable. The principle of hard and soft acids and bases states that hard acids prefer to bond to hard bases and soft acids prefer soft bases; see E. G. Rochow, G. Fleck, and T. R. Blackburn, "Chemistry: Molecules That Matter," p. 189. Holt, New York, 1974.

[5] 1 pm $= 10^{-12}$ m $= 10^{-2}$ Å.

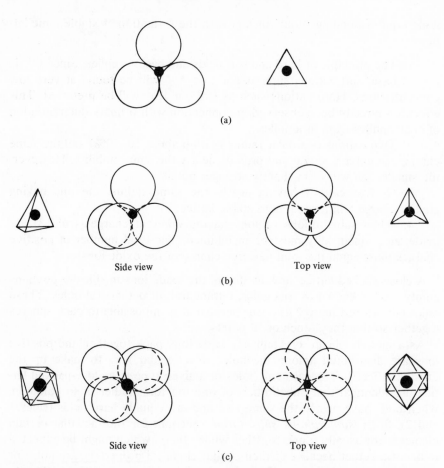

Figure 2.4 Illustration of triangular, tetrahedral, and octahedral holes: (a) triangular hole – three anions in the same plane, (b) tetrahedral hole – three anions in one plane capped by one in the plane above, and (c) octahedral hole – three anions in one plane capped by three in the plane above.

resulted in an initial precipitation of $CaAl_2Si_2O_8$ followed by formation of the solid solution (labradorite and bytownite) and finally precipitation of the remaining $NaAlSi_3O_8$. The latter is more highly concentrated in the upper portion of the mantle and in the crust, whereas $CaAl_2Si_2O_8$ concentrates at greater depths. Note that the much larger size of K^+ ($r = 133$ pm) makes its substitution for Ca^{2+} much more difficult. The mineral orthoclase, $KAlSi_3O_8$, is well known, but its crystal structure has completely different symmetry properties from those of anorthite, labradorite, and albite. The same is true of minerals containing Rb^+ ($r = 145$ pm) and Cs^+ ($r = 167$

TABLE 2.4 Properties of Some Important Minerals

Mineral	Formula	Structure	Melting point (K)	Density (g/cm³)
Olivine	$(Mg,Fe)_2SiO_4$	SiO_4 tetrahedra	2163 (Mg) –1493 (Fe)	3.26–3.40
Pyroxene	$(Mg,Fe)SiO_3$	Single chains	2103 (Mg)	3.2–3.9
Amphibole	$Ca_2Mg_5(Si_4O_{11})_2(OH)_2$	Double chains	—	3.0–3.5
Biotite	$K(Mg,Fe)_3(AlSi_3O_{10})(OH)_2$	Sheets	—	2.69–3.16
Feldspars				
Anorthite	$CaAl_2Si_2O_8$	Framework	1824	2.7–2.8
Bytownite	$NaAlSi_3O_8 \cdot (3–5)CaAl_2Si_2O_8$	Framework	1763–1794	2.75–2.80
Labradorite	$NaAlSi_3O_8 \cdot (1–3)CaAl_2Si_2O_8$	Framework	1723–1763	2.70–2.72
Albite	$NaAlSi_3O_8$	Framework	1373	2.61–2.64
Orthoclase	$KAlSi_3O_8$	Framework	—	2.56
Muscovite	$KAl_2(AlSi_3O_{10})(OH)_2$	Sheets	—	2.76–3.00
Quartz	SiO_2	Framework	<1743	2.59–2.66

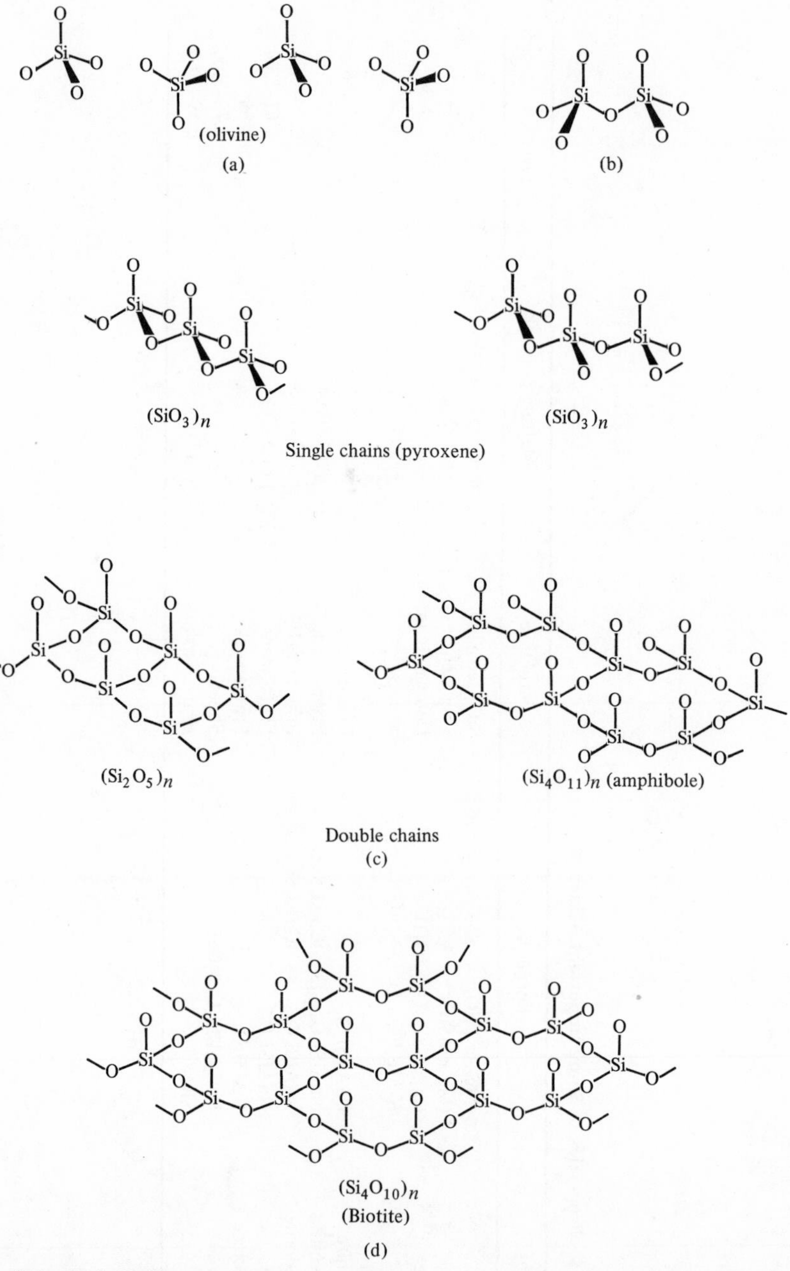

Figure 2.5 Typical silicate structures. Source for e.1, e.2, and e.3: A. F. Wells, "Structural Inorganic Chemistry," 3rd ed. Oxford Univ. Press (Clarendon), London and New York, 1961. (a) Discrete SiO_4 tetrahedra (zero oxygen atoms shared per SiO_4), (b) pyrosilicate ion (one oxygen atom shared per SiO_4), (c) chain structures (two or three oxygen atoms shared per SiO_4), (d) layer structure (three oxygen atoms shared per SiO_4), (e) framework structures

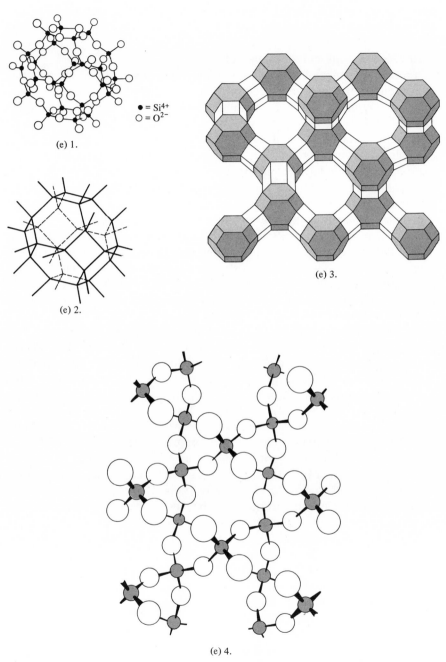

● = Si^{4+}
○ = O^{2-}

(e) 1.

(e) 2.

(e) 3.

(e) 4.

(four oxygen atoms shared per SiO_4): 1. Connection of SiO_4^{4-} tetrahedra to form a basket-like framework, 2. line representation of basket-like framework, 3. linkage of basket-like structures to form the space-filling framework of a zeolite, and 4. framework structure of quartz, SiO_2.

pm). Because lithium ions ($r = 68$ pm) are so small they are able to substitute for Mg^{2+} ($r = 65$ pm) but not for Na^+ or K^+. Despite their chemical similarities the variation in ionic radius of the alkali metals results in their occurrence in quite different minerals.

Another sequence of minerals was probably deposited by fractional crystallization as the hot earth cooled. These are the ferromagnesian substances: olivine, pyroxene, amphibole, and biotite. They form discrete phases having quite different crystal structures which illustrate most of the possibilities available for linking of covalently bonded SiO_4 tetrahedra. For example, in olivine Mg^{2+} or Fe^{2+} ions are found in octahedral holes in the close-packed oxide lattice, and Si^{4+} ions occupy tetrahedral holes in such a way that no single oxide ion is nearest neighbor to two Si^{4+} ions. The structure is described conveniently by saying that it consists of discrete SiO_4^{4-} tetrahedral ions surrounded by divalent positive ions (see Figure 2.5a). In the pyroxene structure the SiO_4 tetrahedra are arranged into long chains by having two oxygen atoms from each one shared between a pair of silicons (see Figure 2.5c). The amphibole structure is similar except that the chain is two SiO_4 tetrahedra wide. These chains are then held together by Mg^{2+}, Fe^{2+}, or Ca^{2+} ions.

Biotite is an aluminosilicate, meaning that one-fourth of the tetrahedral holes in the layer structure shown in Figure 2.5d is occupied by Al^{3+} ions. This produces a sheet whose formula is $(AlSi_3O_{10}^{5-})_n$. The flat layers of AlO_4 and SiO_4 tetrahedra usually form in such a way that the fourth (unshared) oxygen always projects on the same side. These latter oxygen atoms surround Mg^{2+} and Fe^{2+} ions which hold two adjacent layers together in a sort of "sandwich." Vacant sites in the close-packed oxide ion lattice within this sandwich are occupied by OH^- ions. Thus, each sandwich layer has the formula $[Mg_3(AlSi_3O_{10})(OH)_2]_n^-$, $[Fe_3(AlSi_3O_{10})(OH)_2]_n^-$ or some intermediate formula where some Fe^{2+} ions have replaced Mg^{2+} ions. The negative charge on the layers allows them to be bound weakly together by K^+ ions. Biotite is an example of a mica. It is easily cleaved into thin, flat sheets since the K^+ ions do not hold adjacent layers tightly together.

Two other arrangements for the connection of SiO_4 tetrahedra are possible. If only one oxygen from each tetrahedron is shared the pyrosilicate ion, $Si_2O_7^{6-}$, is formed (Figure 2.5b). If all four oxygens are shared a three-dimensional *covalent network* (often called a "framework structure") occurs. One example of such a structure is silica, SiO_2, and others may be formed by isomorphous substitution of Al^{3+} for Si^{4+}. In the latter cases additional positive ions are needed for electroneutrality. These ions are usually found in octahedral holes as in the structures of the feldspars we have already discussed. The family of minerals known as zeolites is another important example of the framework structure. In this case the structure is not derived from a close-packed arrangement of oxide ions and there are many voids or open spaces within it. Substances such as water, carbon dioxide, and ammonia may enter such voids and be loosely held, and the zeolites are often capable of exchanging one type of ion for another.

As a result of isomorphous substitutions, the less common lithophiles generally occur as trace constituents in the same types of crystal lattices described above. Barium ions, for example, are too large ($r = 134$ pm) to replace Ca^{2+} and are found primarily in potassium-containing minerals such as biotite and orthoclase. The lanthanides form tripositive ions and range from 85 pm (Lu) to 114 pm (La) in radius. They, as well as Y^{3+} ($r = 95$ pm), are often found in calcium-containing minerals.

A number of the lithophile elements, however, have ionic radii or charges that are so different from those of the common elements that they prevent isomorphous substitution. Some examples are B^{3+} ($r = 23$ pm) and Be^{2+} (35 pm) which are too small; Th^{4+} (102 pm), U^{4+} (97 pm), and Cs^+ (167 pm) whose radii are much greater than plentiful ions of similar charge; and W^{6+} (62 pm), which has such a large charge that electrical neutrality is difficult to maintain when it replaces another ion. These elements concentrated in the molten phase as others crystallized out. They are usually found in pegmatites, the lowest melting fraction of the silicate melt and the last to crystallize in the crust. Although 90% of pegmatites consist essentially of quartz and feldspar, the remaining 10% are important sources of many rare elements, including those mentioned above. This is the reason that relatively large quantities of uranium are found at the surface of the earth despite its small cosmic abundance.

2.5 CONCENTRATION OF THE ELEMENTS INTO ORES

The cooling of the earth and crystallization of silicates did not end but rather began the evolution of the crust from which we now obtain a majority of our resources. Approximately four billion years have elapsed since the earth's core, mantle, and crust differentiated. During that time a variety of processes have acted to transform the igneous rocks which first crystallized, those which later intruded into the crust from the hotter mantle, or those extruded from the crust by volcanic action. In some cases these processes have served to concentrate compounds containing a particular element to the point where it has become economically feasible to extract the desired element. Such deposits, when readily available with respect to the surface and geographical location, constitute *ores*.

Besides the direct segregation from molten magma described earlier in the case of the Sudbury nickel deposits, processes of metamorphosis, weathering, hydrothermal transport, and sedimentation have contributed to ore formation. Metamorphosis refers to changes induced in existing crustal rock immediately adjacent to a molten magma. In the presence of carbonate minerals such as limestone, iron(II) oxide from the mantle can be partially oxidized (equation 2.3) to form magnetite, an important iron ore.

$$CaCO_3 \xrightarrow{\Delta} CaO + CO_2 \qquad (2.3a)$$

$$CO_2 + 3\ FeO \longrightarrow Fe_3O_4 + CO \qquad (2.3b)$$

This can only occur at locations where the high temperature of an iron-rich magma transforms already existing limestone, and hence does not account for the formation of all iron ore deposits.

Hydrothermal transport is illustrated by the equilibrium 2.4:

$$ZnS + 2\ HCl \rightleftharpoons ZnCl_2 + H_2S \qquad (2.4)$$

Since $ZnCl_2$ is relatively volatile (mp = 535 K, bp = 1005 K) and water soluble, it could be produced in aqueous solution or as a gas by action of HCl. This would allow it to be transported for some distance before the gas cooled or the acid was neutralized by some relatively basic rock such as limestone. In such locations ZnS (mp = 1293 K) might precipitate to form a concentrated deposit. Numerous ores of other chalcophiles such as copper, lead, mercury, and silver probably have formed in similar fashion.

Weathering of silicate rocks by rain or ground water can remove those elements (such as alkali metals) whose compounds are relatively soluble. (This process is described more fully in Section 12.1.). Weathering is extremely important in the development of soil, a resource not usually thought of as an ore, but which has an enormous influence on human well being. In advanced stages of weathering most of the cations (including silicon itself) originally present in silicate minerals are removed. Only those elements (such as aluminum) whose oxides or hydroxides are extremely insoluble remain behind in more concentrated form. Bauxite, our main source of aluminum, is formed this way in tropical regions where high rainfall is common.

Weathering also transports large quantities of dissolved elements to the oceans. Some, such as sodium, chlorine, magnesium, and bromine, can be recovered economically from solution, making the ocean itself an ore. In other cases, over the billions of years of geological evolution, compounds of intermediate and low solubility have formed sedimentary deposits. In localized, oxygen-deficient, less basic oceanic regions conditions are ideal for precipitation of apatite $[Ca_5(PO_4)_3(OH)]$ and fluorapatite $[Ca_5(PO_4)_3F]$. These form the deposits from which phosphate fertilizers (Section 11.1) are obtained, an excellent example of the role of sedimentation.

2.6 RELATIONSHIP OF ORES TO AVERAGE CRUSTAL ABUNDANCE

It should be clear from the foregoing discussion that there is considerable variation in the distribution of the elements throughout the earth's crust. If one analyzes a large number of samples for a given element, only a few will contain enough to be classed as ore. The entire collection of samples will usually exhibit a *lognormal distribution curve* of the type shown in Figure 2.6. The majority of the samples will fall somewhat below the average content of the element. The high concentrations rather than the great number of samples above the average are responsible for that figure being as large as it is. In statistical terms the distribution is *positively skewed*. Since the concentration of a metal in an ore is often more than 1000 times its average crustal abundance, samples from ores would be found at the far right-hand end of curves such as Figure 2.6.

One additional point should be emphasized regarding lognormal distribution curves. The total amount of an element summed over all samples is given by the integral of the distribution function (the area under the curve). Therefore only a very small proportion of the total amount of any element in the crust is available in ores. Suppose, for instance, that the minimum concentration of calcium which constitutes an ore deposit is 2.6% as CaO.

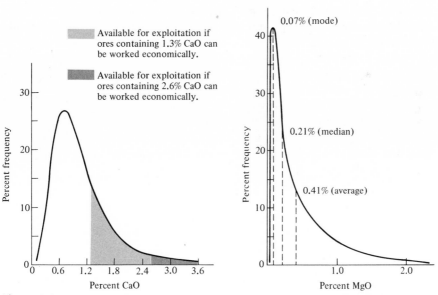

Figure 2.6 Occurrence of various concentrations of calcium and magnesium in large numbers of specimens of granite. Source: L. H. Ahrens, "Distribution of the Elements in Our Planet." McGraw-Hill, New York, 1965.

The quantity of calcium available for exploitation is then given by the double shaded area under the upper curve in Figure 2.6. If a new chemical process were developed or the value of the element were to increase such that deposits containing only half as much calcium (1.3% as CaO) could be worked economically, the amount of calcium available for recovery would increase by a factor of from six to eight (see the lightly shaded area of Figure 2.6).

A small advance in technology or a minor increase in the price of a substance can have a disproportionate effect on the known, exploitable reserves of that substance. This fact makes predictions of the quantities of mineral resources which will be available in the future somewhat difficult. They depend on a knowledge of the technology which will have been developed for extraction and purification, the value of the mineral, and even the geographic location of the deposits. One might also be tempted by the erroneous argument that because we now use such a small fraction of the total available amount of most metals, nearly unlimited supplies await only our ingenuity in finding ways to separate and purify them. Unfortunately, techniques for separation and purification will almost certainly require a nearly unlimited supply of energy, which does not appear to be a reasonable expectation for the future (see Section 7.5). The geological processes which concentrated the elements into ore deposits are extremely slow relative to the time scale of human exploitation, making these resources nonrenewable for all practical purposes. As we shall see in later chapters, the transformation of concentrated ores into dispersed solid waste may be an aspect of industrial civilization which man will eventually come to regret.

EXERCISES

(The first group of exercises is based primarily on material found in this text.)

1. Classify each of the elements below according to the Goldschmidt geochemical categories of atmophile, lithophile, chalcophile, or siderophile: (In cases where an element's properties allow it to fit more than one category list as many as apply.)

   a. Fe   b. N   c. O   d. Ni
   e. Zn   f. Li   g. Ne   h. Cu

2. How is the Goldschmidt geochemical classification of the elements related to the electromotive force series? Explain why metals which are more readily oxidized than iron are almost invariably lithophiles.

3. From the following list of cations select three groups of two or more which

might reasonably be expected to be found in similar minerals. (A table of ionic radii may be useful.)

Li^+ Na^+ K^+ Rb^+ Cu^{2+} Ca^{2+} Zn^{2+} Ba^{2+} Cd^{2+} Al^{3+} La^{3+} Th^{4+}

(Some of the ions in the list may not belong to any of the groups.)

4. What does the term "isomorphous substitution" mean? What rules must be followed if it is to occur?
5. Describe the different ways in which SiO_4 tetrahedra may be linked together in various minerals. Why are the densities of zeolites much less than the density of silica even though both have framework structures?
6. What factors enter into the classification of a mineral deposit as an ore? What are some of the chemical and physical processes by which ore deposits are formed?
7. What is a lognormal distribution curve? Explain its significance in terms of the effects of new technology on exploitable reserves of minerals.
8. The abundance of xenon on earth is nearly 10^7 times less than in the universe as a whole. Give a plausible explanation. How does this fact affect theories of the origin of the earth?

(Subsequent exercises may require more extensive research or thought.)

9. Suppose you had a sample of N_2, a sample of H_2, and a sample of O_2, each at 298 K and 1 atm, and you observed a single molecule from each sample. Can you say with certainty which molecule will be moving fastest? Why or why not? If you were placing a bet on the outcome, which choice would you make?
10. Use equation 2.1 to calculate the average molecular speeds of H_2, He, Ne, N_2, Kr, and Xe at the following temperatures: 400 K, 1000 K, 5000 K, 10 000 K, and 100 000 K. Which gases might escape from planet earth over a geological time scale? Which might have escaped from the moon? (The mass of the moon is 1.2% that of the earth.)
11. Under standard conditions thermodynamic parameters for Fe_2O_3 are

$$\Delta H_f^0 = -822 \text{ kJ}$$
$$\Delta S_f^0 = -0.272 \text{ kJ}$$

Assuming (incorrectly) that these values are independent of temperature, at what T would Fe_2O_3 decompose to its elements? How does this compare with the temperatures at which xenon and krypton would approach escape velocity from the earth (Exercise 10)? Would this estimate be improved if the absolute entropy of Fe(g) were available? How would the results be changed?
12. Although Ni^{2+} is present in the crust in much less abundance than Mg^{2+} the two ions have similar radii (68 and 65 pm, respectively) and occasionally are found in the same mineral. In such cases Ni^{2+} is found preferentially (by a factor of as much as 200) in the portion of the deposit which crystallized first. The same type of observation is made for chromium. What property of these transition elements might account for their preferential stability in the oxide lattice as opposed to the melt? (See L. H. Ahrens, "Distribution of the Elements in Our Planet," pp. 77–78. McGraw-Hill, New York, 1965; J. E. Huheey, "Inorganic Chemistry," pp. 299–316. Harper, New York, 1972.)

3

Chemical Evolution and the Origin of Life

We are not looking into the universe from outside. We are looking at it from inside. Its history is our history; its stuff our stuff. . . . Surely this is a great part of our dignity as men, that we can know, and through us matter can know itself; . . . .

George Wald

The great variety currently observed in the living organisms on the earth's surface (*the biosphere*) leads one to expect a similar variety of biological molecules. Within even a unicellular organism there may be as many as 5000 different substances. Fortunately for those who study biochemistry,[1] most of the important biological molecules are polymeric. Although more than 10^{10} different kinds of molecules are estimated to occur naturally in the biosphere, their structures are derived from combinations of fewer than 10^3 different monomers, and many of the latter (such as the amino acids) are closely related structurally. This limited number of building block molecules together with the fact that they vary little from species to species leads one naturally to the main thesis of biological evolution: all species have evolved from a common ancestor. That common ancestor was very likely an extremely simple organism which itself evolved through aggregation of inanimate substances. Prior to the emergence of this first life-form the biomonomers and polymers necessary for its existence may have been synthesized by processes which were not mediated by any

---

[1] The basic facts required for an understanding of this and later chapters relating to living organisms may be obtained from W. L. Masterton and E. J. Slowinski, "Chemical Principles," 3rd ed. Chapter 23. Saunders, Philadelphia, Pennsylvania, 1973.

living system; that is, there may have been a period of *chemical evolution* which preceded biological evolution in the development of the living environment.

Since 1950 many experiments have been done in the field of chemical evolution, and a fairly clear picture of the processes by which life could have evolved from inanimate matter has begun to appear. Indeed, some have argued that given primitive earth conditions and the nature of the chemical elements such evolution was inevitable. In the process of examining the evolution of the living environment and the types of molecules that are involved in it, it should become evident that each molecule has a particular function which it can fulfill because of its structure and chemical properties. Disruption of such structure and the consequent impairment of function can often occur as a result of environmental pollution.

A second major feature which arises from a consideration of chemical and biological evolution is that living systems have influenced and modified their environments over the geological time scale. This is most readily apparent in the case of the development of the atmosphere, although it is true of the lithosphere and hydrosphere as well. There is considerable evidence that the atmosphere of 20% O_2 and 80% N_2, which is today essential for the propagation of life, was originally very different. Indeed, it is likely that the existence of life was a necessary precursor to the oxidizing atmosphere we take for granted today.

3.1 THE ATMOSPHERE AND HYDROSPHERE OF PRIMITIVE EARTH

According to the scenario presented in Sections 2.1 and 2.2, many volatile substances escaped prior to and during formation of the earth. At this stage of evolution there was no atmosphere or hydrosphere, but eventually, as the earth cooled, liquids and gases which had been physically trapped or chemically bound began to escape through the surface by volcanic action. Because molten, metallic iron had not yet migrated to the core, such gaseous emissions probably were in intimate contact and at equilibrium with zerovalent iron as well as iron oxides and silicates.

Under such conditions strongly oxidizing gases such as O_2 should have reacted with iron. Holland[2] has used thermodynamic parameters to estimate that the partial pressure of dioxygen in volcanic gases at this stage of earth's development must have been on the order of 10^{-12} what it is in today's atmosphere. Equilibrium calculations similar to those in Exercise 12 indicate that the most common compounds released to the atmosphere were probably H_2, H_2O, and CO. Minor amounts of CO_2, N_2, and H_2S may also have been present in the volcanic emissions.

[2] H. D. Holland, *in* "Petrologic Studies" (A. E. G. Engle, H. L. James, and B. F. Leonard, eds.), p. 447. Geol. Soc. Amer., Boulder, Colorado, 1962.

Once gases reached the surface they cooled fairly rapidly since the crust was relatively cool. A number of equilibria involving atmospheric gases were probably affected by this temperature change. Most of the water condensed to form the hydrosphere, leaving H_2O as only a minor atmospheric component. Carbon oxides may have been reduced to methane according to equations 3.1 and 3.2.[3]

$$CO_2(g) + 4\ H_2(g) \rightleftharpoons CH_4(g) + 2\ H_2O(g) \tag{3.1}$$

$$\Delta G^0_{298} = -39.44 - T(-0.04118) = -27.17\ \text{kJ}$$

$$CO(g) + 3\ H_2(g) \rightleftharpoons CH_4(g) + H_2O(g) \tag{3.2}$$

$$\Delta G^0_{298} = -33.97 - T(-0.05132) = -18.68\ \text{kJ}$$

Dinitrogen might also have been partly reduced to form ammonia, or in the hydrosphere, ammonium ions. It was under these seemingly unfavorable conditions that the first molecular steps toward life were probably taken.

3.2 SYNTHESIS OF BUILDING BLOCK MOLECULES

It is possible to simulate in the laboratory the conditions described in the preceding section. The first successful primitive earth simulation was performed by Miller[4] in the early 1950's, using the apparatus shown in Figure 3.1. Urey[5] had just proposed that the original atmosphere of the earth was strongly reducing, so Miller confined a mixture of methane, ammonia, dihydrogen, and water in his apparatus. Energy was supplied by an electric discharge (to simulate lightning in the atmosphere) and by boiling the liquid water (to simulate the hydrologic cycle of evaporation and precipitation). The latter energy input set up a cyclic flow of vapors past the electric discharge.

The usual procedure in primitive earth simulations is to set up the system and then allow it to operate for a few weeks without human intervention. Since the geological time scale for chemical evolution was probably on the order of millions of years, the formation in a few weeks of trace quantities of biological molecules is significant. After one week of sparking with 60 000 V discharge (which supplied about 6000 kJ), Miller observed that the aqueous phase had turned from colorless to yellow and a yellow-brown material had collected on the electrodes. Chromatographic analysis of a portion of the aqueous phase indicated the presence of the compounds shown in Table 3.1.

The most striking aspect of the Miller–Urey experiment was the discovery that amino acids such as glycine, alanine, and glutamic and aspartic

---

[3] For arguments pro and con this hypothesis, see D. H. Kenyon and G. Steinman, "Biochemical Predestination," pp. 90–105. McGraw-Hill, New York, 1969.

[4] S. L. Miller, *Science* **117**, 528 (1953).

[5] H. C. Urey, "The Planets." Yale Univ. Press, New Haven, Connecticut, 1952.

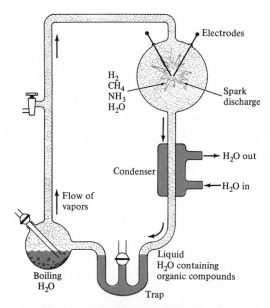

Figure 3.1 Apparatus used by Miller for simulation of primitive earth atmosphere.

acids could have been synthesized on the primitive earth without the intervention of living systems. These compounds are among the "building blocks" from which the protein molecules found in all modern organisms may be constructed by polymerization reactions.

Subsequent to Miller's pioneering work many other investigators have carried out primitive earth simulations. In addition to amino acids, they have synthesized purines and pyrimidines, sugars, nucleosides, nucleotides, and porphyrins. A summary of their results is provided in Table 3.2. Ap-

TABLE 3.1 Structures and Yields of Some Compounds Formed by Electrical Discharge in a Reducing Primitive Atmosphere[a]

Name	Yield (μmol)	% Yield[b]	Structure
Formic acid	233	3.9	HCOOH
Glycine	630	2.1	H_2NCH_2COOH
Glycolic acid	560	1.9	$HOCH_2COOH$
Lactic acid	390	1.8	$CH_3CH(OH)COOH$
Alanine	340	1.7	$CH_3CH(NH_2)COOH$
Propionic acid	126	0.6	CH_3CH_2COOH
Acetic acid	152	0.5	CH_3COOH
Glutamic acid	6	—	$HOOCCH_2CH_2CH(NH_2)COOH$
Aspartic acid	4	—	$HOOCCH_2CH(NH_2)COOH$

[a] S. L. Miller, *J. Amer. Chem. Soc.* **77**, 2351 (1955); *Biochim. Biophys. Acta* **23**, 480 (1957).
[b] Based on initial carbon.

TABLE 3.2 Representative Abiotic Synthesis Experiments[a]

Compound class	Reactants	Energy	Products	Investigators
Amino acids	CH_4, NH_3, H_2, H_2O	Electric discharge	Amino acids, hydroxyl acids, HCN, urea	Miller, 1953, 1955
	Ammonium fumarate	Heat	Aspartic acid	Fox, Johnson, Middlebrook, 1955
	CO_2, NH_3, H_2, H_2O	Electric discharge	Amino acids	Abelson, 1956
	CH_4, NH_3, H_2O, H_2, CO_2, N_2	x Ray	Amino acids	Dose, Rajewsky, 1957
	Ammonium acetate	γ Ray	Glycine, aspartic acid, diaminosuccinic acid	Hasselstrom, Henry, Murr, 1957
	Ammonium carbonate	β Ray	Glycine	Paschke, Chang, Young, 1957
	CH_4, NH_3, H_2O	uv	Glycine, alanine	Groth, von Weyssenhoff, 1957
	NH_3, HCN, H_2	Heat (343 K)	Amino acids	Oró, Kamat, 1961
	CH_4, NH_3, H_2O	Accelerated electron	Glycine, alanine	Palm, Calvin, 1962
	CH_4, NH_3, H_2O	Heat (>1123 K)	Amino acids	Harada, Fox, 1964
	CO, H_2, NH_3 (Ni-Fe, Fe_3O_4, Al_2O_3, SiO_2 catalysts)	Heat (750–1000 K)	Amino acids	Anders, Hayatsu, Studier, 1973
Purines, pyrimidines	$HC{\equiv}C{-}CN$, HCN, NH_3, H_2O	Heat (373 K)	Aspartic acid	Sanchez, Ferris, Orgel, 1966
	HCN, NH_3, H_2O	Heat (343 K)	Adenine	Oró, Kimball, 1961
	Malic acid, urea, polyphosphoric acid	Heat (403 K)	Uracil	Fox, Harada, 1961
	CH_3, NH_3, H_2O	Accelerated electron	Adenine	Ponnamperuma, Lemmon, Mariner, Calvin, 1963
	CO, H_2, NH_3, catalysts	Heat (700–1000 K)	Adenine, guanine	Anders, Hayatsu, Studier, 1973

	Reactants	Conditions	Products	Reference
Sugars	HCHO, CH$_3$CHO; glyceraldehyde, acetaldehyde, Ca(OH)$_2$	Heat (323 K)	2-Deoxyribose 2-deoxyxylose	Oró, Cox, 1962
	HCHO	uv	Ribose, deoxyribose	Ponnamperuma, 1965
Nucleotides	Adenosine, polyphosphate ester	uv	AMP, ADP, ATP	Ponnamperuma, Sagan, Mariner, 1963
	Nucleoside, phosphate	Heat (433 K)	Nucleotides	Ponnamperuma, Mack, 1965
	Nucleosides, polyphosphoric acid	Heat (295 K)	Nucleotides	Waehneldt, Fox, 1967
Hydrocarbons	Methane	Electric discharge	Higher hydrocarbons	Ponnamperuma, Woeller, 1964
	Methane	Heat (1273 K; silica gel)	Higher hydrocarbons	Oró, Han, 1966
	CO, H$_2$, catalysts	Heat (750–1000 K)	Linear alkanes	Anders, Hayatsu, Studier, 1973
Porphyrins	Pyrrole, benzaldehyde	γ Rays	Tetraphenylporphyrin	Szutka, Hazel, Menabb, 1959
	Pyrrole, formaldehyde, Ni^{2+} Cu^{2+}	Electric discharge	Porphyrin	Hodgson, Baker, 1967
	CH$_4$, NH$_3$, H$_2$O		Porphyrin	Hodgson, Ponnamperuma, 1968
	CO, H$_2$, NH$_3$, catalysts	Heat (750–1000 K)	Cyclic or linear pyrrole polymers	Anders, Hayatsu, and Studier, 1973

[a] Source: S. W. Fox, K. Harada, G. Krampitz, and G. Mueller, Chemical origin of cells. *Chem. Eng. News* June 22, p. 86 (1970). Reprinted with permission. Copyright by the American Chemical Society.

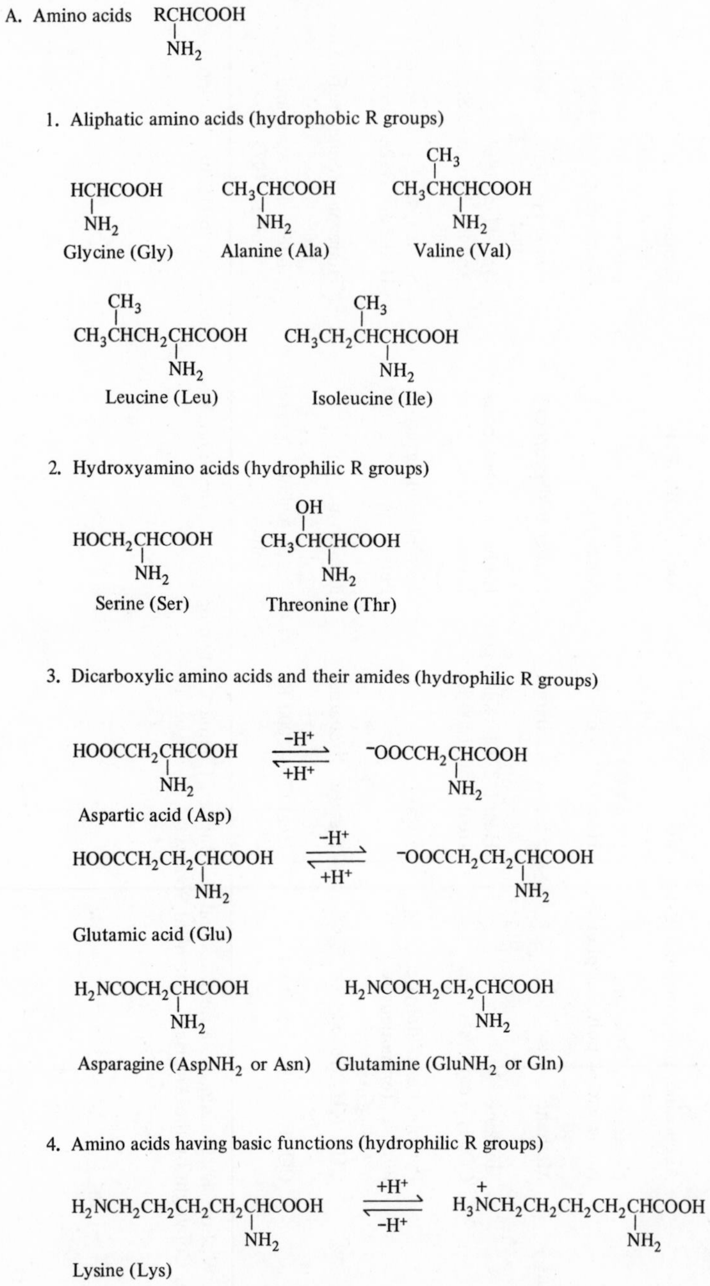

A. Amino acids $\underset{\underset{NH_2}{|}}{R}CHCOOH$

1. Aliphatic amino acids (hydrophobic R groups)

$\underset{\underset{NH_2}{|}}{H}CHCOOH$ $CH_3\underset{\underset{NH_2}{|}}{C}HCOOH$ $CH_3\underset{\underset{NH_2}{|}}{\overset{\overset{CH_3}{|}}{C}}HCHCOOH$

Glycine (Gly)   Alanine (Ala)   Valine (Val)

$CH_3\overset{\overset{CH_3}{|}}{C}HCH_2\underset{\underset{NH_2}{|}}{C}HCOOH$ $CH_3CH_2\underset{\underset{NH_2}{|}}{\overset{\overset{CH_3}{|}}{C}}HCHCOOH$

Leucine (Leu)   Isoleucine (Ile)

2. Hydroxyamino acids (hydrophilic R groups)

$HOCH_2\underset{\underset{NH_2}{|}}{C}HCOOH$ $CH_3\underset{\underset{NH_2}{|}}{\overset{\overset{OH}{|}}{C}}HCHCOOH$

Serine (Ser)   Threonine (Thr)

3. Dicarboxylic amino acids and their amides (hydrophilic R groups)

$HOOCCH_2\underset{\underset{NH_2}{|}}{C}HCOOH \underset{+H^+}{\overset{-H^+}{\rightleftharpoons}} {}^-OOCCH_2\underset{\underset{NH_2}{|}}{C}HCOOH$

Aspartic acid (Asp)

$HOOCCH_2CH_2\underset{\underset{NH_2}{|}}{C}HCOOH \underset{+H^+}{\overset{-H^+}{\rightleftharpoons}} {}^-OOCCH_2CH_2\underset{\underset{NH_2}{|}}{C}HCOOH$

Glutamic acid (Glu)

$H_2NCOCH_2\underset{\underset{NH_2}{|}}{C}HCOOH$ $H_2NCOCH_2CH_2\underset{\underset{NH_2}{|}}{C}HCOOH$

Asparagine (AspNH$_2$ or Asn)   Glutamine (GluNH$_2$ or Gln)

4. Amino acids having basic functions (hydrophilic R groups)

$H_2NCH_2CH_2CH_2CH_2\underset{\underset{NH_2}{|}}{C}HCOOH \underset{-H^+}{\overset{+H^+}{\rightleftharpoons}} \overset{+}{H_3}NCH_2CH_2CH_2CH_2\underset{\underset{NH_2}{|}}{C}HCOOH$

Lysine (Lys)

Figure 3.2   Some important building blocks of biochemistry.

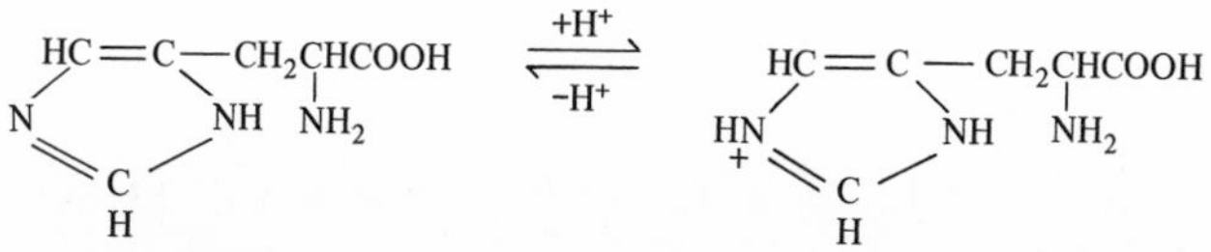

$$HC = C - CH_2CHCOOH$$

Histidine (His)

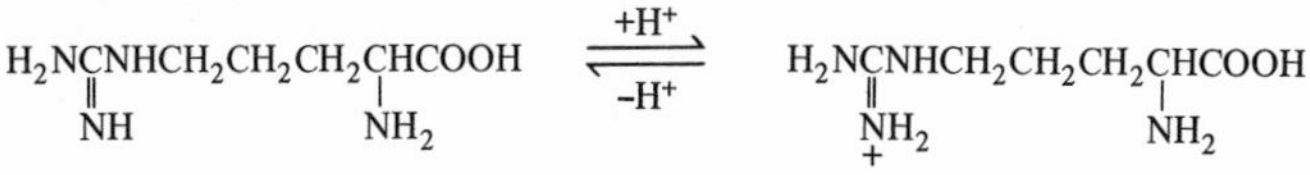

$$H_2NCNHCH_2CH_2CH_2CHCOOH$$

Arginine (Arg)

5. Aromatic amino acids (hydrophobic R groups)

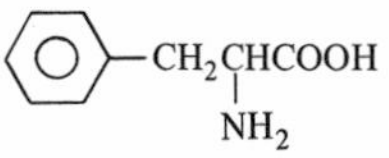

Phenylalanine (Phe)

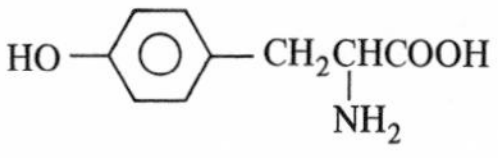

Tryosine (Tyr)

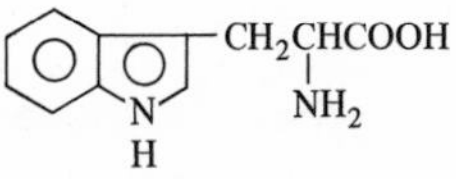

Tryptophan (Trp)

(histidine has been included in the preceding category)

6. Sulfur-containing amino acids (hydrophobic R groups except cysteine)

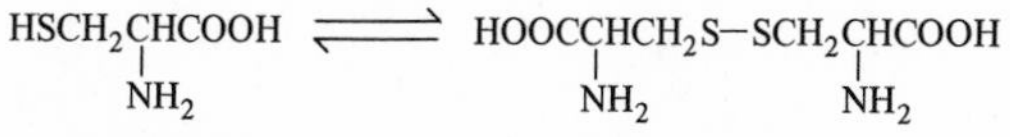

$$HSCH_2CHCOOH \rightleftharpoons HOOCCHCH_2S-SCH_2CHCOOH$$

Cysteine (CySH)   Cystine (CyS-SCy)

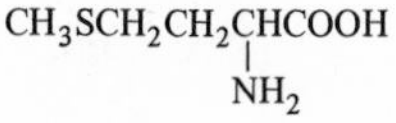

$$CH_3SCH_2CH_2CHCOOH$$

Methionine (Met)

7. Imino acid (hydrophobic R group)

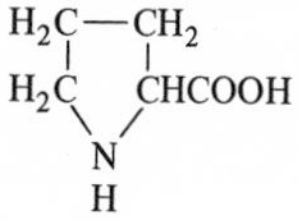

Proline (Pro)

Figure 3.2 (Cont.)

B.   Pyrimidines

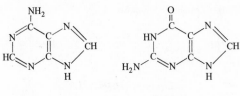

Uracil   Thymine   Cytosine

C.   Purines

Adenine   Guanine

D.   Sugars

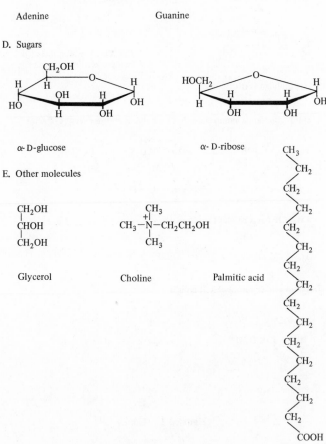

α- D-glucose   α- D-ribose

E.   Other molecules

CH₂OH   CH₃
| +|
CHOH   CH₃—N—CH₂CH₂OH
| |
CH₂OH   CH₃

Glycerol   Choline   Palmitic acid

Figure 3.2   (Cont.)

parently all that is necessary for synthesis of the building block molecules of biochemistry is the interaction of compounds containing the appropriate elements (carbon, hydrogen, nitrogen, oxygen, phosphorus, and sulfur are the most important) with a source of energy. It is not necessary, for example, that carbon be completely reduced to CH_4 or nitrogen to NH_3, so long as these elements *are present*. Under these circumstances many of the biological molecules shown in Figure 3.2 are formed, at least in small yields.

3.3 THE IMPORTANCE OF "REACTIVE COMPOUNDS"

Several so-called "reactive compounds" were apparently quite important both in the initial synthesis of biomonomers described in the preceding section and in the linking of these monomers to form polymers. The most important reactive molecules, formaldehyde and hydrogen cyanide, have also been discovered in interstellar space (Table 1.2), giving further credence to the hypothesis that reasonably large quantities could form on primitive earth.

Formaldehyde and hydrogen cyanide are the simplest, smallest molecules containing respectively, carbon, hydrogen, and oxygen, and carbon, hydrogen, and nitrogen. Carbohydrates, which have the empirical formula CH_2O, may be regarded as polymers of formaldehyde. Polymerization of H_2CO made basic with limestone, calcium oxide, or ammonia yields sugars containing two, three, four, five, or six carbon atoms.[6] Among these reaction products are ribose, 2-deoxyribose, galactose, fructose, and mannose.

The importance of hydrogen cyanide appears at first glance to be an example of what Forrester[7] has termed the "counterintuitive behavior" of complex systems. One does not expect what today is an extremely poisonous substance to have played a major role in the early evolution of life. Upon closer chemical examination, however, one finds a number of properties which make HCN fit for such a role. In organic compounds the —C≡N (nitrile) group undergoes hydrolysis to produce carboxylic acids, providing a reasonable pathway for production of compounds such as the amino acids. Also, the large degree of unsaturation represented by the triple bond of the cyanide ion allows many molecules to undergo addition reactions with HCN, producing higher molecular weight compounds.

Miller proposed that a Strecker synthesis (involving NH_3, H_2CO, HCN, and H_2O) produced glycine and other amino acids in his pioneering experiments. Such a reaction sequence is shown in equations 3.3.

$$\begin{array}{c}H\\ \diagdown \\ \diagup \\ H\end{array}C{=}O \quad + \quad NH_3 \quad \rightleftharpoons \quad \begin{array}{c}H\\ \diagdown \\ \diagup \\ H\end{array}C{=}NH \quad + \quad H_2O \qquad (3.3a)$$

[6] J. Oró, *in* "The Origins of Prebiological Systems" (S. W. Fox. ed.), p. 157. Academic Press, New York, 1965.

[7] J. W. Forrester, *in* "Toward Global Equilibrium: Collected Papers" (D. L. Meadows and D. H. Meadows, eds.), Chapter 1. Wright-Allen Press, Inc., Cambridge, Massachusetts, 1973.

$$\underset{H}{\overset{H}{}}\!\!\diagup\!\!\!C\!=\!NH \;+\; HCN \;\longrightarrow\; H\!-\!\underset{\underset{C\equiv N}{|}}{\overset{\overset{H}{|}}{C}}\!-\!NH_2 \tag{3.3b}$$

$$H\!-\!\underset{\underset{C\equiv N}{|}}{\overset{\overset{H}{|}}{C}}\!-\!NH_2 \;+\; H_2O \;\longrightarrow\; H\!-\!\underset{\underset{O}{\diagdown}{\overset{C}{\diagup}}NH_2}{\overset{\overset{H}{|}}{C}}\!-\!NH_2 \tag{3.3c}$$

$$H\!-\!\underset{O\diagup^{C}\diagdown NH_2}{\overset{\overset{H}{|}}{C}}\!-\!NH_2 \;+\; H_2O \;\longrightarrow\; H\!-\!\underset{O\diagup^{C}\diagdown OH}{\overset{\overset{H}{|}}{C}}\!-\!NH_2 \;+\; NH_3 \tag{3.3d}$$

(Glycine)

More recent work has indicated that polymerization of HCN, apparently by successive additions of HCN across the C≡N triple bond, may be involved in the synthesis of many biologically important molecules. As shown in equations 3.4, oligomers through the tetramer have been characterized. Addition of one more HCN molecule to the photoisomerized tetramer provides a route for the synthesis of adenine which has recently been developed into an industrial process.[8] There is evidence that more than fifty different organic substances observed in primitive earth simulations might have been derived from HCN.

$$H\!-\!C\equiv N \;\xrightarrow{\;CN^-\;}\; H\!-\!\underset{\underset{C\equiv N}{|}}{C}\!=\!N^- \;\xrightarrow{\;H^+\;}\; H\!-\!\underset{\underset{C\equiv N}{|}}{C}\!=\!NH \tag{3.4a}$$

$$H\!-\!\underset{\underset{C\equiv N}{|}}{C}\!=\!NH \;+\; HCN \;\longrightarrow\; H\!-\!\underset{\underset{C\equiv N}{|}}{\overset{\overset{C\equiv N}{|}}{C}}\!-\!NH_2 \tag{3.4b}$$

$$H\!-\!\underset{\underset{C\equiv N}{|}}{\overset{\overset{C\equiv N}{|}}{C}}\!-\!NH_2 \;+\; HCN \;\longrightarrow\; H\!-\!\underset{N\equiv C\diagup^{C}\diagdown NH}{\overset{\overset{C\equiv N}{|}}{C}}\!-\!NH_2 \;\longrightarrow\; \underset{N\equiv C\diagdown_{C}\diagup NH_2}{\overset{N\equiv C\diagdown_{C}\diagup NH_2}{}} \tag{3.4c}$$

$$\underset{N\equiv C\diagdown_{C}\diagup NH_2}{\overset{N\equiv C\diagdown_{C}\diagup NH_2}{}} \;\xrightarrow{\;h\nu\;}\; \underset{H_2N\diagdown_{C}\diagup N}{\overset{N\equiv C\diagdown_{C}\diagup N}{}}\!C\!-\!H \tag{3.4d}$$

$$\underset{H_2N\diagdown_{C}\diagup N}{\overset{N\equiv C\diagdown_{C}\diagup N}{}}\!C\!-\!H \;+\; HCN \;\longrightarrow\; \underset{H\diagdown_{C}\diagup N}{\overset{NH_2}{\underset{N\diagdown_{C}\diagup N}{}}}\!C\!-\!H \tag{3.4e}$$

---

[8] *Chem. & Eng. News* August 8, p. 39 (1966).

3.4 FORMATION OF BIOPOLYMERS

Most of the molecules whose structures are shown in Figure 3.2 are polyfunctional; that is, they have more than one reactive site. For example, all amino acids contain an acid group

$$-C\diagup^{O}_{\diagdown OH}$$

and an α-amino function (—NH$_2$ on a carbon atom adjacent to the COOH):

$$R-\underset{\underset{NH_2}{|}}{\overset{\overset{H}{|}}{C}}-C\diagup^{O}_{\diagdown OH}$$

Different amino acids are distinguished by the different groups, R, which are attached to this characteristic structure. Sugars such as glucose or ribose contain several —OH groups, each of which represents a site of high reactivity. Therefore, once formed, the primordial biomolecules were capable of linking together with other molecules of the same or different structures to form polymers. Because each molecule contained at least two reactive sites it could serve as a bridge between two others and contribute to the formation of a long chain of monomers. At each step in the process of forming such a chain any one of several monomers could become attached to it, providing a means for construction of a tremendous variety of polymers from only a few initial molecules. This variety permitted a natural selection of those polymers whose properties were most nearly ideal and provided a mechanism by which minor changes in these properties could occur and be tested.

A few important biomolecules are monofunctional, and an important class of biological molecules (the lipids) which is derived from them is not polymeric, although molecular weights are rather high. For example, palmitic acid (see Figure 3.2) contains only a single

$$-C\diagup^{O}_{\diagdown OH}\quad\text{group.}$$

Once this group has reacted with one of the —OH groups in glycerol, all reactive sites of the palmitic acid molecule are used up and no polymerization can occur. Nevertheless, glycerol is trifunctional (it has three —OH groups), and the lipids can be built up by reaction of three other molecules with it. Thus, in nearly all cases the important molecules in living systems can be built up by combinations of the simple molecules in Figure 3.2.

It is rather interesting to note that the reactions involved in the linking together of biomonomers all have a common feature. A number of these reactions are shown in Figure 3.3, where it is easily seen that for every pair of monomers which is connected together one molecule of water is given off. A reaction in which a small molecule such as water, CO_2 or HCl is given off when two larger molecules combine is known as a condensation. When the product of a condensation which yields H_2O is brought into contact with excess water, the reverse reaction, hydrolysis, can occur. For all of the reactions in Figure 3.3 it has been found that hydrolysis of the products is extensive. This immediately raises the question of how appreciable quantities of biopolymers could be formed in spite of their lack of stability.

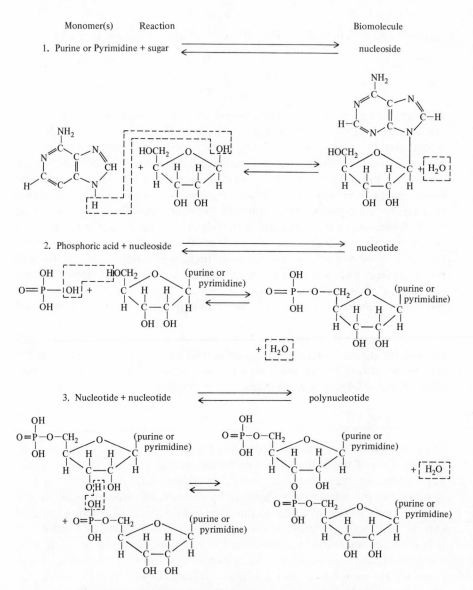

Figure 3.3   Linking of biomonomers to form active biomolecules.

Two means of supplying the work necessary to overcome the positive ΔG's of the reactions in Figure 3.3 come readily to mind. First, if the temperature were high enough, *pyrocondensation* could take place. Water, being relatively volatile compared to the other molecules, would vaporize, and its removal would shift the equilibria to the right. Volcanism on the primitive earth could easily have supplied the heat necessary for pyrocondensation. Reactions such as those in Figure 3.3 could also be forced in a forward direction if they were coupled with other reactions which used up water and had negative free energy changes. This would mean that a compound with a large affinity for water must have been present. Hydrogen

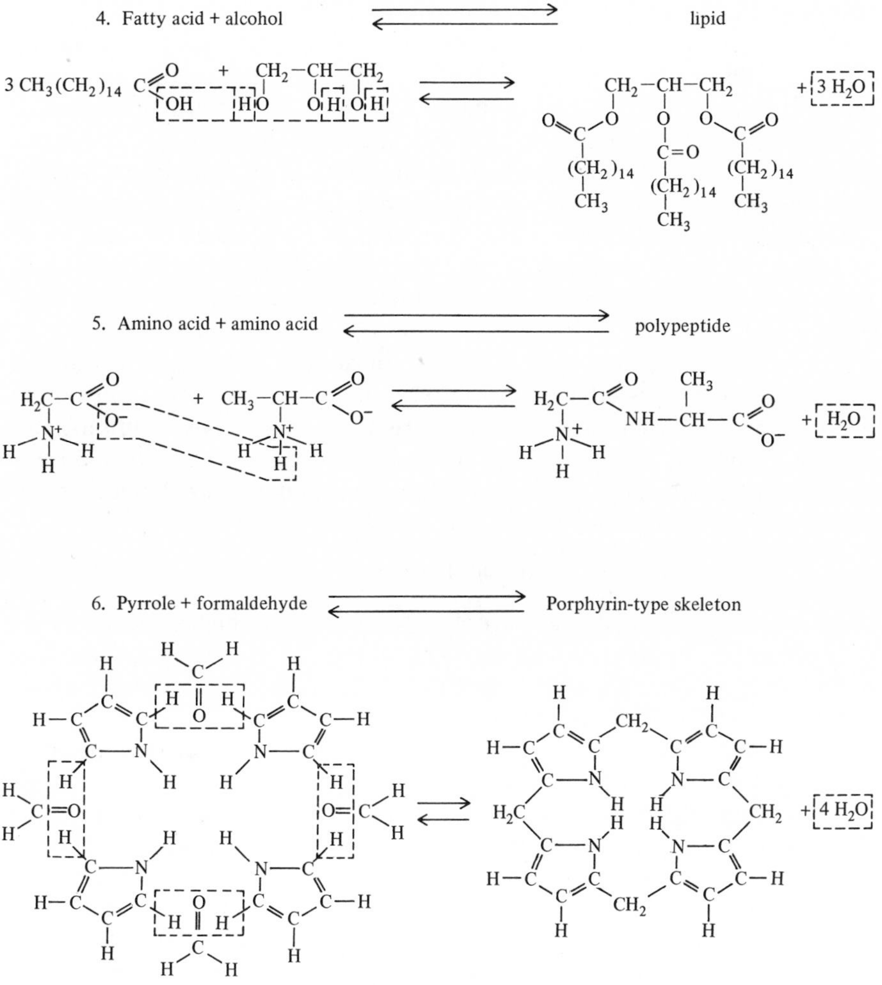

4. Fatty acid + alcohol ⇌ lipid

5. Amino acid + amino acid ⇌ polypeptide

6. Pyrrole + formaldehyde ⇌ Porphyrin-type skeleton

Figure 3.3 (Cont.)

cyanide, some of its derivatives, and a number of polyphosphates may have played such a role since they hydrolyze readily.

3.5 APPEARANCE OF LIVING CELLS

Although the discussion of the chemical evolution of polymeric bio-molecules has necessarily been limited, it should be evident that plausible mechanisms have been found in the laboratory for the conversion of the constituents of the primitive atmosphere and hydrosphere into the types of molecules found in living organisms today. But how did the *cells* from which all modern organisms are made arise? How did molecules formed at low concentrations in the "primeval soup" of the early oceans collect together to form the first living organism?

Contemporary cells contain membranes, a nucleus, exhibit metabolism, and are capable of self-replication and self-regulation. However, it is unlikely that such a cell sprang full-blown from the primordial mixture of biomonomers and biopolymers. The minimum requirement for a primitive "protocell" is that it must have had the capability of *evolving* into a modern cell. The characteristic of a primitive cell which is probably the most favor-able toward such evolution is the formation of membranes because these provide a means of separating and concentrating biopolymers in a region where conditions were better regulated and natural selection could occur more readily.

One of the originators of the concept of abiogenic chemical evolution, A. I. Oparin, suggested that the process known as *coacervation*[9] could account for the formation of cell-like structures in aqueous solutions. Co-acervation occurs when water of hydration is removed from a relatively nonpolar molecule such as a biopolymer. For example, as the concentra-tion of KCl in an aqueous solution of potassium oleate

$$K^{+-}O_2C(CH_2)_7CH = CH(CH_2)_7CH_3$$

is increased, small coacervate droplets which contain high concentrations of the long-chain fatty acid appear. The added salt competes for water of hydration with the oleate ions, and the latter associate, forming a phase separate from the aqueous solution.

Coacervate droplets formed from biologically active macromolecules have been shown to absorb organic materials from their immediate envi-ronment, increasing their mass and volume. Those which could perform this function more rapidly would grow at the expense of their less active neighbors, and an assemblage of biopolymers which could carry out benefi-cial reactions would be favored by natural selection. External forces (wind or waves) or growth beyond the capacity of surface tension to maintain the integrity of the coacervate droplet could be considered as a primitive form

---

[9] A. I. Oparin, "Origin of Life." Dover, New York, 1953.

of cell division. The development of reaction pathways which would make them less dependent on the external environment might lead to specialization in the functions of any catalytic substances included in a droplet, initiating the evolution of enzymes.

Thermal polymerization (pyrocondensation) of amino acids to "proteinoids" has also been shown by Fox and Harada[10] to produce "microspheres" which, upon microscopic examination, closely resemble simple bacterial cells. Other investigators have discovered a variety of ways of generating physical subsystems which can separate and protect the material within them from its immediate environment, selectively transfer molecules in and out, and permit high concentrations of substrates to come in contact with biopolymers having catalytic activity. Coacervate droplets, proteinoid microspheres, and other similar phenomena all appear to have the capability of evolving into the cells we are familiar with today, although such evolutionary development has not yet been demonstrated in the laboratory.

It must be emphasized that prebiotic chemical evolution is a relatively new and rapidly growing area of research. Many of the scenarios suggested in the preceding paragraphs would not be agreed to by all scientists, and some hypotheses regarding the development of life have been neglected altogether. It should be clear, however, that plausible mechanisms, which do not violate the accepted laws of chemistry and physics, do exist for the development of life on planet earth.

The view of chemical evolution set forth so far in this chapter is summarized in Figure 3.4. The next step in the development of the biosphere was conversion of coacervates, microspheres, or other protocells into more complex unicellular organisms. The biological evolution of these organisms into modern plant and animal life was thus initiated.

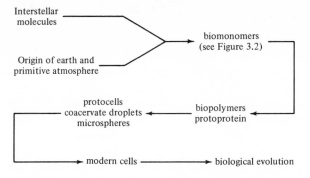

Figure 3.4 Flow sheet for one view of chemical evolution.

[10] S. W. Fox and K. Harada, *Science* **128**, 1214 (1958).

3.6   THE SECOND STAGE OF ATMOSPHERIC EVOLUTION[11]

At about the same time that chemical evolution was occurring, metallic iron was migrating to the earth's core. Since they were no longer in contact with iron, gases emitted to the atmosphere became more highly oxidized. At this stage the principal volcanic gases were probably H_2O, CO_2, and SO_2, with smaller amounts of N_2 and H_2. This composition is little different from contemporary emissions.

The three major components of volcanic vapor were probably removed from the atmosphere rather rapidly. Water continued to condense in the oceans. The acidic oxides CO_2 and SO_2 reacted with relatively basic crustal minerals (equation 3.5).

$$CaSiO_3 + CO_2 \rightleftharpoons CaCO_3 + SiO_2 \tag{3.5}$$

Reduced compounds such as CH_4 and NH_3 began to oxidize as the partial pressure of atmospheric H_2 decreased. Although the quantity of free O_2 increased, it was still only a very minor component. This left N_2 to ac-

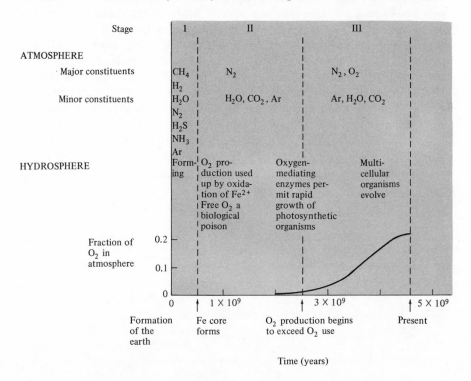

Figure 3.5   Stages in evolution of the atmosphere and hydrosphere.

---

[11] Much of the information in this section in based on P. Cloud, *Amer. Sci.* **62**(1), 54–66 (1974).

cumulate as the principal atmospheric gas. This change is indicated in Figure 3.5 which summarizes the evolution of both the air and water environments.

The first living organisms probably developed in shallow bodies of water or at the edges of the oceans. They must have been *anaerobic* and *heterotrophic*; that is, they obtained their energy from fermentation processes in the absence of dioxygen and they were dependent on externally synthesized molecules for their energy production. Eventually an *autotrophic* organism, one which could utilize an external energy source to manufacture complex molecular structures from simple ones, evolved. It is highly probable that the first autotrophs utilized photosynthesis and produced dioxygen. However, the multiplication of such photoautotrophs may well have been hindered by a biological poison which they themselves produced—dioxygen! Because the compounds from which these first photosynthesizers were made were in a relatively reduced state, the dioxygen or peroxides which they synthesized would have been likely to have burned up the organisms themselves.

One mechanism by which this problem might have been circumvented is for a strong reducing agent to have trapped dioxygen as soon as it formed. A good candidate for such a task is iron(II) ion. So-called *banded iron formations* containing alternate layers of Fe_2O_3-rich and iron-poor silica were deposited between about 2×10^9 and 3×10^9 years ago. Similar sedimentary deposits are not found in rocks of more recent origin, and it seems reasonable to suppose that iron(III) in these formations resulted from oxidation of iron(II) which came into contact with the early photoautotrophs. Because the dioxygen produced by the earliest photosynthetic organisms was trapped in sedimentary deposits there would have been little increase in the partial pressure of dioxygen in the atmosphere during this period.

3.7 THE THIRD STAGE OF ATMOSPHERIC EVOLUTION

The evolution of protein molecules which could transport dioxygen safely away from the sites where it was produced heralded the third stage in atmospheric evolution. Photosynthetic organisms could then spread throughout the hydrosphere, and their increased O_2 production converted the remaining iron(II) compounds in the oceans to less soluble iron(III) species. These precipitated to form additional banded iron formations, and only then did the concentration of O_2 in the atmosphere begin to increase.

Much of this initial dioxygen was consumed by the great number of reduced compounds which still remained in the atmosphere and crust. The rates of such reactions were probably increased by the photolytic formation of the strong oxidants ozone and atomic oxygen:

$$O_2 \xrightarrow[\text{(λ < 242 nm)}]{h\nu} O + O \tag{3.6a}$$

$$O_2 + O + M \longrightarrow O_3 + M \text{ (excited state)} \tag{3.6b}$$

(M is a third molecule, needed to carry off excess energy which might cause O_3 to dissociate.) Red beds, in which iron(III) oxide coats individual grains of sedimentary rocks, are found to have been formed about 2×10^9 years ago. They are apparently the result of atmospheric oxidation of previously deposited material by O or O_3.

Photochemical reactions such as 3.6a do not occur at the surface of the earth today because light in the ultraviolet region of the spectrum is prevented from reaching the surface. The very ozone produced by reaction 3.6b absorbs most of the uv photons having wavelengths less than 340 nm.

$$O_3 \xrightarrow{h\nu} O_2 + O \tag{3.6c}$$

Oxygen atoms produced by equation 3.6c react with dioxygen molecules (see equation 3.6b) to regenerate ozone. At the same time excess energy originally supplied by the ultraviolet radiation is carried off by the third molecule, M, and contributes to the warming of the upper atmosphere. Thus, equations 3.6b and 3.6c are steps in a chain reaction mechanism which is initiated by 3.6a. The chain can be interrupted by combination of ozone with atomic oxygen:

$$O_3 + O \longrightarrow 2 O_2 \tag{3.6d}$$

or by reaction of oxygen with itself:

$$O + O + M \longrightarrow O_2 + M \tag{3.6e}$$

Both processes are relatively slow because the concentrations of O_3 and O are of the order of 10^{-5} and 10^{-10} the total concentration of molecules. The overall effect of this sequence of reactions (equations 3.6a–e) is that ozone serves as a catalyst for the conversion of ultraviolet radiant energy into translational kinetic energy of molecules in the earth's upper atmosphere, maintaining the relatively high temperatures observed in the stratosphere (see Section 8.1).

Prior to the formation of this ozone shield against uv light the entire surface of the earth was bathed with radiation whose photons contained large quantities of energy. A quantum whose wavelength is 200 nm has $E = h\nu = hc/\lambda = (6.62 \times 10^{-34} \text{ J s}) \times (3.0 \times 10^8 \text{ m/s})/(200 \times 10^{-9} \text{ m}) = 9.93 \times 10^{-19}$ J. The energy required to break most chemical bonds is less than 100 kcal/mol or 6.95×10^{-19} J/bond. Thus, a great many photochemical reactions in addition to those already mentioned should have been possible. Until the partial pressure of dioxygen rose to about 5% of its current level and could provide enough ozone to screen out much uv radiation, life on earth was restricted to those habitats (usually under water) which afforded some protection from the sun's high energy rays. The photosynthetic production of O_2 was probably greatly augmented by the formation of the ozone screen because it allowed the photoautotrophic organisms a much broader range (the solid land as well as the hydrosphere) and therefore a much greater population.

When dioxygen is produced by photosynthesis, the oxidation number of each of its atoms is raised by two units from the value of -2 that oxygen normally has in water or the silicate minerals of the crust. A question that comes immediately to mind is: If oxygen is oxidized, what is reduced? The answer is carbon, as equation 3.7 shows:

$$n\ CO_2 + n\ H_2O \underset{\text{release of heat}}{\overset{h\nu}{\rightleftharpoons}} (CH_2O)_n + n\ O_2 \tag{3.7}$$

The fate of this reduced carbon is also of crucial importance to the development of the atmosphere. The carbohydrates, whose general formula is $(CH_2O)_n$, are today almost entirely consumed by living organisms as a source of energy, utilizing the reverse of reaction 3.7. Most dioxygen which is produced by photosynthesis is soon reduced to the -2 state by respiration or decay processes, as indicated in Figure 3.6.

A very small fraction (about four parts in 10^4 at present) of the organic carbon produced by photosynthesis manages to escape reoxidation. This usually occurs by deposition of the organic material as sediments at the bottoms of the oceans where there is little O_2. According to equation 3.7, for every mole of carbon atoms in ocean sediments one mole of O_2 will be left in the atmosphere. At the time when photosynthetic organisms first

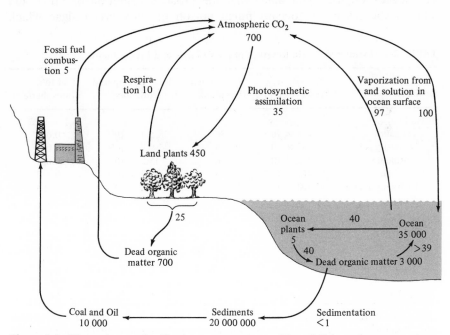

Figure 3.6 The carbon cycle. Numbers represent quantities of C in various reservoirs or annual flows from one reservoir to another. All are in units of 10^9 tonne. Data from: B. Bolin, *Sci. Amer.* **223**(3), 125–132 (1970).

began to proliferate little O_2 was available in the atmosphere. Organisms which could utilize carbohydrates as an energy source probably did not evolve until plentiful supplies of O_2 became available. Thus, the rate at which sedimentation of reduced carbon released dioxygen to the atmosphere may have been greater than it is today, and the formation of an oxidizing atmosphere may have occurred relatively rapidly on the geological time scale.

One other mechanism has been proposed for production of O_2. It involves photodissociation of water followed by escape of hydrogen atoms or molecules from the upper atmosphere. However, this mechanism is limited by a negative feedback system. Ultraviolet radiation required to dissociate H_2O is cut off by the ozone shield once substantial amounts of O_2 begin to accumulate. Apparently photodissociation of water cannot account for more than 1% of the current supply of atmospheric dioxygen.

3.8 THE NONPROBLEM OF CATASTROPHIC DIOXYGEN DEPLETION

Because O_2 is of such great importance in the functioning of the biosphere, some environmentalists have been concerned that our supply might be seriously depleted by some catastrophic effect resulting from pollution. The suggestion has been made that the broad distribution of DDT and other pesticides in the oceans might poison the blue-green algae which

TABLE 3.3 **Estimated World Reserves of Fossil Fuels and Living Plants**[a]

Substance	Estimated reserves (kg)[b]	% Assumed C content	Moles O_2 required for oxidation	% of total atmospheric O_2[c]
Fossil fuel				
Tar sand	75×10^{12}	86	5.4×10^{15}	0.018
Natural gas	166×10^{12}	70	9.7×10^{15}	0.032
Petroleum	212×10^{12}	86	15.2×10^{15}	0.051
Oil shale	470×10^{12}	86	33.7×10^{15}	0.112
Coal and lignite	$15\,300 \times 10^{12}$	75	956.2×10^{15}	3.188
Total			1020.2×10^{15}	3.401
Living organic matter (dry weight)				
Ocean	22×10^{12}	45	0.8×10^{15}	0.003
Land	276×10^{12}	45	10.4×10^{15}	0.034
Total			11.2×10^{15}	0.037
Grand total			1031.4×10^{15}	3.438

[a] Data from E. K. Peterson, *Environment,* **12**(10), 44 (1970); M. K. Hubbert, Energy resources. *In* "Resources and Man," pp. 167–242. Freeman, San Francisco, California, 1969.
[b] Total original reserves, including those already consumed and those not minable.
[c] Total O_2 content of atmosphere is 3×10^{19} mol.

carry out a sizable amount of photosynthesis. A second problem has to do with the burning of coal, petroleum, and natural gas, which consist largely of reduced carbon compounds. This reverses equation 3.7 and could conceivably deplete the supply of atmospheric O_2.

Both of the questions raised in the preceding paragraph may be answered by studying the data in Table 3.3. The total quantity of O_2 available in the atmosphere is approximately 3×10^{19} mol. Green plants produce only 4×10^{15} mol of dioxygen per year, of which all but 1.5×10^{12} mol is consumed by respiration or decay processes. Some of this remainder may be destroyed through oxidation of iron(II), sulfur, or other reduced substances which are brought into contact with the atmosphere by weathering processes. Thus the annual increase resulting from photosynthesis is no more than one part in twenty million of our total store of dioxygen. Even if all photosynthesis were to stop tomorrow and enough dioxygen were consumed to decompose all photosynthetic organisms, atmospheric O_2 would be decreased by only 0.04%. Thousands of years would be required to use up what remained.

The depletion of O_2 by fossil fuel combustion will be somewhat greater, but it certainly does not represent a crisis. Even if all of the projected reserves of fossil fuel and all of the living organic matter now available on earth were burned completely, only about 3.5% of the dioxygen now present in the atmosphere would be required. Apparently most of the reduced carbon in the earth's crust occurs in deposits of such low concentration or at such great depths that man's activities are unlikely to disturb them. A great many environmental problems are created by combustion of fossil fuels, but substantial depletion of O_2 is not one of them.

The Study of Critical Environmental Problems,[12] sponsored by the Massachusetts Institute of Technology, has concluded that depletion of atmospheric dioxygen by man's activities is a "nonproblem." They do recommend, however, that the level of O_2 be measured every decade to be certain that it is remaining constant.

3.9 CONCLUSIONS

The preceding chapters have summarized the origins of the chemical elements, the solar system, and the earth including its lithosphere, atmosphere, hydrosphere, and biosphere. During the course of evolution each of these four "spheres" was affected by interactions with the other three and with energy supplied by solar radiation or by heat escaping from below the crust. As we have seen, there would appear to be no reason to discard the hypothesis that the earth has *evolved slowly* toward its current state over a

---

[12] "Man's Impact on the Global Environment. Report of the Study of Critical Environmental Problems," p. 18. MIT Press, Cambridge, Massachusetts, 1970.

period of approximately 4.6×10^9 years, and that such evolution could have occurred without violating accepted laws of physicochemical behavior.

Until the present no organism has existed which could institute radical changes (occurring over relatively short time scales of tens or hundreds of years) throughout large fractions or perhaps even the whole of the earth's surface. Even the relatively rapid development of the oxidizing atmosphere by photoautotrophic organisms required millions of years. But now the scientific, industrial, and medical revolutions have given the human race the capacity to mobilize resources and increase its numbers to the point where it might be capable of such large-scale modifications of the environment. The pollution problems of the past few decades give convincing evidence that human activities can cause significant deviation from the "natural" conditions achieved through chemical and biological evolution. Although most of these effects currently are observed only in areas where human populations are concentrated, their extrapolation unabated into the future gives cause for concern.

It is the purpose of this text to consider these deviations and compare them where possible with naturally occurring processes. Although the environment has evolved so that small alterations are usually controlled by negative feedback, scientists and citizens must constantly be aware that natural systems can be overloaded and perhaps destroyed by human activities. Indeed, it is precisely this human *consciousness* and *knowledge* of the consequences of individual and collective actions that can serve the feedback function of regulating human impacts on the environment. We may well be at an evolutionary juncture not unlike the period of chemical evolution or the initiation of an oxidizing atmosphere in its importance. For the first time an organism *conscious of its impact* may act to regulate that impact instead of allowing natural forces to run their course.

The remainder of this text has been designed to raise your consciousness and knowledge of environmental problems and of the role which chemistry plays in creating and attacking them. It is organized not too differently from the ancient Greek conception that the universe was made up of four elements; earth, air, fire, and water. To these has been added the subject of life in order to include all of the spheres alluded to earlier. The role of energy flow in driving the spheres is fundamental, and therefore we shall proceed to examine it more carefully in Part II.

EXERCISES

(The first group of exercises is based primarily on material found in this text.)

1. Describe in some detail the evolution of the earth's atmosphere. What reasons can be given for the change from a reducing to an oxidizing atmosphere?

2. Explain why O_2 may have been a major pollutant of the environment at one point in the earth's history. How may its polluting effect have been counteracted?

3. Why is photochemical dissociation of dioxygen no longer an important process at the surface of the earth, even though it is thought to have been important when O_2 first began to accumulate in the atmosphere?

4. Explain why theories of the origin of O_2 in the atmosphere might lead one to fear that our supply of dioxygen could be seriously depleted by man's activities. Is this in fact likely to occur? Explain clearly and give data to support your conclusion.

5. Using the data which accompany equations 3.1 and 3.2, explain why both equilibria would be expected to shift to the left at high temperatures. Obtain data for reaction of N_2 and H_2 to form ammonia. Which way does this latter equilibrium shift as T increases?

6. Explain clearly how O_3 serves as a catalyst for conversion of solar ultraviolet radiation into heat in the atmosphere. Why is only a small concentration of O_3 necessary to intercept a large number of uv photons?

7. Give at least three reasons why HCN is thought to have been an important substance at the time life originated on earth.

8. What attribute of many of the building block molecules whose structures are shown in Figure 3.2 made them especially fit for incorporation in living systems? Explain why this attribute was important.

9. What is a condensation reaction? Give three examples of condensation which may have been important during chemical evolution. How may they have been driven to completion?

(Subsequent exercises may require more extensive research or thought.)

10. Obtain solubility product constants for iron(II) and iron(III) hydroxides from a handbook or a textbook of analytical chemistry. Assuming a solution of pH = 6.0 saturated with $Fe(OH)_2$, what fraction of the iron would precipitate if it were all oxidized to iron(III)? What bearing does this result have on the formation of banded iron formations?

11. Equation 3.5 may be thought of as an acid-base reaction. Label each substance as an acid or a base. Compare the strengths of CO_2, SO_2, and SiO_2 as Lewis acids. Which is strongest and why? How does this affect the equilibrium 3.5? Which would react more readily with $CaSiO_3$, CO_2 or SO_2?

12. Assume that gases escaping from the crust of primitive earth were at equilibrium with magma whose temperature was 1500 K. The equilibrium constant for

$$CO(g) + \tfrac{1}{2} O_2(g) \rightleftharpoons CO_2(g)$$

$K_p = 3.5 \times 10^5$ at this temperature. Show that CO would have been approximately three times as abundant as CO_2 in the early atmosphere, assuming Holland's estimate of O_2 partial pressure. Obtain data for the combination of dihydrogen and dioxygen to form water at the same temperature and show that H_2 would have been twice as abundant as H_2O.

Suggested Readings

1. P. H. Abelson, Chemical Events on Primitive Earth. *Proc. Nat. Acad. Sci. U.S.* **55**, 1365–1372 (1966). A description of the importance of HCN in the chemical evolution of life.

2. L. H. Ahrens, The significance of the chemical bond for controlling the geochemical distribution of the elements. *In* "Physics and Chemistry of the Earth" (L. H. Ahrens, F. Press, and S. K. Runcorn, eds.), Vol. 5, p. 1. Macmillan, New York, 1964. A detailed treatment of the classification of the elements as siderophiles, lithophiles, and chalcophiles. The theoretical chemistry involved is at about the level of Pauling's "The Nature of the Chemical Bond." The faint of heart may wish to stick to Ahrens' smaller book.

3. L. H. Ahrens, "Distribution of the Elements in Our Planet." McGraw-Hill, New York, 1965. An elementary treatment of the distribution of the elements in the solid earth is given in just over ninety pages. Requires no more than the usual general chemistry background and contains much useful information.

4. L. H. Aller, "The Abundance of the Elements." Wiley (Interscience), New York, 1961. A detailed treatise on the abundances of the elements in the cosmos, the solar system, and the earth. Chapter 10 is devoted to theories of the origin of the elements.

5. G. M. Barrow, "Physical Chemistry," 2nd ed. McGraw-Hill, New York, 1966. This or any of several other physical chemistry textbooks may be referred to for a more complete discussion of the kinetic theory of gases and molecular spectroscopy.

6. H. P. Berlage, "The Origin of the Solar System." Pergamon, Oxford, 1968. A brief but fairly detailed treatment of the subject indicated by the title.

7. W. S. Broecker, Man's oxygen reserves. *Science* **168**, 1537 (1970). [Reprinted under the title, Enough air. *Environment* **12**(7), 28 (1970).] This article seems to have laid to rest once and for all the specter of an atmosphere depleted of dioxygen.

8. M. Calvin, "Chemical Evolution." Oxford Univ. Press, London and New York, 1969; Chemical evolution. *Amer. Sci.* **63**(2), 169–177 (1975). Calvin's book is an expression of his work and views and is considerably less general than Kenyon and Steinman (see ref. 18 below). Nevertheless, it contains much useful and interesting information.

9. P. Cloud, Evolution of ecosystems. *Amer. Sci.* **62**(1), 54–66 (1974). An application of geological evidence to the problem of origin of ecosystems.

10. P. Cloud, Atmospheric and hydrospheric evolution of the primitive earth. *Science* **160**, 729 (1968). A chronology of events in the evolution of the atmosphere and hydrosphere is given. Cloud spends considerable time on the problem of lunar capture and its effects on the atmosphere and crust of the earth.

11. S. W. Fox, K. Harada, G. Krampitz, and G. Mueller, *Chem. Eng. News* June 22, pp. 80–94 (1970); Dec. 6, pp. 46–53 (1971). This general description of chemical evolution reflects Fox's predilections. The description of proteinoid microspheres is especially useful.

12. J. C. Giddings, "Chemistry, Man and Environmental Change." Canfield Press, San Francisco, California, 1973. Contains a very readable discussion of the origin of atmospheric O_2 and the possibility of its being depleted.

13. J. Gilluly, A. C. Waters, and A. O. Woodford, "Principles of Geology," 3rd ed. Freeman, San Francisco, California, 1968. Most geology textbooks devote some discussion to the origin of the earth, the evolution of the hydrosphere, and especially the formation of ores. Chapters 21 and 22 of this text are particularly helpful.

14. R. G. Gymer, "Chemistry: An Ecological Approach." Harper, New York, 1973. Chapter 13 gives a useful summary of information regarding the formation of the earth, geochemical classification of the elements, and crystallization of magma.

15. H. D. Holland, Model for the evolution of the earth's atmosphere. *In* "Petrologic Studies" (A. E. J. Engle, H. L. James, and B. F. Leonard, eds.), pp. 447–477. Geol. Soc. Amer., Boulder, Colorado, 1962. A complete account of one scenario for the evolution of the present atmosphere from a primitive reducing atmosphere.

16. R. Jastrow, "Red Giants and White Dwarfs." Harper, New York, 1967. A popularized account of the histories of stars, solar systems, and planet earth, based on a series of CBS television programs. Few details are given but it is a very readable account for the novice.

17. E. Keller and J. A. Wood, Man and the universe. *Chemistry* **45**(7), 4–26 (1972). A discussion of the origin of the universe and the chemical elements, radio astronomy, and the early history of the moon which is accessible by a nonsophisticated reader.

18. D. H. Kenyon and G. Steinman, "Biochemical Predestination." McGraw-Hill, New York, 1969. As Melvin Calvin notes in the foreward: "The book is really the first attempt to produce what might be called a comprehensive essay which could be used as the basic textbook for a systematic discussion of the problem in an academic and scientific environment." The "problem" is the origin of life on earth and chemical evolution.

19. R. M. Lemmon, *Chem. Rev.* **70**(1), 96–109 (1969). A general discussion of the chemical evolution problem.

20. B. Mason, "Principles of Geochemistry," 3rd ed. Wiley, New York, 1966. Discusses all of the topics of this chapter except interstellar molecules. Plenty of details are given and it is assumed that the reader is at least familiar with the subject matter of general chemistry, including elementary thermodynamics. A very useful reference.

21. W. D. Metz, Interstellar molecules: New theory of formation from gases. *Science* **182**, 466–468 (1973). Discusses gas phase kinetics of formation of interstellar molecules.

22. A. I. Oparin, "Genesis and Evolutionary Development of Life." Academic Press, New York, 1969. Chapter 4 contains a good description of the formation of coacervate drops.

23. R. G. Pearson, *J. Chem. Educ.* **45**, 581 and 643 (1968); *Surv. Progr. Chem.* **1**, 1 (1969). Two recent summaries of the application of the principle of hard and soft acids and bases.

24. E. K. Peterson, Letter to the editor. *Environment* **12**(10), 44 (1970). Discusses rates of change of atmospheric O_2 and CO_2 during recent geological history.

25. C. Ponnamperuma and N. W. Gabel, Current status of chemical studies on the origin of life. *Space Life Sci.* **1**, 64–96 (1968). A general review of the status of chemical evolution research in 1968.

26. D. M. Rank, C. H. Townes, and W. J. Welch, Interstellar molecules and dense clouds. *Science* **174**, 1083 (1971). An advanced review of the results of radio astronomical observation.

27. C. Sagan, Interstellar organic chemistry. *Nature (London)* **238**, 77 (1972). Discusses the formation of interstellar molecules and their relation to theories of chemical evolution. Concludes that interstellar biology is unlikely.

28. J. Selbin. The origin of the elements. I. *J. Chem. Educ.* **50**(5) 306 (1973); II. **50**(6), 380 (1973). These articles discuss the origin of the universe, nucleosynthesis of the elements, and stellar evolution.

29. P. M. Solomon, Interstellar molecules. *Phys. Today,* March p. 32 (1973). The most recent and up-to-date catalog of molecules found in interstellar space by means of radioastronomy. Heralds the founding of the new field of interstellar chemistry.

30. B. E. Turner, Interstellar molecules. *Sci. Amer.* **228**(3), 50 (1973). A general treatment containing a beautiful color photograph of the nebula in Orion.
31. H. C. Urey, "The Planets." Yale Univ. Press, New Haven, Connecticut, 1952. This book approaches the origin of the solar system from the standpoint of a physical chemist. It is an important basic reference for anyone doing advanced work in this area, even though discoveries made in the 20 years since its publication have altered some of its conclusions.
32. L. Van Valen, The history and stability of atmospheric oxygen. *Science* **171**, 439 (1971). An ecologist's approach to the problem. Concludes that current theories are not completely adequate.
33. G. Wald, The origins of life. *Proc. Nat. Acad. Sci. U.S.* **52**, 595–611 (1964). A general account of the origin of life problem from the perspective of 1964, this paper contains good descriptions of the properties of elements and compounds which contribute to their fitness.
33. G. Wallerstein, Astronomical evidence for nucleosynthesis in stars. *Science* **162**, 625 (1968). A fairly complete report of the evidence available to support the contention that the elements heavier than hydrogen are manufactured by nuclear reactions in stars.
34. J. C. Wheeler, After the supernova, what? *Amer. Sci.* **61**(1), 42 (1973). An up-to-date account of stellar evolution, nucleosynthesis, and supernovas.
35. P. J. Wyllie, The earth's mantle. *Sci. Amer.* **232**(3), 50–63 (1975). An excellent description of the structure, composition, and dynamics of the mantle.
36. S. W. Fox and K. Dose, "Molecular Evolution and the Origin of Life." Freeman, San Francisco, California, 1972. An up-to-date presentation of this subject.

II
Energy

Energy is the only life and is
from the Body;
and Reason is the bound or outward
circumference of Energy
Energy is Eternal Delight.

William Blake

A QUICK REVIEW OF CHAPTER 3 WILL REVEAL that energy was a necessary ingredient in primitive earth simulation experiments and presumably throughout the evolution of life. A flow of energy is absolutely essential to drive the important *biogeochemical cycles* of matter throughout the lithosphere, atmosphere, hydrosphere, and biosphere, and much of the history of mankind revolves around the discovery and use of sources of energy which could replace human muscle power.

The recent "energy crisis" serves very well to illustrate how intricately energy resources have become woven into the fabric of modern society. An embargo and quadrupling of the price of petroleum resulted in increased prices for food, clothing, and nearly all material goods as well as raising the cost of gasoline and fuel oil. It also produced numerous calls for the repeal, relaxation, or postponement of pollution control regulations. Although many environmentalists would argue that alleviation of the energy shortage need not necessarily imply increased pollution, it is certainly true that a great many technological "solutions" to environmental problems require greater energy consumption. A good example is the control of automobile air emissions, which has significantly increased fuel consumption per mile of travel.

Human conversion of energy from fossil fuel and other resources invariably has negative environmental effects. In 1965 more than 80% of air pollution was the direct result of fuel combustion. Sizable quantities of solid waste, disturbance of land by strip mining, and water pollution by oil spills and acid mine drainage could be attributed to the same cause. The use of nuclear fission reactors to supply projected energy needs of the future will also entail negative consequences, and even solar, tidal, and geothermal energy sources may produce unexpected effects when used on the large scale at which we currently consume fossil fuel.

Why is energy so important and so inextricably intertwined with the environment? Much of the answer to this question becomes evident after a study of the first and second laws of thermodynamics. The limitations which these laws place on the use of energy by man or in the natural world are the subject of Chapter 4. The insights developed here will be used again throughout the remainder of the book. Chapters 5 and 6 consider the energy sources we currently use and those which are being or may be developed for the future. At the close of this part of the text the subject of energy storage and conservation is treated. One lesson which seems fairly evident from both the immediate energy crisis and a longer term view is that the profligate consumption of energy resources currently practiced in United States society cannot be maintained. Chapter 7 will indicate a number of areas where savings might be made.

4

Theoretical Treatment
of Energy

*The law that entropy increases—the Second Law of Thermo-
dynamics—holds, I think, the supreme position among the
laws of nature. If someone points out to you that your pet
theory of the universe is in disagreement with . . . the Second Law of Thermodynamics, I can give you no hope; there
is nothing to do but collapse in deepest humiliation.*

Arthur S. Eddington

The scientific discipline known as *thermodynamics* arose during the
nineteenth century as a result of a great many observations of conversions
of energy from one form to another, especially with regard to transformation of heat to work in steam engines. These studies led to the formulation
of several laws, the first and second of which are the most important for
our understanding of energy and its interaction with the environment.
These two laws might be described by paraphrasing George Orwell: All
forms of matter-energy are equal, but some are more equal than others. Or,
as biologist Garrett Hardin puts it: "First, you can't win; second, you can't
even break even; and what's more, you can't get out of the game." These
laws imply limitations on human capabilities and thus have challenged a
great many persons to test them. More than a century of such testing has
failed to find exceptions, leading to the confidence in their correct description of nature expressed in Eddington's words above.

4.1 THE FIRST AND SECOND LAWS
OF THERMODYNAMICS[1]

The *first law of thermodynamics* states that the total amount of matter-
energy in the universe is constant. Matter can be converted from one form

---

[1] At this point you may wish to review the chapter on thermodynamics in your general chemistry textbook. One reasonable source is F. Brescia *et al.,* "Fundamentals of Chemistry,"
3rd ed., Chapters 5 and 17. Academic Press, New York, 1975.

to another, so can energy, and under the proper circumstances matter can be converted into energy and vice versa. In the latter case Einstein's famous relation $E = mc^2$ applies (see the discussion relating to equation 1.6). Under no circumstances, however, may matter be destroyed or created without a corresponding creation or destruction of energy.

The usual formulation of the first law is

$$\Delta E = q - w \qquad (4.1)$$

where ΔE represents the change in the *internal energy* of some portion of the universe upon which we wish to focus attention (the *system*). The internal energy depends on how many molecules occupy each of the different kinds of energy levels discussed in Section 1.2, and the relative energy of each occupied level.[2] For an ideal gas it is directly proportional to the absolute temperature, so for an *isothermal* (constant temperature) process $\Delta E = 0$. The symbols q and w represent the amount of heat added to the system and the amount of work obtained from the system, respectively. Electrical, mechanical, chemical, or other types of work might be obtained.

The first law says that if heat is added to a system (say the boiler of a steam engine), some combination of two things may happen. The heat may be converted to work ($\Delta E = 0$, $q = w$) or it may be stored as increased energy of the molecules in the boiler ($w = 0$, $\Delta E = q$). All forms of matter-energy are equal and interconvertible, but you can't win (create new matter-energy). Nothing in the first law favors one type of energy over another, and there is no reason to expect that interconversions of work and heat would not occur with 100% efficiency.

The first law does not completely describe mankind's experience with energy, however. Table 4.1 indicates that few conversions even approach 100% efficiency, and many are well below 50%. The transformation of heat (thermal energy) into work is especially inefficient, the highest figure in the table being 45% for a modern steam turbine. These inefficiencies are not entirely the fault of scientists' inability to devise appropriate means of carrying out energy conversions. To a great extent they arise out of the same cosmic evolution described in Part I. The universe is evolving in one direction and events which would reverse that evolution do not occur *spontaneously* (without outside intervention). Some forms of matter-energy are more equal than others; i.e., they have not yet evolved as far and their spontaneous evolution may be harnessed by mankind. If we attempt to convert other forms of matter-energy back to these original forms we can't even break even — some matter-energy must be wasted in the process.

This state of affairs is summarized in the *second law of thermodynamics*, which states that there is an increase in *entropy* for every spontaneous

---

[2] For a complete discussion, see W. G. Davies, "Introduction to Chemical Thermodynamics," Chapter 5. Saunders, Philadelphia, Pennsylvania, 1972.

TABLE 4.1 Efficiency of Energy Converters[a]

Type	Efficiency (%)	Type of conversion	
		From	To
Electric generator	99	Mechanical	Electrical
Electric motor			
Large	92	Electrical	Mechanical
Small	62	Electrical	Mechanical
Dry cell battery	90	Chemical	Electrical
Large steam boiler	88	Chemical	Thermal
Home furnace			
Gas	85	Chemical	Thermal
Oil	65	Chemical	Thermal
Storage battery (lead-acid)	72	Electrical	Chemical
		Chemical	Electrical
Fuel cell	60	Chemical	Electrical
Man on a bicycle	50	Chemical	Mechanical
Liquid fuel rocket (H_2)	47	Chemical	Thermal
Turbine			
Steam	45	Thermal	Mechanical
Gas (aircraft or industrial)	35	Chemical	Thermal
		Thermal	Mechanical
Electric power plant			
Fossil fueled	40	Chemical	Thermal
		Thermal	Mechanical
		Mechanical	Electrical
Nuclear fueled	32	Nuclear	Thermal
		Thermal	Mechanical
		Mechanical	Electrical
Internal combustion engine			
Diesel	37	Chemical	Thermal
Otto cycle (automobile)	25		
Wankel (rotary)	18	Thermal	Mechanical
Lasers	30–40	Electrical	Radiant
Lamps			
High intensity	32	Electrical	Radiant
Fluorescent	20	Electrical	Radiant
Incandescent	4	Electrical	Radiant
Unaided walking man	12	Chemical	Mechanical
Solar cell	10–15	Radiant	Electrical
Steam locomotive	8	Chemical	Thermal
		Thermal	Mechanical
Thermocouple	6	Thermal	Electrical

[a] Much of the data of this table was obtained from C. M. Summers, *Sci. Amer.* **224**(3), 155 (Sept. 1971).

process. The universe, which apparently began as a tightly packed ball of neutrons (Section 1.1), has continually evolved toward a more random, more highly disordered and therefore more probable structure. The quantity of entropy is a measure of the extent of this evolution. For any given *state* of the universe or a smaller system, entropy, S, can be defined quantitatively as

$$S = k \ln W_{max} \qquad (4.2)$$

where k is Boltzmann's constant and W, the *thermodynamic probability*, increases with the number of different ways the particles of the system can be arranged among the energy levels available to them. This definition can be used to show[3] that for a transformation at constant temperature

$$\Delta S = q/T \qquad (4.3)$$

provided the process is carried out reversibly (see Section 4.3).

Matter and energy both tend to be converted from forms having smaller entropies to those exhibiting greater disorder, so long as their surroundings do not undergo an offsetting decrease in entropy. You are probably quite familiar with many such transformations of matter—at appropriate temperatures and pressures solids melt, liquids vaporize, and reactions which produce gaseous products are favored because of positive values of ΔS. These increases may be estimated on the basis of some simple rules[4] or calculated using tables of standard molar entropies which are available in many text or reference books.

A similar situation applies to transformation of energy. Different types of energy may be characterized by a hierarchy of "quality" as shown in Table 4.2. Those forms having little entropy per unit of energy tend to transform into others farther down in the table which have more. Since the total amount of energy must remain constant such transformations result in an entropy increase. Objects at high temperatures tend to transfer energy to those that are cooler. Gravitational potential energy (of an elevated object) tends to be converted into low temperature thermal energy (once the object has fallen and come to rest). At the surface of the earth the lowest quality energy is usually the "thermal surroundings"—objects at "normal" temperatures on the order of 250–300 K. The lowest quality energy in the universe is the cosmic microwave radiation referred to at the beginning of Part I. Its color temperature is 4 K, and it provides the ultimate heat sink to which other energy forms will eventually be degraded.

---

[3] W. G. Davies, "Introduction to Chemical Thermodynamics," pp. 136–140. Saunders, Philadelphia, Pennsylvania, 1972.

[4] W. G. Davies, "Introduction to Chemical Thermodynamics," pp. 125–132. Saunders, Philadelphia, Pennsylvania, 1972.

TABLE 4.2 Quality of Various Forms of Energy[a]

Form of energy	Entropy per unit energy
Gravitational	0
Nuclear	10^{-6}
Thermal	
Stars (10^8 K)	10^{-3}
Earth (10^2 K)	10^2
Chemical	1–10
Radiant	
Sunlight (visible)	1–10
Cosmic microwave	10^4

[a] Data adapted from F. J. Dyson, *Sci. Amer.* **224**(3), 50–59 (Sept. 1971).

4.2 ENTROPY POLLUTION AND FREE ENERGY

The term "energy resource" implies more than just a given quantity of energy. The energy must be of high quality; that is, it must have low entropy per unit energy so that its subsequent, spontaneous evolution into a higher entropy form may be harnessed to do useful work. As indicated in Table 4.3 there are three major energy resources available at the surface of the earth whose quality (Table 4.2) and quantity are high enough to make them attractive. In order of decreasing quality they are nuclear, chemical (fossil fuels), and solar. Small quantities of gravitational and medium temperature thermal energy are also available from the tides and geothermal sites. Each of these resources will be discussed further in the next two chapters.

The energy crisis is not the result of a lack of energy—energy cannot be created, destroyed, or "used up." Rather it has resulted from the production of entropy associated with the conversion of high-grade energy sources into less desirable forms. The same problem arises when we consider matter instead of energy—the most useful forms (ores) contain individual elements in high concentrations. They tend spontaneously to be transformed into less concentrated materials having greater amounts of entropy. This production of entropy has been referred to as *entropy pollution* since it is a measure of the extent to which the universe has been *irreversibly degraded.*[5] The second law assures us that it cannot be avoided, but there are techniques by which it can be minimized.

The use of entropy increase as a criterion of spontaneous change requires that we consider the entire universe, i.e., the system we are interested in plus all surroundings that may exchange matter or energy with it. Usually it is more convenient if we can limit our attention to the system

---

[5] H. A. Bent, *Chemistry* **44**(9), 6 (1971).

TABLE 4.3   Estimated Energy Resources[a]

Type	Total world supply[b]	Economically available (at no more than double current costs)	
		World	U. S.
Depletable supplies (10^{21} J)			
Fossil (chemical)			
Tar sands	2.1	—	—
Natural gas	3.8	2.2–3.8	0.6–1.1
Petroleum	6.0	3.2–6.0	0.6–1.1
Oil shale	13.3	—	—
Coal and lignite	185.	21.1–31.5	5.0–7.2
Total fossil	210.2	26.5–41.3	6.2–9.4
Nuclear			
Ordinary fission	$2 \times 10^{4\,c}$	14.	1.4
Breeder fission	$6 \times 10^{6\,c}$	4000.	400.
Fusion (D-T)	215.	—	—
Fusion (D-D)	1×10^{10}	—	—
Continuous supplies (10^{21} J/year)			
Solar	899.	—	50.[d]
Tidal	0.094	—	0.0094[d]
Geothermal	0.010	—	0.0010[d]

[a] Current worldwide rate of consumption is 5×10^{12} W or 0.16×10^{21} J/year. Data adapted from C. Starr, *Sci. Amer.* **224**(3), 43 (Sept. 1971).

[b] Total supply including amount consumed to date.

[c] Estimated from total quantity of uranium and thorium within 1 mile of land surface assuming only 1% to be minable [*Int. At. Energy Ag. Bull.* **14**(4), 11 (1972)].

[d] Total supply is listed since no cost figures are available.

only, because changes in the surroundings are often less obvious. This can be done in the following way. Assume that the process we wish to study takes place at constant temperature and pressure. The entropy of the universe will increase for a spontaneous process or remain constant if the system is at equilibrium:

$$\Delta S_{\text{universe}} = \Delta S_{\text{system}} + \Delta S_{\text{surroundings}}$$

$$= \Delta S_{\text{system}} + \frac{q_{\text{surroundings}}}{T} \geqslant 0 \qquad (4.4)$$

(Equation 4.3 was used to obtain the last part of 4.4.)

If the system undergoes a chemical reaction in which heat is absorbed or given off, the change in enthalpy[6] corresponds to the quantity of heat ab-

---

[6] Your general chemistry text should be consulted for the definition of ΔH, the enthalpy, and its relationship to ΔE.

sorbed: $\Delta H_{system} = q_{system}$. Since energy cannot be destroyed the quantity of heat absorbed by the system must be given off by the surroundings: $\Delta H_{system} = q_{system} = -q_{surroundings}$. If we assume that the nature of the surroundings is such that this transfer can be accomplished reversibly, equation 4.4 may be rewritten as

$$\Delta S_{universe} = \Delta S_{system} - \frac{\Delta H_{system}}{T} \geqslant 0 \qquad (4.5)$$

Multiplying both sides of 4.5 by $-T$ (the minus sign changes the direction of the inequality), the equation takes a form familiar to everyone who has studied elementary chemical thermodynamics:

$$-T \, \Delta S_{universe} = \Delta H_{system} - T \, \Delta S_{system} = \Delta G_{system} \leqslant 0 \qquad (4.6)$$

ΔG_{system} is defined by equation 4.6 and is known as the change in *Gibbs free energy* of the system. It is a measure of the increase in entropy of the universe at a given temperature and pressure and is convenient to use because it requires information about the system only—not the surroundings.

Equation 4.6 indicates that there are two contributing factors to the negative ΔG required in any spontaneous process. The system tends to transfer heat to the surroundings, lowering its energy (negative value of ΔH_{system}), and its entropy tends to increase (positive ΔS_{system}). Either or both of these changes will make ΔG_{system} negative. At low temperatures ΔH_{system} is the most important factor (if T is small, $-T \, \Delta S_{system}$ is usually negligible), while at high temperatures the entropy term controls the sign of ΔG_{system}. Although they do vary with temperature (especially if reactants or products undergo phase changes), ΔH_{system} and ΔS_{system} usually do not change sign as T changes unless they were originally nearly equal to zero. Therefore, values calculated at one temperature (usually 298 K) may be used to predict *qualitatively*[7] the behavior of a chemical reaction at other temperatures. You should already have used this "low energy—high entropy" rule in answering Exercise 5 in Chapter 3.

4.3 CONVERTING HEAT TO WORK— REVERSIBLE PROCESSES

Although the foregoing section may have clarified the fact that our crisis of consumption of energy resources is actually one of production of entropy, it still has not indicated why the efficiencies of some of the energy

---

[7] W. G. Davies, "Introduction to Chemical Thermodynamics," pp. 147–150. Saunders, Philadelphia, Pennsylvania, 1972.

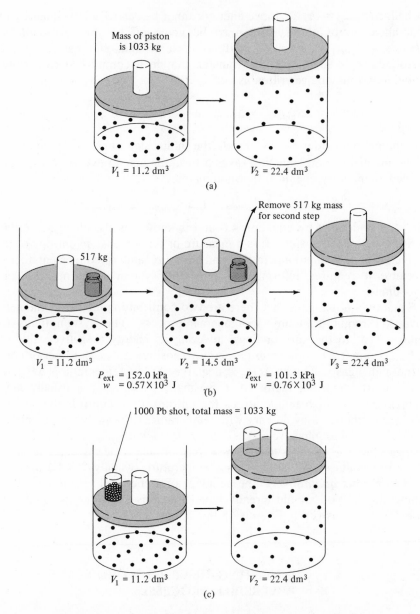

Mass of piston is 1033 kg

$V_1 = 11.2$ dm³ $V_2 = 22.4$ dm³

(a)

Remove 517 kg mass for second step

517 kg

$V_1 = 11.2$ dm³ $V_2 = 14.5$ dm³ $V_3 = 22.4$ dm³

$P_{ext} = 152.0$ kPa $P_{ext} = 101.3$ kPa
$w = 0.57 \times 10^3$ J $w = 0.76 \times 10^3$ J

(b)

1000 Pb shot, total mass = 1033 kg

$V_1 = 11.2$ dm³ $V_2 = 22.4$ dm³

(c)

Figure 4.1 Conversion of heat to work by expansion of an ideal gas.

transformations in Table 4.1 are so low. This can perhaps be done best by examining one of the less efficient conversions — thermal to mechanical — in greater detail.

One system from which energy may be extracted in a fairly simple way consists of an ideal gas which is allowed to expand against some external pressure, perhaps forcing a piston along a cylinder as shown in Fig. 4.1. If the external pressure remains constant the amount of work done is $P_{ext} \Delta V$. Depending on how the expansion of the gas is carried out different amounts of work may be obtained. Assume that one mole of gas at a temperature of 273 K is confined in a cylinder whose cross-sectional area is 1000 cm² by a piston 11.2 cm above the bottom. The volume of this sample of gas is 11 200 cm³ or 11.2 dm³ (11.2 l) and the pressure can be calculated from the ideal gas law ($PV = nRT$) to be 202.6 kPa (2.0 atm). If the gas is permitted to expand at constant temperature to a volume of 22.4 dm³ (by allowing the piston to move 11.2 cm along the cylinder), the new pressure of the gas will be 101.3 kPa. If the cylinder were projecting from a spaceship (zero external pressure and negligible gravitational force on the piston), the total work done by the expansion of the gas would be zero [$P_{ext} \Delta V = (0 \text{ kPa}) \times (11.2 \text{ dm}^3) = 0 \text{ J}$]. On the other hand, if the cylinder were located at the surface of the earth the external pressure would be 101.3 kPa and 1.13×10^3 J of work would be accomplished by the expansion. Note that in this latter case heat would have to be added to the system to compensate for the work done.

The quantity of work obtained and the quantity of heat that must be added to maintain the internal energy (and therefore the temperature of the gas) both increase with the external pressure. But if a pressure greater than 101.3 kPa is applied to the piston the expansion of the gas will stop before the final volume of 22.4 dm³ is reached. The internal pressure must exceed the external pressure in order for the piston to be raised. Suppose that enough mass were added to the piston to increase the external pressure to 152.0 kPa. The piston would rise until the internal pressure dropped to 152.0 kPa (a distance of 3.73 cm). The amount of work done would be $w = (152.0 \text{ kPa}) \times (3.73 \text{ dm}^3) = 0.57 \times 10^3$ J. The expansion of the gas could then be completed by removing the extra mass. The pressure would drop to 101.3 kPa, the piston would rise an additional 7.47 cm ($\Delta V = 7.47 \text{ dm}^3$) and a quantity of work, $w = (101.3 \text{ kPa}) \times (7.47 \text{ dm}^3) = 0.76 \times 10^3$ J, would be extracted from the system. Although the total volume change of 11.2 dm³ did not differ from the previous two examples, the total work done in this case was even greater. Again, in order to maintain the internal energy of the system (keep the temperature constant), a quantity of heat equivalent to 1.33×10^3 J would have to be added.

If the expansion were carried out in five steps (by beginning with an external pressure of 182.34 kPa and then decreasing P_{ext} by increments of

20.26 to 101.3 kPa), the total work done would be 1.47×10^3 J. As a general rule, if the expansion is carried out in n steps

$$w = \sum_{i=1}^{n} (P_{ext})_i \, (\Delta V)_i \tag{4.7}$$

The smaller the volume increments (the greater the number of steps) the larger the quantity of work obtained from the gas. In the idealized mathematical limit of an infinite number of steps and infinitesimal volume increments,

$$w = \int_{V_{initial}}^{V_{final}} P_{ext} \, dV \tag{4.8}$$

If the external pressure is allowed to vary continuously so that it is always just infinitesimally less than the internal pressure, the expansion process is called *reversible*; only an infinitesimal increase in the external pressure is required to stop it. Since the external and internal pressures differ by only an infinitesimal amount,

$$P_{ext} = P_{gas} = \frac{nRT}{V} \quad \text{(ideal gas law)} \tag{4.9}$$

Equation 4.9 may be substituted into 4.8, and since n, R, and T remain constant,

$$w_{rev} = \int_{V_{initial}}^{V_{final}} nRT \, \frac{dV}{V} = nRT \ln \frac{V_{final}}{V_{initial}} \tag{4.10}$$

Substituting values of $n = 1$ mol, $R = 8.3144$ J K^{-1} mol^{-1}, $T = 273$ K, $V_{final} = 22.4$ dm^3, $V_{initial} = 11.2$ dm^3, w is calculated to be 1.57×10^3 J. This is considerably larger than for either the two-step or five-step expansion described above.

The generalization that the maximum amount of work is obtained from a process which is carried out reversibly is an important one, and it applies to the other types of energy conversion listed in Table 4.1 as well. Usually reversibility is approached when a process can be divided into a number of small steps, each of which must be carried out against some resistance. Getting more and more work out of any process requires that we take longer and longer to do it, as in the case of the 1000 lead shot in Figure 4.1c , and this requires that we give up some *power*. Power is the rate at which work can be done. It is usually measured in watts (joule/second), although for easy comparison the power available from continuous energy supplies was reported as joule/year in Table 4.3.

Maximum power is not obtained from a reversible process because an infinite number of infinitesimal steps would require an infinite time to complete. The relationship between power, the quantity of useful work obtained, and the time (number of steps) required is shown schematically in

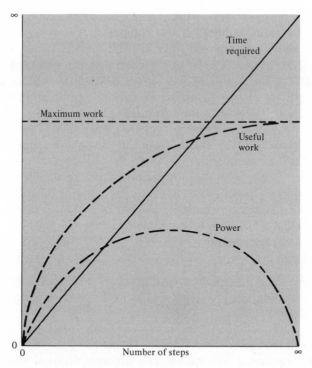

Figure 4.2 Schematic relationship of work, power, and time required for some process.

Figure 4.2. Generally this confirms our common sense expectation that, for example, a large, powerful automobile engine will require more fuel to move a given mass a given distance than will a smaller, less powerful one.

4.4 CONVERTING HEAT TO WORK — CARNOT EFFICIENCY

The expanding-gas energy conversion system described in the preceding section contains a major flaw. Work is obtained from it only as long as the piston continues to move along the cylinder. We have not yet considered how this process might be repeated. If the system is to convert heat into work at 100% efficiency, ΔE must be zero so that $q = w$, according to the first law. One simple way to insure that $\Delta E = 0$ is to use a cyclic process. Returning the system exactly to its initial state at the end of the transformation requires that the initial and final internal energies be the same.

Suppose a gas is compressed (still at a temperature of 273 K) back to a volume of 11.2 dm³ to complete a cycle. This could be done in a single step by applying an external pressure of 202.6 kPa. The work done on the system (negative value of w) is given by $w = P_{ext} \Delta V = (202.6 \text{ kPa}) \times (-11.2 \text{ dm}^3) = -2.50 \times 10^3$ J. If the compression is carried out stepwise it is possible to use less work. The minimum amount of work (-1.57×10^3 joule) is required if the compression is reversible. It is the same in magnitude but opposite in sign from the maximum work obtainable from the expansion. Unless one is careful to carry out both processes reversibly, the result is the opposite of the desired one—a net quantity of work will be done by the surroundings on the system and an equivalent amount of heat will appear in the surroundings.

There is a way of transforming heat into work using our cylinder and piston. Less work will be required from the surroundings during compression of the gas if the temperature is lower than during expansion (equation 4.10). This can be accomplished by means of four reversible steps, known collectively as the *Carnot cycle*. Because each step is reversible it can be shown that the quantity of work obtained per unit of heat supplied from the surroundings in the first step (the *efficiency*) of the Carnot cycle cannot be exceeded in any real system.

The four steps are

1. *Reversible, isothermal expansion*—This has been described in detail in the preceding section. Heat added to the system is converted into mechanical work.

2. *Reversible, adiabatic expansion*—In an adiabatic process the system is insulated against heat transfer so that $q = 0$. Some of the internal (translational, rotational, vibrational, and perhaps electronic) energy of the gas molecules will be converted to work ($\Delta E = -w$) and the temperature will fall.

3. *Reversible, isothermal compression*—The work required is calculated from equation 4.10. Unless $T = 0$ K, so that the gas molecules are motionless and exert no resistance, some of the work obtained in step 1 will be transformed back into heat, but *at a lower temperature* and hence of lower quality. This *waste heat* is not counted when the efficiency is calculated because it is at the same temperature as the thermal surroundings and cannot do useful work.

4. *Reversible, adiabatic compression*—The amount of work yielded in step 2 is converted back into internal energy, restoring the gas to its initial condition.

A more quantitative analysis of the Carnot cycle[8] confirms the importance of the temperature of step 3 in determining Carnot efficiency:

$$\text{Efficiency} = \frac{w}{-q_{\text{hot reservoir}}} = \frac{T_1 - T_2}{T_1} = \frac{T_{\text{high}} - T_{\text{low}}}{T_{\text{high}}} \qquad (4.11)$$

At the surface of the earth, where the thermal surroundings average 288 K, efficient conversion of heat to work requires a large T_{high} in step one of the

---

[8] A. W. Adamson, "A Textbook of Physical Chemistry," pp. 200–207. Academic Press, New York, 1973; or W. J. Moore, "Physical Chemistry," 3rd ed., pp. 70–75. Prentice-Hall, Englewood Cliffs, New Jersey, 1962.

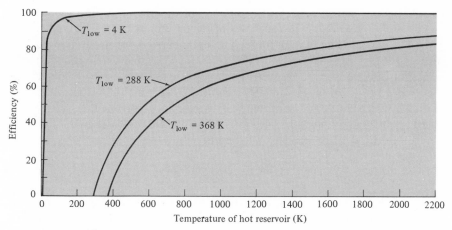

Figure 4.3 Theoretical (Carnot) efficiencies of heat engines.

cycle. As Figure 4.3 indicates, the efficiency of a heat engine (such as an electric power plant) whose maximum boiler temperature reaches 600 K drops from 52 to 38% as T_{low} increases from 288 to 368 K. Thus, power plants are usually constructed along rivers or lakes which can absorb their waste heat. When adverse environmental effects occur this heat is termed *thermal pollution.*

4.5 CONVERSION OF MATTER INTO MORE USEFUL FORMS—ALUMINUM FROM BAUXITE

Desirable chemical conversions, such as the separation of aluminum from bauxite ore ($Al_2O_3 \cdot 2\ H_2O$), are often found to have positive free energy changes. If such a process is to be carried out, it must be *coupled* to another which has a ΔG greater in magnitude but of negative sign. The coupled pair of processes can thus exhibit an overall increase in entropy. A *positive ΔG indicates the minimum useful work which must be obtained from a coupled reaction in order to force the desired one to occur.*

The fundamental chemical change required in the Hall process for recovery of aluminum is

$$Al_2O_3 \longrightarrow 2\ Al + \tfrac{3}{2}\ O_2 \qquad \Delta G^0_{298} = 1670 - T(0.314) = 1576\ kJ \tag{4.12}$$

Work equal to at least 1576 kJ must be done on the system for every 2 mol of aluminum produced. This work could in principle be obtained from any spontaneous process, but in practice it is supplied electrically. The required current is generated by burning coal or allowing water to fall through a turbine in a power plant. To keep this example as chemical as possible we shall assume the former source.

If the free energy stored in dioxygen and coal could be transferred by an infinite number of infinitesimal steps into the aluminum and dioxygen, there would be zero net increase in the entropy of the universe. But, as we have seen, such a reversible process would occur infinitely slowly, and this is too long to wait. The Hall process is actually carried out in such a way that only 17% of the free energy in the coal and O_2 is used. (Since other raw materials are consumed as well, the overall efficiency is even lower; see Section 11.5.) For every mole of refined aluminum 3847 kJ of energy which could have done useful work is irreversibly degraded.

By rearranging equation 4.6 it is possible to relate this wasted free energy to the entropy increase of the universe:

$$\Delta S_{universe} = -\frac{\Delta G_{system}}{T} \tag{4.13}$$

Since two systems are coupled together in the aluminum refining example, we can write

$$\Delta S_{universe} = \left(-\frac{\Delta G_{coal}}{T}\right) + \left(-\frac{\Delta G_{bauxite}}{T}\right) \tag{4.14}$$

(In what follows we shall assume that both coupled processes occur at the same temperature, although of course they usually do not.) Since $\Delta G_{bauxite}$ is positive, some of the entropy created by burning the coal will be destroyed, but only 17% of it; that is, the free energy change of the bauxite system must be 0.17 as large as and opposite in sign from that of coal:

$$\Delta G_{bauxite} = -0.17 \, \Delta G_{coal} \tag{4.15}$$

Therefore,

$$\Delta S_{universe} = \left(-\frac{\Delta G_{coal}}{T}\right) + \left(\frac{0.17 \, \Delta G_{coal}}{T}\right) \tag{4.16}$$

This result can be generalized to any fuel and any efficiency as follows:

$$\Delta S_{universe} = (1 - \text{efficiency}) \left(\frac{-\Delta G_{fuel}}{T}\right) \tag{4.17}$$

Three factors act to increase the entropy of the universe:

1. A large negative molar ΔG for combustion of the fuel.
2. Combustion of large amounts of fuel.
3. Low efficiency of conversion.

The third factor is perhaps the most important. When a product such as aluminum is produced by a highly efficient process nearly all of the free energy remains stored and available to do useful work. Eventually, of course, the aluminum will react spontaneously, releasing the free energy, and completing the increase in entropy begun when the coal was burned, but at this later time useful work can still be obtained.

It should be pointed out that when we talk about "energy consumption" or "energy conservation," our language is a bit loose. What we really consume or want to conserve is *free energy* — that which can do useful work. Conservation and wise consumption of free energy will minimize the increases in entropy which necessarily occur in the course of human existence.

4.6 THE IMPROBABILITY OF LIFE IN A UNIVERSE AT EQUILIBRIUM

When a molecule dissociates to form atoms there is invariably an increase in entropy due to the tendency of rotational and vibrational energy to be converted to translational energy.[9] Dissociation is prevented at ordinary temperatures because the potential energy of the atoms in a molecule is lower than the sum of their separate potential energies when dissociated. This reduced potential energy, sometimes referred to as "chemical energy," can be calculated by summing the bond energies of the molecule. As the temperature rises the effect of entropy becomes more important (the "low energy — high entropy" rule discussed in Section 4.2), and at sufficiently high temperatures all molecules become unstable.

A very simple example of this type of effect is the dissociation of dihydrogen:

$$H_2 \longrightarrow H + H \qquad \Delta G^0_{298} = 435 - T(0.099) = 464 \text{ kJ} \qquad (4.18)$$

There is an entropy increase of 99 J/K resulting from the conversion of more widely spaced (less randomly occupied) vibrational and rotational levels of H_2 to translational levels of the hydrogen atoms. The process does not occur spontaneously at 298 K because of the positive ΔH which requires that 435 kJ be transferred from the surroundings. (This entails an entropy decrease, $\Delta S_{\text{surroundings}} = q/T = -435 \text{ kJ}/298 \text{ K} = -1.46 \text{ kJ/K}$, more than 14 times the effect of the entropy increase in the system.)

As the complexity of a molecule increases the entropy effect illustrated in the preceding paragraph becomes more important. At 298 K all hydrocarbon compounds containing more than five carbon atoms are unstable with respect to dissociation into graphite and dihydrogen, even though many have negative enthalpies of formation. The same is true of biologically important molecules such as amino acids. They are invariably unstable with respect to decomposition into water, carbon dioxide, dinitrogen, and other small, stable molecules. Hence, they should exist only at very small concentrations, if the surface of the earth were at thermodynamic equilibrium. The degree of order represented by a protein molecule or even the simplest organism is exceedingly improbable. This is indicated in Table

---

[9] See W. G. Davies, "Introduction to Chemical Thermodynamics," Chapters 2 and 3. Saunders, Philadelphia, Pennsylvania, 1972.

TABLE 4.4   Probability of a System Occurring as a Spontaneous Fluctuation from Equilibrium[a]

System	Maximum probability of system	W Maximum probability that system would have occurred during history of universe
Escherichia coli	$10^{-10^{11}}$	$10^{-10^{11}}$
Mycoplasma hominis H39	$10^{-10^{10}}$	$10^{-10^{10}}$
Hemoglobin molecule	$10^{-40\,000}$	$10^{-39\,866}$
Amino acid	10^{-60}	10^{+74}

[a] Data from H. J. Morowitz, "Energy Flow in Biology," p. 67. Academic Press, New York, 1968.

4.4, which gives estimates of the thermodynamic probabilities of several important biological systems. The larger and more highly organized the system the farther it is from equilibrium and the smaller the probability that it could occur in a universe at equilibrium. Living systems have extremely low entropies.

Of course the universe has been in existence for a very long time (about 10 to 20×10^9 years according to cosmologists). Even a highly improbable event like the organization of the single state represented by a protein molecule from the myriad possible arrangements of its thousands of individual atoms might occur occasionally over such a time scale. To calculate the probability that a hemoglobin molecule might be formed in a universe at thermodynamic equilibrium and as old as ours four factors must be multiplied:

1. The maximum thermodynamic probability of the system (column two of Table 4.4).

2. The total number of systems of any kind present in the universe at a given instant. This cannot exceed the number of atoms in the universe, estimated to be less than 10^{100}.

3. The rate at which one system can be transformed into another. Since atomic and molecular collisions or excitations always require longer than 10^{-16} s, this cannot exceed 10^{16} s^{-1}.

4. The lifetime of the universe. Twenty billion years gives 10^{18} s.

The last column in Table 4.4 has been obtained by multiplying all four factors. Of the systems in the table only the relatively simple amino acids appear to have any chance of being found in a universe at equilibrium.

This type of argument poses a fundamental dilemma for thermodynamics. We know that life exists and biologists tell us that more and more complicated life-forms have been evolving during the last billion years or so of the earth's existence. But thermodynamics seems to predict exactly the opposite—the entropy of the universe should increase. Less and less

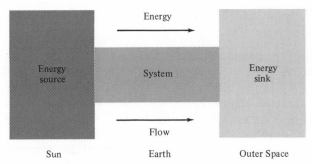

Figure 4.4 Steady-state nonequilibrium thermodynamics. Source: H. J. Morowitz, "Energy Flow in Biology," p. 17. Academic Press, New York, 1968.

order should evolve as time goes on. Some persons have used this seeming contradiction to argue that the entire theory of evolution is erroneous, although, of course, it could just as well mean that thermodynamics does not apply to living systems.

Actually, neither of these alternatives is required by the logic of thermodynamics. In calculating the probability of occurrence of living systems it was assumed that the universe has been an equilibrium ensemble of systems throughout its lifetime. The existence of life contradicts this assumption. Therefore the universe can more reasonably be assumed to be evolving from a state of high order toward one of greater disorder. During this evolutionary process some small portions of the universe (the earth, for example) may become segregated from the general trend of increasing entropy. They become hung up, as it were, as *steady-state, nonequilibrium systems*. Such systems are characterized by a flow of energy from a source to a sink (Figure 4.4). On earth this source and sink are often, but not invariably, the sun and outer space. Remember that the primitive earth simulation experiments and indeed the entire course of chemical and biological evolution described in Chapter 3 were dependent on energy from sunlight or some other source.

4.7 STEADY-STATE, NONEQUILIBRIUM SYSTEMS

Four generalizations can be made[10] from a thermodynamic study of steady-state, nonequilibrium systems. These are

1. The flow of energy from a source to a sink through a system can lead to an internal organization—a decrease in entropy—of that system.
2. Such a system will be in a steady state, not at equilibrium.
3. The flow of energy can lead to cyclic flows of matter in the system.

---

[10] H. J. Morowitz, "Energy Flow in Biology," Chapter 2. Academic Press, New York, 1968.

4. There is at least one maximum in the degree of order in such a nonequilibrium system.

Let us proceed to examine each one.

It is relatively easy to work out the thermodynamics of a steady-state system of a very simple type. Assume that the source of energy is a reservoir of heat at a temperature T_1 and the sink is a reservoir at a temperature T_2, where T_1 is larger than T_2. In order for a spontaneous process to occur, the entropy change of the system plus the entropy change of the surroundings must be positive. The entropy change of the surroundings can be calculated as shown in equation 4.19.

$$\Delta S_{\text{surroundings}} = \frac{-Q}{T_1} + \frac{Q}{T_2} > 0 \qquad (4.19)$$

Because we have chosen a steady-state system, the quantity of heat transferred from the source to the system (which we call Q) must be equal to the quantity of heat transferred from the system to the sink. Since T_1 is larger than T_2, the quantity Q/T_1 will be smaller than Q/T_2, and the entropy change in the surroundings will be positive. As a result, the entropy change of the system can be negative and its order can be increased as long as the change is not greater than the decreased order of the surroundings.

The mechanism by which the earth's surface can become more highly ordered at the expense of its surroundings depends on the energy per quantum of radiation coming from the sun and the energy per quantum radiated to outer space from the dark side of the earth. Energy comes from the sun largely in the form of visible radiation whose quanta are highly energetic. A relatively small number of photons need to be received to transfer a certain amount of energy to earth. On the other hand, the radiation to outer space is mostly in the infrared region of the spectrum and contains a considerably smaller amount of energy per quantum. If the earth is to remain in a steady state, the amount of energy radiated must equal the amount of energy received. Therefore a larger number of photons must be radiated than were absorbed, and they can be distributed more randomly with respect to energy and position. The randomness of the radiation increases and the entropy of the earth can decrease as a result of the energy flow (Figure 4.5).

Turning to the second generalization about steady-state, nonequilibrium systems, it is clear from Table 4.4 that the earth is not at equilibrium but in a very low entropy state. According to the second law of thermodynamics, a very strong tendency exists for this system to react toward equilibrium. In fact, the system is constantly attempting to achieve equilibrium. Free energy must be supplied to do the work necessary to reverse this spontaneous process. This is similar to our previous example in which free energy from burning coal was used to reverse the spontaneous oxidation of aluminum to bauxite. Without a constant supply of free energy the earth system

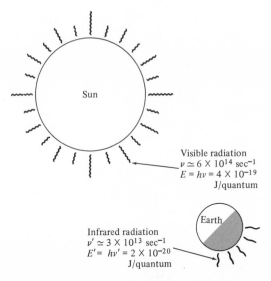

Figure 4.5 The sun as a source of high-grade energy on earth.

would spontaneously approach equilibrium — organisms would die, large molecules would decompose to smaller ones, and all of the order built up over the course of biological evolution would disappear. This is not observed to be happening because the flow of solar energy maintains the system under nearly constant conditions far from equilibrium. This is what we refer to as a steady state.

4.8 MATERIAL FLOWS — BIOGEOCHEMICAL CYCLES

That energy flow can induce cycling of matter can be seen by analyzing the simple system of three isomeric compounds diagramed in Figure 4.6a. Chemical reactions can convert any of these compounds into any of the others. For instance, A can react to form B by a process having a rate constant k_1. Under equilibrium conditions, according to the principle of microscopic reversibility,[11] the rate at which A reacts to form B and the rate at which B reacts to form A must be equal. The same applies for transformation of B to C and C to B, and for transformation of C to A and A to C. Under such conditions, there is no net change in the concentrations of A, B, and C.

An energy versus reaction coordinate diagram for the system might look like the one in Figure 4.6b. Let us suppose that the equilibrium is disturbed by allowing radiation to strike the system such that some of the radiation is absorbed. Eventually, the energy of this radiation will be converted to heat and must flow into some heat sink, as shown in Figure 4.6c. In order for

---

[11] L. Onsager, *Phys. Rev.* **38**, 405–426 (1930).

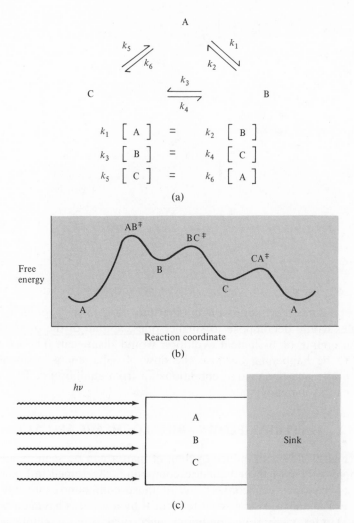

Figure 4.6 System of three isomeric compounds: (a) Kinetic parameters and equilibrium conditions, (b) energy-reaction coordinate diagram, and (c) possible experimental setup for steady-state system.

the radiation to be absorbed, its wavelength must be such that $h\nu$ exactly matches the difference in energy of two states of the system. For example, assume that the energy of radiation is exactly equal to the difference in energy between compound A and the transition state between compound A and compound B:

$$E_{\text{photon}} = h\nu = E_{\text{AB}^\ddagger} - E_{\text{A}} \tag{4.20}$$

Radiation of this frequency can speed up the reaction of A to form B, but it will not affect the other reactions because it does not exactly match any other energy difference in the energy-reaction coordinate diagram. Because the radiation can increase the rate at which A reacts to form B, it can induce a net cycling of material from A to B to C and back to A again. *Energy flowing through the system from a source to a sink has induced a cyclic flow of material through the system.* Once the cyclic flow of matter reaches a steady state, the concentrations of A, B, and C will remain constant over time, but they will be *different* from those in the equilibrium state.

All so-called biogeochemical cycles operate in the same way as this oversimplified example. For instance, the production of dioxygen by photosynthesis was described in Chapter 3. The basic process in the photosynthetic "carbon cycle" is the conversion by sunlight of CO_2 and H_2O into carbohydrates which later are transformed into a great many other compounds by living organisms. As higher concentrations of these compounds are achieved, their rates of reaction with dioxygen are increased and they are recycled to CO_2 and H_2O. The heat evolved in oxidation of the compounds eventually is radiated to the sink of outer space.

4.9 MAXIMIZING THE DEGREE OF ORDER IN A STEADY-STATE, NONEQUILIBRIUM SYSTEM

The fourth generalization states that *there must be at least one maximum in the degree of order of the nonequilibrium system.* This can be seen heuristically by considering the fact that as greater and greater quantities of energy are put into a system there is a greater probability that molecules will receive enough energy to ionize or to dissociate. Once ionization or dissociation begins, a rapid increase in entropy occurs since complex molecules break down into a larger and larger number of simple units. Thus, too large a flow of energy will create an increase in entropy of the system instead of a decrease.

In practical terms this fourth generalization means that on earth, at least, there is no such thing as an unlimited supply of energy. Too much energy input is just as bad for the system as too little. At this point it is hard to say what an optimum supply might be, but there is some evidence[12] that the relative amounts of solar energy received on Venus, Earth, and Mars may have had profound effects on the development of the rather different atmospheres of each planet; thus, the optimal energy flow may not differ drastically from our current solar input. We shall return to this problem toward the end of Chapter 7.

The basic laws of thermodynamics tell us that to create a more highly ordered system, any living organism must have available a supply of free

---

[12] S. I. Rasool, *in* "Exobiology" (C. Ponnamperuma, ed.), Chapter 11. North-Holland Publ., Amsterdam, 1972.

energy. For the earth as a whole, the supply is radiation from the sun. However, man, in order to replace his own and animals' muscle power, has supplemented this energy supply with a variety of other types of fuels. Today the most important of these are the fossil fuels—coal, petroleum, and natural gas. They are the subject of the next chapter.

EXERCISES

(The first group of exercises is based primarily on material found in this text.)

1. Verify the result obtained in Section 4.3 that a five-step expansion of one mole of an ideal gas from 11.2 to 22.4 dm^3 would yield 1.47×10^3 J.
2. List the steps in a Carnot cycle. Why is each one necessary? Explain how the efficiency of a heat engine is calculated and why it cannot reach 100%.
3. Arrange the following types of energy in order of decreasing entropy per unit energy:

   Nuclear   Cosmic microwave radiation
   Chemical   Thermal (10^6 K)
   Gravitational   Thermal (10^2 K)

4. Which of the following processes would produce the smallest increase in the entropy of the universe? The largest?

   a. Burning one ton of coal.
   b. Burning one ton of coal, converting the heat to electrical energy to refine aluminum from bauxite.
   c. Burning one ton of coal, converting the heat to electricity, and using the electrical energy to heat 100 homes.

5. Explain clearly the relationship between the change in Gibbs free energy for a chemical reaction and the increase in entropy of the universe. In a spontaneous, exothermic reaction, why is the increase in entropy of the universe often much greater than the increase in entropy of the chemical system?
6. Some devices are said to "waste" energy. Discuss energy waste in terms of the first and second laws of thermodynamics. Where does wasted energy go?
7. "Any living organism is a system very far from thermodynamic equilibrium." Do you agree? What arguments can you supply to support or disaffirm the above statement?
8. Name four characteristics of steady-state nonequilibrium thermodynamic systems. Explain how these apply if the earth is considered thermodynamically.
9. How much water would have to fall through the turbines at Hoover Dam to make 1 tonne of aluminum? (Assume 95% efficiency for generation of electricity and 18% efficiency for the Hall process. Hoover Dam is 772 feet high and the acceleration of gravity is 9.8 m/s^2.)

(Subsequent exercises may require more extensive research or thought.)

10. Calculate the standard enthalpy, entropy, and free energy changes for formation of any five of the compounds whose structures are shown in Figure 3.2. In each case note whether the enthalpy change or the entropy change favors dissociation. Data may be obtained from a handbook or general chemistry text.

11. Calculate *quantitatively* the heat absorbed at T_1, and the work done in a Carnot cycle and show that the efficiency is given by equation 4.11.

12. Explain why the entropy change of a chemical system where energy is transferred from vibrational levels of the reactants to translational levels of the products would be positive.

13. F. L. Curzon and B. Ahlborn [*Amer. J. Phys.* **43**(1), 22–24 (1975)] give a formula for calculating the efficiency of a Carnot engine operating at maximum power. Calculate this efficiency for an engine operating between 800 and 288 K. This may be compared with the actual efficiencies of power plants which will be discussed in the next two chapters.

5

Fossil Fuels

World resources of petroleum and natural gas . . . will be substantially consumed by the first quarter of the twenty-first century if world trends of production and consumption continue.

U. S. National Academy of Sciences, 1975

*You can make anything
from a salve to a star,
if you only know how
from black coal tar.*

Cornish, Organic Chemistry, 1905

The combustion of fossil fuels — coal, oil, and natural gas — provides by far the largest portion of the current United States supply of free energy. The production of synthetic organic chemicals is also based almost entirely on these raw materials. Because of the large investment of economic capital and chemical free energy in machinery and industrial plants which have been designed around fossil fuels, it would be unwise and perhaps impossible to switch instantaneously to another resource base. Moreover, the manpower and materials to construct the plants or other devices necessary if solar or nuclear energy were to replace fossil fuels could not possibly be mobilized immediately. Past experience (based largely on the lifetimes of industrial installations and the rate of new construction) suggests that a period of 20–40 years is required for one energy technology to replace another.

The foregoing does not imply that rapid development of newer free energy sources is unimportant. Indeed just the opposite is true, because considerable lead time must be allowed for alternative resources to supplant dwindling supplies of fossil fuel. It does emphasize, however, that environmental degradation associated with fossil fuels will persist for a number of decades even if "limitless" supplies of "clean" energy become available.

5.1 ORIGIN AND DEVELOPMENT OF COAL

A small fraction of the total amount of carbon reduced by photosynthesis has been accumulating in sedimentary deposits for the past 0.6 to 1.0×10^9 years. Formation of coal deposits requires special conditions; thus, only a minute portion of this total reduced carbon is involved. Some forms of lignite are apparently no older than 2×10^6 years, bituminous coals range from 100 to 300×10^6 years, and some anthracite coals may be as much as 380×10^6 years old. It is generally agreed that coal has been formed from the remains of terrestrial plants which were deposited in swamps and covered rapidly enough so that they were not oxidized by the atmosphere. Biochemical and physical processes then effected the transformation of this plant debris into various types of coal.

It seems reasonable to suppose that terrestrial plants of the carboniferous period (some 300 million years ago), during which many bituminous coals were laid down, had much the same composition as today's vegetation. If so, the material from which coal has evolved consisted of 0–16% protein, 20–55% cellulose, 10–35% lignin, and ~ 10% inorganic salts. The first stage in coal formation was conversion of this plant material to peat by aerobic and anaerobic degradation. Much of the protein apparently was incorporated in the body structure of the microorganisms and the cellulose was converted to CO_2 and H_2O by their metabolism, leaving behind lignin, a complex polymer of coniferyl, sinapyl, and *p*-coumaryl alcohols, which is difficult to decompose.

Coniferyl alcohol   Sinapyl alcohol   *p*-Coumaryl alcohol

The second stage of coal formation involved geological metamorphism. After the lignin and the remains of the bacteria (which probably account for the increase in nitrogen content at this stage of the process) had been buried under many other layers of sediment, high temperatures and pressures drove off additional H_2O and CO_2, leaving behind mostly carbon with smaller quantities of hydrogen, oxygen, nitrogen, iron, and sulfur. The *rank* of a coal indicates the extent to which this process has occurred, as shown in Table 5.1.

TABLE 5.1   Rank, Composition, and Fuel Value of Various United States Coals[a]

Fuel	Rank	State of origin	Analysis, weight % before drying				Heat content (MJ/kg)
			Moisture	Volatile matter	Carbon	Ash	
Anthracite	High	Penn.	4.4	4.8	81.8	9.0	30.5
Bituminous							
Low volatile		Md.	2.3	19.6	65.8	12.3	30.7
Medium volatile		Ala.	3.1	23.4	63.6	9.9	31.4
High volatile		Ohio	5.9	43.8	46.5	3.8	30.6
Subbituminous		Wash.	13.9	34.2	41.0	10.9	24.0
		Colo.	25.8	31.1	34.8	4.7	19.9
Lignite (brown coal)		N. Dak.	36.8	27.8	30.2	5.2	16.2
Peat		Miss.	—	—	—	—	13.
Wood[b]	Low	—	—	—	—	—	10.4–14.1

[a] Most data from A. L. Hammond, W. D. Metz, and T. H. Maugh, II, "Energy and the Future," Table 1. Amer. Ass. Advan. Sci., Washington, D. C., 1973.

[b] Includes waste; see G. C. Szego and C. C. Kemp, *Chem. Technol.* **4**(5), 275 (1973), Table 4.

The higher the rank of a coal the more nearly it approaches both the composition and the structure of graphite.[1] This structural change involved *aromatization,* the formation and dehydrogenation of six-membered rings of carbon atoms. An oversimplified example of aromatization is indicated by the formation of benzene from *n*-hexane in equations 5.1:

$$CH_3-CH_2-CH_2-CH_2-CH_2-CH_3 \longrightarrow \text{Cyclohexane} + H_2 \qquad (5.1a)$$

n-Hexane

$$\text{Cyclohexane} \longrightarrow \text{Benzene} \left(\text{or} \; \bigcirc \right) + 3H_2 \qquad (5.1b)$$

Cyclohexane   Benzene

Benzene is only the simplest of a whole series of aromatic compounds, many of which, like naphthalene, anthracene, and benzo[*a*]pyrene (Figure 5.1a) involve several benzene rings fused together. (Such polynuclear aromatic hydrocarbons are often found to be carcinogens — cancer-causing agents.) In graphite aromatization is as extensive as possible. The structure consists of planar sheets of six-membered rings as shown in Figure 5.1b.

[1] Recent work by S. Chakrabartty of the Research Council of Alberta, Edmonton, indicates that the amount of aromatization in coal may be less than has been supposed.

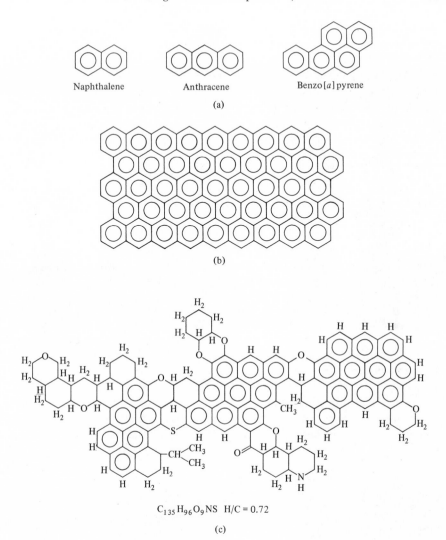

Naphthalene   Anthracene   Benzo[a]pyrene

(a)

(b)

$C_{135}H_{96}O_9NS$ H/C = 0.72

(c)

Figure 5.1 Comparison of a proposed coal structure with a single layer of graphite: (a) polynuclear aromatic hydrocarbons, (b) molecular structure of a single layer of graphite, and (c) proposed molecular structure of coal. Coal structure redrawn from S. M. Manskaya and T. V. Drozdova, "Geochemistry of Organic Substances," p. 131. Pergamon, Oxford, 1968.

The close relationship between a proposed structure of coal and the layers of graphite is shown in Figure 5.1c.

That coal is not a pure substance can be seen from the data in Table 5.1 and the formula in Figure 5.1c. The composition of different coals depends upon the geographical regions and specific mines from which they are obtained and can even vary within a given mine. When moisture is removed most coals contain from 70 to 90% carbon, ~4 to 5% hydrogen, 1 or 2% nitrogen, from 5 to 15% oxygen, and from 0.5 to 5% sulfur. As a result of the combustion of coal, all of these elements are oxidized and appear as reaction products in their oxidized forms; for instance, sulfur is oxidized to sulfur dioxide, nitrogen to nitric oxide, and cɑ ʾbon to carbon dioxide. The *ash* reported in Table 5.1 represents those elements (such as iron and other metals) which are converted into solid oxides. From 5 to 15% of the initial weight of coal appears as ash in the gaseous emissions (fly ash) or the solid waste of a coal-burning plant.

The sulfur content of coals has suddenly become extremely important because of restrictions on sulfur dioxide emissions (Section 9.3) from power plants. In the United States the bituminous coals of the East and Midwest have relatively higher sulfur content on a weight percent basis than the subbituminous and lignite deposits of the western states (Table 5.2).

The origin of sulfur in coal is not completely understood, but in many of the lower rank, more recent deposits there may have been enough sulfur in the plant and animal material originally laid down to account for what is now found. It seems almost certain that sulfur which is covalently bonded in

$$\overset{\backslash}{\underset{\diagup}{C}}-SH, \quad -\overset{\backslash}{\underset{\diagup}{C}}-S-S-\overset{\diagup}{\underset{\backslash}{C}}-, \quad \overset{\backslash}{\underset{\diagup}{C}}H-SH, \quad \text{and} \quad \overset{\backslash}{\underset{\diagup}{C}}H-S-\overset{\diagup}{\underset{\backslash}{C}}H$$

groups originated in this way. However, almost half of the sulfur in older bituminous deposits is in the form of pyrite (FeS_2) or other inorganic compounds. This may have arisen from the same source as the organic sulfur, but it might also be the product of bacterial reduction of dissolved sulfate ions. In some cases FeS_2, ZnS, and other sulfides are found in cracks or fissures, almost certainly the result of hydrothermal deposition after the coal had formed.

Pyritic and other inorganic sulfur can be removed rather easily by grinding coal to a fine powder and washing or screening the lower density carboniferous material away from the heavier inorganics. Since coal is usually pulverized (to improve the kinetics and hence the efficiency of combustion) in modern installations, little additional cost is required for such

TABLE 5.2 Sulfur Content of Coals in the United States

Type of coal	Percentage of total resources		
	Low S (0–1% by weight)	Medium S (1–3% by weight)	High S (>3% by weight)
Anthracite	97.1	2.9	—
Bituminous	29.8	26.8	43.4
Subbituminous	99.6	0.4	—
Lignite	90.7	9.3	—
All ranks	65.0	15.0	20.0

treatment. Unfortunately, it usually does not remove covalently bonded, organic sulfur.

Recently, there has been a rush to exploit deposits of subbituminous coal and lignite in Montana and Wyoming. These readily strip-minable resources have begun to be shipped 1600 km or more to midwestern electric generating plants because their lower weight percent sulfur could meet air quality regulations. There are a number of reasons why such a shift in the source of coal may be misguided. Because they are of more recent origin and hence lower rank the western coals have only one-half to three-quarters the heat value (Table 5.1) of eastern bituminous, and most of their sulfur is covalently bound. Since most coal is consumed to produce energy, the proper basis for measurement of sulfur is per joule, not per kilogram. Recent air quality standards (< 0.517 kg SO_2/kJ) have been written with this in mind. On such a basis little of the western coal appears to be effectively low sulfur,[2] especially when the relatively easy removal of inorganic sulfur from eastern bituminous is accounted for. These two factors combine to raise the effective sulfur content of western coals to as much as four times its nominal value. Moreover, long distance transportation takes an additional toll both in energy and human lives (see Section 5.2).

Because deposition of coal occurred in freshwater sediments and not under the oceans, geologists are fairly confident that all the major coal basins of the world have been discovered. A reasonable estimate[3] of minable amounts of coal is 7.6×10^{12} tonnes (1 tonne $= 10^3$ kg), most of which occurs in the northern hemisphere. As with other depletable energy sources, coal production can be expected to go through a cycle reaching a maximum and then decreasing as the price climbs higher, supplies become more difficult to find, and quality decreases. Hubbert[4] has calculated the curves shown in Figure 5.2 for the cycle of world coal production. Note that halving or doubling world reserves has little effect on the time required for consumption of the resource, and that after the year 2000 neither supply curve can possibly keep up with an extrapolation of the current rate of increase in demand. Nevertheless, reserves of coal are large enough that with prudent use it should remain a major factor in supplying both free energy and reduced carbon for several centuries.

---

[2] R. Curry and S. Andersen, "Some Western Coal Sulfur Problems," Res. Office Bull., Sierra Club, October 17, 1974; M. Rieber, "Low Sulfur Coal. A Revision of Reserve and Supply Estimates," Doc. No. 88. Center for Advanced Computation, University of Illinois, Urbana, 1973 (Stock No PB235-464 from National Technical Information Service, Springfield, Virginia).

[3] P. Averitt, *U. S., Geol. Surv., Bull.* **1275**, Table 8, p. 82 (1969).

[4] M. K. Hubbert, Energy resources. *In* "Resources and Man," pp. 167–242. Freeman, San Francisco, California, 1969; see also M. K. Hubbert, "U. S. Energy Sources as of 1972," pp. 56–64, Committee Print Serial No. 93-40 (92-75), Committee on Interior and Insular Affairs, U. S. Govt. Printing Office, Washington, D. C., 1974, in which new, *lower* estimates of coal reserves are given.

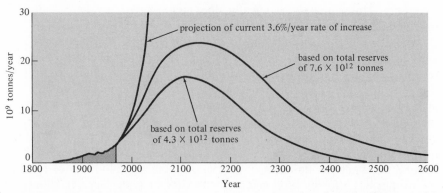

Figure 5.2 Cycle of world coal production, based on estimated total supply of between 4.3 and 7.6×10^{12} tonnes. Source: M. King Hubbert, The energy resources of the earth. *Sci. Amer.* **224** (3), 69 (1971).

5.2   COAL-FIRED POWER PLANTS

Most of the coal which is currently consumed in the United States is used to produce heat, steam, and often electric power for industrial and domestic use. All of these applications require large-scale power plants. A schematic diagram of the flow of material and energy through a plant which can produce electric energy at a rate of 1000 MW (10^9 J/s) is shown in Figure 5.3. The plant is assumed to operate at an average of 75% of its rated capacity. Such a plant is about the same size as many that are under construction today, although it is considerably larger than the average of currently operating plants. It would produce 2.36×10^{16} J each year—enough to meet the demands of a city of 900 000 people. Environmental problems (and the social costs which accompany them) may occur throughout the system represented in Figure 5.3, and any assessment of the true cost of electric power produced from coal must consider the entire system.

In one year our representative power plant produces an output of 2.36×10^{16} J electrical energy (2.16×10^{16} J actually delivered to customers, allowing for transmission losses) from an input of coal (corrected for the quantity lost in transportation, processing, and storage) equivalent to 6.24×10^{16} J. The efficiency of the power plant itself is only 38%, largely because thermal energy must be converted to mechanical during its operation. The basic design of a fossil fuel steam power plant is shown schematically in Figure 5.4. The heat released by burning coal is used to boil water and superheat the steam which drives a turbine connected to an electrical generator. The steam is then passed through a condenser to cool it below 310 K, and recycled to the boiler. The maximum temperature in the cycle is determined largely by the fact that water becomes progressively more reactive as it gets hotter. Above 800 K special, expensive alloys are

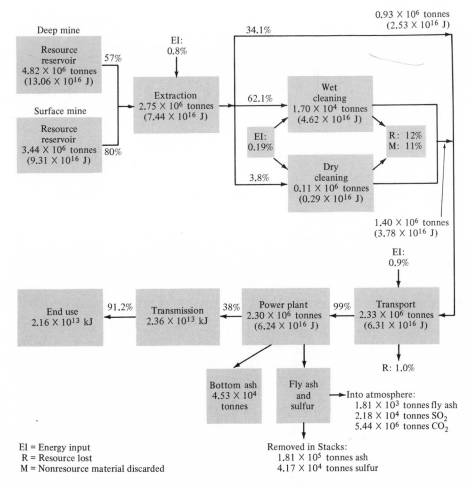

Figure 5.3 Flow of material and energy in a 1000-MW$_e$ coal-fired electric power plant operating annually at 0.75 load factor. Adapted from Council on Environmental Quality, "Energy and the Environment: Electric Power," No. 4111-00019, Superintendent of Documents, Washington, D. C., 1973. (The subscript e in MW$_e$ indicates 1000 MW of electrical power as opposed to the 2632 MW$_t$ of thermal energy needed to generate the electricity at 38% efficiency.)

required to prevent rapid corrosion of the piping. If such a system could be operated completely reversibly as a Carnot cycle, the maximum theoretical efficiency would be $(800 - 310)/800 = 0.61$ or 61%. The fact that the power plant does not follow a Carnot cycle exactly and does not always operate at the maximum possible temperature reduces theoretical efficiency to about 50%. Furthermore, the steam turbine is only 89% efficient, the boiler is ~88% efficient, and the generator ~99% efficient, giving an overall value of about 38%.

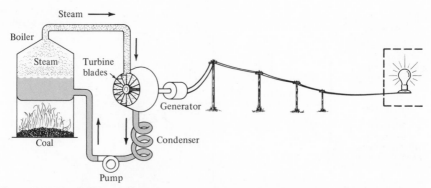

Figure 5.4 Fossil fuel steam power plant (schematic). Redrawn from R. Wilson and W. J. Jones, "Energy, Ecology, and the Environment," p. 20. Academic Press, New York, 1974.

Older plants operate at lower maximum temperatures and therefore are even less efficient; thus, the average of all fossil fuel plants was about 33% in 1970. This average has improved steadily from about 5% in 1910, but as actual efficiencies approach the theoretical limit additional gains become more difficult. For the next decade or two it is almost certain that coal-fired electric power plants will reject nearly two-thirds of the heat content of their fuel to their neighboring environment. They must do so or their efficiencies would become even smaller as the lowest temperature in the cycle increased because of retention of waste heat. The usual solution to the problem of keeping the low temperature portion of the steam turbine power cycle as cool as possible has been to transfer waste heat to the nearest body of water. Because it has a relatively large heat capacity, good heat transfer properties, and is a liquid under normal conditions of temperature and pressure water is seemingly ideal for disposal of unwanted thermal energy. However, it has been estimated[5] that if power consumption continues to grow at its current rate and there is no increase in efficiency, electric power plants of all types will produce enough waste heat in the year 2000 to raise the average temperature of all river water flowing over the lower 48 United States by more than 11 K.

The two major consequences of elevating the temperature of natural waters are increased chemical reaction rates and decreased solubility of essential gases such as dioxygen. As rates of metabolism of water-dwelling organisms go up so does the demand for dioxygen, but the potential supply of this and other gases moves in the wrong direction. Because of the balance between these two effects, the optimal temperature range for most aquatic organisms is not far below the maximum at which they can survive. Elevated temperatures can also destroy enzyme structure and function, and rapid temperature changes which would be encountered at the edge of a power plant discharge may affect cell membranes. All of these are specific

[5] D. E. Abrahamson, "Environmental Cost of Electric Power." Scientists' Institute for Public Information, New York, 1970.

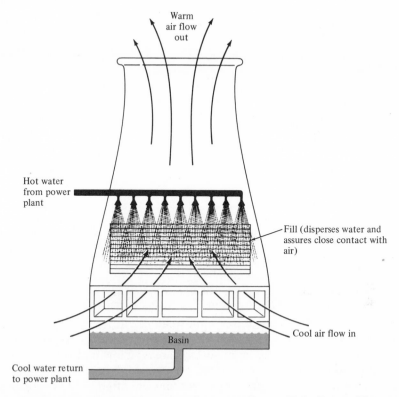

Warm
air flow
out

Hot water
from power
plant

Fill (disperses water and
assures close contact with
air)

Basin

Cool air flow in

Cool water return
to power plant

Figure 5.5 Schematic diagram of a wet cooling tower. Source: T. L. Brown, "Energy and the Environment," p. 78. Merrill, Columbus, Ohio, 1971.

instances of the generalization (Section 4.9) that too great an energy flow may disrupt the steady-state ordinarily maintained by aquatic organisms.

While the foregoing clearly indicates the rationale for calling waste heat discharges *thermal pollution,* not all of the effects are undesirable. If enough O_2 is available, increased temperatures speed growth and development of aquatic life. This sometimes results in better fishing or increased harvests, especially in areas such as large lakes or oceans where the volume of water is sufficient to minimize thermal inpact. On the other hand, increased rates of growth of blue-green algae may accelerate eutrophication (Section 14.3) and a power plant sited on a small, deep lake may destroy the layered structure necessary for survival of many species which spend the summer in deeper regions whose temperature seldom rises above 280 K.

Most technologies for dealing with thermal pollution transfer waste heat to the atmosphere, from which it can be radiated to outer space. The simplest are cooling ponds — large, shallow pools which allow water to cool before it is returned to a stream or small lake. More efficient transfer to the atmosphere can be attained by means of cooling towers. In a *wet cooling tower* (Figure 5.5) hot water is sprayed down through cool air. Some of the

TABLE 5.3   Energy and Dollar Costs of Thermal Pollution Abatement

	Power requirement[a]		% Cost increase to consumer[b]		
	(% of output)		Class of consumer		
Type of system	Fossil	Nuclear	Industrial	Commercial	Residential
Once-through, fresh H_2O	0.4	0.6	—	—	—
Once-through, salt H_2O	0.4	0.6	0.34	0.16	0.14
Cooling pond	—	—	0.94	0.43	0.39
Wet cooling tower	0.9–1.0	1.3–1.7	3.17	1.41	1.28
Dry cooling tower	3.0	4.8	Uncertain (could be in the range of $1\frac{1}{2}$–3% for residential)		

[a] Based on data in M. Eisenbud and G. Gleason, eds., "Electric Power and Thermal Discharge," p. 372. Gordon & Breach, New York, 1969.

[b] D. E. Abramson, "Environmental Cost of Electric Power," p. 9. Scientists' Institute for Public Information, New York, 1970.

water evaporates allowing its heat of vaporization (as well as the heat lost upon cooling) to be transferred to the atmosphere. This more effective heat transfer has the disadvantage that mini-weather effects such as fog or rain may occur downwind. In the *dry cooling tower* water is not sprayed but flows through pipes having large surface areas, allowing heat transfer but no evaporation. The natural tendency of warm air to rise provides sufficient air flow in a wet cooling tower but fans are often used to supplement it. Dry cooling towers almost invariably have mechanical draft. The relative costs in energy and dollars for dispersion of thermal pollution are given in Table 5.3.

A wide variety of environmental costs other than thermal pollution are involved in producing electric power by the combustion of coal. Many of these are listed in Table 5.4, an estimate by scientists at Argonne National Laboratory[6] of the effects of a 1000-MW_e coal-fired electrical generating plant in 1980. The equivalent of nearly 150 of such plants is expected to be in operation by that time. The figures in the table are based on the assumption that federal regulations relating to environmental quality and worker safety will be obeyed. For example, only 1% of total fly ash and 20% of total SO_2 are included in the figures for air emissions. The remainder is assumed to be removed (see Section 9.5) and disposed of as solid or liquid waste.

Underground mining of coal is a hazardous occupation which exposes workers to dangers of disease (black lung), explosion, fire, and entrapment because of collapse of shafts. The injury rate is currently about three times

---

[6] "A Study of Social Costs for Alternate Means of Electric Power Generation for 1980 and 1990," Summary Report, Draft Version. Argonne Nat. Lab., Argonne, Illinois, 1973.

TABLE 5.4 Annual Environmental Effects of 1000 MW$_e$ Coal-Fired Power Plant (Estimated 1980)[a]

Environmental effects	Mining	Transportation	Power plant	Total
Occupational accidents				
Deaths	0.98[b]	0.055	0.03	1.1
Nonfatal injuries	40.5[b]	5.1	1.5	47.1
Mandays lost	8 330[b]	570	350	9 250
Mining[b]				
Land disturbance by stripping (acres)	300			
Land subsidence, underground mining (acres)	200			
Mine drainage, tons	10 000			
Sulfuric acid in drainage, tons	80			
Dissolved iron in drainage, tons	20			
Rail transportation				
Public death		0.55		
Injury		1.17		
Days lost		3 500		
Transportation and handling loss, tons		10 000		
Ash collected, tons			250 000	
Sulfur retained, tons			46 000	
Waste storage area, acres			5	
Thermal discharge, 10^{16} J			3.2	
Stack discharge, 10^{16} J			0.58	
Air emissions, tons:				
Fly ash			2 000	
Sulfur dioxide			24 000	
Carbon dioxide			6 000 000	
Carbon monoxide			700	
Nitrogen oxides (as NO_2)			20 000	
Mercury			5	
Beryllium			0.4	
Arsenic			5	
Cadmium			0.001	
Lead			0.2	
Nickel			0.5	
Radium 226, Ci			0.02	
Radium 228, Ci			0.006	
Facilities land use (acres)[c]				150

[a] Source: "A Study of Social Costs for Alternate Means of Electrical Power Generation for 1980 and 1990," p. 79. Argonne National Laboratory, Argonne, Illinois, 1973.

[b] Fifty percent of production from strip mining and 50% from underground mining.

[c] Facilities are for two generating units at a site and include fuel preparation but exclude transportation area.

the national average for all manufacturing industries. Because they are safer, require less specialized equipment, and cost less, strip and augur mining have grown more rapidly than underground methods in recent years, but they can be used to mine only about 10% of the United States coal reserves. Coal fields in Appalachia follow the contours of the mountains and in many western states they occur in areas of extremely dry climate and sparse vegetation, factors which make restoration of stripped land to its former condition extremely difficult.

Both underground and surface mining can result in acid or alkaline drainage as various compounds are exposed to the atmosphere and become dissolved in natural waters. *Acid mine drainage,* by far the most detrimental to water quality, is initiated by air oxidation of pyrite (equation 5.2a). Equations 5.2b and 5.2c constitute a cyclic process in

$$FeS_2 + \tfrac{7}{2} O_2 + H_2O \longrightarrow Fe^{2+} + 2\ SO_4^{2-} + 2\ H^+ \tag{5.2a}$$

$$Fe^{2+} + \tfrac{1}{4} O_2 + H^+ \longrightarrow Fe^{3+} + \tfrac{1}{2} H_2O \tag{5.2b}$$

$$FeS_2 + 14\ Fe^{3+} + 8\ H_2O \longrightarrow 15\ Fe^{2+} + 2\ SO_4^{2-} + 16\ H^+ \tag{5.2c}$$

$$Fe^{3+} + 3\ H_2O \rightleftharpoons \underline{Fe(OH)_3} + 3\ H^+ \tag{5.2d}$$

which iron(II) is air oxidized to iron(III) which then oxidizes more pyrite into solution. Note that the oxidation of S_2^{2-} by either dioxygen or iron(III) (equations 5.2a and 5.2c) and the hydrolysis of Fe^{3+} (equation 5.2d) all produce hydrogen ions. In some cases pH values as low as 2.5 have been found. The rate-limiting step in this sequence is apparently[7] reaction 5.2b, for which the rate law is

$$\text{rate} = -\frac{d\,[\text{Fe(II)}]}{dt} = k[Fe^{2+}][OH^-]^2\, p_{O_2} \qquad (\text{pH} \geqslant 5.0) \tag{5.3}$$

As more hydrogen ions are produced $[OH^-]$ goes down and the sequence of reactions 5.2 is slowed. Acid mine drainage would probably be much less serious if it were not for the unfortunate fact that as the pH drops below 5.0 the dependence of rate on hydroxide ion concentration disappears from equation 5.3. Negative feedback is no longer provided by decreasing $[OH^-]$. Moreover, field observations have demonstrated that 5.2b occurs more rapidly than would be predicted on the basis of laboratory studies. This apparently results from catalysis by microorganisms which can accelerate the reaction as much as a million times.

Control of acid mine drainage might be effected by halting or slowing down reaction 5.2b. The former process would require elimination of dioxygen from the mine, an expensive and probably impractical measure. Bactericides might be used to eliminate catalysis of 5.2b, and at least one, chromate, might also act to inhibit reaction 5.2a.[8] This would slow the

[7] P. C. Singer and W. Stumm, *Science* **167**, 1121–1123 (1970).

[8] W. H. Hartford, *Science* **169**, 504 (1970).

cyclic process and halt the initiating step at the same time. Nevertheless, streams and mines already polluted by acid drainage are usually coated with reddish brown $Fe(OH)_3$ (or $Fe_2O_3 \cdot nH_2O$). The reversal of 5.2d regenerates $Fe^{3+}(aq)$ from this precipitate, providing reactant for 5.2c even in the absence of O_2. Thus, abatement procedures cannot be expected to take effect immediately.

In addition to thermal pollution and acid mine drainage a number of air pollutants are produced by coal-fired power plants. These include carbon monoxide, nitrogen oxides, poisonous metals (such as lead, beryllium, mercury, arsenic, and cadmium), some radioactive isotopes of radium as well as SO_2, fly ash, and large amounts of carbon dioxide. Trace quantities of nickel in coal may be converted to nickel tetracarbonyl

$$Ni + 4\,CO \longrightarrow Ni(CO)_4 \tag{5.4}$$

by carbon monoxide in the stack gases. Nickel tetracarbonyl is considerably more poisonous than CO and is suspected of causing lung cancer. Incomplete combustion of coal can release benzo[a]pyrene and other compounds having fused aromatic rings, but modern power plants burn coal so efficiently that very few such molecules remain unoxidized.

There is one other environmental cost of electric power generation that has not been listed in Table 5.4. About ten acres of land area is required for every mile of high voltage transmission line. In the northeastern United States, where the concentration of power plants is greatest, the problem of finding suitable rights-of-way for transmission lines is becoming acute. Because such rights-of-way must be kept clear for service vehicles, large quantities of herbicides are often used and erosion sometimes causes extensive damage.

5.3 CLEANER COAL COMBUSTION

Five techniques can be envisioned for alleviating environmental degradation associated with coal combustion: (a) supplies of fuel (low sulfur coal, natural gas) which do not contain the precursors (sulfur, metals) of pollutants may be sought, or entirely new free energy sources may be developed; (b) contaminants (such as FeS_2) may be removed prior to combustion; (c) pollutants may be extracted from effluents after combustion; (d) pollutants may be diluted by taller stacks and greater dispersion; (e) fuel consumption may be reduced by increased efficiency or programs of energy conservation. Of these alternatives only (a) and (d) have been used extensively so far, and both are reaching their limits. Point (c) will be discussed in greater detail in Sections 9.2 and 9.5 because it is applicable to all forms of air pollution, not just from coal combustion. A number of approaches to points (b) and (e) will be considered in this section.

Nearly all of the sulfur (and perhaps other impurities as well) can be removed from coal during processes known as *gasification* and *liquefac-*

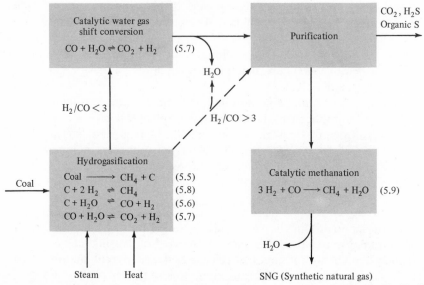

Figure 5.6 Generalized flow chart for gasification of coal. Source: T. H. Maugh, II, Gasification: A rediscovered source of clean fuel. *Science* **178**, 45 (1972). Copyright 1972 by the American Association for the Advancement of Science.

tion. These techniques are currently in pilot plant development stages but will surely become more important during the next two decades as reserves of natural gas and petroleum dwindle. In favorable cases they may even be carried out underground, eliminating some exposure of miners to hazard and simplifying extraction of the coal. Although liquefaction usually refers to production of hydrocarbon fuel similar to petroleum, other liquids such as methanol[9] may be obtained as well.

Since liquefaction and gasification are closely related we shall consider only the latter in detail. A generalized flow chart of the chemical reactions involved is shown in Figure 5.6. The first step in the process is similar to the formation of coke and is usually referred to as *devolatilization:*

$$\text{Coal} \xrightarrow{\text{800–1100 K}} \text{C} + \text{CH}_4 \tag{5.5}$$

Although a relatively high temperature is required for devolatilization to occur rapidly, it is exothermic and represents a continuation of the spontaneous loss of hydrogen and oxygen referred to in the earlier description of the formation of coal. Because most coals contain only a small percentage of hydrogen (the number of hydrogen atoms present is usually less than the number of carbon atoms), less than one-fourth of the total carbon in the coal can be converted to methane by reaction 5.5. Although this is fine if

---

[9] G. A. Mills and B. M. Harney, *Chem. Technol.* **4**(1), 26 (1974).

one is producing coke, an external source of hydrogen is needed if a complete conversion of coal to gaseous or liquid fuel is to be carried out.

The most readily available source of hydrogen is water, which can be reacted directly with the carbon in the char left behind by reaction 5.5 (see Figure 5.6).

$$C + H_2O \underset{}{\overset{1200\ K}{\rightleftharpoons}} CO + H_2 \qquad \Delta G^0_{298} = 131.4 - T(-0.01825) = 136.8\ kJ \qquad (5.6)$$

This so-called *water−gas reaction* produces a combustible vapor which is known as *power gas* or *low BTU gas*. The heat released per unit volume of power gas burned is relatively small−about 15% as much as from natural gas. Power gas is adequate for use in a power plant at the site of coal gasification, but its low energy per unit volume makes it uneconomical to ship it over long distances.

Further chemical processing can convert power gas into *synthetic natural gas* (SNG). The reactions are the *water−gas shift* (equation 5.7) and *hydrogasification* (equation 5.8). The resulting fuel is primarily methane and can adequately replace natural gas.

$$CO + H_2O \rightleftharpoons H_2 + CO_2 \qquad \Delta G^0_{298} = -41.15 - T(-0.04245) = 28.5\ kJ \qquad (5.7)$$

$$C + 2\ H_2 \underset{}{\overset{1200\ K}{\rightleftharpoons}} CH_4 \qquad \Delta G^0_{298} = -74.88 - T(-0.2331) = -5.43\ kJ \qquad (5.8)$$

It is preferable to carry out all four of the reactions (5.5, 5.6, 5.7, and 5.8) in the same reactor so that heat can be provided to reaction 5.6 by the three exothermic steps. This minimizes the amount of heat which must be provided from external sources, and should maximize the thermodynamic efficiency of gasification.

A final reaction, *catalytic hydrogenation* of carbon monoxide to form methane, must be carried out at a lower temperature in a separate reactor. Because three moles of dihydrogen are required for each mole of carbon monoxide in 5.9,

$$3\ H_2 + CO \underset{650\ K}{\overset{Ni}{\rightleftharpoons}} CH_4 + H_2O \qquad (5.9)$$

the water−gas shift reaction may need to be employed outside the principal reactor in order to adjust the H_2/CO ratio (see the upper left box in Figure 5.6). After removal of water the product of the gasification process is almost identical in composition to natural gas.

From an environmental standpoint gasification is advantageous since sulfur must be removed from the gas in order to prevent poisoning of the nickel catalyst for reaction 5.9. The sulfur will be present as H_2S because of the excess of dihydrogen in the first steps of gasification. This compound is more easily reacted with alkalis than the SO_2 which is usually present in stack gases following coal combustion, and its concentration is relatively large because the gasification process is carried out at elevated pressures. Both of these factors aid in its removal. One procedure uses a bed of granular, half-calcined dolomite, which is obtained by gentle heating of dolomite:

$$CaCO_3 \cdot MgCO_3 \xrightarrow{\text{heat}} CaCO_3 \cdot MgO + CO_2 \qquad (5.10a)$$
$$\text{Dolomite} \qquad \text{Half-calcined dolomite}$$

$$CaCO_3 \cdot MgO + H_2S \rightleftharpoons CaS \cdot MgO + H_2O + CO_2 \qquad (5.10b)$$

Apparently the crystal structure of the half-calcined dolomite is such that an extremely large surface area is available to the gaseous mixture, making the dolomite extremely effective for H_2S removal.

Because equation 5.10b is reversible, the H_2S may be recovered at high concentration and purity outside the reactor by adding CO_2 and H_2O; dolomite is recovered at the same time and recycled to the purifier. The hydrogen sulfide is usually converted to

$$H_2S + \tfrac{3}{2} O_2 \longrightarrow SO_2 + H_2O \qquad (5.11a)$$
$$2\,H_2S + SO_2 \xrightarrow[\text{S}]{\text{molten}} 3\,S + 2\,H_2O \qquad (5.11b)$$

elemental sulfur for storage or shipment, since the solid form occupies less space and costs less to ship.

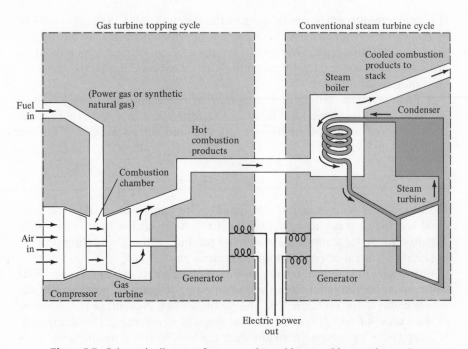

Figure 5.7 Schematic diagram of a power plant with gas turbine topping cycle.

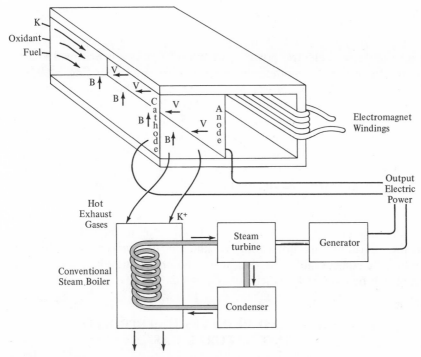

Figure 5.8 Principles of a magnetohydrodynamic electric generator. B, magnetic field (vertical); V, induced voltage (horizontal).

The limitations on efficiency imposed by high temperature reactivity of steam may be partially circumvented by means of gas turbine or magnetohydrodynamic power-generating systems. These *topping cycles,* which operate in the temperature range of 1500–2500 K and whose exhaust gases can then fire a conventional steam boiler, are especially compatible with power gas or synthetic natural gas (SNG) from coal gasification. Both topping cycles require that combustion products pass through turbines or a generator, and large quantities of ash, which would be present in coal prior to gasification, might clog such devices.

A *gas turbine* topping cycle employs an engine roughly equivalent to that of a jet aircraft. Figure 5.7 indicates schematically how it might be added on top of the conventional boiler and steam turbine cycle. A *magnetohydrodynamic (MHD) generator* operates on a completely different principle. Fuel is combined with preheated air in a pressurized combustion chamber and the hot combustion gases are ionized by the addition of an alkali metal such as potassium. The ionized gas is passed through a duct surrounded by coils that produce a very strong magnetic field (see Figure 5.8), inducing a voltage perpendicular both to the direction of motion of the ions and to the direction of the magnetic field. This voltage causes current to flow in an external electric circuit. As was true of the gas turbine, the exhaust gases from an MHD generator are still sufficiently hot to fire a conventional steam boiler.

Both gas turbine and MHD topping cycles appear to be able to increase overall efficiency of electric power generation from its current level to 50% or more. Nevertheless, neither has been employed as a topping cycle in a commercial United States power plant so far. Gas turbines have been used separately to meet demand during peak load periods, and a pilot MHD generator has operated successfully in the Soviet Union. When combined with coal gasification either of these topping cycles would help to offset the fact that the chemical processes required may be only 65–75% efficient.

Conversion of coal into more convenient liquid and gaseous fuels and cleansing it of pollution-causing impurities obviously take their toll in terms of the free energy available. The same is true of the requirement that, because it is deep in the ground, most coal will have to be obtained from shaft mines. While new technologies will enable us to make better use of coal, they certainly are not panaceas, and the *net energy* available from them may be considerably less than would be imagined at first glance. This concept will be explored further in Section 7.3.

5.4   ORIGIN AND RESERVES OF PETROLEUM AND NATURAL GAS

Petroleum originated through the deposition of organic matter from plants and animals at the bottoms of ancient seas. Such organic matter was decomposed by anaerobic bacterial action which removed much of the oxygen, sulfur, and nitrogen. The hydrocarbon compounds produced in this way were probably concentrated by being dissolved in water and transported through sedimentary rocks; the deposits were then trapped in dome-shaped chambers from which they can be obtained by drilling wells.

Because it is relatively easy to recover, handle, and transport, and more recently because it is relatively free of pollution-causing impurities, the use of natural gas (which consists almost entirely of methane) has grown more rapidly than any other fuel during the past few decades. This tremendous consumption is already outpacing the discovery of new reserves in the United States. Figure 5.9 indicates that domestic production of both natural gas and petroleum has already peaked. New discoveries, even those as large as the Alaska north slope, can only flatten the sharp peaks in Figure 5.9 slightly, postponing consumption of the resource by a few years at best.

For the near-term future the United States will have to import large quantities of petroleum and natural gas, but over a longer time scale even imports will become hard to obtain. It is estimated that 50% of the world's recoverable reserves will have been consumed by the year 2000. One might argue that because currently known oil fields are concentrated in a

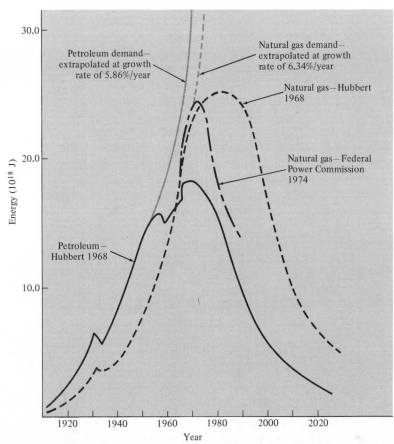

Figure 5.9 Cycle of production of petroleum and natural gas in the United States. Data from M. King Hubbert, Energy resources. *In* "Resources and Man," pp. 167–242. Freeman, San Francisco, California, 1969: "A Realistic View of U. S. Natural Gas Supply," Bureau of Natural Gas Staff Report, p. 11. Washington, D. C., 1974.

relatively small portion of the earth's surface (the Persian Gulf, North Africa, and the Gulf of Mexico), all that is required is further exploration to discover much more. However, formation and concentration of petroleum deposits probably required highly specific conditions of temperature, mineral nutrients for the plankton whose remains would be converted to oil, and appropriate tectonic factors controlling formation of oil-bearing basins and preservation of the resource, once formed.[10] Thus it would not be accurate to extrapolate from known oil fields to the entire globe. Indeed, most large oil fields have probably already been discovered, and returns of gas or oil per meter of new drilling have been decreasing for some time.

Two resources related to petroleum which have not yet been exploited

[10] E. Irving, F. K. North, and R. Couillard, *Can. J. Earth Sci.* **11**(1), 1–17 (1974).

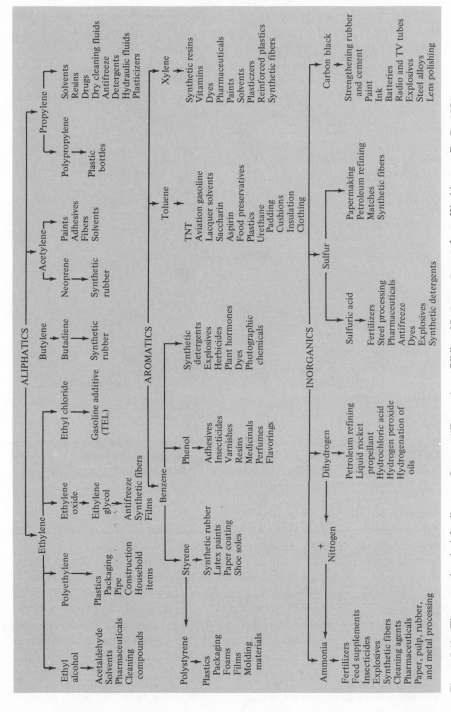

Figure 5.10 The petrochemical family tree. Source: "Facts about Oil," p. 27. Amer. Petrol. Inst., Washington, D. C., 1971.

are tar sands and shale oil. Table 4.3 indicated that the latter especially constitutes a large reservoir of free energy, although both resources are small by comparison with coal. Tar sands consist of extremely viscous, high-boiling liquid hydrocarbons found in association with sand deposits. Oil shale contains about 20% solid hydrocarbon (kerogen) and 80% ordinary shale. Recovery techniques now being tested involve heating in large retorts to drive off the hydrocarbons. Because the treated shale occupies greater volume than the material originally strip mined, and since large quantities of water are required for the retorting process, considerable disturbance of the environment is inevitable. In the western United States, where large deposits of oil shale are available, there is some concern about the adequacy of water supplies. The free energy consumed in retorting will also greatly reduce the net energy available from oil shale.

One other aspect of the use of fossil fuels ought to be mentioned at this point—they constitute the majority of the easily recoverable supply of *reduced carbon* on earth (see Figure 5.10). A variety of useful materials (including plastics and perhaps protein for use as food) can be manufactured from hydrocarbons or coal. The difficulty of synthesizing these substances from oxidized forms such as atmospheric CO_2 or carbonate minerals makes our current practice of burning large quantities of reduced carbon very questionable.

5.5 COMPOSITION AND CLASSIFICATION OF PETROLEUM

Crude petroleum is a mixture of a wide variety of hydrocarbon compounds which can be classified on the basis of structural type or number of carbon atoms. *Paraffins* have the general formula C_nH_{2n+2} with straight or branched carbon chains but no rings or multiple bonds. *Naphthenes* contain primarily five- and six-membered rings. Taken together paraffins and naphthenes are classed as *aliphatic hydrocarbons*. *Aromatic* compounds contain benzene rings.

Hydrocarbons may be separated on the basis of molecular weight by distillation. Because vapor pressures of these nonpolar substances depend on intermolecular attractions resulting from London forces which in turn are determined by the number of electrons in a molecule, hydrocarbons having similar numbers of carbon atoms have similar boiling points. The simplest and most common method of *refining petroleum* involves a fractionating tower like the one illustrated in Figure 5.11. Distillation separates crude petroleum into the fractions shown in Table 5.5.

Before considering other techniques of petroleum refining there is one other factor which should be noted. Within the gasoline fraction some compounds are far better than others for use in an automobile engine because of the conditions under which the fuel is burned. Figure 5.12 diagrams the

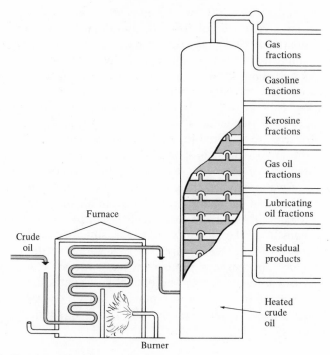

Figure 5.11 Fractionating tower used for petroleum distillation. Source "Facts about Oil," p. 26. Amer. Petrol. Inst., Washington, D. C., 1971.

four steps in operation of such an *Otto cycle* engine. A mixture of air and gasoline is provided to a closed cylinder, compressed by a piston, and ignited by a spark. It burns rapidly, forcing the piston down the cylinder and producing mechanical work. The piston, driven by the inertia of a flywheel, then returns, exhausting combustion products from the cylinder.

TABLE 5.5 Approximate Compositions and Boiling Points of Various Petroleum Fractions

Name of fraction	Representative hydrocarbons	Approximate bp (°C)
Natural gas	CH_4	−161
Liquefied gas (LP gas)	C_3H_8, C_4H_{10}	−44 to +1
Petroleum ether	C_5H_{12}, C_6H_{14}	30–60
Aviation gasoline	C_5 to C_9	32–150
Auto gasoline	C_5 to C_{12}	32–210
Naphtha	C_7 to C_{12}	100–200
Kerosene	C_{10} to C_{16}	177–290
Fuel oil	C_{12} to C_{18}	205–316
Lubricating oils	C_{15} to C_{24}	250–400

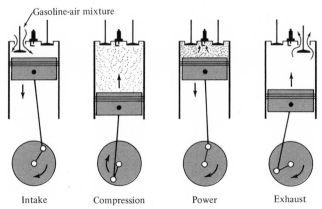

Gasoline-air mixture

Intake   Compression   Power   Exhaust

Figure 5.12 Schematic diagram of the four stages—intake, compression, power, and exhaust—of an internal combustion engine. This is the type of engine used in practically all automobiles and is described technically as a four-stroke (for the four stages) Otto cycle.

Some hydrocarbons have a greater tendency to ignite spontaneously, either because of increased temperature and pressure during compression, hot spots such as lead or other deposits on the valves or cylinder walls, or reaction induced by light, heat, and pressure from the spark-ignited flame. Such pre-ignition or multiple ignitions of the fuel–air mixture produce "knocking," "pinging," and sometimes "dieseling" or "running-on." These reduce power output and efficiency and increase engine wear, making the compounds responsible undesirable components of automobile fuel. As a general rule, paraffin hydrocarbons having the carbon atoms connected in a straight chain, such as *n*-heptane or *n*-octane, have the greatest tendency to produce knocking. Paraffins consisting of branched chain molecules are more desirable, naphthenes are intermediate, and aromatic hydrocarbons have the best antiknock characteristics.

The tendency of different gasolines to produce engine knocking is indicated by the octane rating. On this scale, as shown in Figure 5.13, the compound *n*-heptane is given a rating of zero and the compound 2,2,4-trimethylpentane (also called isooctane) has been given the rating of 100. Any compound or mixture of compounds whose tendency to produce knocking in a test engine is equivalent to a mixture containing 9% *n*-heptane and 91% isooctane would be rated 91-octane. Note that some compounds can have octane ratings below 0 (*n*-octane) and others can have octane ratings above 100 (toluene).

The extent to which the air–fuel mixture in an internal combustion engine is compressed is called the compression ratio and is determined by the engine design. For instance, an engine which has a compression ratio of 8 : 1 would compress the air–fuel mixture to one-eighth of its former volume before the mixture is ignited. As a general rule, the efficiency of an internal combustion engine can be increased by increasing the compression

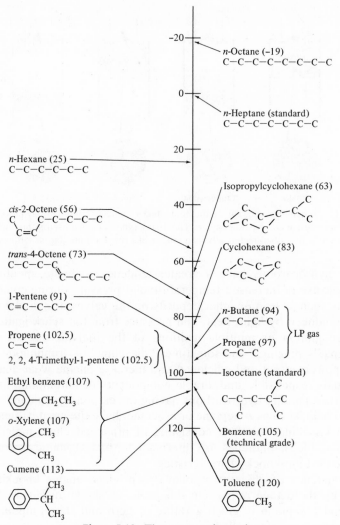

Figure 5.13   The octane rating scale.

ratio. However, heating of the fuel–air mixture and the tendency for knocking become greater also. Therefore, higher octane gasolines are needed for high compression engines.

5.6   PETROLEUM REFINING

In refining petroleum for the present automobile-oriented economy, it is necessary to solve two basic problems. First, in order to make efficient use of the petroleum, compounds which are not part of the gasoline fraction (hydrocarbons having more than ten or fewer than five carbon atoms) should be converted into gasoline. Second, within the gasoline fraction as many high octane compounds as possible should be produced. This means converting straight chain paraffins to branched chain, naphthenic, or aromatic compounds. The most common petroleum refining processes can be divided into five categories: cracking, polymerization, alkylation, catalytic reforming, and hydrocracking. Figure 5.14 contains a generalized equation for each type of process.

Cracking

This oldest refining process consists of breaking down large molecules into smaller fragments, producing greater numbers whose molecular

1. Cracking

$$C_nH_{2n+2} \longrightarrow C_nH_{2n} + H_2$$

$\Delta G = 132.3 - T(0.1407)$

$$C_{(m+n)}H_{2(m+n)+2} \longrightarrow C_mH_{2m} + C_nH_{2n+2}$$

    (Alkane)   (Alkene) (Alkane)

$\Delta G = 83.7 - T(0.1407)$

  a. Thermal cracking—1000 – 1100 K

  b. Catalytic cracking—725 K on SiO_2, Al_2O_3

2. Polymerization

$$m\,C_nH_{2n} \longrightarrow C_{m \cdot n}H_{2m \cdot n}$$

3. Alkylation—strong acid catalyst

$$RH + CH_2{=}C\overset{R'}{\underset{R''}{\diagup}} \xrightarrow{\text{acid}} R{-}CH_2{-}\overset{R'}{\underset{R''}{CH}}$$

4. Catalytic reforming

$$CH_3CH_2CH_2CH_2CH_2CH_3 \longrightarrow \text{(Cyclohexane)} + H_2$$

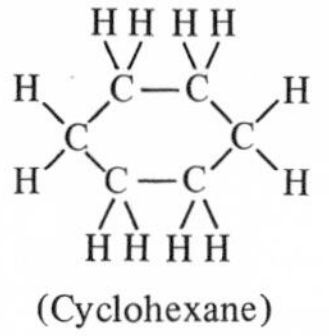

$$C_6H_{12} \longrightarrow C_6H_6 + 3\,H_2$$
(Cyclohexane)   (Benzene)

Catalyst is MoO_3 on Al_2O_3 or Pt on clay

5. Hydrocracking

$$C_{(m+n)}H_{2(m+n)+2} + H_2 \xrightarrow[P]{\text{high}} C_mH_{2m+2} + C_nH_{2n+2}$$

Figure 5.14 Summary of petroleum refining processes.

weights are in the gasoline fraction. An example of a cracking reaction is shown in equation 5.12:

$$CH_3CH_2CH_2CH_2CH_2CH_2CH_2CH_2CH_2CH_2CH_2CH_3 \longrightarrow$$
$$CH_3CH_2CH_2CH_2CH_3 + CH_3CH_2CH_2CH_2CH_2CH_2CH_3 \qquad (5.12)$$

A second type of reaction where one dihydrogen molecule is lost and a double bond is formed in the original compound competes with equation 5.12:

$$CH_3CH_2CH_2CH_2CH_2CH_2CH_2CH_2CH_2CH_2CH_2CH_3 \longrightarrow$$
$$H_2 + CH_3CH_2CH_2CH{=}CHCH_2CH_2CH_2CH_2CH_2CH_2CH_3 \qquad (5.13)$$

At high temperatures thermodynamics favors both of these reactions (see the free-energy changes given in Figure 5.14), but formation of a double bond requires a somewhat higher temperature than breakdown into two smaller fragments. Thus, cracking was originally done by heating the compounds to a temperature (1000–1100 K) just high enough to favor their breakdown and to permit the reactions to proceed at a reasonable rate. However, it is possible to use a catalyst (usually silicon dioxide and aluminum oxide) to speed up the reaction so that a temperature around 725 K can be used. An advantage of catalytic cracking is that at the lower temperature less double bond formation occurs.

Polymerization

Hydrocarbons having fewer than five carbon atoms can be connected together by a process of polymerization to produce molecules in the gasoline fraction. An example is shown in equation 5.14:

$$
CH_2{=}\underset{\underset{CH_3}{|}}{\overset{\overset{CH_3}{|}}{C}} \; + \; CH_3{=}\underset{\underset{CH_3}{|}}{\overset{\overset{CH_3}{|}}{C}} \; \longrightarrow \; CH_3{-}\underset{\underset{CH_3}{|}}{\overset{\overset{CH_3}{|}}{C}}{-}CH_2{-}\underset{\underset{CH_3}{|}}{\overset{\overset{CH_2}{\|}}{C}} \qquad (5.14)
$$

(Isobutene)   (2,4,4-Trimethyl-1-pentene)

Another example of polymerization is the trimerization of propene to form a branch chain compound which should have a relatively high octane rating (equation 5.15).

$$
CH_2{=}\underset{\underset{CH_3}{|}}{\overset{\overset{CH_3}{|}}{CH}} + CH_2{=}\underset{\underset{CH_3}{|}}{\overset{\overset{CH_3}{|}}{CH}} + CH_2{=}\underset{\underset{CH_3}{|}}{\overset{\overset{CH_3}{|}}{CH}} \longrightarrow CH_3{-}\underset{\underset{CH_3}{|}}{\overset{\overset{CH_3}{|}}{CH}}{-}CH_2{-}\underset{\underset{CH_3}{|}}{\overset{\overset{CH_3}{|}}{CH}}{-}CH_2{-}\underset{}{\overset{\overset{CH_2}{\|}}{CH}} \qquad (5.15)
$$

Polymerization involving more than just two or three monomer molecules can also occur, producing much higher molecular weight materials which are not volatile enough to be satisfactory for use in gasoline. Thermodynamic calculations are often used to determine conditions of polymerization so that the process provides the desired mixture of compounds.

Alkylation

This refining process requires catalysts which are strong acids: sulfuric acid, hydrofluoric acid, or aluminum trichloride (a strong Lewis acid) are typical. In the alkylation process a compound having a double bond reacts with a carbon-hydrogen bond to form a somewhat higher molecular weight product. Ethylbenzene can be produced from ethene and benzene and isooctane can be formed from isobutane and isobutene. Alkylation processes are generally capable of producing fairly large increases in octane rating of the compounds involved.

$$\text{benzene} + CH_2{=}CH_2 \xrightarrow{AlCl_3} \text{ethylbenzene} \tag{5.16}$$

$$CH_3{-}\underset{\underset{CH_3}{|}}{\overset{\overset{CH_3}{|}}{C}}{-}H + CH_2{=}\underset{}{\overset{\overset{CH_3}{|}}{C}}{-}CH_3 \xrightarrow{H_2SO_4} CH_3{-}\underset{\underset{CH_3}{|}}{\overset{\overset{CH_3}{|}}{C}}{-}CH_2{-}\underset{\underset{H}{|}}{\overset{\overset{CH_3}{|}}{C}}{-}CH_3 \tag{5.17}$$

Catalytic Reforming

The process known as catalytic reforming can convert paraffin hydrocarbons first to naphthenes and then to aromatic compounds. Since aromatics generally have high octane ratings this is very useful in producing high test gasoline. A good example is shown in equation 5.1 in the discussion of aromatization of coal where *n*-hexane is converted first to cyclohexane and finally to benzene. Catalysts for reforming often consist of molybdenum oxide or platinum on fluoride-containing clay.

Hydrocracking

Note that when 1 mol of benzene is produced from 1 mol of *n*-hexane in the reforming reaction, 4 mol of dihydrogen are produced as a by-product. Some of this dihydrogen can be used in a process known as hydrocracking to add hydrogen to the double bonds produced in the cracking reaction, thus producing smaller amounts of tar and other less desirable high molecular weight compounds which might occur as a result of polymerization of compounds containing double bonds. A general equation for the hydrocracking process can be found in Figure 5.14.

Another advantage of hydrocracking is that sulfur contained in the petroleum will be converted through reaction with dihydrogen to hydrogen sulfide. As previously mentioned in the discussion of coal gasification (see

Section 5.3), there are processes for separating hydrogen sulfide from the remaining hydrocarbons, thus making hydrocracking an important source of sulfur. By 1972 the amount of sulfur produced by mining had dropped to less than 50% of the total United States consumption.

Gasoline Additives

During the 1920's it was discovered that the octane rating of gasoline could be raised by very small quantities of a variety of different compounds. Addition of heavy metals, particularly tetraethyllead [$Pb(C_2H_5)_4$], will greatly increase octane. This was originally done at the service station, but problems arose because of the toxicity of lead and its handling by a large number of inexperienced service station operators. This resulted in the decision to add tetraethyllead at the oil refinery. Although other compounds such as ethyl alcohol can be used to increase octane ratings, they all suffer from deficiencies either because of their chemical properties or because of the high cost of producing them. Tetraethyllead, the cheapest and most effective substance for increasing octane ratings, came to be used in nearly all gasoline.

Tetraethyllead is synthesized by reacting ethane with chlorine to give chloroethane. This is then reacted with a sodium-lead alloy to produce tetraethyllead:

$$C_2H_6 + Cl_2 \longrightarrow C_2H_5Cl + HCl \tag{5.18a}$$

$$Na + Pb \longrightarrow NaPb \text{ (alloy)} \tag{5.18b}$$

$$4 C_2H_5Cl + 4 NaPb \longrightarrow Pb(C_2H_5)_4 + 3 Pb + 4 NaCl \tag{5.18c}$$

When tetraethyllead was first used to increase octane ratings, it was discovered that deposits of lead and lead oxide would build up inside an engine unless something was added to remove them. The compounds used for this purpose today are dichloroethane and dibromoethane. They react

H   H   H   H
| | | |
H—C—C—H   H—C—C—H
| | | |
Cl Cl   Br Br
Dichloroethane   Dibromoethane

with lead to give $PbCl_2$, $PbClBr$, and $PbBr_2$, all of which have relatively low melting and boiling points and are easily vaporized by the heat of the engine. The vapors then exhaust through the tail pipe of the car into the atmosphere where they are converted to finely divided particles of $PbClBr$, $NH_4Cl \cdot 2\ PbClBr$, and $2\ NH_4Cl \cdot PbClBr$. These particles are in a size range ($<2\ \mu m$) which is easily retained by human lungs. As an indication of the magnitude of this problem, in 1968 there was a total of 167 560 tonnes of lead emitted into the atmosphere; it is estimated that 98.2% came from gasoline combustion.

The octane ratings of different hydrocarbons are not affected equally by the addition of tetraethyllead. As a general rule, compounds having a low octane number can be improved a good deal more than those of higher octane. The manufacture of gasolines having an adequate octane rating, but not containing tetraethyllead, is made more difficult by this characteristic since considerably more refining must be done to convert the lower octane paraffins to higher octane aromatic or naphthenic compounds. This requires additional equipment to carry out catalytic reforming or alkylation processes. A further problem is introduced because many aromatic compounds and some of their partially oxidized forms are carcinogenic. The effect on the environment of using increased quantities of these compounds in gasoline, as tetraethyllead is phased out, is not known.

In 1971, General Motors and other United States automobile manufacturers agreed to limit the compression ratios of nearly all automobile engines to approximately 8:1. This permitted use of relatively low octane gasoline — 91 or 92 research octane. Regular leaded gasoline of 93 octane is produced by adding 2.5 to 4 g of lead per gallon to an initial mixture of about 87 or 88 octane. Premium gasoline is raised from a lead-free rating of 93 to 98 or 99 by the same quantity of lead. Since cars with lower compression ratios could burn 93 octane gasoline, the same facilities used to refine premium gasoline could produce a lead-free regular of 93 octane. This is especially important since 1975 automobiles are equipped with catalytic converters which otherwise would be poisoned by the lead. As more and more cars begin to use lead-free gasoline, more new refining equipment must come into use, but such a changeover can be brought about in an orderly way.

On the other hand, Otto cycle engines having lower compression ratios are less efficient, and the limitation of octane ratings to 93 will mean that more gasoline must be consumed for the same amount of travel. An estimate by the Environmental Protection Agency[11] indicates that complete removal of lead will result in an increase of 3.5% in gasoline consumption. Thus, petroleum reserves will be depleted somewhat more rapidly by a changeover to lead-free fuel. A full discussion of the effects of airborne lead, other pollution from automobiles, and possible abatement procedures will be found in Section 10.4.

5.7 ENVIRONMENTAL PROBLEMS ASSOCIATED WITH PETROLEUM

Environmental degradation resulting from petroleum combustion consists mainly of thermal pollution, which has already been discussed, and air

---

[11] "A Report on Automotive Fuel Economy." U. S. Environ. Protect. Ag., Washington, D. C., 1974.

pollution, which will be treated in detail in Chapter 10. However the production, refining, and shipping of petroleum also have significant environmental consequences. Off-shore drilling, supertankers, and loading and unloading procedures at refineries contribute to the magnitude of the oil spill problem, and construction of pipelines in fragile areas such as Alaska has been the subject of much debate. Effluents from petroleum refineries include carbon monoxide (12×10^9 kg/year worldwide), sulfur oxides (3.5×10^9 kg/year), particulate matter (10^9 kg/year), nitrogen oxides, hydrocarbons (11×10^9 kg/year), hydrogen sulfide and other air pollutants, aqueous solutions of sulfuric acid, sodium hydroxide, ammonia, phenols and cresols and naphthalene, and solid wastes such as spent catalysts and clay. The chief hazard from leaks and spills of natural gas is explosions since methane is common in the natural environment and not highly poisonous.

Many of these effects will be treated in later sections on air and water pollution, but it is worthwhile to consider oil spills in more detail now. The total influx of oil into the oceans is estimated at five to ten million tonnes annually from human activities. These include pumping of seawater ballast from tankers, pumping of bilges of ordinary cargo ships, losses during loading and unloading operations, and accidents involving vessels or offshore drilling. In addition, approximately two million tonnes of lubricating oil is "consumed" every year in the United States alone. Much of this probably finds its way into the freshwater environment. The effects of such oil pollution include reduction of light transmission across the water surface and consequent slowing of photosynthesis by aquatic algae, depletion of dissolved dioxygen, damage to water birds whose feathers become coated, smothering of shoreline plants, and toxicity of many petroluem components. In this last respect low molecular weight, relatively soluble aromatic substances are the most dangerous. In addition to short-term degradation and toxicity it is not unreasonable to expect chronic effects resulting from low levels of petroleum constituents. These may include confusion of marine predators by blockage of odor and taste receptors as well as carcinogenic effects. Nonpolar hydrocarbons are readily concentrated along food chains (Sections 16.3 and 17.1) and may also serve as a concentrating medium for other fat-soluble poisons such as pesticides.

Despite numerous attempts to devise cleanup techniques, the best cure for oil spills still appears to be prevention. Because its nonpolar molecules do not mix readily with water, oil may be trapped by floating booms or air bubble barriers and skimmed from the surface. Such techniques do not work well in the rough water which often accompanies large-scale spills, and they cannot remove the more water-soluble aromatics which present greatest problems of toxicity. The use of chemical detergents to emulsify oil and water and thus disperse a spill so that greater contact with microorganisms and dioxygen can decompose the hydrocarbons also has drawbacks. More petroleum constituents dissolve in the water and the small oil droplets produced are easily ingested by organisms at the low end of the food chain. Detergents also remove essential oils from waterfowl feathers and in some cases have proved more harmful than the oil spill itself. After nearly a decade of research no entirely satisfactory cleanup technology has been developed.

EXERCISES

(The first group of exercises is based primarily on material found in this text.)

1. What type of chemical process is characteristic of the formation of coal? Describe qualitatively how composition of coal varies with rank.
2. In what two forms is sulfur found in coal? Why is this distinction important for techniques of pollution control? What types of United States coal are low in sulfur? Where are they located? How does this affect the costs (both economic and social) of energy production from coal? Why is the distinction between weight percent sulfur and gram per joule an important one?
3. What are the factors which reduce the efficiency of coal-fired electric power plants from 100% to only 40%?
4. Why is discharge of waste heat (to water or air) a necessity for a steam electric power plant? How can "thermal pollution" of water by power plants be reduced?
5. Make a list of as many of the environmental costs of electric power production from coal as you can think of. Compare your list with Table 5.4.
6. Write balanced equations for the reactions involved in acid mine drainage. Why does sealing of old mine entrances or pumping N_2 into old shafts help alleviate the problem of acid drainage?
7. What is a topping cycle? Describe in detail how at least one type of topping cycle works. Why is gaseous fuel advantageous if topping cycles are to be used?
8. Write equations for the reactions involved in coal gasification. What is the difference between power gas and SNG?
9. Why do schemes for coal gasification invariably involve removal of sulfur? What reactions are involved in sulfur removal?
10. What are the main objectives (in terms of chemical reactions and structures) of petroleum refining? List at least three types of refining processes and give an example of a reaction that might occur in each.
11. How was the octane scale originally defined? What types of compounds result in high octane ratings? What types give low octane?
12. Give equations for synthesis of tetraethyllead. Why must lead be exhausted from the cylinders of automobile engines? What compounds are added to gasoline to insure that lead is exhausted?
13. What are some of the problems associated with removal of lead from gasoline? Do you think it is a good idea? Why or why not?

14. Why does the presence of $Fe(OH)_3$ in streams which have been exposed to acid mine drainage make abatement more difficult?

15. Using data from this chapter compare the efficiencies of:

    a. A conventional coal-fired electric power plant with dry cooling tower.

    b. Coal gasification plus a gas turbine topping cycle on a power plant with a cooling pond.

(The following exercise requires more extensive research.)

16. It is proposed that supertankers carrying liquefied natural gas (LNG) in cryogenic tanks transport this resource from the Middle East to the United States. Suppose such a tanker were to spill, releasing LNG onto the ocean. Calculate the density of methane vapor at its boiling point (111 K). To what temperature would a cloud of pure CH_4 have to be heated in order that its density become less than that of air at 293 K? Why would this be important? [see W. W. Crouch and J. C. Hillyer, *Chem. Technol.* **2**(4), 210 (1972)]. Repeat the above calculation for H_2 at its boiling point.

6

Newly Developing Energy Sources

We nuclear people have made a Faustian compact with society: we offer . . . an inexhaustible energy source . . . tainted with potential side effects that, if uncontrolled, could spell disaster.

Alvin Weinberg

. . . it is important to re-examine all the ways in which science and technology can help to make direct use of the sun practical.

Farrington Daniels

Much of the sponsorship and planning of energy research and development during the past three decades has been predicated upon the assumption that nuclear power would fill the gap between ever-growing energy demand and dwindling fossil fuel supplies. As you can demonstrate by working Exercise 1, on an atom-for-atom comparison nuclear reactions produce a million times as much energy as the chemical reactions of fossil fuel combustion. A seemingly limitless resource such as this could hardly fail to captivate the imaginations of scientists, engineers, and technologists and to appropriate many of their best research and development efforts.

Referring back to Figure 1.3 one should recall that there is a minimum in energy per nucleon in the vicinity of $_{26}^{56}$Fe. If a heavy nucleus, such as $_{92}^{238}$U, could be split into two fragments, such as $_{36}^{84}$Kr and $_{56}^{138}$Ba, which lie lower on the curve, approximately 1.5×10^{-11} J/nucleon or 2×10^{15} J/mol would be released. Such a *fission* process provides the energy in the explosion of an atomic bomb. Similarly, *fusion* of two $_{1}^{2}$H nuclei to form $_{2}^{4}$He results in increased stability and hence release of energy. Such a reaction occurs during a hydrogen bomb explosion. The fact that energy can be released very rapidly in a nuclear explosion does not guarantee that the same reaction can be carried out under controlled conditions to produce electric power or other useful work. Fission reactions have been developed to this stage, however, and the first two sections of this chapter will be devoted to them. Fusion power is less well developed, but one section will be devoted to it as well.

As Weinberg implied in the quotation at the outset of this chapter, many of the problems associated with nuclear power are more social than technical. Many observers feel that we would do well to maximize the development of solar, geothermal, and tidal power in order to minimize the impact of nuclear technology. These energy sources will be considered in Sections 6.4 through 6.7.

6.1 NUCLEAR FISSION REACTORS

Although $^{238}_{92}U$ was used in a hypothetical example above, it does not undergo spontaneous fission. The reaction used in most currently operating nuclear power plants is

$$^{235}_{92}U + ^1_0n \text{ (slow or } thermal\text{)} \longrightarrow ^{142}_{56}Ba + ^{91}_{36}Kr + 3\,^1_0n \text{ (fast)} \tag{6.1}$$

In the equation barium and krypton are indicated but a variety of other pairs of products is possible. The important aspect is that on the average three neutrons are produced whereas only one is required to initiate fission. This permits a *chain reaction* in which neutrons which are the product of one reaction initiate the fission of another uranium 235 nucleus. $^{233}_{92}U$ and $^{239}_{94}Pu$ also undergo fission by thermal neutrons and are suitable for chain reactions.

One very important aspect of the nuclear reaction process is the speed of the neutrons that are involved. Relatively slow moving (thermal) neutrons split ^{235}U nuclei more efficiently than high velocity (fast) neutrons. From equation 6.1 it can be seen that the neutrons which are products of the reaction have large energies and velocities. They must be slowed down if they are to cause more ^{235}U nuclei to break apart. Substances that can reduce the energy of neutrons are called *moderators*. They must contain light elements and therefore usually consist of carbon, hydrogen, or oxygen; graphite or water is often used.

A second problem of controlling a nuclear fission reaction is regulation of the number of neutrons which are available within the fissionable material. The larger the neutron flux the faster the reaction and therefore the higher the temperature. Unless the number of neutrons is reduced by means of cadmium *control rods* to a level where just enough are available to maintain but not augment the chain reaction, the reactor material may become hot enough to melt. Control rods alternate with rods of fissionable material in the reactor core. In order to start the reaction the control rods

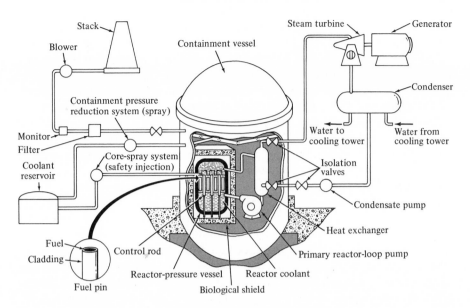

Figure 6.1 Schematic diagram of a nuclear fission reactor for electric power generation; the various safeguards can also be seen. Source: W. H. Jordan, Nuclear energy: Benefits versus risks. *Phys. Today* May, 35 (1970). © American Institute of Physics.

are partially removed, increasing the number of neutrons within the core; to stop the reaction they are reinserted so that they will absorb nearly all of the neutrons produced by the fission reaction. Figure 6.1 shows schematically this and other components of a nuclear fission reactor for electric power generation.

To absorb the heat generated by a nuclear reaction and transfer it outside the reactor core where it can be converted to useful work, a *cooling system* must be provided. Most reactors use either water or liquid sodium as coolants because of their high thermal conductivity and large heat capacity, but inert gases such as helium may also be employed. Some provision is also made for emergency cooling in the event that the nuclear reaction begins to go faster than expected.

There must be sone kind of *shielding material* to prevent neutrons, γ rays, and other radioactivity from escaping from the reactor core. Usually this is provided by a reinforced concrete shield to which salts of lead may be added to absorb more radiation. A final line of defense is the *containment vessel* which surrounds all portions of the power plant which might be contaminated by radioactivity (see Figure 6.1). Even if a fairly serious accident were to occur, the containment vessel should not allow radioactive material to escape.

Light Water Reactors (LWR's)

Two types of controlled fission reactors are now being used to generate electricity in the United States. The first type is named the *boiling water reactor (BWR)* because the heat of the reaction boils the cooling water. This produces steam which is fed directly to turbines which turn generators, producing electrical current. A slightly newer system is called the *pressurized water reactor (PWR)*; here the cooling system water is superheated under pressure and then passed through a heat exchanger (see Figure 6.1) which transfers the heat to a secondary water circuit. Steam in the outer circuit operates turbines and generators to produce electricity. The primary cooling system in this type of reactor is completely enclosed, making the escape of radioactive material from the core much less likely.

Most BWR's and PWR's use either uranium oxide, U_3O_8, or pellets of uranium metal as fuel. Both types of fuel must be greatly enriched in ^{235}U, the fissionable isotope of uranium. This is usually done by gaseous diffusion of UF_6 through a great many slightly permeable barriers. The fuel is enclosed in stainless steel or zirconium alloy *cladding* tubes which are usually about 3 m long and 2 cm in diameter. A 500-MW_e reactor, an average size for electric power generation, might have 20 000 such tubes.

High Temperature Gas-Cooled Reactors (HTGR's)

A third type of nuclear reactor uses an inert gas, usually helium, in the primary coolant circuit instead of water. Because higher temperatures are involved the efficiency of electricity generation by HTGR's can reach 39% — about the same as a modern coal-fired plant. Otherwise the construction and operation of a HTGR is similar to that of a LWR. A few HTGR's are planned or under construction in the United States today, and by 1990 they will probably be producing significant amounts of power.

The amount of energy which can be produced by LWR's and HTGR's is limited by the amount of ^{235}U fuel that is available in the surface of the earth. The uranium 235 isotope amounts to only about seven-tenths of one percent of all uranium. Present supplies of uranium are such that the amount of energy available at reasonable cost from this type of reactor is only about one-half the amount available from coal reserves.

Breeder Reactors

The uranium fuel supply can be multiplied by a factor of more than 100 by using what is known as a "catalytic" or *breeder reactor*. Here neutron flux is used to produce other fissionable isotopes — to "breed" additional nuclear fuel which can be used in a second reactor. By means of the nuclear reactions indicated in equations 6.2 and 6.3, both ^{238}U and ^{232}Th

can be converted into fissionable fuels: plutonium of mass 239 can be produced from ^{238}U using fast neutrons; ^{233}U can be produced from ^{232}Th if neutrons whose velocities have been reduced by a moderator are used.

$$
\begin{array}{c}
{}^{238}_{92}\text{U} + {}^{1}_{0}n \text{ (fast)} \longrightarrow {}^{239}_{92}\text{U} + \gamma \\
\downarrow t_{1/2} = 24 \text{ min} \\
{}^{239}_{93}\text{Np} + \beta^{-} \\
\downarrow t_{1/2} = 2.3 \text{ days} \\
{}^{239}_{94}\text{Pu} + \beta^{-}
\end{array}
\tag{6.2}
$$

$$
\begin{array}{c}
{}^{232}_{90}\text{Th} + {}^{1}_{0}n \text{ (moderate)} \longrightarrow {}^{233}_{90}\text{Th} + \gamma \\
\downarrow t_{1/2} = 22 \text{ min} \\
{}^{233}_{91}\text{Pa} + \beta^{-} \\
\downarrow t_{1/2} = 27.4 \text{ days} \\
{}^{233}_{92}\text{U} + \beta^{-}
\end{array}
\tag{6.3}
$$

Remember that when one ^{235}U nucleus undergoes fission, an average of three neutrons are given off. One is required to maintain the self-sustaining fission reaction, but the other two can be used to produce more fuel. Thus in a breeder reactor it is possible to produce more fissionable fuel than is being used and at the same time obtain useful energy, which seems almost too good to be true. Note, also, that either of the two fuels (^{233}U or ^{239}Pu) produced by breeder reactors can be used in a second breeder reactor. Thus the amount of nuclear fuel available through the use of breeder reactors depends only on the total supply of all isotopes of uranium and thorium; this is well over 100 times the supply of ^{235}U alone. Because of the limited supply of ^{235}U it is generally agreed that obtaining adequate energy from fission will require the use of breeders.

6.2 FISSION POWER AND THE ENVIRONMENT

Environmental problems resulting from the operation of nuclear power plants can be divided into four classes: thermal pollution (common to all power generating plants); the discharge into the environment of some radioactive materials during the normal operation of the plant; the possibility of catastrophic destruction of the nuclear plant with the subsequent release of large quantities of radioactivity; and the problem of storing, safeguarding, and disposing of radioactive wastes which are a normal product of the operation of any nuclear power plant.

Thermal Effects

The nuclear power plants constructed to date are slightly less efficient than similar coal-fired plants. Boilers of BWR's operate at lower temperatures and pressures and heat transfer to the secondary cooling systems in PWR's cancels most of the advantage gained by pressurizing the primary system. These lower efficiencies result at least in part from the necessity of preventing loss of radioactive materials. Light water reactors transfer to the immediate environment as thermal pollution 65–70% of the total energy content of their fuel. Thermal pollution of the water environment is much worse than for fossil-fueled plants since there are no stacks for the emission into the atmosphere of part of this heat loss. The same means of alleviating thermal pollution may be applied to nuclear plants as to coal-fired or oil-fired power stations.

Discharge of Radioactivity

Although great precautions are taken to prevent the discharge of radioactive isotopes from a nuclear power plant during its normal operation, small quantities of material are lost. The main problem area is the cladding of the fuel elements, which often develops pinhole leaks because of the large number of neutrons, their ability to transmute the cladding alloy, and the high temperature in the reactor core. Many of the products of fission reactors are radioactive and can escape in liquid or gaseous effluents. Some of these fission products, their half-lives, and environmental effects are summarized in Table 6.1. Some reduction of radioactivity loss can be effected by monitoring each of the 20 000 fuel elements in an average reactor and removing those in which pinhole leaks have developed when the plant is shut down for periodic replenishment of the fuel. However, monitoring is not a universal practice at the present time.

It is technologically feasible to improve the treatment of gaseous effluents from nuclear plants. A typical offgas system monitors the radioactivity of the effluents before they pass from the plant, and has sufficient volume to hold the gases for 30 min between monitoring and emission into the atmosphere. If the radioactivity monitor detects a high level of radiation, an isolation valve can be closed, but this also requires a complete shutdown of the plant. Since it takes several days to get a plant running after a shutdown, there is considerable resistance to lowering the permissible level of radiation in effluent gases. An improved control system would provide

TABLE 6.1   Environmental Effects of Some Fission Reactor Emissions

Radioisotope	Half-life	Environmental effect
^{90}Sr	25 years	Collects in bones, may cause leukemia
^{131}I	8 days	Concentrates in thyroid
^{85}Kr	9.4 years	Not chemically reactive so not concentrated
^{3}H (^{3}T)	12.5 years	Hard to separate T_2O from H_2O

TABLE 6.2 Atomic Energy Commission Radiation Emission Guidelines for LWR's[a]

Type of emission	Guideline
Noble gases	10 mrem yearly integrated dose at any point on the site boundary
Liquid wastes	5 mrem maximum calculated dose to any person over a year's time
Long-lived isotopes (half-life of 8 days or more)	5 mrem maximum annual dose to any organ (this is a reduction of previous guidelines by a factor of 10^5)

[a] *Federal Register,* **36,** no. 111, 11113–11117 (1971).

large storage tanks to allow time for much of the radioactive material to decay. Activated carbon filters could adsorb substances such as krypton which are not chemically reactive. The flexibility of such a system would permit stricter radiation requirements to be implemented without causing numerous plant shutdowns.

Because of the many safeguards built into nuclear power plants, discharge of radioactive material during normal operation appears not to be a major problem. The Atomic Energy Commission [now the Nuclear Regulatory Commission (NRC)] has issued guidelines (see Table 6.2) for light water reactors. Exposure is listed in millirem, a measure of radiation dose which takes account of the *quantity* and *type of radiation* as well as the *biological effect* it produces. One rem corresponds to the deposition in living tissue of 10^{-5} J of energy, corrected by appropriate factors for type of radiation and biological damage. If the guidelines are met the maximum dose to any individual from a single nuclear plant would be 20 mrem, less than one-fourth of the most recent[1] estimate of average background radiation. Nuclear power plants generally operate below the guidelines and as indicated in Table 6.3 the radiation received by the average person from nuclear power plants in 1970 was much smaller than that from medical sources, fallout, or natural background (terrestrial radioactivity plus cosmic rays). Even allowing a factor of one hundred or more for the increasing number of nuclear plants, this source of radiation is expected to account for only a minor fraction of the total dose received by the average citizen of the United States in the year 2000.

Catastrophies and Nuclear Plants

A third problem relating to nuclear power plants is the possibility of some catastrophic destruction to the plant itself. Although many people seem to think that having a nuclear power plant in the vicinity is akin to living next door to an atomic bomb, there is nearly zero probability of this

---

[1] D. T. Oakley, "Natural Radiation Exposure in the United States." U. S. Environ. Protect. Ag., Washington, D. C., 1972.

TABLE 6.3   Estimated Annual Whole-Body Radiation Dose Rates in the United States (1970)[a]

Source of radiation	Average[b] dose rate (mrem/year)
Environmental	
Natural background	102[c]
Global fallout	4
Nuclear power	0.003
Subtotal	106
Medical	
Diagnostic (x rays, etc.)	72[d]
Radiopharmaceuticals	1
Subtotal	73
Occupational	0.8
Miscellaneous (TV sets, etc.)	2
Total	182

[a] Source: Advisory Committee on the Biological Effects of Ionizing Radiations, ''The Effects on Populations of Exposure to Low Levels of Ionizing Radiation,'' p. 19. Nat. Acad. Sci.— Nat. Res. Counc., Washington, D. C., 1972.

[b] The numbers shown are averages only. Certain segments of the population may experience dose rates considerably greater than these.

[c] A value of 84 mrem/year is given by the estimate referred to in the text.

[d] Based on the abdominal dose. More recent information indicates that this figure may be too high by as much as a factor of two [*Sci. Trends* Dec. 23, p. 76 (1974)].

type of explosion occurring in a light water reactor—the fuel elements would be separated from each other before explosive force could be reached. An atomic bomb requires chemical explosives to drive together into a fairly dense mass several pieces of radioactive material; such densities could not be achieved under any conceivable conditions in an LWR.

Nevertheless, circumstances can be imagined under which large amounts of radiation could escape from a nuclear power plant. These usually involve failure of the plant's cooling system with the subsequent melting of the core. In water-cooled reactors where the water serves as a moderator to slow down neutrons, it was first assumed that loss of cooling water would simply stop the nuclear reaction. While this is true, considerable energy remains in excited nuclear states of fission products, and this could melt the fuel elements and their cladding. At the high temperatures involved (between 1300 and 1800 K) many metals are much more reactive with water than they are at room temperature; thus, violent chemical reactions might also contribute to the destruction of the plant. Large amounts of steam might be produced and the gas pressure could destroy the containment building, the last line of defense against the radioactive materials. It has even been suggested that sufficient heat could be produced to melt the concrete and steel floor of the containment building. While the Atomic Energy Commission discounts such suggestions, one AEC report has indicated that catastrophic reactor failure could result in "subsequent deposition at undesirable locations" of fission product material. Some recent small-scale experiments sponsored by the AEC imply that steam pressure inside the containment room might prevent emergency cooling water from entering.

Cooling is especially important in fast breeder reactors, most of which use liquid sodium as the coolant and hence are called *liquid metal fast breeder reactors* (LMFBR's). These have three separate cooling circuits in tandem since sodium can react with neutrons to form a radioactive isotope:

$$\ce{^{23}_{11}Na} + \ce{^{1}_{0}n} \longrightarrow \ce{^{24}_{11}Na} \xrightarrow{\text{14.9 hours}} \ce{^{24}_{12}Mg} + \beta^- + \gamma \tag{6.4}$$

Thus the primary coolant (sodium) becomes radioactive. Heat is transferred from this primary cooling system to a second liquid sodium system which is used to boil water as shown schematically in Figure 6.2. Although the core of the fast breeder operates at a higher temperature than a light water reactor, overall thermal efficiency is no higher because of the three-fold cooling system. Also, liquid sodium is very reactive chemically so any breach in the cooling system could cause catastrophic chemical reactions to occur. To date the only fast breeder reactor that has been used to produce electric power is the Enrico Fermi I plant in Monroe, Michigan.

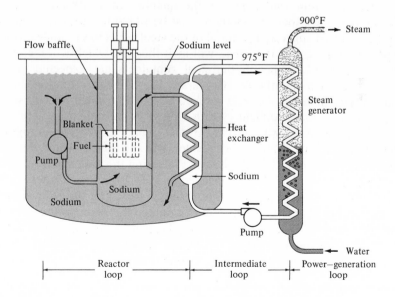

Figure 6.2 Schematic diagram of a liquid metal fast breeder reactor. Source: M. Eisenbud, "Environmental Radioactivity," 2nd ed., p. 238. Academic Press, New York, 1973.

The first time this plant was operated at full capacity a failure of the cooling system very nearly caused melting of the core; it has never generated electricity economically and has been dismantled.[2] A second demonstration breeder plant is now being built on the Clinch River near Oak Ridge, Tennessee.

Radioactive Wastes

Efficient operation of a nuclear power plant demands that fuel elements be replaced before all of the fissionable material has been depleted. Spent fuel rods are then taken to a reprocessing plant where the remaining uranium is separated from the products of the nuclear reaction and recycled. Many fission products and materials transmuted by the high neutron flux of a reactor are highly radioactive and have relatively long half-lives. Some plutonium isotopes will have dangerous levels of radioactivity for up to half a million years. As of 1971 the Atomic Energy Commission was storing more than 30 000 m³ of highly radioactive liquid waste, and the quantity of this material is growing at an increasingly rapid rate as more nuclear power plants are constructed. It is estimated that by the year 2000 commercial power plants will be producing 22 000 m³ of radioactive waste annually. In some cases the aqueous solutions of these wastes contain enough radioactive material to keep them boiling for 2 years.

Radioactive waste materials are now stored in steel containers in underground concrete vaults, but as the quantity of waste grows, a more efficient and less expensive method must be found for disposal. The technique most often suggested is burial of the steel tanks in salt mines. It is expected that the heat of the waste will melt the salt, which will then tightly enclose the tanks. Although the steel containers are expected to disintegrate in from 6 months to 10 years, salt has good radiation-shielding properites, is quite strong, and is usually stable to disturbances such as earthquakes. A major difficulty in this program is finding an appropriate site for disposal of the wastes. The Atomic Energy Commission proposed a location near Lyons, Kansas, but residents of the area objected that water might be diverted into the salt deposits, releasing the radioactive wastes by dissolving the salt. The AEC conceded in May 1972 that further study of such problems was necessary. The Energy Research and Development Administration (ERDA) is currently conducting a comprehensive environmental review of all techniques for disposition of radioactive wastes.

Several other environmental problems are related to nuclear power. Reprocessing plants will very likely emit more gaseous waste than the

---

[2] For a thorough account of the mishaps at the Monroe plant, the reader may consult R. L. Scott, Jr., *Nucl. Safety* **12** (2), 123 (1971); or M. Eisenbud, "Environmental Radioactivity," 2nd ed., p. 248. Academic Press, New York, 1973.

reactors themselves since the cladding of each fuel element must be broken open to separate fissionable material from the radioactive waste. Although this is done in a completely enclosed hot lab, the extremely high level of radiation can result in the loss of some gaseous radioactive materials. (Reprocessing plants are privately owned but operate under NRC guidelines.) Mining and milling of uranium to produce fuel rods also results in land disturbance and considerable radiation hazard to workers. In Colorado some houses have been built on uranium mine tailings from which radioactive radon gas is emitted. This can diffuse up through concrete floors and may cause chromosome damage in children.

As more and more power plants are constructed, especially if they are breeder reactors based on a plutonium fuel cycle, transportation of radioactive materials will become a greater problem. Hijacking, sabotage, and diversion of fissionable material from power plants, reprocessing plants, or in transit may pose very real problems. About 5 kg of plutonium is needed to produce a bomb, and it has been demonstrated that a bright chemistry graduate student can learn how to do so from published, unclassified U. S. Government documents. Moreover, the high toxicity of plutonium will make it an ideal target for terrorist activities. The area of *nuclear safeguarding* is concerned with prevention of such hazards.

Assessing the Risks of Nuclear Power

The so-called "Rasmussen report,"[3] a three million dollar study of light water reactor safety, concluded that deaths and injuries from accidents at nuclear power plants would be much less probable than those from other natural and man-made disasters. It found the probability of a core meltdown serious enough to produce 2300 fatalities and 6×10^9 in property damage to be only one in 10^7 for example. The report has a number of limitations, however, perhaps the most important of which is that it considered LWR power plants only. Hazards involving LMFBR's, transportation, reprocessing, storing, and safeguarding of radioactive wastes and fissionable materials were ignored. Possibilities of human error were factored into the equations, but it was deemed impossible to determine the probability that fundamental design errors might be present in safety systems.

Fission power plants are very complicated, high technology systems. A high level of training and vigilance will be required of those who operate *each* such installation. Even when they become commonplace and familiar there will be no room for complacency by nuclear technologists or the general public. Weinberg's[4] characterization of fission power as a "Faust-

---

[3] "An Assessment of Accident Risks in U.S. Commercial Nuclear Power Plants," WASH-1400. U. S. At. Energy Comm., Washington, D. C., 1974.

[4] For more details, see A. M. Weinberg, *Science* **177**, 27 (1972).

ian compact with society" is a true one which should not be ignored by scientists or citizens.

6.3 FUSION POWER

A second method of obtaining energy from nuclear processes — still in the experimental stages — is to fuse two or more light nuclei together to form a heavier element. The hydrogen-burning reaction of the sun is one example of *nuclear fusion,* but, as we have seen, such reactions have very high activation energies. The two reactions below have relatively low activation energies and seem reasonable for controlled fusion power plants.

$$\ce{^2_1D + ^2_1D ->[4 \times 10^8 \text{ K}] ^3_2He + ^1_0n}$$ (6.5a)
$$\ce{-> ^3_1T + ^1_1H}$$

$$\ce{^2_1D + ^3_1T ->[4 \times 10^7 \text{ K}] ^4_2He + ^1_0n}$$ (6.5b)

(Deuterium, ^2_1D, and tritium, ^3_1T, are isotopes of hydrogen containing one and two neutrons, respectively.)

Although deuterium occurs in nature (there are about 156 deuterium atoms for every 100 000 000 hydrogen atoms in the universe), tritium does not. It must be manufactured (or "bred") from lithium as shown in the following reactions:

$$\ce{^6_3Li + ^1_0n \text{ (thermal)} -> ^4_2He + ^3_1T}$$ (6.6a)

$$\ce{^7_3Li + ^1_0n \text{ (fast)} -> ^3_1T + ^4_2He + ^1_0n \text{ (thermal)}}$$ (6.6b)

The quantity of lithium on the surface of the earth is limited and thus equation 6.5b could only supply energy roughly equal to the amount now available from fossil fuels. The fusion of deuterium nuclei (6.5a) must be carried out at a somewhat higher temperature than the deuterium–tritium fusion, but there is enough deuterium available in the oceans to supply this reaction essentially forever. The quantity of energy which could be obtained from deuterium fusion is about 50 000 000 times the supply of energy available from fossil fuels.

As of this writing, power has not been obtained from a controlled fusion reaction. The basic problem is to hold the nuclei together long enough at a high enough temperature so that the activation barrier to fusion can be overcome. At temperatures in the range of 10^7 to 10^8 K no molecules can exist; most atoms ionize to form positive ions and electrons. The description of materials as solids, liquids, or gases no longer applies; an unusual state of matter called a plasma is formed. This plasma behaves much like the universal solvent of the alchemists by converting any solid materials which it contacts to vapor. The plasma is significantly cooled as well, reducing the kinetic energy of the ions and consequently decreasing the probability that fusion of two nuclei will occur. Containing the plasma is therefore very difficult.

Magnetic Containment

The usual approach to containment has been to construct what is known as a magnetic bottle. Because the charged particles of the plasma move along magnetic lines of force, a strong magnetic field can hold the plasma together. Unfortunately, a very large amount of electric power is required to produce a magnetic field sufficiently strong to contain a plasma. One device for doing this, the Soviet Tokamak, is shown schematically in Figure 6.3. At the current stage of Tokamak development a plasma can be contained for about 0.03 s, not quite long enough for usable power to be withdrawn. It is expected that a workable laboratory model will be available by 1980.

Laser Fusion

A second technique for producing controlled fusion is to rapidly heat and compress a small pellet of deuterium and tritium by means of a sharply focused laser beam. Single path or multipath (see Figure 6.4) devices have been suggested. The main problem appears to be delivery of a large enough amount of energy rapidly enough so that the target pellet does not have time to expand significantly. Most experimental work in the laser fusion field is aimed at developing high power lasers whose energy can be delivered within a period of a few nanoseconds. It will probably be several years before the feasibility of laser fusion can be adequately tested.

Environmental Effects

Many proponents of fusion reactors as a power source have claimed that no environmental problems would result from their use. While this is not completely true, it does appear that the environmental effects would be considerably less than those of fission reactors. The quantity of radioactive material produced in a fusion reactor would be only about one-ten-thousandth of that produced in a fission reactor, but most of this radioactivity would be in the form of tritium, which is very difficult to contain within the reactor itself. Tritium, just like ordinary hydrogen, can diffuse through metals from the reactor to the atmosphere; it can also react to form T_2O which has almost identical chemical properties with normal water, H_2O, making it especially susceptible to incorporation in living systems.

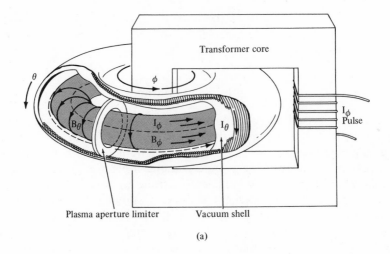

(a)

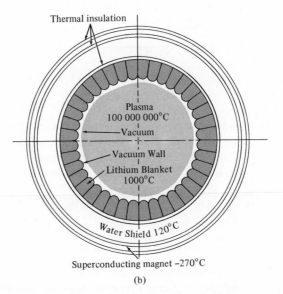

(b)

Figure 6.3 Schematic diagrams of magnetic fusion reactor: (a) Perspective view of To-kamak reactor. Source: D. J. Rose, Controlled nuclear fusion: Status and outlook. *Science* **172,** 799 (1971). Copyright 1971 by the American Association for the Advancement of Science. (b) Cross section of Tokamak reactor. Source: U. S. Atomic Energy Commission, "Fusion Power: An Assessment of Ultimate Potential," WASH-1239, p. 12. USAEC, Washington, D. C., 1973.

Fusion reactors would also have a very high flux of high energy neutrons which could cause structural damage to their metal components. The purpose of the liquid lithium moderator shown in Figure 6.3 is to absorb such neutrons and to breed more tritium by reactions such as 6.6a and 6.6b.

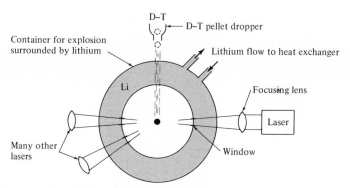

Figure 6.4 Conceptual diagram of a laser fusion device. Source: R. Wilson and W. J. Jones, "Energy, Ecology, and the Environment," p. 168. Academic Press, New York, 1973.

There could also be other hazards associated with a fusion reactor of which we are not yet aware. Once workable laboratory-scale fusion reactors are actually constructed, it may be possible to evaluate their environmental impact more accurately.

Recently, a concept called "Pacer" has been advanced[5] in which thermonuclear explosives would be used to generate high pressure steam in salt deposits along the coast of the Gulf of Mexico. This would involve 700 detonations per plant per year and planners have projected production of 100 000 hydrogen bombs per year should the concept be realized. Neutrons from the explosions could be used to breed $^{239}_{94}Pu$ and $^{233}_{92}U$, which would then be used in fission power plants. Such a concept flies in the face of fusion power as a clean, limitless energy source, but is probably the only way that significant quantities of fusion power could be produced in the next decade. One hopes that the Energy Research and Development Administration will stick to its old timetable of commercial feasibility for magnetic or laser fusion by 2000 and significant power production by 2020.

6.4 SOLAR ENERGY—THERMAL COLLECTION

In the rough order of energy quality shown in Table 4.2 solar radiation is similar or somewhat superior to the potential energy stored in chemical bonds. This is near the average of all forms of energy but considerably better than the thermal energy of the earth's surface. The quality of solar energy is due to the fact that the sun behaves almost exactly as a blackbody radiator having a color temperature of 6000 K (compare Figure 1.7 with Figure 6.5). This solar radiation drives, directly or indirectly, nearly every process in the biosphere.

---

[5] W. D. Metz, *Science* **188,** 136–137 (1975).

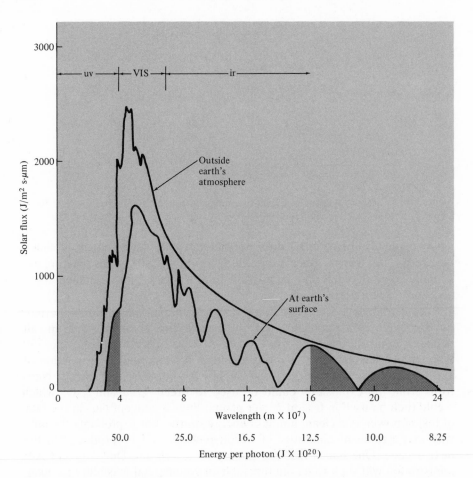

Figure 6.5 Solar flux as a function of wavelength. Adapted from "Handbook of Geophysics," rev. ed. Macmillan, New York, 1960.

Figure 6.6 indicates the partitioning of solar radiation (as well as other natural, continuous energy supplies) as it encounters the earth's atmosphere, hydrosphere, and lithosphere. About 30% is reflected by clouds or light surface areas, while nearly half serves to raise the temperature of land, air, and water. Most of the remainder is stored as heat of vaporization or melting of water, serving to transport this vital substance from hydrosphere to atmosphere. A small fraction is converted to convection currents such as the Gulf Stream or trade winds, and about 0.02% is trapped by the photosynthetic reactions of plants. The supply of solar energy available for human use (Table 4.3) is reduced by corrections for direct reflection and the fact that much of the earth's surface is uninhabitable ocean, but in just one year it amounts to more than four times the total estimated fossil fuel

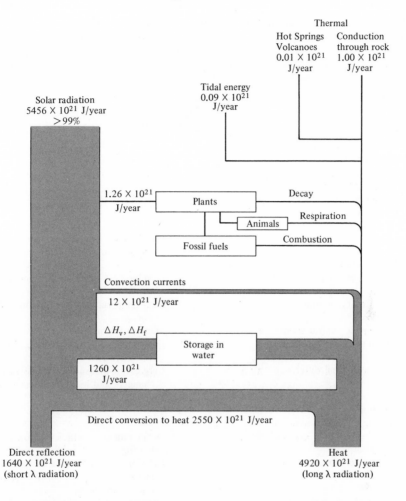

Figure 6.6 Continuous energy supplies and their flow through the biosphere (width of bars indicates quantity of flow). Data from M. K. Hubbert, The energy resources of the earth. *Sci. Amer.* **224**(3), 62–63 (1971).

supply. The lower 48 United States receive from the sun between 500 and 700 times their annual free energy consumption. One might well ask why this resource is not more widely developed.

The basic difficulty which faces any scheme for utilization of solar radiation is the large area over which it is spread and its intermittent nature. On the average in the United States solar energy arrives at the rate of 180 J/m² s (180 W/m²). This is only one five-hundredth the rate at which heat passes into a typical steam boiler of a fossil-fueled power plant. Even if solar power could be converted to a usable form with 100% efficiency,

TABLE 6.4   Present Status of Solar Utilization Techniques[a]

Application	Research	Development	Systems test	Pilot plant	Prototype plant	Commercial	Major technical problems to be solved
Thermal energy for buildings							
Water heating	x	x	x	x	x	x	—
Building heating	x	x	x				Development of air conditioning and integration of heating and cooling
Building cooling	x	x					
Combined system	x	x					
Electric power generation							
Thermal conversion	x						Development of collector, heat transfer, and storage subsystems
Photovoltaic							
On buildings	x						Development of low-cost long-life solar arrays
Central station	x						High temperature operation and energy storage
Space central station	x						Many problems
Wind energy conversion	x	x	x				Integration with energy storage and delivery systems
Ocean thermal difference	x	x	x				Large, low-pressure turbines, heat exchangers and long, deep-water intake pipe
Hydroelectric	x	x	x	x	x	x	
Renewable clean fuel sources							
Combustion of organic matter	x	x	x	x			Development of efficient growth, harvesting
Bioconversion of organic materials to methane	x	x	x	x			Development of efficient conversion processes and economical sources of organic materials
Pyrolysis of organic materials to gas, liquid, and solid fuel	x	x	x	x	x		Optimization of fuel production for different feed materials
Chemical reduction of organic materials to oil	x	x	x				Optimization of organic feed systems and oil separation process
Photolysis or pyrolysis of water	x						Discovery of appropriate sensitizers

[a] Source: NSF/NASA Solar Energy Panel, "An Assessment of Solar Energy as a National Energy Resource," pp. 8–9. University of Maryland, College Park, 1972.

an area of about two square miles (5.18×10^6 m^2 = 518 hectares) would be required for collection of 1000 MW. The scale of equipment necessary requires a large initial investment of labor, capital, and perhaps free energy. This short-term loss often causes prospective users to choose other energy supplies even though solar might be favorable in the long run. The present status of most direct and indirect applications which are thought to be worth pursuing is indicated in Table 6.4. Of these only hydroelectric plants, windmills, and hot water heating (which has found only limited acceptance) are currently available on a commercial scale.

Solar energy applications may also be classified according to the type of process by which the energy of solar photons is transformed. In the case of *thermal collection* relatively closely spaced translational, rotational, and vibrational energy levels store the energy. Photons of nearly every wavelength can excite transitions among such states. When electronic energy levels are involved, however, not all wavelengths of the blackbody radiation of the sun are sufficiently energetic to contribute to the excitation. In this and the next section we shall compare the relative theoretical efficiencies of solar energy conversion involving electronic (photochemical or photoelectric) and thermal processes.

Solar Collectors

Conversion to thermal energy amounts approximately to transfer of radiation from one blackbody (the sun at 6000 K) to another (the *solar collector*). The quantity of energy emitted by a blackbody is proportional to the fourth power of its temperature. As the collector absorbs thermal energy its temperature (and therefore its rate of emission) increases, as does its rate of heat loss to the surroundings by conduction and by convection of the surrounding air. A steady state is eventually attained when the quantity of heat lost equals that absorbed. Under the most favorable conditions collector temperature might reach 550 K,[6] but in practice temperatures above 400 K are not easily attained in heat transfer fluids. A blackbody in this temperature range exhibits maximum emission at wavelengths between 8 and 10 μm (equation 1.9). Performance of a collector can be greatly improved if it is enclosed by several layers of a material such as glass which is transparent to most visible and some ultraviolet wavelengths but opaque to the infrared rays emitted by the black collector surface. The temperatures achieved by *flat plate collectors* (Figure 6.7a) are adequate for water and space heating, and probably for air conditioning. The latter application has the definite advantage that the amount of solar energy is usually at a maximum when it is needed.

Heat Storage

Because of interruptions by clouds or nightfall, solar heating schemes must include some heat storage device. Two different methods have been proposed. The first involves raising the temperature of a material with a

[6] H. C. Hottel and J. B. Howard, "New Energy Technology — Some Facts and Assessments," p. 338ff. MIT Press, Cambridge, Massachusetts, 1971.

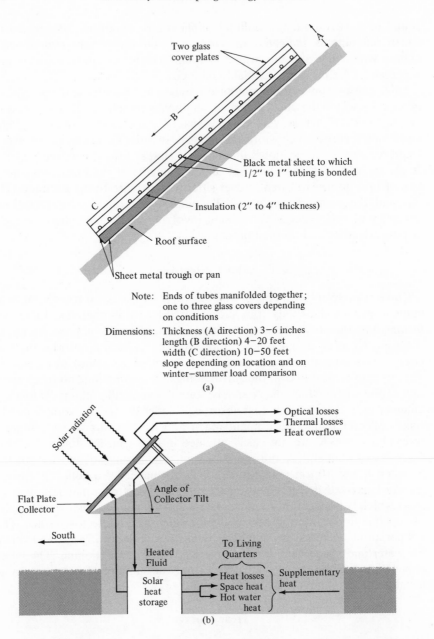

Figure 6.7 (a) Flat plate solar collector (cross section). Source: NSF/NASA Solar Energy Panel, "An Assessment of Solar Energy as a National Energy Resource," p. 15. University of Maryland, College Park, 1972. (b) Solar house heating system using flat plate collector. Some system variables are indicated. Source: R. A. Tybout and G. O. G. Lof, Solar house heating. *Nat. Res. J.* **10**(2), 268–326 (1970). The same figure has also appeared in: Cost of house heating with solar energy, *Solar Energy* **14**, 255 (1973).

TABLE 6.5   Heat Storage in Phase Change

Compound	Transition temp. (K)	Heat (J/g)
$CaCl_2 \cdot 6 H_2O$	302–312	174
$Na_2CO_3 \cdot 10 H_2O$	305–309	267
$Na_2HPO_4 \cdot 12 H_2O$	309	265
$Ca(NO_3)_2 \cdot 4 H_2O$	313–315	209
$Na_2SO_4 \cdot 10 H_2O$	305	241
$Na_2S_2O_3 \cdot 5 H_2O$	322–324	209

high heat capacity such as water or crushed rocks. Air passed through the storage area, warmed, and then circulated through a building could retrieve the heat later. A second possibly more efficient technique would employ the heat of fusion or other *phase change* of some substance. To be most useful, a large amount of heat should be absorbed and the temperature of the phase change should be approximately 10–15 K above normal room temperature. Examples of compounds which could be used to store heat in this way are given in Table 6.5. Note, however, that to provide for the heat loss of 2 to 4×10^8 J/day expected for an average home in freezing weather, anywhere from one to two tonnes of material would be necessary. Although it is possible to operate a completely solar-heated house in most areas of the United States, the large initial investment in solar collectors and heat reservoirs usually makes systems involving supplementary fossil fuel energy more favorable economically. Such systems are already competitive with electric heating, and rising costs of fossil fuels may soon make them even more attractive.

Concentrating Collectors

While heat can be collected fairly efficiently by a flat plate collector, even in hazy or partly cloudy weather, temperatures are not high enough for efficient operation of a heat engine. For a Carnot cycle operating between 400 and 288 K, the theoretical efficiency would be 28%. Since the collectors cannot absorb all of the energy striking them and some heat loss from the transfer fluid is inevitable, it is unlikely that an overall efficiency greater than about 10% can be attained for electric power generation. The hypothetical 1000-MW electric power plant mentioned earlier would thus occupy twenty (rather than two) square miles if thermal conversion were to be employed.

It is possible, by means of *spherical* or *parabolic mirrors* or with *lenses,* to *concentrate* solar radiation, achieving higher temperatures. While this would improve the efficiency of a solar-powered heat engine, it would also greatly increase the cost and complexity of the collection device. Because a distinct image of the sun is required if mirrors or lenses are to focus it onto a small area, concentrating collectors must be designed to follow the apparent daily and yearly motions of the sun. Moreover, on a cloudy day no energy could be collected at all. Despite the problems involved in concentrating it, thermally collected solar energy seems ideal for applications such as space heating and cooling and water heating whose low temperatures do not require extremely high energy quality.

6.5 SOLAR ENERGY–ELECTRONIC COLLECTION

In the case of *electronic collection* of solar energy the effect of the incoming radiation is to excite a molecule from one electronic state to another. Electrons in a higher energy orbital may provide the work necessary to drive nonspontaneous chemical reactions or to force an electric current to flow. An oversimplified example of the enhanced chemical reactivity[7] of an excited electronic state is provided in Figure 6.8. The higher energy electron is more easily donated to another species, making the excited molecule a better reducing agent, and there is a lower energy unfilled orbital which allows the excited molecule to accept an electron and serve as an oxidizing agent.

In order to excite a molecule the energy per quantum of solar radiation must be at least as great as the difference between two energy states. This implies that photons whose wavelengths are greater than some maximum value (λ_{max}) will be ineffective. Only those of shorter wavelength contain sufficient energy to excite the molecule from E_1 to E_2. For the purpose of calculating maximum efficiency we shall assume that all of the latter photons can be made to contribute the portion of their energy corre-

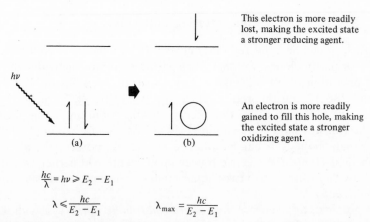

$$\frac{hc}{\lambda} = h\nu \geqslant E_2 - E_1$$

$$\lambda \leqslant \frac{hc}{E_2 - E_1} \qquad \lambda_{max} = \frac{hc}{E_2 - E_1}$$

Figure 6.8 Enhancement of chemical reactivity by means of electronic excitation. (a) Ground state molecule, E_1; and (b) excited state molecule, E_2.

[7] J. H. Wang, *Accounts Chem. Res.* **3**, 90 (1970).

sponding to $E_2 - E_1$. However, any excess over this value will be assumed not to be collected since a photon can only be absorbed once and therefore cannot excite two or more molecules.

If the energy difference $E_2 - E_1$ is large, say, 50.0×10^{-20} J, only those photons represented by the shaded area at the left of Figure 6.5 will have sufficient energy to produce electronic excitation. The energy of the vast majority of photons can only be converted to heat. A small value of $E_2 - E_1$, say, 12.5×10^{-20} J, would allow a large λ_{max} (1.6×10^{-6} m) and all photons except those represented by the shaded area on the right of Figure 6.5 could be collected. In this case, however, most photons would contain excess energy. (A photon of $\lambda = 8 \times 10^{-7}$ m contains 25.0×10^{-20} J, double the required amount.) Since this excess cannot electronically excite another molecule, it will be converted to heat. Obviously, there must be an optimum energy difference between these two extremes. It is $E_2 - E_1 = 18.1 \times 10^{-20}$ J or $\lambda_{max} = 1.1$ μm for blackbody radiation of 6000 K color temperature. The corresponding theoretical efficiency of collection is 44%.[8]

Photovoltaic Cells

The most efficient *photovoltaic cells* in use today are constructed using semiconductors based on silicon. In this case the "molecule" being excited is an entire silicon crystal in which the atoms are arranged in the same covalent network structure as diamond. Such a crystal may be "doped" by adding one part per million of arsenic to give an N-type or a similar concentration of boron to give a P-type semiconductor. (The letters "N" and "P" refer to electrical conduction by negative charges—the extra, fifth valence electron of arsenic—or positive charges—the "holes" created by the deficiency of one valence electron in boron.) If an N-type semiconductor is placed in contact with a P-type (Figure 6.9) the extra electrons and the positive holes neutralize each other, creating a barrier layer which is a fairly effective insulator.

When photons strike the semiconductor "sandwich" electrons are excited to the conduction band, and positive holes are created. Since neither electrons nor holes can pass through the barrier layer, a potential of 0.5–0.6 V is set up and electrons flow around the external circuit to neutralize the positive holes. For silicon all photons having $\lambda < 1.15$ μm can excite electrons to the conduction band, but the 44% theoretical efficiency is reduced to about 20% by losses of reflected light at the surface of the cell, recombination of electrons and holes and leakage of current across the barrier layer. Most cells in commercial production are 10–15% efficient,

---

[8] D. Trivich and P. A. Flinn, *in* "Solar Energy Research" (F. Daniels and J. A. Duffie, eds.), p. 143. Univ. of Wisconsin Press, Madison, 1955.

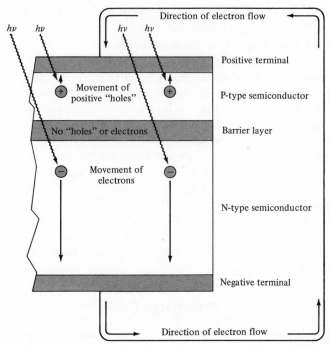

Figure 6.9 Schematic diagram of silicon photovoltaic cells. Free electrons and positive holes produced by photoionization yield a one-way current flow.

and at their current state of development high dollar and free-energy costs make them an unlikely choice as a panacea for the energy crisis. Recently developed techniques for growing single silicon crystals[9] give some hope that these problems may be solved in the future.

Photochemical Fuels

The concept of converting solar energy directly into a clean-burning chemical fuel which can be shipped wherever it is needed is a very appealing one. Many chemists have attempted to find photochemical processes which would efficiently produce this type of fuel in reasonable quantities. One example is the *cerium(IV)—cerium(III) system* shown in equations 6.7:

$$Ce^{4+} + \tfrac{1}{2} H_2O + \text{light or heat} \longrightarrow Ce^{3+} + \tfrac{1}{4} O_2 + H^+ \tag{6.7a}$$
$$\underline{Ce^{3+} + H_2O + \text{light } (\lambda < 350 \text{ nm}) \longrightarrow Ce^{4+} + \tfrac{1}{2} H_2 + OH^-} \tag{6.7b}$$
$$H_2O + \text{light} \longrightarrow H_2 + \tfrac{1}{2} O_2 \tag{6.7}$$

[9] *Chemical and Engineering News,* July 29, pp. 16–17 (1974).

Cerium(IV) ions are reduced, either photochemically or thermally, to produce both cerium(III) ions and dioxygen. The cerium(III) ions can be photochemically oxidized to produce cerium(IV) ions and dihydrogen. The overall process (equations 6.7) is the photolysis of water to produce dihydrogen and dioxygen. The former could easily be shipped as a chemical fuel while the latter might find numerous industrial applications – steelmaking, for example. (For further discussion of dihydrogen fuel see Section 7.1.) Unfortunately, because of the requirement of very short wavelengths in equation 6.7b, overall efficiency of solar energy collection by the cerium(IV)–cerium(III) system is less than 0.1%. One means of improving this value would be the addition of *sensitizers* – compounds which absorb longer wavelength photons and could transfer their energy to Ce^{3+}, helping water to oxidize it.

6.6 PHOTOSYNTHESIS

As described in Section 3.7, the nonspontaneous reaction of water with carbon dioxide may be driven by solar energy, producing carbohydrates. Because of the ubiquity of polymeric forms such as cellulose and starch there is probably more carbohydrate in the biosphere than all other organic substances combined. *Photosynthesis* involves oxidation and reduction, and the overall process may be separated into half reactions:

$$2 H_2O \longrightarrow O_2 + 4 H^+ + 4 e^- \qquad E^{0\prime} = -0.82 \text{ V} \qquad (6.8a)$$
$$CO_2 + 4 H^+ + 4 e^- \longrightarrow [CH_2O] + H_2O \qquad E^{0\prime} = -0.40 \text{ V} \qquad (6.8b)$$
$$\overline{H_2O + CO_2 \longrightarrow O_2 + [CH_2O]} \qquad \text{emf}\prime = -1.22 \text{ V} \qquad (6.8)$$

Since the electromotive force (emf) of a cell constructed from these two half-reactions is negative, free energy must be supplied to force the reaction from left to right. The amount needed can be calculated as

$$\Delta G^{0\prime} = -nFE^{0\prime} = -(4)(96.53 \text{ J/V})(-1.22 \text{ V}) = 471 \text{ J/mol } [CH_2O] \qquad (6.9)$$

(In equations 6.8 and 6.9 $[CH_2O]$ represents the empirical formula of carbohydrate. The primes on $E^{0\prime}$ and $\Delta G^{0\prime}$ indicate that these values apply under conditions of 10^{-7} molar hydrogen ion instead of the usual 1 M.) This free energy must be supplied by solar photons.

A molecule which can absorb visible light and whose excited state lasts long enough for it to undergo oxidation and/or reduction (recall Figure 6.8) is needed to initiate the photochemical process. Chlorophylls and several accessory pigments (sensitizers) such as β-carotene serve this function in most photosynthetic systems.[10] Chlorophyll contains a Mg^{2+} ion surrounded by a complex organic ring system with numerous, delocalized

---

[10] Structures of important compounds in photosynthesis and a somewhat more detailed treatment may be found in R. Gymer, "Chemistry: An Ecological Approach," pp. 658–669. Harper, New York, 1973.

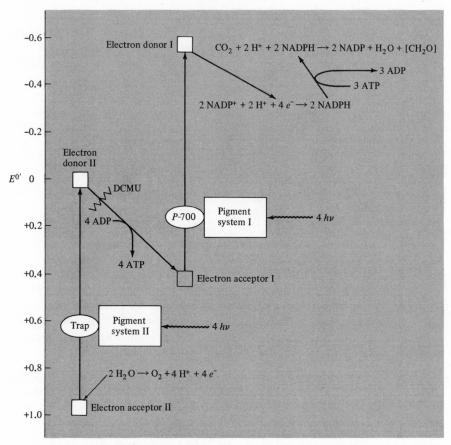

Figure 6.10 Schematic mechanism for photosynthesis.

double bonds. Evolution of this structure has resulted in "fine tuning" to the wavelengths of visible light, increasing the efficiency of collection. Energy absorbed by these pigments is apparently transferred to a single specialized chlorophyll molecule, designated *P*-700 because its absorption maximum is at 700 nm, which can carry out redox processes with the aid of enzymes and special energy carrier molecules in its vicinity.

It has been observed that the efficiency of photosynthesis is much greater if light whose wavelength is above 680 nm is combined with light whose wavelength is below that figure. This suggests that there is more than one system for absorbing photons and more than one redox couple driven by absorbed radiation, as indicated in Figure 6.10. The energy of photons striking pigment system II is trapped by a molecule similar to *P*-700 but whose structure is not yet known. This molecule (called "trap" in the figure) can only be excited by photons of wavelength less than 680 nm, and produces electron acceptor II whose $E^{0'}$ is more positive than that of the water–oxygen half-reaction. This generates O_2 and a flow of four electrons into acceptor II. Each electron is raised to the excited state, from which it can be transferred to electron donor II at $E^{0\prime} \simeq 0$. Donor II begins a chain of reduction reactions during the course of which four molecules of adenosine diphosphate (ADP) react with $H_2PO_4^-$ to form adenosine triphosphate (ATP). The ATP stores some of the free energy released by the spontaneous reactions between electron donor II and electron acceptor I.

In pigment system I the collection of radiant energy is again used to "pump" electrons to a more negative $E^{0\prime}$, producing electron donor I. This molecule initiates another chain of spontaneous reductions which culminates in the reduction of two molecules of nicotinamide adenine dinucleotide phosphate to form NADPH. The final step in the process, known as the *dark reaction* because it does not involve a photochemical excitation, is the reaction of NADPH with CO_2 to yield carbohydrate. This process has a positive ΔG (involves upward movement on Figure 6.10), but chemical energy (stored in ATP molecules) is used to drive it forward instead of radiant energy.

Any substance which interferes with the flow of electrons along the pathway of Figure 6.10 can decrease the efficiency of photosynthesis. For example, the effectiveness of the weed killer DCMU (dichlorophenyl-dimethylurea) is a result of its ability to react with one of the substances in the chain of spontaneous reductions leading from electron donor II to electron acceptor I. This blocks electron flow and prevents pigment system II from reducing substances which have been oxidized by acceptor I. There are probably at least five different redox couples along this chain, providing numerous opportunities for interference. In addition there are two or three substances between electron donor I and NADPH. Knowledge of intermediates between oxidation of water and electron acceptor II is very sparse, but opportunities for foreign compounds to block electron flow probably exist here as well.

Photosynthesis can also be interrupted if the supply of reactants (CO_2 and/or H_2O) is cut off. In green plants, for example, the stomata (openings in the leaves) which allow gases to come in contact with the photosynthetic centers (chloroplasts) may become clogged by particulate matter from air pollution, eliminating the supply of CO_2. Finally, the death of photosynthetic organisms on a large scale could interrupt photosynthetic production of O_2 and carbohydrate. Although one usually thinks of trees and other terrestrial green plants as the most important sites for photosynthesis, it is estimated that over half the carbohydrate on earth is produced by aquatic microorganisms such as blue-green algae, diatoms, and dinoflagellates. It was concern over the effect of DDT on such organisms that sparked the controversy over world supplies of dioxygen which was discussed at the end of Chapter 3.

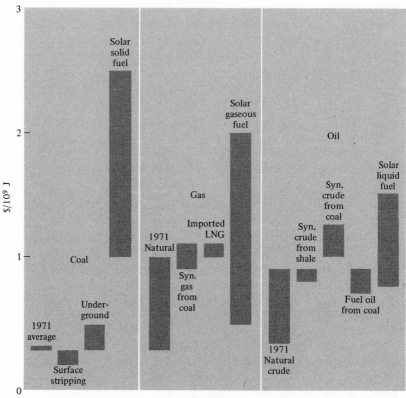

Figure 6.11 Costs of fossil and solar renewable fuels (1971). Source: NSF/NASA Solar Energy Panel, "An Assessment of Solar Energy as a National Resource," p. 44. University of Maryland, College Park, 1972.

Although the efficiency of most photosynthetic processes is below 1% based on the total light available to green plants, it is estimated that a maximum efficiency of about $12\frac{1}{2}\%$ could be reached by providing optimum concentrations of carbon dioxide and water. Blue-green algae might be grown in a culture medium of water saturated with carbon dioxide. After they had stored considerable solar energy, the water would be removed, first mechanically, then by heat drying (perhaps using solar radiation), and, finally, by either a pyrolysis or fermentation process. This would convert the algae into usable fuels containing carbon monoxide, methane, or ethylene as well as other liquid and solid carbon compounds. Similar processes have been applied to convert organic waste materials such as sewage sludge into useful fuels. Estimates of the cost of such fuels are shown in Figure 6.11. There has been considerable increase in fossil fuel prices since 1971, making solar photosynthetic fuels even more attractive than the figure might imply.

Photosynthesis and Dihydrogen

It seems possible that the photosynthetic process might successfully be modified to produce dihydrogen rather than carbohydrate.[11] If NADPH produced by electron donor I (Figure 6.10) is reduced by a bacterial enzyme (hydrogenase) before it can react with CO_2, H_2 can be evolved. Such a reaction has been demonstrated in the laboratory using triphosphopyridine nucleotide in place of $NADP^+$ and hydrogenase from the bacteria *Clostridium kluyveri,* but efficiencies are very low so far.

6.7 TIDAL AND GEOTHERMAL ENERGY

The movement of ocean water over the earth's surface, mainly resulting from the gravitational influence of the moon, can produce useful energy at sites where especially high tides occur. One of these, the Rance River estuary in France, has been developed to yield 240 MW of electrical power. Unfortunately, estimates of total energy available from tides are low (Table 4.3), and the construction of dams and turbines cannot help but affect the ecology of estuarine areas which often are responsible for much marine food production.

Geothermal energy is listed in Table 4.3 as a continuous supply, under the assumption that heat would not be withdrawn at rates greater than it is supplied to earth's surface by radioactive decay and conduction from the core and mantle. However, such a supply is small compared to current consumption. Even if favorable geothermal sites are developed nonrenewably this energy resource cannot replace more than a fraction of current fossil fuel consumption. Moreover, because of the low temperature of geothermal steam, nine times more waste heat is produced than by a fossil-fueled power plant, and a larger area is required. Problems of small earthquakes, contamination of aquifers, and release of sulfur, arsenic, and mercury compounds also make geothermal development less desirable. At carefully selected sites both tidal and geothermal plants may make environmentally safe contributions to United States and world free-energy supply, but neither appears to be a panacea.

EXERCISES

(The first group of exercises is based primarily on material found in this text.)

1. Calculate the energy available from one gram of hydrogen, assuming the following techniques for its conversion:

   a. The entire 1 g of hydrogen is converted into energy.

---

[11] American Chemical Society, "What's Happening in Chemistry," pp. 16–21. ACS, Washington, D. C., 1975.

    b. The nuclear reaction in equation 1.1 is used to react all of the hydrogen.
    c. The natural abundance of deuterium is recovered and reaction 6.5a carried
       out.
    d. The hydrogen is burned chemically: $H_2 + \frac{1}{2} O_2 \longrightarrow H_2O$.

2. Explain how both fission and fusion of atomic nuclei can result in production of
   energy. Write equations for the reactions which are being used or have been
   proposed for controlled fission and fusion reactors.
3. What are the major differences among BWR's, PWR's, HTGR's, and
   LMFBR's? What components and safeguards are included in current designs
   for LWR's?
4. Breeder reactors can produce energy and at the same time yield more fuel than
   they burn. Does this not violate the second (and perhaps the first) law of thermodynamics?
5. List environmental problems which might be associated with conversion of the
   United States energy system to nuclear power. Which do you think are the
   most important?
6. Why has the deuterium–tritium reaction been chosen for laboratory studies of
   the scientific feasibility of fusion power? If such experiments are successful
   does this guarantee that fusion reactors will be commercially successful?
7. What quantity of each of the salts in Table 6.5 would be required to provide for
   storage of 4×10^8 J? Using a handbook to obtain densities calculate the *volume*
   required for heat storage. Do the same for water as a storage medium.
8. Calculate the maximum wavelength that could excite a molecule between two
   states differing in energy by

    a. 50×10^{-20} J
    b. 12.5×10^{-20} J

9. How much excess energy would be contained in a photon of $\lambda = 5 \times 10^{-7}$ m
   over that required to excite the molecule in Exercise 8b?
10. United States energy consumption is estimated (Table 7.4) to be approximately
    140×10^{18} J in the year 2000. If all of this were to be collected by silicon pho-
    tocells, what area would be required for collection?

(Subsequent exercises may require more extensive research or thought.)

11. Obtain data from the weather bureau on the average number of successive
    cloudy days to be expected in your locale. Combine this with the results of Ex-
    ercise 7 to estimate the volume of heat storage material for an all-solar house in
    your area.
12. If you are considering chemical research as a career, to which of the techniques
    of solar energy collection do you think you might make the greatest contribu-
    tion? How would you proceed to attack this problem?
13. The Rance River tidal power plant has the following characteristics:

    a. Tidal range: 14 m.
    b. Time between successive high tides: 12 hours 20 min. (This allows two
       cycles of the plant, one on the incoming and one on the outgoing tide.)
    c. Area of reservoir 23 km^2.

Assuming a net efficiency of conversion of 90% calculate the power rating of the plant. Compare your result with that given in the text.

14. Assuming that the earth's core contains iron at an average density of 11.0 g/cm³ and an average temperature of 3300 K, how long could the current yearly energy consumption of 0.16×10^{21} J be sustained by geothermal energy (assuming that energy could be withdrawn at that rate)?

7

Storage, Distribution, and Conservation of Energy

Industrialism is the systematic exploitation of wasting assets
. . . progress is merely an acceleration in the rate of that
exploitation. Such prosperity as we have known up to the
present is the consequence of rapidly spending the planet's ir-
replaceable capital.

Aldous Huxley

Of the developing energy supplies discussed in the preceding chapter only solar is not tied to generation of electric power. Nuclear, tidal, and geothermal sources all require complicated, large-scale plants, technically trained personnel to operate them and carefully chosen sites, factors which make large plants more economical than small ones. Most predictions have naturally assumed that such plants would generate electricity, although in some cases agricultural or industrial uses of waste heat have been included in proposed designs. Evolution toward such systems carries with it two interrelated problems: there are few good ways to store electric power, and its generation and transportation are relatively inefficient.

Figure 7.1 indicates that 78.2% of the United States gross energy consumption in 1971 was in the form of fluid fossil fuels — petroleum or natural gas. As we have seen (Section 5.4), it is just these fuels which are in shortest supply and must be replaced, and we are likely to encounter numerous difficulties attempting to adapt electric power to the end uses (especially transportation) which they now serve. Fluid fuels are readily portable, easy to handle, and can store a large quantity of energy in a small mass and volume. One proposed means of adapting nuclear fission power to such uses is generation of dihydrogen fuel, especially if it is later used in conjunction with fuel cells. This will be discussed in the next two sections.

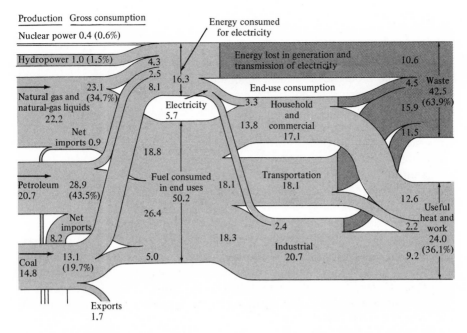

Figure 7.1 Flow of energy through the United States system, 1971, showing sources, uses, and amount/percent utilization versus waste (in units of 10^{18} J). Data from E. Cook, Texas A & M University, College Station.

Interconversion of one energy source to another, construction of power and processing plants, and even the mining and transportation of primary energy resources all consume some free energy. Although past availability of high quality resources has permitted us to ignore this fact, it is constantly increasing in importance as more advanced technologies such as fission power or coal gasification and lower grade ores such as lignite and oil shale are developed. The *net energy* available is often much less than might be expected, as we shall see in the third section.

One very successful means of increasing net free energy is conservation and more efficient consumption. That current efficiencies are low should be obvious from Figure 7.1, which shows that only 36% of the primary United States energy consumption in 1971 went into end-use work. This is probably a gross overestimate of efficiency since it fails to consider alternative means of accomplishing the same end—bicycles in place of automobiles and the like. More details will be given in Section 7.4. Finally, we shall return to the question raised in Section 4.9 regarding optimal energy flow through the earth system and demonstrate that even if a limitless source of free energy were available, there would be limitations on its use.

7.1 SYNTHETIC CHEMICAL FUELS – THE H_2 ECONOMY

Because of the difficulty of storing electrical energy all power plants cannot be run at full load except during the few hours each day when individuals and industries are using maximum power. Should demand exceed generating capacity, however, brownouts and blackouts would occur; thus, numerous small-scale, flexible, low-efficiency gas- and oil-fired "peak shaving" power plants are necessary. If large, base-load power plants could operate around the clock, storing energy generated early in the morning for use during peak afternoon hours, efficiencies might be improved. This is currently being done by pumping water to elevated reservoirs during off-peak hours and later releasing it through turbines, but the number of suitable sites for such *pumped storage* is not large enough.

Energy can also be stored by mechanical devices such as springs, compressed gases, or flywheels. Although the energy content of a compressed steel spring is rather low per unit mass, synthetic materials such as polyurethane plastics can approach the energy density of a lead-acid storage battery. Compression of air to 10^4 kPa or more stores a good deal of energy, but if the mass of an adequate container is included energy density is low again. Flywheels constructed of high-strength carbon or glass filaments and operating at extremely high speeds can store more energy per unit mass, but ancillary equipment to transfer work to and from such a flywheel reduces its utility. Most experience with energy storage leads one to the conclusion that the potential energy of chemical bonds is the most efficient, most convenient, and safest means of accomplishing this essential task. The activation barriers to chemical reactions prevent energy from being released until it is wanted, and liquid or gaseous substances are relatively easy to handle and transport.

In addition to higher overall efficiency of power plants if dihydrogen or other chemical fuel were synthesized during off-peak periods, the transmission of energy as gaseous fuel is from three to six times more efficient than transmission of electric power. Public objection to the siting of nuclear plants in heavily populated areas will require transmission of power over greater distances in the future, making this savings even more important.

Production of Dihydrogen

Most H_2 is currently produced from hydrocarbon fuels by reactions such as catalytic reforming (see Section 5.6, Figure 5.14), but the fact that these fuels are in short supply has created the need for substitute means of synthesizing dihydrogen. The only other method commercially available is electrolysis of water:

$$H_2O(l) \xrightarrow{\text{electrolysis}} H_2 + \tfrac{1}{2} O_2 \qquad \Delta G^0_{298} = 286.0 - T(0.1632) = 237.4 \text{ kJ} \qquad (7.1)$$

At present electrolyzers operate at efficiencies between 60 and 70%, but some prototypes have reached 85%. In theory the maximum *electrical* efficiency of such a cell is nearly 120% — because the reaction is endothermic heat absorbed from the surroundings can also contribute to overcoming the positive free-energy change.

No matter what efficiency may be reached in the electrolyzer, however, one must still contend with the 35–45% maximum efficiency expected for the generation of the electricity. Therefore some attention has been turned to as yet undeveloped methods for decomposition of water by direct use of thermal energy from nuclear plants. The increase in entropy for reaction 7.1 implies that it will occur spontaneously at sufficiently elevated temperatures. Appreciable quantities of dihydrogen may be obtained only above 2800 K, however, and nuclear reactors are not expected to exceed 1200 K.

Several multistep reaction sequences have been proposed in order to accomplish thermal decomposition of water at lower temperatures.[1] Perhaps the most thoroughly studied is the following:

$$CaBr_2 + 2\ H_2O \longrightarrow Ca(OH)_2 + 2\ HBr \qquad (7.2a)$$

$$Hg + 2\ HBr \longrightarrow HgBr_2 + H_2 \qquad (7.2b)$$

$$HgBr_2 + Ca(OH)_2 \longrightarrow CaBr_2 + HgO + H_2O \qquad (7.2c)$$

$$\frac{HgO \longrightarrow Hg + \tfrac{1}{2}O_2 \qquad\qquad\qquad}{H_2O \longrightarrow H_2 + \tfrac{1}{2}O_2 \qquad\qquad\qquad (7.2)}$$

All steps may be carried out at 1000 K or less, and the thermal efficiency is reported to exceed 50%. So far this sequence has been tested only under laboratory conditions, and considerable development work remains before its feasibility can be established definitely.

Dihydrogen Storage

Although it seems reasonably certain that efficient methods for generating dihydrogen will be available in the future, the problem of a convenient means of storing it may be more difficult to solve. The critical temperature of dihydrogen is only 33 K (at a pressure of 1300 kPa); thus, cryogenic tanks are required if it is to be stored as a liquid. In gaseous form H_2 has a relatively low density, even at high pressures, and bulky, heavy containers are required. This compares unfavorably with methane, which has a critical temperature of 191 K, a critical pressure of 4641 kPa, and a liquid density more than five times that of liquid dihydrogen.

---

[1] G. DeBeni and C. Marchetti, *Eur. Spectra* p. 46 (1970); see also R. H. Wentorf, Jr. and R. E. Hanneman, *Science* **185**, 311–319 (1974); B. M. Abraham and F. Schreiner, *ibid.* **180**, 959–960 (1973); **182**, 1372–1373 (1973).

A more efficient means of storage would employ *metal hydrides* as shown in equations 7.3.

$$H_2 + Ti \rightleftharpoons TiH_2 \qquad (7.3a)$$

$$H_2 + 4 Pd \rightleftharpoons 2 Pd_2H \qquad (7.3b)$$

$$3 H_2 + LaNi_5 \rightleftharpoons LaNi_5H_6 \qquad (7.3c)$$

In TiH_2 the density of hydrogen is about four times that of liquid H_2. Hydrogen atoms dissolve in the titanium by occupying all of the tetrahedral holes in the cubic close-packed array of titanium atoms. (The formation of Pd_2H is similar except that only one-fourth of the tetrahedral holes are occupied by hydrogen atoms.) Titanium hydride is especially useful because reaction 7.3a is exothermic; if the temperature of TiH_2 is raised, the equilibrium shifts to the left and H_2 is released. This would be ideal if dihydrogen were to be used to operate automobiles since waste heat from the engine could be used to generate fuel pressure as needed.

Possible Environmental Impacts
of Hydrogen Economy

The properties of dihydrogen would have to be considered carefully in any eventual plan for its use as an energy storage and transmission medium. Because of its low density the energy content per unit volume of gaseous H_2 is less than one-third that of methane; it also has about one-third the viscosity of methane. This would permit increased flow rates in pipelines which are already being used for CH_4, provided larger capacity pumps were installed, making the dollar and energy transmission costs of the two gases similar. The diffusion of hydrogen into structural materials causing their subsequent embrittlement might complicate storage and transmission of H_2, but plastic or other coatings could protect exposed metal.

Fears concerning the safety of liquid or gaseous dihydrogen as a fuel are often referred to as the "Hindenburg syndrome" because of the spectacular fire involving that H_2-filled dirigible at Lakehurst, New Jersey, in 1937. Because of its low molecular weight dihydrogen diffuses more rapidly than any other gas, making leakage from even "airtight" containers a possibility. It forms an explosive mixture with air over a range of 4–75% H_2 by volume (compared with a range of 5–15% for methane and 1–10% for gasoline), and the amount of energy necessary to ignite a flame is small enough that a minor discharge of static electricity can provide it. Nevertheless, numerous shipments of H_2 are made without incident in the United States every day, and at least one case is known where the entire contents of a liquid H_2 tank truck accidentally spilled but no fire ensued. The facts that dihydrogen is a gas and it has an extremely high rate of diffusion mean that it will be removed rapidly from the scene of a spill, minimizing the *time* during which ignition could occur.

The low energy necessary for ignition of H_2–O_2 mixtures has a useful feature. The low activation energy for combustion means that dihydrogen can be burned at low temperatures in catalytic heaters, providing that boon to the advertisers of all-electric systems, flameless heat. The need for a flue would disppear, as would the loss of the 25–50% of the fuel heating value which usually goes up the chimney because of furnace inefficiency, and home heating could be individualized on a room-to-room basis. Prior to the development of long-distance natural gas pipelines many cities were supplied with water gas manufactured from coal. As noted in Section 5.3, water gas contains up to 50% H_2 by volume and much of the remainder is CO. The techniques used for handling water gas should carry over to pure dihydrogen. Certainly it will be necessary to add an odorant (as is currently done with natural gas) to make leaks more readily detectable, and an illuminant to make the normally colorless dihydrogen flame visible.

One further possible drawback relating to some of the above applications of dihydrogen fuel is the possibility of production of H_2O_2 upon combustion. Griffith[2] has observed as much as 230 ppm H_2O_2 in the effluent from a H_2-burning internal combustion engine. Thermodynamics favors production of both H_2O and H_2O_2 (compare the reverse of equation 7.1 with 7.4):

$$H_2 + O_2 \longrightarrow H_2O_2 \qquad \Delta G^0_{298} = -136.2 - T(-0.1101) = -103.4 \text{ kJ} \qquad (7.4)$$

and the known ability of H_2O_2 to generate free radicals might make it a health hazard as well as contributing to photochemical smog reactions. On the other hand, the low temperature of H_2 combustion produces almost none of the nitrogen oxide emissions associated with high temperature combustion of most fossil fuels.

7.2 ELECTROCHEMICAL ENERGY CONVERSION

Conversion to a hydrogen economy would probably result in an increased adoption of *fuel cells* to convert chemical energy to electrical. A shift toward primary energy sources associated with electric power generation without production of synthetic chemical fuel would greatly increase the importance of galvanic cells as free energy storage devices. A fuel cell operates in the same manner as a galvanic cell with the exception that the chemical reactants are supplied continuously from an external reservoir. Consequently, a battery of fuel cells does not run down or require recharging as a battery of galvanic cells would, and it can supply uninterrupted electric power until its electrodes or other structural components wear out. A schematic diagram of a single fuel cell is shown in Figure 7.2.

---

[2] E. J. Griffith, *Nature (London)* **248**, 458 (1974).

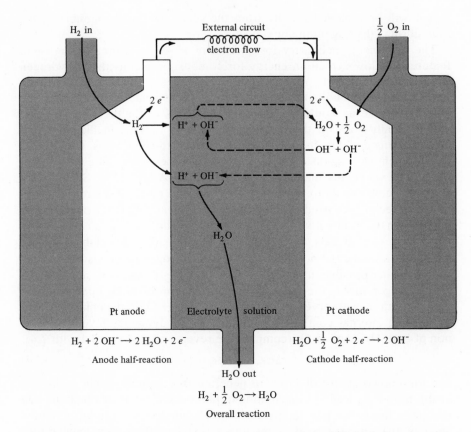

Figure 7.2 Schematic diagram of a dihydrogen–dioxygen fuel cell with alkaline electrolyte.

The principles of operation and the basis on which comparisons of operating efficiency can be made are the same for fuel cells and galvanic cells. In both cases an oxidation half-reaction and a reduction are carried out simultaneously at two different locations. The two half-cells are connected internally by conduction of ions through a salt bridge or electrolyte solution. By connecting them externally an electric circuit is completed, electrons released by the oxidation half-reaction flow through the circuit to the reduction half-cells, and useful work may be extracted from the chemical reaction. Because there is no intermediate heat engine in the conversion of chemical to electrical energy the Carnot cycle dependence of efficiency on temperature does not apply. In principle 100% efficiency can be attained, and if the cell reaction is endothermic heat absorbed from the surroundings can also be converted to electrical energy, permitting electrical efficiencies in excess of 100%.

In order to achieve maximum efficiency, however, an electrochemical cell must be operated reversibly. As in the case of an expanding gas (Section 4.3) the maximum work can be obtained from a reversible discharge and a minimum amount of work is required to recharge a galvanic cell reversibly. This reversible work equals the change in Gibbs free energy, and may be calculated as shown previously in equation 6.9. The reason that electrochemical cells are not used more extensively for chemical-electrical energy interconversions is that methods have not yet been devised by which they may be operated nearly reversibly. Complete reversibility of any reaction implies an infinitesimal rate, which in this case means that zero current could be drawn from a cell. As more current is drawn various irreversible processes combine to reduce the voltage (V) from the value (E) that it would have under reversible (zero current) conditions. The electrochemical efficiency of the cell may be expressed as V/E since the actual voltage under load, V is proportional to the actual work done and E is proportional to the maximum work (equation 6.9).

Polarization

Some of the reasons for *polarization* (the reduction of the actual cell voltage from E when current is drawn) may be understood by considering the reactions of an H_2–O_2 fuel cell in more detail (Figure 7.2). The half-reactions found in a table of electrode potentials are

Anode:	$H_2 + 2\ OH^- \longrightarrow 2\ H_2O + 2\ e^-$	$E^0 = 0.828$	(7.5a)
Cathode:	$\frac{1}{2}\ O_2 + H_2O + 2\ e^- \longrightarrow 2\ OH^-$	$E^0 = 0.401$	(7.5b)
Overall:	$H_2 + \frac{1}{2}\ O_2 \longrightarrow H_2O$	$E^0 = 1.229$	(7.5)

but the actual processes occurring at an electrode surface are considerably more complicated. One major source of irreversibility, called *activation polarization,* is the activation energy for the half-reactions since they require cleavage of strong bonds in both dioxygen and dihydrogen. Metals such as platinum, palladium, and nickel which catalyze hydrogenation of organic double bonds have been found to be useful in dihydrogen cleavage. Metals such as gold, silver, platinum, and nickel, which form relatively unstable hydroxides and are not readily oxidized, are suitable catalysts for formation of OH^- ions from O_2. All of the electrode materials mentioned so far, with the possible exception of nickel, are relatively expensive because they are not very abundant. A second approach to overcoming activation polarization—raising the cell temperature—is inconvenient in many of the situations where fuel cells or galvanic cells might be used and might require a supplementary energy supply.

Two other polarization processes have to do with the fact that cell reactions occur at an electrode surface. The thin layer at the solid-solution interface may have considerable electrical resistance, which usually obeys Ohm's law and is referred to as *IR drop.* The second process, *concentration polarization,* occurs when high current flow depletes the concentra-

TABLE 7.1 **Ratings for Various Rechargeable Batteries**[a]

System	Energy density	Power density	Ability to recharge or refuel	Service life	Material availability	Cost	Other problems
Pb-acid	E	G	G	G	E	E	—
$PbO_2 + Pb + 2 H_2SO_4 \rightleftharpoons 2 PbSO_4 + 2 H_2O$							
Ni-Fe	P	F	G	E	G	F	—
$Fe + 2 NiO(OH) + 2 H_2O \rightleftharpoons Fe(OH)_2 + 2 Ni(OH)_2$							
Ni-Cd	P	E	E	E	P	P	—
$Cd + 2 NiO(OH) + 2 H_2O \rightleftharpoons Cd(OH)_2 + 2 Ni(OH)_2$							
Ag-Zn	F	E	P	P	P	P	—
$Zn + Ag_2O + 2 OH^- + H_2O \rightleftharpoons Zn(OH)_4^{2-} + 2 Ag$							
Zn-air[b]	G	F	G	F(?)	E	F[c]	Complexity, maintenance(?)
$2 Zn + O_2 \rightleftharpoons 2 ZnO$							
Mg-air[b]	E	F	F(?)	?	E	E[c]	Refueling is mechanical
$2 Mg + O_2 \rightleftharpoons 2 MgO$							
Li-organic[b]	E	P	F-G	F(?)	G	G[c]	—
$2 Li + NiF_2 \underset{\text{electrolyte}}{\overset{\text{nonaqueous}}{\rightleftharpoons}} 2 LiF + Ni$							
LiCl[b]	E	E	E	E	G	G	Safety, toxicity, complexity, startup(?)
$2 Li(l) + Cl_2(g) \underset{(900\text{ K})}{\overset{\text{molten LiCl}}{\rightleftharpoons}} 2 LiCl$							
Li-Te[b]	E	E	E	E	P(?)	F(?)	Temperature, toxicity, startup(?)
$2 Li + Te \underset{(700\text{ K})}{\overset{\text{molten LiCl}}{\rightleftharpoons}} Li_2Te$							
Na-S[b]	G-E	E	E	E	E	E[c]	Safety, high temperature, startup
$2 Na + 2 S \underset{(600\text{ K})}{\overset{\text{ceramic electrolyte}}{\rightleftharpoons}} Na_2S_2$							

[a] Data from: R. U. Ayres and R. P. McKenna, "Alternatives to the Internal Combustion Engine," p. 107. Johns Hopkins Univ. Press, Baltimore, Maryland, 1972. E stands for excellent; G, good; F, fair; and P, poor.
[b] Estimates.
[c] In production.

tions of reactants at the electrode surface and ions or molecules in the bulk of the solution cannot diffuse to the electrode rapidly enough to replace them. Because of the reduced concentrations of the reacting substances the voltage of the cell drops.

A great many combinations of half-cell reactions are possible for use in fuel or galvanic cells, but only a few have been found to be suitable for energy storage and regeneration. In addition to efficiency, *three basic characteristics* are required. First, the reversible voltage of the cell ought to be relatively large and should be approachable at normal temperatures. Some types of polarization are roughly constant no matter what the voltage and will therefore cause a greater loss of efficiency in a low voltage cell. This has led to considerable study of the alkali and alkaline earth metals for the oxidation half-reaction and F_2, Cl_2, O_2, or S for reduction upon discharge. Second, since many applications are expected to involve transportation, a large ratio of total energy and power to weight is desirable. This favors elements with low atomic weights (such as carbon, hydrogen, and nitrogen) in the fuel, electrodes, and electrolyte, and is the main reason why the familiar lead-acid battery is not used much more extensively. Finally, the materials from which a cell is constructed must not be so rare or costly as to make large-scale production economically impractical. A rechargeable cell must be able to undergo a great many charge–discharge cycles, and both types of cell must have reasonably long service lives. Table 7.1 rates various rechargeable batteries on the basis of the above criteria.

Fuel Cells and the H_2 Economy

The fact that gaseous fuel is more economically transportable than electric power has led to the suggestion that electricity might be generated from natural gas in individual homes or at local power substations by means of fuel cells. Considerable research in this area has already been supported by gas utilities, and fuel cells based on the H_2–O_2 couple which are 45% efficient at power outputs between 100 kW and 100 MW have been developed. This is well in excess of the efficiencies of diesel or Otto cycle engines or steam and gas turbines. Such fuel cells require a *chemical reformer* to convert hydrocarbon fuels into carbon dioxide and dihydrogen. The latter can then react with O_2 from the air yielding electric energy. The relationship between fuel cells and a hydrogen economy should now be apparent. If our chief fluid fuel were H_2, the reforming step in fuel cell power generation could be eliminated and efficiencies might reach 60%. For example, an automobile powered by fuel cells and electric motors could far exceed the 25% efficiency of today's internal combustion engine and would have the advantage of drawing no energy at all when at rest in heavy traffic.

7.3 NET ENERGY

Throughout the history of development of energy resources each new technology has built upon the old. For example, coal mining required machines, such as steam engines and railroad cars, whose materials and operation were dependent upon energy which was originally supplied by combustion of wood and charcoal. After significant quantities of coal were available the energy necessary for mining and processing began to be supplied by coal itself, but in either case the *net energy* available from coal was much less than would have been calculated on the basis of the heat of combustion of the total resource present as ore in the ground.

As any resource is consumed more effort must be expended in locating new deposits, ores are usually of lower concentration and quality, and their accessibility relative to geographical areas of maximum consumption and the surface of the earth is usually poorer. In the case of petroleum many more feet of exploratory drilling are required to find an equivalent new field

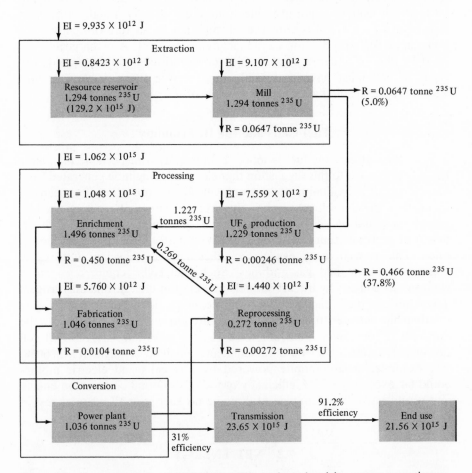

Figure 7.3 Energy analysis of 1000 MW$_e$ LWR nuclear electricity system operating annually at 0.75 load factor; EI, energy input; R, resource lost. Source: Council on Environmental Quality, "Energy and the Environment: Electric Power," No. 4111-00019, p. 56. Superintendent of Documents, Washington, D. C., 1973.

today than 30 years ago, the field usually is at considerably greater depth and, like the Alaskan north slope, may be far from centers of population. All of these factors imply that a growing fraction of the total energy available from new energy resources must be expended to recover, process, and transport the resources themselves. While this fraction does "useful work" in the physicochemical sense, it does not contribute correspondingly to the quality of life of the final consumer, and in a social sense is "wasted." An evaluation of the extent of this social waste is inherent in studies of net energy. These require consideration of an entire energy *system,* from resource in the ground to end use.

As a sample application of the net energy concept consider a LWR system capable of producing 1000 MW electric power and operating for one year at 75% of maximum capacity.[3] Figure 7.3 shows each step in mining, processing, and converting ^{235}U into electric power, together with external inputs of energy and the amounts of ^{235}U lost. One way of looking at the data is to note that the original resource, 1294 tonnes ^{235}U, contained 129.2×10^{15} J, but only 21.6×10^{15} J were delivered to end-use consumers—a system efficiency of 16.7%. However, this ignores energy inputs necessary for mining and milling uranium ore, converting it to UF_6 and concentrating ^{235}U, fabrication of fuel rods, and reprocessing spent fuel. These total 1.07×10^{15} J and must be subtracted from the 21.56×10^{15} J available for end uses, yielding an overall efficiency of 15.8%. Another way of looking at the system is to calculate the ratio of end-use energy output to external energy inputs required for extraction,

TABLE 7.2 **Efficiencies and Energy Ratios of Electric Power Systems**[a]

	% Efficiency					Energy ratio[c]
System	Extraction	Processing	Transport	Conversion	Total[b]	
Coal						
Deep mine	56	92	98	38	18	17.2
Strip mine	79	92	98	38	25	17.2
Oil						
Onshore	30	88	98	38	10	2.8
Offshore	40	88	98	38	13	2.8
Natural gas	73	97	95	38	24	4.7
Nuclear	95	57	100	31	16	20.1

[a] Data from Council on Environmental Quality, "Energy and the Environment: Electric Power." US Govt. Printing Office, Stock No. 4111-00019, Washington, D. C., 1973.

[b] Assumes 91% efficiency of transmission to end users.

[c] Energy ratio = (end-use energy)/(external energy input to the system).

---

[3] Data on this and other electric power generating systems are available in Council on Environmental Quality, "Energy and the Environment: Electric Power." US Govt. Printing Office, Stock No. 4111-00019, Washington, D. C., 1973.

processing, and conversion of the fuel. For our sample LWR the result is $21.56/1.07 = 20.1$. Approximately 5% of end-use output *must* be expended to provide a continuous supply of enriched uranium to the power plants. Table 7.2 gives comparable data for other methods of generating electric power. It should be obvious from these numbers why petroleum production and refining *consumed* nearly 5% of the total United States energy supply in 1968.

Of course, the energy ratios in Table 7.2 are especially low because of the inefficiency of electric power generation. For end uses such as direct (nonelectric) home or hot-water heating, where much higher efficiencies of conversion can be attained, these numbers might be higher by nearly a factor of three for natural gas or oil. Nevertheless, a sizable fraction of total energy (5–10%) would still be required for processing. Moreover, the analysis in Figure 7.3 and Table 7.2 has completely ignored what might be termed the "capital energy investment" required for construction of pipelines, power plants, and the like. One estimate[4] indicates that when amortized over its 30-year life span the annual energy investment required for construction of a nuclear power plant and ancillary facilities is comparable to the approximately 10^{15} J required for enrichment of ^{235}U by gaseous diffusion. Since the latter accounts for 98% of the external energy inputs in Figure 7.3 this would effectively halve the energy ratio to a value of 10. Furthermore, it implies that there is an upper limit on the rate of expansion of nuclear power (or any other energy resource) because the capital investment of free energy in the materials necessary for a new plant's construction must come from the end-use output of existing ones. Since that output is limited, so is the number of new plants which can be constructed during a given period.

7.4 CONSERVATION OF FREE ENERGY

The rush to "solve" our energy problems by developing new sources of supply must be tempered by considerations of net energy. In some cases (shale oil, perhaps) energy ratios may be less than 1.0 so that no net energy will be obtained. In every case far less than 100% of the energy ore can be made to do the work we want. Perhaps the most successful strategy will be to attack the tail instead of the head of an energy system like the one in Figure 7.3. Reduction of end-use consumption of electricity by 1 J can save from 4 to 10 J of fossil or nuclear fuel resources based on the overall efficiencies already tabulated. While efficiencies are higher in some other sectors, *free energy conservation* still has an impact many times that of development of new supplies. The thermodynamic basis for such a pro-

---

[4] M. Rieber, Center for Advanced Computation, University of Illinois, Urbana (private communication).

TABLE 7.3 Energy and Price Data for Various Modes of Transportation[a]

Mode	Energy		Price	
	kJ/ tonne-km	kJ/ passenger-km	cents/ tonne-km	cents/ passenger-km
Bicycle	—	131	—	~0.0
Walking	—	197	—	0
Pipeline	325	—	0.19	—
Bus (intercity)	—	1050	—	2.2
Railroad	484	1900	0.96	2.5
Waterway	492	—	0.21	—
Truck	2 020	—	5.1	—
Automobile (intercity)	—	2230	—	2.5
Bus (urban)	—	2430	—	—
Mass transit (urban)	—	2490	—	5.2
Automobile (urban)	—	5310	—	6.0
Airplane	30 400	5510	15.0	3.7

[a] Based on: E. Hirst and J. C. Moyers, Efficiency of energy use in the U. S. *Science* **179**, 1299 (1973).

gram has already been developed in Section 4.5, equation 4.17, and may be summarized as follows: If we wish to achieve the same end result but consume less free energy, the *efficiency* of conversion must be improved.

A review of Figure 7.1 reveals that the gross United States energy consumption in 1971 could be divided into four roughly equal categories: electric power, transportation, industrial, and household and commercial. The first of these has already been discussed extensively, and there is room for improvement in the others as well. The weighted average efficiency of all modes of transportation is estimated to have been 15% in 1971, and current trends are away from the more efficient modes shown at the top of Table 7.3. This is at least partly because prices for more rapid and convenient air and automobile travel do not increase in proportion to their energy consumption.

Savings of free energy in the transportation sector could be effected by reduction in traffic levels, increasing the efficiency of individual modes or shifts to those of higher efficiency. The first two of these have been implemented to some extent as a result of increased fuel prices, but the third will require significant changes in attitudes and life-styles. Because of the large investment, economic and thermodynamic, in highways and dispersed living patterns, shifts to less energy intensive transportation modes will probably be slow.

In the household and commercial sector, the most important energy-consuming function is space heating. A new gas or oil furnace can deliver 75% or more of the heat content of its fuel to the interior of a building, but the average efficiency of units in use today is estimated to be between 35 and

60%, primarily because of soot on heat exchange surfaces and the fact that furnaces have not been designed to be easy to clean. Electric resistance heating requires 50–100% more primary fuel than gas or oil because of the low efficiencies of power generating plants. Better insulation seems capable of providing more fuel savings than any other change in home heating practice. It can be shown, for example, that in a climate like that of New York upgrading insulation to the economic optimum could save the average homeowner from $32 to $155 per year and reduce energy consumption by nearly 50%.

In the longer run the use of solar energy for space and water heating (and perhaps air conditioning) as well as redesign of homes to take advantage of solar radiation in the winter and protect from it in summer could reduce the need for external fuels. Building design also should minimize the use of energy intensive materials such as aluminum, glass, concrete, and steel. (It should be noted that the glass and copper used in flat plate collectors require considerable energy for manufacture, reducing somewhat the net energy available from the sun.) Air conditioners currently available range over nearly a factor of three in efficiency and are especially important because they create the peak summer loads which lead to brownouts and blackouts. A 70% increase in average efficiency appears to be readily achievable and could save 17 800 MW.

A number of other measures could be taken to reduce household energy consumption. Outdoor gas lights and appliance pilot lights use surprisingly large amounts of fuel and could be eliminated or replaced by electronic igniters. Frost-free refrigerators and freezers could be replaced by manually defrosted units which use two-thirds as much energy, and supermarket freezer chests could be covered. Incandescent lamps could be replaced by fluorescents which are four times as efficient. Finally, mobile homes, which account for one out of four new dwellings in the United States, could be made to adhere to stricter standards with regard to insulation. As currently constructed they are extremely high users of energy.

The effectiveness with which energy is used in industry is quite variable and hard data on efficiencies are often difficult to find. Some industries (electric power and chemical processing) have always been concerned about efficient energy use, but low prices for energy have permitted others (machine manufacturing and metal forming) to give energy conservation low priority. Shortages of natural gas (which has been the fastest growing industrial energy source, accounting for nearly half of the total) and restrictions on permissible levels of air pollution will increase energy costs to industry in the near future, and this will certainly cause adoption of numerous energy-saving techniques. There is not space here for analysis of specific industrial processes, but their energy intensiveness will receive considerable attention in Part IV. For now it suffices to say that energy savings of about 30% are estimated to be possible in the industrial sector.

A recent study[5] suggests that the most significant steps toward energy conservation which have a reasonable probability of being carried out in the near future are as follows:

Improve insulation in homes
Adopt more efficient air conditioning
Shift intercity freight traffic from highway to rail
Shift intercity passenger traffic from air to ground
Shift urban passenger traffic from automobile to mass transit and consolidate urban freight movement
Introduce more efficient industrial processes and equipment

It is estimated that by 1980 these measures could save about two-thirds of projected imports of petroleum to the United States, a result whose economic, political, and environmental advantages would be enormous.

7.5 THE ENERGY BALANCE OF THE EARTH

In Section 4.7 it was emphasized that the high degree of order on the earth's surface is made possible by a flow of energy from the sun to earth and thence to outer space. Radiation reaching the earth is distributed over approximately those wavelengths expected for a blackbody at 6000 K (Figure 6.5), although some uv and ir wavelengths are absorbed by the atmosphere. The average temperature of the earth's surface is between 255 and 287 K, and it too radiates as a blackbody, with maximum intensity around 10 μm in the ir (Figure 7.4).

The Greenhouse Effect

Compounds in the atmosphere can absorb much of the earth's infrared radiation as shown in the figure. Since they reradiate approximately half of it back toward the surface this has a significant effect on world climate. The most important of these compounds is water. It is plentiful and absorbs wavelengths below 8 μm and above 20 μm, leaving a "window" at just the point of maximum earth flux. Carbon dioxide, which has a strong absorption band between 13 and 18 μm, partially blocks this window. Taken together these two gases serve much the same function as the glass cover on a flat plate solar collector (Section 6.4). Their interception of infrared radiation is referred to as the *greenhouse effect,* although this is partially a misnomer since the glass on a greenhouse prevents convection and wind cooling as well as reradiation.

---

[5] Executive Office of the President, Office of Emergency Preparedness, "The Potential for Energy Conservation," US Gov. Printing Office, Stock No. 4102-00009, Washington, D. C., 1972.

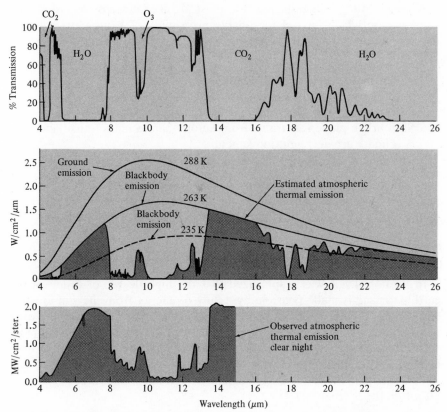

Figure 7.4 Absorption of earth's blackbody radiation by water, carbon dioxide, and ozone. Source: D. M. Gates, Radiant energy, its receipt and disposal. *Meteorol. Monogr.* **6,** No. 28 (Agricultural Meteorology), 16 (1965).

Since human activities (primarily the combustion of fossil fuels) have increased natural levels of carbon dioxide, some climatologists have predicted that the greenhouse effect should have increased average annual temperatures somewhat. From the beginning of the Industrial Revolution to about 1950 this appeared to be true (Figure 7.5), but during the past 20 years temperatures have fallen drastically—almost back to the level of 1900. Some have attributed this drop to increased reflection of incoming energy as a result of particulate matter air pollution, but the mathematical models developed to predict effects of CO_2 and particles are very sensitive to the size and altitude of the particles. Depending on the choice of these parameters, widely different conclusions may be drawn. There is general agreement, however, that variations in average surface temperature as small as 2–3 K could trigger drastic changes in the environment (on the scale of an ice age). Such a change corresponds to a 2% variation in the energy reaching or being radiated from the earth's surface.

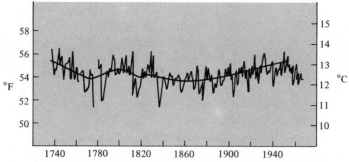

Figure 7.5   Annual average temperatures—Philadelphia. Source: H. E. Landsberg, Man-made climatic changes. *Science* **170**, 1265 (1970). Copyright 1970 by the American Association for the Advancement of Science.

Heat Balance

Regardless of the outcome of the greenhouse effect arguments, there is another long-term problem involved in the energy balance of the earth. The second law of thermodynamics implies that high quality energy provided to the surface of the earth eventually will be converted into heat. Thus, an energy source such as fossil fuel or nuclear power will add to the input of solar energy. To maintain the average temperature of the earth, this additional energy must be radiated away into outer space.

The solar energy reaching the United States in a year is estimated at 5.7×10^9 J/m^2 of surface. Table 7.4 compares past and projected energy consumption in the United States with the total amount of solar energy which must be reradiated. For instance, in 1950 our energy consumption added about 4.6×10^6 J/m^2 to the 5.7×10^9 J/m^2 provided by the sun; this level of energy consumption was about 0.08% of the available solar radiation. However, by the year 2050 the energy consumption in the United States might reach 0.62% of the total solar energy input. This is getting dangerously close to the 2% figure at which global changes in the climate

TABLE 7.4   Comparison of United States Energy Consumption with Solar Input

Year	Annual energy consumption (J/m²)		% Solar	
1900	1.3×10^6		0.02	
1950	4.6×10^6		0.08	
2000	$21 \times 10^{6\,a}$	$23 \times 10^{6\,b}$	0.37^a	0.40^b
2050	$30 \times 10^{6\,a}$	$35.6 \times 10^{6\,b}$	0.53^a	0.62^b

[a] Assumes per capita daily consumption stable after 2000, population stable at 400×10^6 in 2100.

[b] Assumes daily consumption stable after 2000, population growing according to current projections.

might be expected to occur. On a worldwide basis, assuming a per capita energy consumption approximately twice the current United States average and reradiation of heat from land areas of the globe only, this constraint would limit human population to about 20×10^9.[6]

In the densely populated, highly industrialized northeastern United States energy consumption per unit area is far above the national average. One estimate indicates that by the year 2000 this region will be producing heat at a rate of 15% of solar input in summer and 50% in winter. Large urban areas thus constitute *heat islands,* in which local effects on weather and climate have already been detected. Both energy consumption and population are rapidly approaching the point where thermal effects will be of concern on a regional and perhaps even a global scale.

It should be pointed out that utilization of solar, tidal, or renewable geothermal energy does not add to the global thermal pollution just described. In some cases (for example, conversion of solar energy into chemical free energy of a refined substance such as aluminum) the heat load on the earth might even be decreased. The best approach to the energy crisis would appear to be a policy of energy conservation and attention to efficiency of conversion, combined with the development of new, nonpolluting energy resources (such as solar) which are ideally matched with particular end uses (such as space or water heating). This can give rise to a diversified energy base which will be much less vulnerable to a large-scale interruption than the current system of monolithic power plants. The era of unrestrained growth in energy consumption and frantic exploitation of new energy resources to meet a demand which is encouraged by minimal fuel prices has come to an end. Inefficient, unnecessary energy consumption is responsible for too much pollution, too much depletion of resources, and too many social problems to be permitted to continue.

EXERCISES

(The first group of exercises is based primarily on material found in this text.)

1. Using the data in Figure 5.3, verify the overall efficiency and energy ratio for coal-fired electric power in Table 7.2. What changes would occur in these figures if the coal were gasified prior to combustion?

2. Using data from Table 4.1, calculate the change in net energy (assuming a constant primary supply available in the necessary mix of fluid fuel and electricity) should the United States instantaneously switch from gasoline-powered automobiles to electric cars based on lead-acid batteries.

---

[6] A. M. Weinberg and R. P. Hammond, *Amer. Sci.* **58**, 412–418 (1970).

3. Repeat Exercise 2 assuming that cars are to be powered by H_2–O_2 fuel cells and H_2 can be generated at 75% efficiency by power plants.
4. It was estimated in Section 7.4 that 17 800 MW could be conserved by technologically feasible improvements in air conditioner efficiencies. Assuming an average operating time of 886 hours/year, translate this saving into

  a. Tons of coal needed to generate electric power
  b. Acres of strip-mined land (Table 5.4)
  c. Human deaths
  d. Nonfatal injuries

5. Mass transit currently accounts for 3% of intracity travel while automobiles account for 97%. Using the data in Table 7.3 and the total travel of 7.1×10^{11} passenger miles, calculate the annual energy savings if 50% of travel were via mass transit. Could such savings be realized instantaneously? Why not?
6. The average occupancy of urban automobiles is 28% of capacity while for mass transit the figure is 20%. How much energy could be saved if both figures were increased to 60%?
7. Some automobiles obtain more than three times the fuel economy of others. Assuming equal occupancy, what fraction of total United States energy consumption could be saved by switching 100% to such cars for urban trips? How would this compare with a complete switch to mass transit?

(Subsequent exercises may require more extensive thought or research.)

8. Apply the net energy concept to each of the following energy sources:

  a. Shale oil
  b. Solar space heating
  c. Tidal power
  d. Geothermal power
  e. Solar electric power using silicon photocells

  In each case first try to analyze the components of the system which require energy inputs or lose part of the resource. Then try to find data on those inputs and losses. (Some of the latter may be available in Part IV of this book.)
9. Using references given in Section 7.1, look up at least three schemes for thermal cracking of water to give H_2. Evaluate each with respect to its feasibility for use with

  a. Nuclear reactors
  b. Solar collectors

10. Look up as many of the properties of TiH_2 as you can find. How would these properties affect its use as a storage medium for H_2? Using the density of TiH_2 calculate the mass and volume of a storage cell which could contain enough H_2 to drive an automobile 100 miles. (Assume that the car would have gotten 15 miles/gallon of gasoline, 1 gallon $= 1.46 \times 10^8$ J, and H_2 can be burned at 50% efficiency in an Otto cycle engine.)

Suggested Readings

General Energy Concepts

1. AAAS, *Science* **184**, no. 4134, April 19 (1974). This entire issue is devoted to excellent articles on the energy crisis.
2. A. W. Adamson, "A Textbook of Physical Chemistry." Academic Press, New York, 1973. This or another physical chemistry textbook may be consulted for a more advanced treatment of chemical aspects of thermodynamics.
3. W. G. Davies, "Introduction to Chemical Thermodynamics." Saunders, Philadelphia, Pennsylvania, 1972. A noncalculus approach to chemical equilibrium from a molecular point of view. The statistical aspect is emphasized.
4. D. Hafemeister, Science and society test for scientists: The energy crisis. *Amer. J. Phys.* **42**(8), 625–641 (1974). Some of the exercises at the end of chapters are derived from this excellent set of "back-of-the-envelope" calculations.
5. A. Hammond, W. Metz, and T. Maugh, II, "Energy and the Future." Amer. Ass. Advan. Sci., Washington, D. C., 1973. This book is based on a series of articles which appeared in *Science*. All aspects, current and future, of energy generation are included.
6. R. S. Lewis and B. I. Spinrad, eds., "The Energy Crisis." Education Foundation for Nuclear Science, Chicago, Illinois, 1972. This Science and Public Affairs Book of the Bulletin of Atomic Scientists contains 30 articles on the energy crisis, radiation hazards, economic aspects, technical alternatives, international aspects, and protests against energy-related projects.
7. G. T. Miller, Jr., "Energy and Environment: Four Energy Crises." Wadsworth, Belmont, California, 1975; "Energetics, Kinetics, and Life." Wadsworth, Belmont, California, 1971. Both Miller's earlier, higher level book and his most recent effort apply scientific and moral principles to energy problems.
8. H. J. Morowitz, "Energy Flow in Biology." Academic Press, New York, 1968. The primary concern of this book is resolution of the seeming conflict between the second law of thermodynamics and biological evolution. The nonequilibrium steady state is emphasized.
9. National Academy of Sciences, "Resources and Man." Committee on Resources and Man, Freeman, San Francisco, California, 1969. Chapter 8 on energy resources is especially useful.
10. *Scientific American* **224**(3) (1971). This entire issue of *Scientific American* (11 articles) was devoted to energy.
11. C. E. Steinhart and J. S. Steinhart, "Energy," Duxbury Press, North Scituate, Massachusetts, 1974. An excellent overview of energy problems.
12. H. S. Stoker, S. L. Seager, and R. L. Capener, "Energy." Scott, Foresman & Co., Glenview, Illinois, 1975. An overview of energy from supplies through current and developing sources – written by chemists.

13. P. K. Theobald, S. P. Schweinfurth, and D. C. Duncan, Energy resources of the United States. *U. S., Geol. Sur., Cir.* **650** (1972). An up-to-date 27-page summary of energy resources in the United States.

14. R. Wilson and W. J. Jones, "Energy, Ecology and the Environment." Academic Press, New York, 1974. Two physicists' approach which contains many interesting problems and much useful data.

Fossil Fuels

15. Argonne National Laboratory, "A Study of Social Costs for Alternative Means of Electrical Power Generation for 1980 and 1990," Summary Report, February 1973 draft. ANL, Argonne, Illinois, 1973. A careful study of the internal and external costs of electric power generation during the next two decades.

16. D. F. Boesch, C. H. Hershner, and J. H. Milgram, "Oil Spills and the Marine Environment." Ballinger, Cambridge, Massachusetts, 1974. This reports a portion of the work of the Ford Foundation Energy Policy Project.

17. Council on Environmental Quality, "Energy and the Environment: Electric Power," Stock No. 4111-00019. US Govt. Printing Office, Washington, D. C., 1973. This study applies the net energy concept to electric power plants.

18. R. F. Goldstein and A. L. Waddams, "The Petroleum Chemical Industry." Spon, London, 1967. Chapters 1 and 2 give general material on sources and structures of petroleum hydrocarbons as well as refining processes.

19. J. P. Henry, Jr. and B. M. Louks, An economic study of pipeline gas from coal. *Chem. Technol.* **2**(4), 238 (1971). Emphasizes that the heat evolved in the exothermic steps of coal gasification must be available to the endothermic steps for maximum efficiency.

20. B. Mason, "Principles of Geochemistry," 3rd ed. Wiley, New York, 1966. Chapter 9, "The Biosphere," discusses the origin of coal and petroleum and the geochemical carbon cycle.

21. D. Murchison and T. S. Westoll, eds., "Coal and Coal-Bearing Strata." Amer. Elsevier, New York, 1968. A compendium of papers contributed to the Thirteenth Inter-University Geological Congress, 1965. Discusses coal formation and origin of sulfur in coal.

22. A. M. Squires, Clean power from dirty fuels. *Sci. Amer.* **227**(4), 26 (1972). Gas turbine topping cycles and coal gasification are discussed.

23. W. Stumm and J. J. Morgan, "Aquatic Chemistry." Wiley (Interscience), New York, 1970. Section 10.3 discusses the kinetics of oxidation of iron(II) by O_2 and application to acid mine drainage.

24. D. W. VanKrevelen, "Coal." Elsevier, Amsterdam, 1961. The subtitle is "Typology — Chemistry — Physics — Constitution."

Nuclear and Solar Energy

25. R. C. Axtmann, Environmental impact of a geothermal power plant. *Science* **187**, 795–803 (1975). Some negative features of a supposedly "clean" energy source.

26. J. Barnea, Geothermal power. *Sci. Amer.* **226**(1), 70 (1972). A more optimistic view.

27. B. Coppi and J. Rem, The Tokamak approach in fusion research. *Sci. Amer.* **227**(1), 65 (1972). Thorough treatment on the most nearly successful plasma confinement scheme.

28. F. L. Culler, Jr. and W. O. Harms, Energy from breeder reactors. *Phys. Today,* May, p. 28 (1972). A thorough treatment of breeder reactors by two scientists from Oak Ridge National Laboratory.

29. F. Daniels, "Direct Use of the Sun's Energy." Yale Univ. Press, New Haven, Connect-

icut, 1964. To quote the *Whole Earth Catalog:* "The best book on solar energy that I know of." This applies whether your interest is in a commune or a chemistry lab.

30. M. Eisenbud, "Environmental Radioactivity," 2nd ed. Academic Press, New York, 1973. A classic treatise providing a thorough treatment of all aspects of nuclear energy and radioactivity with the exception of fusion power.

31. R. Gillette, Nuclear reactor safety. *Science* **172,** 918 (1971); **176,** 492 (1972); **177,** 771, 867, 970, and 1080 (1972); **178,** 482 (1972); **179,** 360 (1973). This series of seven news articles spanning 18 months points up apparent flaws and omissions in Atomic Energy Commission studies of power reactor accidents.

32. H. C. Hottel and J. B. Howard, "New Energy Technology — Some Facts and Assessments." MIT Press, Cambridge, Massachusetts, 1971. A critical look at energy technology. The section on solar energy should be required reading for those whose proposals tend to be overoptimistic.

33. W. H. Jordan, Nuclear energy: Risks *versus* benefits. *Phys. Today* May, p. 32 (1970). Compares the risks associated with nuclear energy to the benefits to be obtained from the elimination of pollution from other energy sources.

34. NSF/NASA Solar Energy Panel, "An Assessment of Solar Energy as a National Energy Resource." University of Maryland, College Park, 1972. A report on the status of solar energy conversion and proposed actions to develop it to commercial feasibility. Perhaps somewhat overoptimistic.

35. D. J. Rose, Controlled nuclear fusion: Status and outlook. *Science* **172,** 797 (1971). Comprehensive treatment of the pros and cons of various methods for magnetic containment of plasma. Also considers environmental aspects of fusion reactors.

36. E. C. Tsivoglou, Nuclear power: The social conflict. *Environ. Sci. Technol.* **5**(5), 404 (1971). A general discussion of the hazards associated with nuclear power plants and their effects on the general public.

37. U. S. Atomic Energy Commission, "Environmental Survey of the Uranium Fuel Cycle," WASH-1248. USAEC, Washington, D. C., 1974.

Energy Storage and Conservation

38. R. U. Ayres and R. P. McKenna, "Alternatives to the Internal Combustion Engine." Johns Hopkins Univ. Press, Baltimore, Maryland, 1972. Evaluates energy storage and conversion devices with regard to their suitability for use in automotive vehicles.

39. C. A. Berg, Energy conservation through effective utilization. *Science* **181,** 128 (1973).

40. T. L. Brown, "Energy and the Environment." Merrill, Columbus, Ohio, 1971. Devoted to the greenhouse effect, ultimate thermal pollution, and energy conservation.

41. D. P. Gregory, The hydrogen economy. *Sci. Amer.* **228**(1), 13 (1973). A strong argument for conversion of all primary energy sources to H_2.

42. A. L. Hammond, Conservation of energy: The potential for more efficient use. *Science* **178,** 1079 (1972).

43. A. B. Hart and G. J. Womack, "Fuel Cells." Chapman & Hall, London, 1967. Basic principles.

44. E. Hirst and J. C. Moyers, Efficiency of energy use in the United States. *Science* **179,** 1299 (1973); see also E. Hirst, "Energy Consumption for Transportation," ORNL-NSF-EP-15 (1972); "Energy Intensiveness of Passenger and Freight Transport Modes, 1950–1970," ORNL-NSF-EP-44 (1973). Oak Ridge Nat. Lab., Oak Ridge, Tennessee. Opportunities for energy conservation in transportation, space heating, and air conditioning are discussed.

45. L. W. Jones, Liquid hydrogen as a fuel for the future. *Science* **174,** 367 (1971); W. E. Winsche, K. C. Hoffman, and F. J. Salzano, Hydrogen: Its future role in the nation's energy economy. *ibid.* **180,** 1325 (1973). Two articles from *Science* on the proposed

hydrogen economy. The latter estimates costs for various means of incorporating H_2 into the United States fuel supply.

46. H. E. Landsberg, Man-made climatic changes. *Science* **170**, 1265 (1970). A thorough treatment of man-made effects on local, regional, and global climate.
47. T. H. Maugh, II, Fuel cells: Dispersed generation of electricity. *Science* **178**, 1273 (1972).
48. Office of Emergency Preparedness, "The Potential for Energy Conservation," OEP, Washington, D. C., 1972. A detailed treatment of practical means that can be implemented over the short, medium, and long range to decrease energy consumption in all sectors.
49. D. W. Rabenhorst, Super flywheel configurations form the heart of mechanical-powered drives. *Prod. Eng.* April 12 (1971). Proposed energy storage in "super flywheels" constructed from high-strength filaments.
50. T. A. Robertson, Systems of energy and the energy of systems. *Sierra Club Bull.* **60**(3), 21–23 (1975). The principle of net energy is developed. Concentrates on the work of Howard Odum to the exclusion of others in the field such as E. Hirst, R. S. Berry, and B. Hannon.
51. Stanford Research Institute, "Patterns of Energy Consumption in the United States." SRI, Menlo Park, California, 1971, prepared for the Office of Science and Technology, Washington, D. C., 1972.
52. *Technology Review,* **76**(4) (1974). An entire issue devoted to energy conservation and energy systems studies.

For a more extensive bibliography on environment and energy topics see J. W. Moore and E. A. Moore, *J. Chem. Educ.* **52**(5), 288–295 (1975).

III

Air

... [*London's*] Inhabitants *breathe nothing but an impure and thick Mist, accompanied with a fuliginous and filthy vapor,* ... *corrupting the* Lungs *and disordering the entire habit of their Bodies;* ....

John Evelyn, *Fumifugium*, 1661

AS THE PRECEDING QUOTATION IMPLIES THE history of air pollution is lengthy, perhaps going back to the first caveman (or woman) who brought fire into human living quarters. The invention of the chimney followed much later, allowing its discoverer the benefits of fire while foisting environmental costs on the community at large. As communities grew so did these external costs, and the first documentation of the problem may have been Roman complaints of the foulness of their city's air. Coal burning was prohibited in London as early as 1273, and one man was beheaded for ignoring Edward I's proclamation enjoining the combustion of "sea-coal" in 1306. By 1578 Elizabeth I still found "hersealfe greately greved and anoyed with the taste and smoke of the sea-cooles," and she prohibited coal burning while Parliament was in session. John Evelyn, seventeenth century diarist and environmentalist, was instrumental in reversing the progressive deforestation of England but unsuccessful in reducing air pollution from the fuel which replaced the uncut wood.

During the mid-nineteenth century Parliament again attempted to cleanse the air, but alternatives to smoke were not obvious and enforcement was lax. A fog in 1873 killed a number of prize cattle (and is now realized to have hastened several hundred human deaths as well), and near the turn of the twentieth century smoke abatement societies became somewhat popular. One of these was headed by Dr. H. A. Des Voeux, who amalgamated the words smoke and fog into smog. In the United States the first air pollution ordinance, enacted in St. Louis in 1876 merely required factory chimneys to be at least 20 feet higher than adjacent buildings. "Dense smoke" was declared a nuisance, punishable by fines of five to fifty dollars, in Chicago in 1881, and 12 years later St. Louis declared its right to set standards defining these two words quantitatively.

In the absence of strictly enforced controls smog "episodes," during which stagnant air trapped pollutants over a city or industrial area, became more prevalent, occurring in London, Glasgow, the Meuse Valley in Belgium, and Donora, Pennsylvania. The most serious of these occurred in London during the first week of December 1952, causing an estimated 4000 deaths in excess of the normal rate for that month. Despite the passage of clean air legislation in the United States in 1955 and in England in 1956 such killer smogs have continued, occurring in London during 1956, 1957, and 1963 and in New York in 1953, 1963, and 1966. In August 1969 an episode of high air pollution potential caused alerts across much of the eastern United States, from the Great Lakes and Gulf Coast to the Atlantic Seaboard.[1] On the first "Earth Day" in 1970 atmospheric degradation was perhaps the environmental problem of greatest concern, although more recent crises relating to energy and drinking water quality have displaced it by now.

In the next three chapters we shall examine the structure and properties of the atmosphere to determine whether and how pollutants can be removed from it as well as the conditions which lead to air pollution episodes. Each of the five substances emitted from United States sources in the largest quantity will be examined from the standpoint of its cycling through the atmosphere, its detection, health and other effects, and available control technology. Several other air pollutants which may become more infamous as the "big five" are brought under control will be treated as well.

---

[1] V. Brodine, *Environment* **13**(1), 3–27 (1971).

8

The Atmosphere

This most excellent canopy, the air, look you, this brave o'er-hanging firmament, this majestic roof fretted with golden fire, why, it appears no other thing to me than a foul and pestilent congregation of vapours.

Shakespeare

8.1 STRUCTURE AND PROPERTIES OF THE ATMOSPHERE

Some 99.999% of the 5.1×10^{18} kg mass of the atmosphere is within 80 km of the earth's surface, accounting for just over 1% of the diameter but only $10^{-8}\%$ of the mass of the planet. There is considerable variation of temperature with height, and this forms the basis for one classification of the atmosphere into layers as shown in Figure 8.1. Another scheme terms the regions above the stratosphere the chemosphere and ionosphere because they contain molecular and atomic ionic species, respectively. These species are formed by very energetic, short wavelength solar uv radiation. Man's only penetration of these upper levels has been by balloon or spacecraft, and so far they appear not to have been modified in any way which can affect the biosphere.

The temperature profile of the atmosphere is controlled mainly by the degree of absorption of solar radiation in the various regions. For example, the increasing temperature of the thermosphere is the result of kinetic energy imparted to oxygen atoms by the reaction:

$$O_2 \xrightarrow[135-176 \text{ nm}]{h\nu} O + O \tag{8.1}$$

183

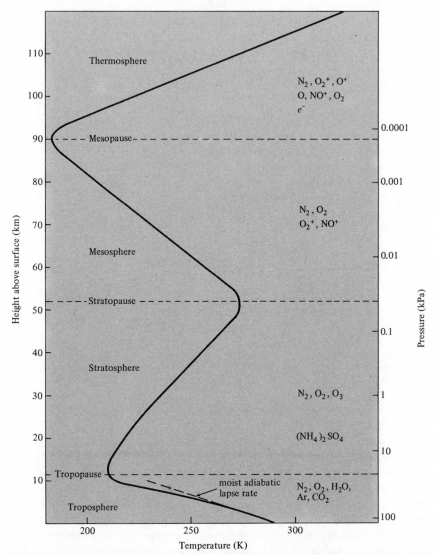

Figure 8.1 Temperature distribution and important chemical species in earth's atmosphere.

Ozone contributes to the warming of the stratosphere as described at the beginning of Section 3.7 (equations 3.6b and 3.6c). Another important source of heat is absorption of visible radiation by the earth's surface and by clouds in the lower troposphere. This leads to the profile of decreasing temperature with height observed up to an altitude of 10 km.

The temperature profile is of prime importance in determining the stability and degree of mixing of the atmosphere as a whole as well as those small parts of it in contact with cities or other high pollution areas. Everyone is familiar with the generalization that, because its density is smaller, warm air rises above cold at the same pressure. In the atmosphere, however, pressure decreases significantly with altitude, allowing rising warm air to expand. Because air is a poor conductor of heat its expansion is usually adiabatic (Section 4.4) and its temperature must decrease by an amount equivalent to the work it does by expanding. If the rising air cools sufficiently it will reach the same temperature and density as its surroundings and its rise will therefore be halted. If, however, the temperature profile of the surroundings is such that at higher altitudes they are cooler than the rising body of air, the latter can continue upward.

The *lapse rate,* defined as the change in temperature with increasing altitude, is easily calculated (Exercise 9) as $-gM/C_p$ for an ideal gas, where $g = 9.8$ m/s^2 is the gravitational acceleration, M the molecular weight of the gas, and C_p its molar heat capacity at constant pressure. Averaging over the components of dry air gives a value of -9.8 K/km. In the troposphere much water vapor is present which can condense and release heat as rising air expands. This reduces the cooling effect and consequently the *moist adiabatic lapse rate* is on the order of -6 K/km—smaller than the dry rate.

To determine the stability and hence the degree of vertical mixing of a portion of the atmosphere the actual lapse rate must be compared with the moist adiabatic value. If the former is more negative, cooling of a rising parcel of air will not be rapid enough to increase its density above that of the surroundings and it will continue to rise. On the other hand, if the actual lapse rate is more positive than the moist adiabatic value, the atmosphere is in a stable condition and vertical motion cannot occur.

The troposphere is (on the average) an example of the former situation. As shown in Figure 8.1 the slope of its temperature–altitude curve is slightly more negative than the moist adiabatic lapse rate. Although vertical mixing may be absent at specific times and places, the general behavior of the troposphere is such that pollutants will be dispersed and diluted. On the other hand, the stratosphere exhibits stable atmospheric behavior. Because its temperature–altitude profile has a positive slope vertical mixing is very slow. Material injected into the stratosphere tends to remain at a given altitude for years, even though it may be carried many times around the globe by jet-stream winds. A good example of this is a thin band of $(NH_4)_2SO_4$ aerosol which has been discovered at an altitude of 20 km.

8.2 TEMPERATURE INVERSION AND AIR POLLUTION EPISODES

A temperature–altitude profile similar to that of the troposphere and stratosphere combined can occasionally occur on a smaller scale within the troposphere itself. When the lapse rate is positive and temperature increases with altitude, even for only a short distance, a *temperature inversion* is said to exist. Stable atmospheric conditions prevent vertical motion within the inversion layer; thus, gases and suspended matter become trapped and cannot rise past the altitude at which temperature begins to increase. If an inversion occurs above a city or industrial region, pollutants which would normally disperse throughout the troposphere cannot do so. This is especially important in a valley or among skyscrapers in a city where horizontal transport of hazardous substances is also cut off. All of the air pollution episodes discussed in the introduction to Part III were associated with temperature inversions which lasted from several days to a week.

Inversion layers may develop as a result of several circumstances. The meeting of two large bodies of air at different temperatures can create a *frontal inversion* since the colder air generally slides under the warmer at the front where they meet. Occasionally such a frontal system will remain stationary long enough for pollution to collect, although moving fronts break up inversions more often than they create them. On a smaller scale cold air often flows down the sides of valleys at night, leading to lower temperatures and a localized inversion at the bottom. In a small valley early in the morning one often observes a layer of fog a few meters thick which dissipates as the sun warms the ground and low-lying air. In many of the industrialized areas of the Appalachian Mountains this effect combines with surface cooling resulting from loss of infrared radiation on clear nights to yield a significant depth of cool air trapped by an inversion layer above. Not only do pollutants collect during the night but also when solar heating breaks up the inversion layer the following day, mixing of the first few hundred meters of the atmosphere may bring even higher concentrations to breathing level. This latter process is known as *fumigation*.

Should such an inversion layer be deep enough and the accompanying fog dense enough to prevent solar heating of the ground, it might persist for several days until broken up by a change in weather. This atmospheric condition was the cause of the Meuse Valley and Donora, Pennsylvania air pollution disasters already mentioned. An even larger geographical area may be affected by anticyclones—regions of high pressure which often are characterized by light winds and upper layers which descend to the ground, contracting and becoming warmer. The upper levels of such a weather system generally remain warm enough to maintain stable atmospheric conditions, and in the absence of wind horizontal dispersion is impossible. When an anticyclone slows or halts in its easterly movement across the United States the *subsidence inversion* accompanying it may raise pollution levels in half a dozen or more states, as in the August 1969 episode mentioned earlier.

There is little that can be done to prevent a temperature inversion from forming or to destroy one that already exists. To remove an inversion layer

TABLE 8.1 United States Alert, Warning, and Emergency Level Air Pollution Criteria[a]

| | Level (μg/m^3) | | | Averaging |
Pollutant	Alert	Warning	Emergency	period (hours)
SO$_2$	800	1 600	2 100	24
Particulates	375	625	875	24
Combined SO$_2$ and particulate[b]	65 000	261 000	393 000	24
CO	17 000	34 000	46 000	8
Oxidant	200	800	1 200	1
NO$_2$	1 130	2 260	3 000	1

[a] *Federal Register* **36** (206), 15593 (1971).
[b] Product of SO$_2$ level and particulates level. Units are (μg/m^3)^2.

from even a limited area such as the Los Angeles Basin by heating the air beneath it[1] would require energy equivalent to the explosion of a hydrogen bomb or approximately 1000 times the daily output of electricity from Hoover Dam. In a few cases extra tall stacks may enable pollutants to be emitted above the inversion, but, in general, pollution episodes must be dealt with by halting the output of harmful substances. Until controls are instituted which reduce emissions to the point where air pollution alerts become unnecessary, the criteria in Table 8.1 are being used to determine the actions of control officials. At each successive stage more stringent restrictions must be imposed on polluting activities, up to a 12-hour emergency situation where only those industrial operations essential to prevent equipment damage may be permitted to continue.

8.3 ATMOSPHERIC PHOTOCHEMISTRY

Because the atmosphere is nearly transparent its constituent molecules are exposed to a large quantity of visible radiation. At high altitudes, in the mesosphere and thermosphere, uv photons of even higher energy are also prevalent. Therefore, *photochemistry* is extremely important in understanding atmospheric interactions.

The absorption of radiation by atoms and molecules and its ability to induce chemical reactions have been touched upon in Sections 1.1 and 6.6. Once a molecule has been raised to an excited state—whether by absorption of a photon, chemical reaction, or other interaction with its neighbors—there are a variety of pathways by which the excitation energy may be dissipated.[2] Taking excited dioxygen (O$_2$*, where the asterisk in-

---

[1] M. Neiberger, *Science* **126**, 637–645 (1957).
[2] A more thorough discussion is given in A. W. Adamson, "A Textbook of Physical Chemistry," Chapter 18. Academic Press, New York, 1973.

dicates that the molecule is not in its ground electronic state) as an example, a photon might be emitted:

$$O_2^* \longrightarrow O_2 + h\nu \qquad (8.2)$$

returning the molecule to the lowest electronic energy level. This process is called *luminescence*. (Fluorescence and phosphorescence are special cases involving different types of excited states.) Luminescence is responsible for faint atmospheric light emission known as airglow.

Energy may also be transferred from electronic levels to more closely spaced vibrational (and eventually rotational and translational) ones of the same or a different molecule. In such *radiationless deactivation* the net result is an increase in temperature of the region surrounding the absorbing species. Transfer of energy to electronic levels of an adjacent molecule, M:

$$O_2^* + M \longrightarrow O_2 + M^* \qquad (8.3)$$

may also result in radiationless deactivation, but if M^* undergoes chemical reaction the process is known as *photosensitization*. This is extremely important in photosynthesis where energy collected by a variety of sensitizers is funneled to a single reactive site such as *P*-700 (Section 6.6).

The excited molecule itself may undergo *photoionization:*

$$O_2^* \longrightarrow O_2^+ + e^- \qquad (8.4)$$

or *photodissociation:*

$$O_2^* \longrightarrow O + O \qquad (8.5)$$

Both produce reactive species which often induce additional ordinary chemical reactions. We have already seen (Figure 6.8) that electronic excitation can induce redox reactions, and since excited states generally are associated with weakened chemical bonds, displacement and isomerization reactions can also proceed more readily. Thus the visible and especially the uv components of sunlight, which can produce excited electronic states, are capable of inducing or accelerating a tremendous variety of chemical reactions.

The rate of an elementary photochemical process depends on the fraction of absorptions which lead to the desired products (φ, the *quantum yield*) and the number of photons absorbed by the primary photoreactive species. The latter quantity varies with the intensity, I_0, of light, the molar absorptivity (or extinction coefficient), α, and the concentration of the sample. It should be noted that φ and α both change as a function of wavelength—the former depends on the type of state which can be excited by a photon of a given energy and the latter on the absorption spectrum of the substance whose molecules are to be excited. I_0 is dependent on the blackbody radiation curve on the sun (Figure 6.5), the amount of solar radiation of different wavelengths absorbed by the atmosphere at a given altitude, and the amount of scattered radiation incident on a sample. Concentrations of reactive species are also dependent on altitude. Thus any calculation of atmospheric photochemical reaction rate must be integrated over all visible and uv wavelengths and over altitude.

This greatly complicates the use of laboratory data to predict the rates of atmospheric photochemical reactions. Numerous assumptions or approximation (especially with respect to I_0) may have to be made in lieu of missing empirical data when models of atmospheric photochemistry are constructed, and the situation is made even more complex by the fact that one is never absolutely certain that all relevant reactions have been included in the model. Usually data and reaction schemes are fed to a computer which simulates the effects of changing concentrations of existing species or the introduction of new species into the atmosphere.

8.4 POSSIBLE DEPLETION OF STRATOSPHERIC OZONE

An excellent example of such applications of chemical kinetics and photochemistry to the atmosphere is evaluation of possible mechanisms by which the stratospheric ozone layer might be depleted. The role of O_3 in blocking short wavelength uv radiation from the earth's surface and warming the stratosphere was delineated in Section 3.7, and equations given there will not be repeated here. In addition to the multitude of effects on the biosphere to be expected upon removal of a substances as crucial as O_3 for the evolution of life on earth, there is specific statistical evidence from epidemiological surveys which links human skin cancer with increased levels of uv radiation.[3] Clearly, destruction of stratospheric ozone would be an environmental catastrophe of the first order.

A quantitative study of the rates of reactions 3.6 shows[4] that 3.6d does not destroy O_3 rapidly enough to account for its small concentration (< 10 ppm) in the stratosphere. Reasonable estimates are that 3.6d consumes about 18% of the O_3 formed by 3.6b, implying the existence of other mechanisms by which O_3 can be decomposed or 3.6d can be catalyzed. One of these involves reaction of O_3 with hydroxide radicals ($\cdot \ddot{O}-H$) derived from stratospheric water vapor:

$$O + H_2O \longrightarrow 2\ OH \qquad (8.6a)$$

$$OH + O_3 \longrightarrow HOO + O_2 \qquad (8.6b)$$

accounting for about 11% of ozone depletion. A second depends on catalysis of 3.6d by the sequence:

$$NO + O_3 \longrightarrow NO_2 + O_2 \qquad (8.7a)$$

$$\underline{NO_2 + O \longrightarrow NO + O_2} \qquad (8.7b)$$

$$O_3 + O \longrightarrow 2\ O_2 \qquad (8.7\ \text{or}\ 3.6d)$$

[3] A. J. Grobecker, S. C. Coronti, and R. H. Cannon, Jr., "The Effects of Stratospheric Pollution by Aircraft," p. xx. U. S. Dept of Transportation, Washington, D. C., 1974.

[4] H. Johnston, *Science* **173**, 517–522 (1971).

It is estimated that under normal circumstances reaction 8.7 consumes 50–70% as much O_3 as would be necessary to maintain a steady-state concentration. Since NO serves a catalytic function only a small quantity is required. It is apparently derived from tropospheric N_2O which is formed by denitrifying soil bacteria, slowly diffuses to the stratosphere, and reacts with atomic oxygen. The uncertainties in the above percentages for natural ozone destruction allow room for other mechanisms, but no significant ones have been elucidated so far.

Plans for a fleet of supersonic transport (SST) aircraft which would cruise in the stratosphere raised the possibility of interference with the stratospheric ozone balance. Combustion of any hydrocarbon fuel yields water, which would increase the rate of 8.6a, and elevated temperatures (such as would be encountered in jet engines) favor combination of dinitrogen and dioxygen:

$$N_2 + O_2 \longrightarrow 2\,NO \qquad \Delta G^0_{298} = 90.42 - T(0.0247) = 83.06 \text{ kJ} \tag{8.8}$$

The NO so produced might contribute significantly to the catalytic sequence 8.7. In 1971, Johnston[4] considered the collective effect of thirty-one different elementary processes, including the sequences mentioned above, and concluded that nitric oxide represented a much greater threat to the ozone shield than water. His results suggested that if Federal Aviation Administration projections (500 four-engine and 166 two-engine SST's flying an average of 7 hours/day in the stratosphere by 1985) were realized, increased stratospheric NO could reduce the ozone shield in the northern hemisphere by one-half. A more recent study[5] has confirmed these results, although its more reassuring conclusion that the currently operating fleet of French Concorde and Soviet TU-144 aircraft poses no immediate threat has received greater publicity.

Freons

Because of uncertainties about the completeness of Johnston's model there has been considerable research since 1971 on other mechanisms for consumption of O_3 which might contribute to maintenance of a steady-state concentration in the stratosphere. One such mechanism, first reported by Stolarski and Cicerone,[6] involves catalysis of reaction 3.6d by chlorine atoms:

$$Cl + O_3 \longrightarrow ClO + O_2 \tag{8.9a}$$

$$\underline{ClO + O \longrightarrow Cl + O_2} \tag{8.9b}$$

$$O_3 + O \longrightarrow 2\,O_2 \tag{8.9 or 3.6d}$$

---

[5] A. J. Grobecker, S. C. Coronti, and R. H. Cannon, Jr., "The Effects of Stratospheric Pollution by Aircraft," p. xvi. U. S. Dept. of Transportation, Washington, D. C., 1974.

[6] R. S. Stolarski and R. J. Cicerone, *Can. J. Chem.* **52**, 1610 (1974).

(Note that the overall reaction 8.9 is identical to both 8.7 and 3.6d.) It was originally thought that atomic chlorine might arise from photolytic dissociation of HCl, but Molina and Rowland[7] have suggested that freon-11 ($CFCl_3$) and freon-12 (CF_2Cl_2) are much more important sources of stratospheric chlorine. Production of these two compounds was respectively 0.3 and 0.5 Tg in 1972, and their use as aerosol spray propellants and refrigerants has been increasing with a doubling time of 6 years or less for the past 20 years.

Because they are readily liquefied under pressure and are almost completely inert, the chlorofluoromethanes are ideal aerosol propellants; but the latter property, combined with the fact that they originate entirely from man-made processes and have no natural routes of microbial decomposition, appears to make their tropospheric lifetimes (> 10 years for $CFCl_3$ and > 30 years for CF_2Cl_2) long enough that they can diffuse into the lower stratosphere. Only when the freons reach an altitude where there is significant flux of uv radiation do they decompose photolytically, yielding atomic chlorine. In 1974, Molina and Rowland estimated that O_3 had been depleted by 1%. Because of the induction period required for diffusion to the stratosphere, they concluded that even immediate cessation of release of the compounds would not stop depletion of O_3 until 2% was gone. Assuming no growth beyond 1974 annual freon consumption, a steady-state involving about 10–15% ozone reduction was expected to be reached during the 1990's. Other investigators have published similar conclusions.[8]

Of course, none of the above-mentioned laboratory-based studies *proves* that freons affect stratospheric ozone levels—conclusions based on correlations of freon and ozone concentrations actually measured in the stratosphere would be more convincing. Without scientific proof of their harm the U. S. Government concluded that banning freon propellants would be premature, and in late 1974 it formed an Interagency Task Force on Inadvertant Modification of the Stratosphere to study the matter further. However, the lag time of 20–30 years inherent in diffusion of chlorofluoromethanes to the stratosphere requires rapid resolution of the uncertainties which remain. If indeed a real problem exists, continued consumption of freon propellants at the current rate of growth will double its severity in about 5 years. Should sufficient scientific data not soon become available, prudence would seem to dictate a moratorium on the use of freon-propelled aerosols. Most such products are available in other packaging, and the strong *possibility* of irreversible environmental damage ought to weigh heavier than consumer convenience and the profits of a few manufacturers.

---

[7] M. J. Molina and F. S. Rowland, *Nature (London)* **249**, 810–812 (1974).
[8] R. J. Cicerone, R. S. Stolarski, and S. Walters, *Science* **185**, 1165–1167 (1974); S. C. Wofsy, M. B. McElroy, and N. D. Sze, *ibid.*, **187**, 535–537 (1975).

8.5   NATURAL VERSUS POLLUTED AIR

An important aspect of the freon controversy is the apparent absence of natural mechanisms by which these synthetic chemicals may be removed from the troposphere. Man has added a new substance which is not amenable to participation in already existing biogeochemical cycles (Section 4.8). Fortunately, this is not true of most air pollutants—nearly all have natural as well as man-made *sources*. Therefore natural *sinks*—processes which can convert the pollutants into less harmful forms—also are available. In some cases, such as that of carbon monoxide, we have only discovered those sources and sinks within the past few years, and little is known about them. Indeed, because analytical techniques capable of detecting trace atmospheric substituents only became available after man's population, affluence, and technology grew large enough to pollute the atmosphere, it is not a simple matter to determine the composition of *natural air*. (Others have used the terms "clean" or "pure," but these imply the total absence of any substances normally considered to be pollutants, whether of natural origin or not. They also put a negative connotation,

TABLE 8.2   Composition of Dry, Natural Air near Sea Level

Component	Concentration[a]		Total mass (10^6 tonnes)
	Volume %	ppm	
Major components			
Dinitrogen	78.09	780 900	3 850 000 000
Dioxygen	20.94	209 400	1 180 000 000
Minor components			
Argon	0.93	9 300	65 000 000
Carbon dioxide	0.0318	318	2 500 000
Trace components			
Neon	—	18	64 000
Helium	—	5.2	3 700
Methane	—	1.3	3 700
Krypton	—	1	15 000
Dihydrogen	—	0.5	180
Nitrous oxide	—	0.25	1 900
Carbon monoxide	—	0.1	500
Ozone	—	0.02	200
Ammonia	—	0.01	30
Nitrogen dioxide	—	0.001	8
Sulfur dioxide	—	0.0002	2

[a] Data from "Cleaning Our Environment: The Chemical Basis for Action," p. 24. Amer. Chem. Soc., Washington, D. C., 1969.

which may not necessarily be deserved, on any deviation from natural concentrations.) One estimate of the composition of a contemporary atmosphere unaffected by human activities is given in Table 8.2. The data are given for dry air because of variability of water vapor concentration, which can range from 0.1 to 5% with an average of 3.1%.

The alert reader will have noted that different units have been used for expressing the concentrations in Tables 8.1 and 8.2. Parts per million (ppm) is 10^6 times the ratio of volume of a constituent gas to total volume of air. It is assumed that all gases behave ideally in calculating such volumes.[9] Micrograms per cubic meter is derived as its units would imply, with the gas volume adjusted to standard temperature and pressure. In the case of suspended solid particles ppm does not make sense and only $\mu g/m^3$ is used.

Emissions of the five major air pollutants within the United States in 1970 are given in Table 8.3. It is evident that in some cases (sulfur oxides, SO_x, and nitrogen oxides, NO_x) United States emissions in one year greatly exceed the amount of substance present in the worldwide natural environment. Fortunately, there are natural sinks which rapidly remove sulfur and nitrogen oxides, but yearly injection to the atmosphere of more than 15 times the natural abundance of a substance such as SO_2 emphasizes the need for emission controls.

The cost of air pollution in the United States are also large. Estimates of yearly damage range from 6.1×10^9 to 18.5×10^9 with the average at

TABLE 8.3 **Estimates of United States Air Pollution Emissions and Costs, 1970**[a]

Source category	Substance emitted (10^6 tonnes)					
	CO	Particulates	SO_x	HC	NO_x	Total
Transportation	100.6	0.6	0.9	17.7	10.6	130.4
Fuel combustion in stationary sources	0.7	6.2	24.0	0.5	9.1	40.5
Industrial processes	10.3	12.1	5.4	5.0	0.2	33.0
Solid waste disposal	6.5	1.3	0.1	1.8	0.4	10.1
Agricultural burning	12.5	2.2	~0	2.5	0.3	17.5
Miscellaneous	4.1	1.4	0.3	4.1	0.2	10.1
Total emissions	134.7	23.8	30.7	31.6	20.8	241.6
Total costs (10^9 \$)	?[b]	5.8	5.4	1.1[c]		12.3

[a] Data from T. E. Waddell, "The Economic Damages of Air Pollution," pp. 127–131. U. S. Environ. Protect. Ag., Washington, D. C., 1974.
[b] Costs of effects on human health and the natural environment are indeterminate.
[c] Cost of total oxidant level.

[9] In the case of water pollution concentrations ppm is expressed as a ratio of *weights* instead of volumes.

12.3×10^9. It should be emphasized that health and other costs are related to physical and chemical characteristics of a substance as well as to the amount emitted. Thus, although more carbon monoxide is emitted than any other pollutant, considerably greater costs are attributed to the more poisonous sulfur oxides.

Four major factors are of concern to persons involved in the abatement and control of air pollution:

1. What natural and man-made sources and sinks are available for the substance in question? How does the material fit into natural biogeochemical cycles?

2. What means are available for detection and accurate quantitative analysis of the pollutant?

3. What are its effects on human health, animal life, plants, and property? Are there synergistic interactions (Section 9.4) with other substances?

4. What techniques are available for abatement? Do they perhaps create unintended effects such as increases in other forms of environmental degradation?

The next two chapters deal with the five major air pollutants as well as several other less common ones. In each case the discussion will be organized as much as possible in terms of the four factors above.

EXERCISES

(The first group of exercises is based primarily on material found in this text.)

1. Explain clearly what a temperature inversion is and how one might arise. Why are temperature inversions of importance with respect to air pollution episodes?

2. Name at least two ways in which SST flights might affect the stratosphere. In each case write equations for the reactions involved.

3. Nitrogen dioxide is much more reactive chemically and photochemically in the atmosphere than CO_2. Draw Lewis diagrams for both molecules. Give at least two properties of NO_2 which may account for its greater reactivity.

4. Obtain conversion factors relating $\mu g/m^3$ to ppm for each of the four major gaseous pollutants in Table 8.3 (see H. C. Perkins, "Air Pollution," Appendix A. McGraw-Hill, New York, 1974).

5. Suppose a molecule is raised to a higher electronic energy level by absorption of a photon. By what processes may this excitation energy be lost?

6. Why does the temperature of the atmosphere pass through a maximum at the boundary between stratosphere and mesosphere (Figure 8.1)?

(Subsequent problems may require more extensive research or thought.)

7. Given the experimentally determined data tabulated below, determine the maximum altitude that could be reached by a sample of dry air beginning at 0 m and 303 K. (A graphical solution is appropriate, see W. Bach, "Atmospheric Pollution," p. 17. McGraw-Hill, New York, 1972.) Repeat the exercise for a sample of moist air.

Height (m)	T (K)
0	303
500	291
750	288
1000	288
1500	303

8. Using values of C_p obtained from a handbook and the proportions given in Table 8.2, show that the dry adiabatic lapse rate for air is -9.8 K/km.

9. Using the ideal gas law, the hydrostatic equation $\partial P/\partial h = -g\rho$ (where P is pressure, h, height; g, 9.8 m/s²; and ρ is the average density of air) and the law for adiabatic expansion $T/T_0 = (P/P_0)R/Cp$, derive the equation given in the text for the dry adiabatic lapse rate (see L. Hodges, "Environmental Pollution," p. 54. Holt, New York, 1973).

10. It is a well-known fact that the minimum altitude of the ionosphere is higher at night than during the day, allowing longer range radio broadcasts. Why might this be true? (see S. Manahan, "Environmental Chemistry," p. 291. Willard Grant Press, Boston, 1972).

11. Suppose you were the government official in charge of the Interagency Task Force on Inadvertant Modification of the Stratosphere (Section 8.4). How would you set up a program of research to determine scientifically whether SST's, freons, and possibly other hazards to stratospheric ozone pose a real threat? What kinds of chemists and other specialists would you include on the task force? What procedures would you follow? What aspects of the problem would have to be taken into account?

9

Air Pollution—Industry and Energy Related

We used to think that if we knew one, we knew two, because one and one are two. We are finding that we must learn a great deal more about "and".

Arthur S. Eddington

The data in Table 8.3 indicate that fuel combustion in stationary sources (mainly power plants producing electricity and industrial process heat) and industrial processes were the second and third largest sources of air pollution in 1970. Closer examination reveals that the particulate matter and sulfur oxides emitted from such sources account for the majority of air pollution costs, making their control especially important.

9.1 PARTICULATE MATTER

Historically the oldest air pollution problem is that of smoke and soot—*particulate matter*. Although particulates may contain a variety of chemical compounds depending on the sources from which they are emitted, all have the common property of consisting of finely divided solid particles suspended in the atmosphere. This permits a number of generalizations related to *particle size*. Diameters of particulates range from several centimeters down to 10^{-7} cm, just a bit larger than a small molecule. Effects related to particle size are summarized in Figure 9.1.

The main sink for atmospheric particulates is precipitation, and its rate depends on the density and diameter of the particles according to Stokes' law:

$$v_0 = gd^2(\rho_1 - \rho_2)/18\eta \qquad (9.1)$$

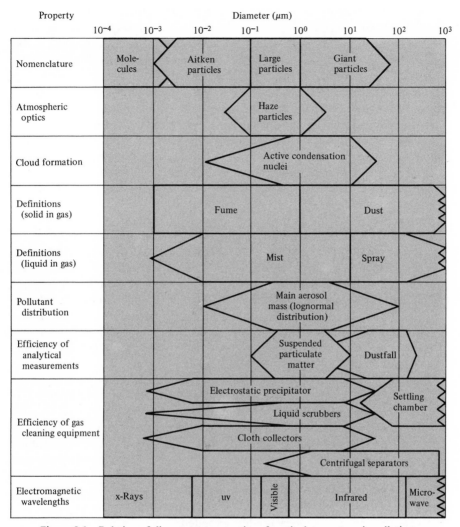

Figure 9.1   Relation of diameter to properties of particulate matter air pollution.

The terminal velocity, v_0, for a spherical particle of density ρ_1, and diameter d, falling through air of density ρ_2 and viscosity η, is rather slow—less than 10^{-6} m/s for particles smaller than 0.1 μm whose densities are close to that of water. This corresponds to a descent of 32 m ($\sim$ 100 feet) in a year. Although deviations from Stokes' law occur as a result of irregular particle shapes, turbulent rather than viscous flow around large particles, random, Brownian motions of very small particles, and weather effects such as wind or precipitation, it is clear that small-diameter particulates must aggregate to form larger ones if they are to be removed rapidly from the air.

Particle size affects other atmospheric properties as well. For a given quantity of pollutant the smaller the diameter the greater the surface area. Adsorption of gaseous or liquid materials and subsequent catalysis of chemical reactions are both expedited by large surface areas. This may be one reason for increased toxic effects in London smog (Section 9.4). Particles smaller than about 0.1 μm behave much like molecules, producing Rayleigh scattering (which has little effect on visibility). Particles in the range between 0.1 and 1 μm cause the greatest effect because their diameters are comparable to wavelengths of visible radiation and interference phenomena are important. In the size range > 1 μm scattering is proportional to diameter. The general effect of particulates is to reduce visibility as well as the amount of solar radiation reaching the ground. As mentioned in Section 7.5 this latter property may be reducing the impact of the greenhouse effect.

Appropriate concentrations of particles in the size range 0.1–10 μm can serve as nucleation centers for formation of raindrops, increasing cloud cover and rainfall over cities. A well-known, though disputed, example of this phenomenon is the excess precipitation in LaPorte,[1] an Indiana city 48 km (30 miles) downwind of the Chicago–Gary industrial complex, which had 31% more precipitation, 38% more thunderstorms and 246% more hail days than surrounding areas during the period 1951–1965. If too many nuclei are available, water droplets too small to precipitate may form. Under such circumstances particulates decrease rainfall but greatly increase cloud cover.[2]

Biological impact and particle size are also related. Impairment of circulation of CO_2 and O_2 through plant leaf stomata has already been mentioned (Section 6.6). In the case of animals, the respiratory tract is the site of most damage from particulates. Those larger than 10 μm are almost entirely trapped in the nose or pharynx and cause fewer problems because they are generally swallowed. Absorption through this gastrointestinal tract is approximately 80 times less efficient than through the lungs. Particles smaller than 0.1 μm have better than a 50% chance of being deposited in the lungs. They can often cause more damage than gaseous pollutants because the latter are more readily expelled. There is not space here to discuss toxic effects of specific substances associated with particulate matter, but in some cases this will be done on subsequent pages.

Sources of Particulates

There is a variety of natural sources of atmospheric suspended solids. Spray from ocean waves evaporates, leaving salt particles whose average diameter usually lies between 1 and 10 μm. These may reach altitudes well

[1] S. A. Changnon, Jr., *Bull. Amer. Meteorol. Soc.* **49**(1), 4–11 (1968).

[2] W. Bach, "Atmospheric Pollution," pp. 31–33, and references therein. McGraw-Hill, New York, 1972.

above 500 m over the oceans and have been detected as much as 1500 km inland. Volcanoes are another important, if intermittent, source. In some explosive eruptions particles have been injected into the stratosphere as well as the troposphere. Even relatively minor eruptions may produce haze 3000 km away, and there is evidence that the 1883 explosion of Krakatoa in the East Indies may have affected climate for several years.

A third important, though less massive, source of particles is biological. Spores, pollen, viruses, or bacteria have been found as high as the stratosphere, and photochemical decomposition of terpenes derived from trees results in haze of the type that gives the Great Smoky Mountains of the southeastern United States their name. Forest, grass, and brush fires, some of man-made origin, produce as many as 5×10^{18} particles per square meter, and smoke from major conflagrations can travel around the world. Dust and sand storms, which produce mainly large and giant particles (Figure 9.1), are the final component in nature's array of sources.

Man's activities, unfortunately, are comparable in magnitude to the sources listed above. In the United States fuel combustion and industrial processes both emit more suspended solids than forest fires. The former produces large quantities of *fly ash* containing carbon, oxides of silicon, iron, aluminum, calcium, nickel or phosphorus, and sulfates. Composition of industrial particles depends on the type of process (see Chapter 11 for some of the more important ones). These and other human activities yield particles whose diameters follow a lognormal distribution (Section 2.6) whose maximum lies in the vicinity of 1 μm.

9.2 ANALYSIS AND CONTROL OF PARTICULATES

Analysis and control of particulates are connected in two ways. The former is necessary to prove the need for the latter, and many methods of collecting samples for analysis are scaled-down versions of techniques by which suspended particles may be removed from an effluent gas stream.

The oldest, least expensive, and simplest method of measuring particulate emission levels is the *Ringlemann scale* (Figure 9.2). The degree to which light is attenuated by smoke is compared with six gray scales ranging from 0% black (#0) to 100% black (#5). Another technique which depends on the absorption of light by particulates is *densitometry.* Here air is passed through a filter, usually a continuous strip of paper which is automatically moved past the collection site at constant intervals. The particles collected during each time increment appear on the filter as a dark spot whose opacity or reflectance is proportional to the pollutant concentration.

Both the Ringlemann chart and densitometry err when quantities of light colored particulates must be compared with darker ones. A *high volume sampler,* which uses an air pump similar to a vacuum cleaner, can collect sufficient solid material over a 24-hour period for accurate weighing by

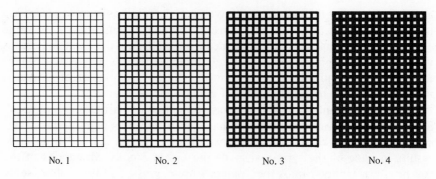

Spacing of lines on Ringelmann chart

Ringelmann chart no.	Width of black lines (mm)	Width of white spaces (mm)	Percent black
0	All white		0
1	1	9	20
2	2.3	7.7	40
3	3.7	6.3	60
4	5.5	4.5	80
5	All black		100

Figure 9.2 Ringelmann smoke chart. Source: A. C. Stern, H. C. Wohlers, R. W. Boubel, and W. P. Lowry, "Fundamentals of Air Pollution," p. 428. Academic Press, New York, 1973).

means of an analytical balance. With such a large sample further analysis may also be done to determine chemical properties of the substance. One of the most common is *benzene extraction,* which separates nonpolar, organic compounds from metal oxides and other inorganic materials. A good example of further analysis for a specific compound is provided by benzo[a]pyrene (Figure 5.1a). Solid residue from evaporation of benzene extract is dissolved in hexane and the benzo[a]pyrene separated by thin layer chromatography on silica-alumina plates. Quantitative analysis is achieved by scraping the appropriate spot from the plate, dissolving it in hexane, and measuring absorbance or fluroescence in the ultraviolet. Obviously beyond the benzene extraction step methods of separation and analysis must be developed for each compound among the many hundreds which may be present.

Filtration cannot easily separate particles by diameter because as pores become clogged by larger particles smaller ones may be trapped. A train of *impingers* may be used for this purpose, however. Collection by an impinger (Figure 9.3) depends on the inertia of the particles, which causes them to travel in more nearly linear paths than the less massive gas molecules. Greater gas velocities are therefore required to collect progressively smaller sizes. Occasionally a large particle will break up upon impact, and in some cases impingers yield spuriously large estimates of the amounts of small diameter solids.

Control Devices

Devices for trapping the millions of tonnes of particulates generated by human activities each year fall into five categories:

1. Gravitational settling chambers
2. Centrifugal separators (cyclones)
3. Wet scrubbers
4. Filters
5. Electrostatic precipitators

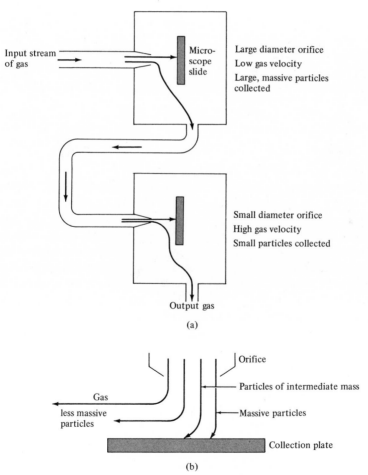

Figure 9.3 A train of impingers for collection of suspended particulates as a function of diameter: (a) two-stage impinger train and (b) principle of impinger operation.

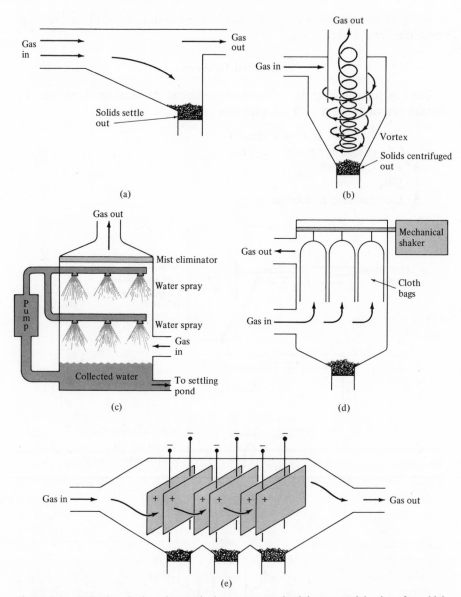

Figure 9.4 Collection devices for particulate matter and minimum particle sizes for which they give efficient removal: (a) settling chamber ($d > 50 \ \mu$m), (b) centrifugal separator ($d > 1 \ \mu$m), (c) wet scrubber (several other designs also used: $d > 0.05 \ \mu$m; can also collect water-soluble gases), (d) baghouse filter ($d > 0.01 \ \mu$m), and (e) electrostatic precipitator ($d > 0.005 \ \mu$m).

In some cases *ultrasonic agglomeration,* in which sound waves cause small particles to aggregate into larger ·ones, may be useful as a first step in treating particulate-laden gas. Schematic diagrams of each type of collection device are shown in Figure 9.4.

In a *gravitational settling chamber* reduced gas velocities allow particles to settle out. Only large ones can be collected because of the limitation on settling velocity imposed by Stokes' law. The *cyclone collector* uses the centrifugal force imposed by a vortex to increase settling rates in much the same way that a laboratory centrifuge speeds separation of solids from liquids. *Wet scrubbers* may be based on a simple counterflow spray (Figure 9.4c), the spray tower may be packed with low-density spheres or baffles, liquid may be introduced at a venturi throat which produces maximum gas velocity, or scrubbing may be combined with a cyclone to give increased efficiency. In all scrubbers impingement on liquid droplets and consequent entrapment of particulates serves to clean the gas. The operation of *filters* (bag houses) should be obvious. The selection of fiber for filter material is especially important for high-temperature gas streams since most commonly used fabrics are combustible.

The most efficient method for the removal of particles smaller than 1 μm is the *electrostatic (Cottrell) precipitator.* A very high dc voltage (40–60 kV) is applied between discharge wires and collector plates. The former ionize gas molecules and the charges so produced collect on the particulates, attracting them to the collector plates. Above about 1100 K thermal ionization begins to produce gas of sufficiently high conductivity that maintenance of the discharge voltage becomes difficult. This places an upper limit on the temperature range for efficient electrostatic collection. Efficiency of collection depends on collector plate area—a tenfold reduction in emissions corresponding to doubling the area—and in some cases on the composition of the gas stream. Reduction of SO_2 emissions by combustion of low sulfur coal has been found to increase particulate levels from precipitator-controlled power plants, for example, because of the high resistivity of the ash from this type of coal. To maintain high efficiency in any precipator it is necessary to shake or "rap" the discharge electrodes regularly to remove collected dust.

All of the collection devices mentioned in preceding paragraphs convert suspended particulate matter into solid (or in the case of the wet scrubber, liquid) wastes. Proper disposal of such materials (usually in landfills at present) is essential if the collection devices are to be considered real solutions to the particulate air pollution problem. However, all of these methods have been demonstrated to be both technologically and economically feasible. Once proper regulations are enacted and enforced there appears to be no reason why anthropogenic particulate emissions cannot be reduced by 95–99% from their uncontrolled levels.

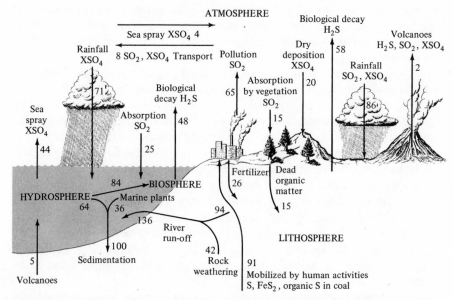

Figure 9.5   The global sulfur cycle. Arrows represent transport of sulfur in units of Tg. Note that because of lack of precise data, balance of lithosphere and hydrosphere sulfur transport is not exact. Data from S. I. Rasool, ed., "Chemistry of the Lower Atmosphere," Chapter 4. Plenum, New York, 1973.

9.3   SULFUR OXIDES

Sulfur in the environment provides an excellent example of a biogeo-chemical cycle involving the lithosphere, hydrosphere, and biosphere as well as the atmosphere. It also affords an opportunity to compare human mobilization of an element with natural sources and sinks. Figure 9.5 shows schematically recently derived estimates of the magnitude of transport of sulfur among the various spheres. It is obvious that SO_2 and sulfates derived from air pollution sources and fertilizers (Section 11.11) constitute significant fractions of total sulfur transport on earth. One estimate[3] suggests that the total amount of sulfur mobilized from the lithosphere by human activities such as coal or sulfur mining is nearly twice that freed by natural processes like rock weathering or volcanism. Furthermore, this ratio is increasing and may approach 10:1 in the near future.

The data in Figure 9.5 demonstrate the considerable potential for over-loading the biogeochemical sulfur cycle via addition of anthropogenic mobilization. Fortunately, there are a variety of fairly rapid pathways for transport and removal of sulfur from the atmosphere (in which it is present as

---

[3] J. P. Friend, The global sulfur cycle. *In* "Chemistry of the Lower Atmosphere" (S. I. Rasool, ed.), p. 177. Plenum, New York, 1973.

the poisonous gases H_2S, SO_2, and SO_3, and to which further discussion in this section will be confined). The half-life (making the not-fully justified assumption of first-order removal) for atmospheric sulfur averages less than 3 days.[4] In general, the mechanism for its removal depends on oxidation followed by precipitation in the form of H_2SO_4 or sulfates.

Hydrogen sulfide can undergo oxidation by oxygen, dioxygen, or ozone:

$$H_2S + O \longrightarrow HS + HO \xrightarrow{O_2} SO_2 + H_2O \tag{9.2a}$$

$$H_2S + \tfrac{3}{2} O_2 \longrightarrow SO_2 + H_2O \tag{9.2b}$$

$$H_2S + O_3 \longrightarrow SO_2 + H_2O \tag{9.2c}$$

Reaction 9.2c requires catalysis by aerosol particles to proceed at an appreciable rate in the troposphere. All of these reactions can occur more rapidly in photochemical smog (Section 10.3), since O, O_3, and particulates are all present in greater concentration.

Sulfur dioxide may occasionally be reduced to H_2S, but most is oxidized to SO_3, which forms sulfuric acid mist within seconds by reacting with water:

$$SO_3 + H_2O \longrightarrow H_2SO_4 \tag{9.3}$$

Sulfur dioxide may be oxidized by oxygen or dioxygen and perhaps O_3, although the latter requires near-uv radiation and few studies have been made of the mechanism.

$$SO_2 + O + M \longrightarrow SO_3 + M \tag{9.4a}$$

$$SO_2 + \tfrac{1}{2} O_2 \xrightarrow[\text{particulate}]{h\nu \text{ or}} SO_3 \tag{9.4b}$$

$$SO_2 + O_3 \xrightarrow{h\nu} SO_3 + O_2 \tag{9.4c}$$

Reaction 9.4b catalyzed by particulate matter provides the most important pathway for gas phase reaction of SO_2. Oxidation of sulfurous acid formed by reaction of SO_2 with water is also quite rapid, especially in the presence of metal salts supplied by particles which served to nucleate raindrops.

$$SO_2 + H_2O \longrightarrow H_2SO_3 \tag{9.5a}$$

$$H_2SO_3 + \tfrac{1}{2} O_2 \longrightarrow H_2SO_4 \tag{9.5b}$$

Once sulfuric acid has formed it may react with ammonia or metal salts (such as NaCl from sea salt particulates) to produce sulfates (XSO_4):

$$H_2SO_4 + 2 NH_3 \longrightarrow (NH_4)_2SO_4 \tag{9.6a}$$

$$H_2SO_4 + 2 NaCl \longrightarrow Na_2SO_4 + 2 HCl \tag{9.6b}$$

---

[4] J. P. Friend, The global sulfur cycle. *In* "Chemistry of the Lower Atmosphere" (S. I. Rasool, ed.), p. 197. Plenum, New York, 1973.

Sulfur dioxide is also known to form sulfate rapidly in the presence of ammonia and moist air:

$$SO_2 + 2\ NH_3 + H_2O + \tfrac{1}{2}\ O_2 \longrightarrow (NH_4)_2SO_4 \tag{9.6c}$$

Sulfuric acid or sulfates then precipitate with rain.

Acid Rain

Sulfurous and sulfuric acids, HCl produced by reaction 9.6b, and hydrolysis of ammonium sulfate all reduce the pH of rainwater. In both Europe

$$(NH_4)_2SO_4 \longrightarrow 2\ NH_4{}^+ + SO_4{}^{2-}$$
$$\Updownarrow \tag{9.6d}$$
$$2\ NH_3 + 2\ H^+$$

and the United States *acid rain,* with pH values ranging from 5.0 to as low as 2.1, is commonplace. (Because of dissolution of atmospheric CO_2 to form H_2CO_3 the pH of natural rain is somewhat below 7.0, averaging around 5.7.) There is also some speculation that pollution abatement technology currently in use—tall smokestacks especially—contributes to the long-range (up to 1000 km) transport of sulfur, making the acid rain problem a regional one which can cross state or national boundaries.[5] There is convincing evidence that rainfall in the industrialized northeastern United States has a significantly lower pH than elsewhere in the country.

The ecological effects of acid rain have not yet been evaluated quantitatively and are undoubtedly complex. They probably include increased leaching of nutrients from plant foliage, removal of basic nutrients such as calcium from soil, changes in rates of metabolism of organisms which depend on acid or base catalysis, and acidification of lakes or rivers. Corrosion of basic materials such as limestone or marble (which consist of $CaCO_3$ or $CaCO_3 \cdot MgCO_3$) is also accelerated. Such "stone leprosy," in which the carbonates are converted to sulfates that later dissolve has accounted for considerable damage to structural and artistic (statuary) stone during the twentieth century.[6]

$$CaCO_3 + H_2SO_4 \longrightarrow CaSO_4 + H_2O + CO_2 \tag{9.7a}$$
$$CaCO_3 + SO_2 + \tfrac{1}{2}\ O_2 \longrightarrow CaSO_4 + CO_2 \tag{9.7b}$$

Monitoring of Sulfur Oxides

In Europe the standard method for collection and analysis of sulfur oxides is to bubble a gas sample through 0.3 M H_2O_2 at a pH of 5. This converts SO_2 and SO_3 to H_2SO_4 which can be titrated with base. The pres-

---

[5] G. E. Likens and F. H. Bormann, *Science* **184,** 1176–1179 (1974); H. Reiquam, *ibid.* **170,** 318–320 (1970).

[6] E. M. Winkler, "Stone. Properties, Durability in Man's Environment." Springer-Verlag, Berlin and New York, 1973.

ence of other acidic (HCl; NO_2 to a slight extent) or alkaline (NH_3) gases in the sample can lead to erroneous results, but the method is simple and inexpensive. In the United States collection in H_2O_2 solution is occasionally employed, but analysis is usually by measurement of conductivity of the H_2SO_4 solution. The same interferences apply.

The most common method of analysis [adopted as standard by the U. S. Environmental Protection Agency (EPA)] was devised by West and Gaeke.[7] The sample is collected by bubbling through a dilute aqueous solution of sodium tetrachloromercurate(II), which can trap SO_2 quantitatively at concentrations as low as 0.002 ppm:

$$SO_2 + HgCl_4^{2-} + H_2O \longrightarrow HgCl_2SO_3^{2-} + 2\,H^+ + 2\,Cl^- \qquad (9.8)$$

After collection of a 1-hour or 24-hour average sample in the field, the solution is returned to the laboratory. Sulfamic acid is added to reduce nitrogen dioxide, which otherwise would destroy some of the color reagent. Then the dichlorosulfitomercurate(II) is reacted with formaldehyde and pararosaniline in a solution whose pH has been controlled using H_3PO_4. The pararosaniline methylsulfonic acid formed has an absorption maximum at 575 nm at pH = 1.

$$\lambda_{max} = 575 \text{ nm} \qquad (9.9)$$

If the concentration of $HgCl_2SO_3^{2-}$ is too high more than one site on the pararosaniline dye may react, and Beer's law will no longer be obeyed. This limits the method to solution concentrations of SO_2 between 0.05 and 5 mg/dm³. The range of atmospheric concentrations may be extended by appropriate dilutions of concentrated samples from 0.002 to 5 ppm.

Using the West-Gaeke method adapted for automated analysis the EPA

[7] P. W. West and G. C. Gaeke, Jr., *Anal. Chem.* **28**, 1816 (1955); see also J. B. Pate, J. P. Lodge, Jr., and A. F. Wartburg, *ibid.* **34**, 1660 (1962); F. P. Scaringelli, B. E. Saltzman, and S. A. Frey, *ibid.* **39**, 1709 (1967).

monitors SO_2 levels throughout the United States as part of its National Air Sampling Network. Particulates, nitrogen oxides, carbon monoxide, and photochemical oxidants are also sampled regularly in rural and urban environments. Urban areas such as Pittsburgh (Allegheny County, Pennsylvania), Chicago, St. Louis, and New York as well as states (Washington, New Jersey, and New York) have set up their own regional networks, also.

9.4   EFFECTS OF SULFUR OXIDES AND PARTICULATES

Toxic and other effects of air pollution (or of environmental contamination in general) may be divided into two classes. *Acute effects* manifest themselves immediately upon short-term exposure (a few days at most). *Chronic effects* only become evident after much longer periods of continu-

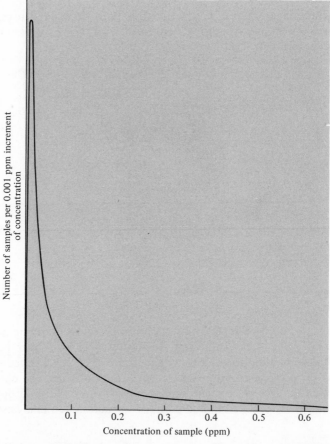

Figure 9.6   Lognormal distribution of 5-min average sulfur oxide concentrations, Chicago, 1962–1967. Based on data in National Air Pollution Control Administration, "Air Quality Criteria for Sulfur Oxides," p. 35. US Govt. Printing Office, Washington, D. C., 1970.

ous exposure and are consequently less obvious. (In the case of chemical carcinogens the distinction between acute and chronic toxicity is less easily made, see Section 17.5).

Because of the different time scales of the two types of effect and the large fluctuations of pollutant concentrations which may occur, different ways of averaging such concentrations are appropriate. In some cases averaging is inherent in the method of analysis. The West-Gaeke technique usually requires collection times of at least 1 hour to obtain sufficient material, and any fluctuations occurring over that period are therefore averaged out. High volume particulate samplers usually operate on a 24-hour collection cycle. Thus a sudden, large pulse of pollution, whose concentration might be large enough to cause severe acute effects for a few minutes, might not be detected by such methods.

On the other hand, exposure periods of a year or more are required for evaluation of chronic effects, and no sampling method has a collection time that long. Therefore, mathematical averaging must be done. Because the distribution of number of analyses versus concentration is usually log-normal (Figure 9.6), the appropriate method of averaging is the *geometric mean* rather than the arithmetic. The latter would give excessive weight to the more concentrated samples (see Exercise 10). Based on geometric averaging of typical United States results the 1-day maximum SO_x concentration can be expected to be from four to seven times and the 1-hour maximum as high as 20 times the annual mean. In the vicinity of a point source (such as an electric power plant), where the smoke plume would change with wind direction and velocity, fluctuations up to 160 times the annual mean have been observed.

Because of such fluctuations the type of averaging as well as the concentration of pollutant is important in establishing air quality criteria, and such information is given in Figure 9.7, where various concentrations of SO_2 are related to effects on materials, vegetation, visibility, and human health. In addition to hastening dissolution of stone, sulfur oxides contribute to corrosion of metals, deterioration of electric equipment, bleaching and weakening of fabric and leather, and discoloration and embrittlement of paper. Plants are damaged by both acute injury, in which sulfurous acid attacks cells, and chronic effects related to disruption of chlorophyll synthesis. Low levels of SO_x may significantly decrease crop yields without visible damage. Since about 80% of sulfuric acid and sulfate aerosol falls in the size range (0.1–1 μm) that causes greatest loss of visibility, sulfur oxide air pollution is capable of curtailing activities (such as scheduled airline flights) which depend on clear lines of sight.

Synergistic Effects

In Figure 9.7 levels of particulate matter are often reported in conjunction with those of SO_x when a specific effect is described. This is necessary because of a *synergistic effect* between these two pollutants. The total influence of both combined is greater than the sum of the negative consequences each would produce alone. In such a case one and one are not necessarily two—it becomes very important to consider the nature of "and" as Eddington's words at the beginning of this chapter suggested. Although the mechanisms by which synergism operates between sulfur oxides and particulates have not been completely elucidated, the general picture has begun to emerge. This is shown in Figure 9.8, using London smog as an example.

London smog involves three major components: sulfur oxides and particulates from smoke, and water droplets from fog. Particulates serve as nucleation sites for formation of fog droplets and their large surface area allows catalysis of oxidation of SO_2 (equation 9.4b). Metal ions such as iron(II) and manganese(II), which might pass into solution from a nucleating particle, also catalyze oxidation of aqueous SO_2 (equations 9.5).

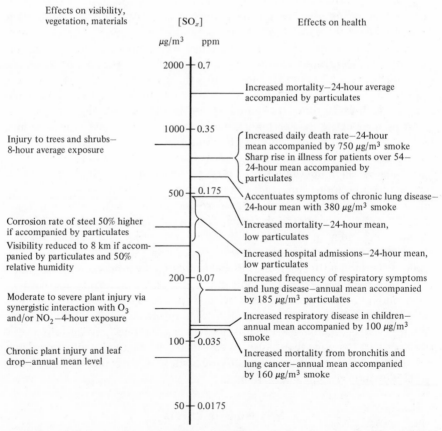

Figure 9.7 Effects of sulfur oxides (accompanied by particulates in some cases) on visibility, vegetation, materials, and health. Data from National Air Pollution Control Administration, "Air Quality Criteria for Sulfur Oxides," pp. 161–162. US Govt. Printing Office, Washington, D. C., 1970.

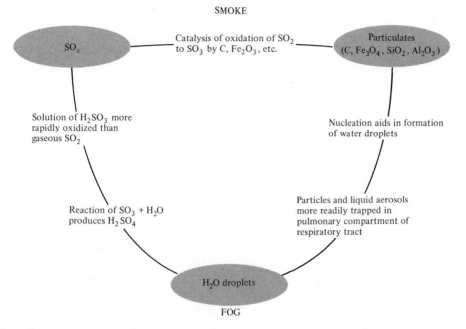

Figure 9.8 Synergism in London smog.

Therefore rapid production of sulfuric acid aerosols is likely when all three components are present. Moreover, relatively concentrated solutions of H_2SO_4 are often adsorbed on the surfaces of particles whose diameters fall in precisely the range necessary for deposition deep in the lungs (in the bronchioles and alveoli). From the latter soluble material may enter the bloodstream, the lymphatic system, or connective lung tissue. High concentrations of H_2SO_4 cause large amounts of fluid to be released as a diluent, and bronchioles may contract spasmodically in an attempt to prevent admission of the irritant. Both of these natural defense mechanisms make breathing more difficult and strain the victim's heart, leading to both acute and chronic health damage.

The possibility of synergistic effects is a very real one both with respect to air pollution and other areas of environmental contamination. It makes the task of setting safe levels for materials released into the environment extremely difficult because it is impossible to check for synergism all combinations of the large number of substances dealt with by modern technology. Prior to recognition of the causes of death in London-type smogs there were numerous examples of exposure of industrial workers *to SO₂ alone* at levels well above the 1.7 ppm reached in the 1952 episode, and no ill effects were recorded. The folly of assuming without proof that data obtained under controlled conditions of exposure to a single substance are applicable to the complex mixtures found in urban atmospheres should be evident.

9.5 CONTROL TECHNOLOGY FOR SULFUR OXIDES

A number of means by which SO_x emissions might be alleviated have already been mentioned (Sections 5.1 to 5.3). Most of them were based on use of low-sulfur fuels in place of coal, removal of sulfur prior to coal combustion or conversion to nuclear power. So far the major responses to EPA efforts to reduce pollution associated with coal combustion have involved taller smokestacks and use of low-sulfur coal, fuel oil, or natural gas. The latter two fuels are in short supply, low-sulfur coal is not invariably low on a per unit energy basis and is not unlimited in supply, and taller stacks are a stopgap which eventually result in regional rather than local problems. Moreover, SO_2 is produced by ore smelting (Section 11.6) and sulfuric acid manufacture (Section 11.10), processes which are not affected by the removal of sulfur from fuel. Therefore, techniques by which sulfur oxides may be removed from gaseous effluents are and will remain important.

Emissions from ore smelters are usually sufficiently concentrated (4–8% SO_2) to make feasible catalytic oxidation and manufacture of sulfuric acid via the reaction sequence: 9.4b, 9.3. Several such processes are available.[8] That effluent gases are permitted to escape, often denuding the adjoining landscape of trees, instead of being fed to sulfuric acid manufacturing plants is a consequence of low prices and lack of local markets for the acid—not lack of technologically feasible systems for SO_2 removal. Similarly, emissions from sulfuric acid plants may be controlled by combining more efficient conversion of SO_2 to H_2SO_4 with lime scrubbing (see equations 9.11), but capital investment in added equipment increases costs 10% (about one dollar per tonne).[9] Therefore legislation has been essential in bringing about such reductions of emissions.

In the case of coal-fired electric power plants, which produce 60% of United States sulfur oxide emissions, at least ten removal processes have been or are being installed on a test basis in large (> 100 MW_e) plants.[10] Space does not permit a detailed description of each one, but a recent study sponsored jointly by EPA, Council on Environmental Quality, Department of Commerce, and the Federal Power Commission[11] concluded that the four processes described below were sufficiently developed for full-scale commercial use by 1978.

In *wet lime—limestone scrubbing,* powdered limestone is injected into the power plant boiler along with powdered coal. At the high temperature

---

[8] L. P. Argenbright and B. Preble, *Environ. Sci. Technol.* **4**(7), 554–561 (1970); F. A. Ferguson, K. T. Semrau, and D. R. Monti, *ibid.* pp. 562–568.

[9] W. R. King, Air pollution control. *In* "Riegel's Handbook of Industrial Chemistry" (J. A. Kent, ed.), 7th ed., p. 874. Van Nostrand-Reinhold, Princeton, New Jersey, 1974.

[10] A. V. Slack, *Environ. Sci. Technol.* **7**(2), 110–119 (1973).

[11] *Chem. & Eng. News* July 16, p. 15 (1973).

of the boiler $CaCO_3$ decomposes:

$$CaCO_3 \xrightarrow{\Delta} CaO + CO_2\uparrow \qquad (9.10)$$
$$\text{Limestone} \qquad \text{Lime}$$

Calcium oxide then reacts with SO_2 to form solid $CaSO_3$ or $CaSO_4$, both of which are somewhat water soluble:

$$CaO + SO_2 \longrightarrow CaSO_3 \qquad (9.11a)$$
$$CaO + SO_2 + \tfrac{1}{2} O_2 \longrightarrow CaSO_4 \qquad (9.11b)$$

The overall process is the same one involved in stone leprosy. Calcium sulfite and sulfate can be removed from stack gases by scrubbing with a slurry of limewater:

$$CaO + H_2O \xrightarrow[H_2O]{excess} Ca(OH)_2 \qquad (9.12)$$
$$\text{Lime} \qquad \text{Limewater}$$

The main problems of the wet limestone process are the extremely large quantities of $CaCO_3$ required, the fact that only half of the $CaCO_3$ actually reacts with SO_2 (apparently because only the surface of CaO particles is in contact with the gas phase in the stack), the possibility of water pollution by scrubber liquid or runoff from solid wastes of $CaSO_3$ and $CaSO_4$, and the difficulty of disposing of such solids, for which no commercial use has been found. Nevertheless, the wet limestone process can remove 90% of the SO_2 and 98% of the fly ash from power plant effluents and is suitable as an "add-on" to existing plants.

In *magnesium oxide scrubbing* a slurry of $Mg(OH)_2$ is used in much the same way as $Ca(OH)_2$ in the limestone process. In this case the $MgSO_3$ which forms can be calcined (heated) to regenerate SO_2 at a concentration acceptable for sulfuric acid manufacture:

$$MgSO_3 \xrightarrow{\Delta} MgO + SO_2\uparrow \qquad (9.13)$$

MgO can be recycled to the scrubber. The main disadvantages of this process are large energy requirements for the calcining operation and the question of profitable sale of H_2SO_4.

A third category of removal processes, *sodium-based scrubbing,* involves formation of bisulfite (HSO_3^-) ions in the scrubbing liquor. This can be done using sodium salts such as the sulfite or citrate:

$$Na_2SO_3 + SO_2 + H_2O \rightleftharpoons 2\,NaHSO_3 \qquad (9.14a)$$

$$NaH_2Cit + SO_2 + H_2O \longrightarrow NaH[HSO_3 \cdot H_2Cit] \qquad (9.14b)$$

$$\left(\;\; Cit = H_2C\!\!-\!\!\overset{\displaystyle\overset{CO_2^-}{|}}{\underset{\displaystyle\underset{OH}{|}}{C}}\!\!-\!\!CH_2, \text{ the citrate ion} \;\;\right)$$

Reaction 9.14a can be reversed, yielding SO_2, which can be combined with H_2S (equation 5.11b) to give elemental sulfur. Similarly, the bisulfite-dihydrogen citrate complex from 9.14b can be converted to sulfur:

$$NaH[HSO_3 \cdot H_2Cit] + 2\ H_2S \longrightarrow 3\ S + NaH_2Cit + 3\ H_2O \qquad (9.15)$$

regenerating the absorbent, NaH_2Cit, for recycle to the scrubber. H_2S may be manufactured as needed by reduction of sulfur:

$$4\ S + CH_4 + 2\ H_2O \xrightarrow{Al_2O_3} 4\ H_2S + CO_2 \qquad (9.16)$$

The sodium-based scrubbing techniques have the advantage that elemental sulfur is much cheaper to ship than H_2SO_4, considerably widening the market area available to such a plant.

Removal of SO_2 by *catalytic oxidation* is an application of the same chemistry used in sulfuric acid manufacture (equation 9.4b followed by 9.3) to the less concentrated SO_2 emerging from a power plant. A vanadium pentoxide catalyst speeds oxidation of SO_2 to SO_3, and the latter reacts with water to form sulfuric acid mist which can be removed by filtration. The equipment is fairly simple, requiring no recycling of reagents, but a high-temperature electrostatic precipitator is required upstream from the catalyst bed to remove particulates which might clog and poison it. Corrosion by the dilute acid is also a problem. The process can only be installed on new plants, and it is generally agreed that sales of H_2SO_4 would not bring enough profit to pay for the added equipment.

Despite the drawbacks already mentioned and considerable public debate, which has seen different branches of government arguing on opposite sides,[12] the technological feasibility of currently available SO_x removal systems seems assured. A limestone scrubber at the Mitsui Aluminum Company in Japan was available over 90% of the time during 6 months of power plant operation as early as 1972, and EPA has concluded that scrubbers are "permanent, reliable and effective."[13] Debate has turned now to economic feasibility, with costs for SO_x removal estimated at 1.1 to 3.0 mills/kW-hour of electric energy generated. This corresponds to a cost increase of from 6 to 17% for electricity, a price which is comparable to the more than three billion dollars of damage done by sulfur oxides from power plants in 1970 (Table 8.3).

9.6 OTHER INDUSTRIAL AIR POLLUTANTS

A variety of substances is emitted to the atmosphere by industrial processes in addition to the five major air pollutants listed in Table 8.3. Some of them are reaching levels high enough to affect materials, plants, or

---

[12] J. A. Noone, Great scrubber debate pits EPA against electric utilities. *Nat. J. Rep.* pp. 1103–1114 (1974).

[13] *Power Eng.* September, p. 51 (1972); Scrubbers: permanent, reliable, effective. *Environ. Midwest* November, p. 2 (1974).

TABLE 9.1 Sources, Annual Emissions, and Health Effects of Minor Air Pollutants (United States, 1970)

Substance	Sources[a]	Emissions[a] (tonnes/year)	Health effects[b]
Lead	Auto exhaust, industry, solid waste disposal, coal combustion, paint	208 250	Brain damage, behavioral disorders, convulsions, death
Fluorides	Industry, coal combustion	150 400	Mottled teeth, weakening of bone, weight loss, thyroid and kidney injury, death
Vanadium	Coal and petroleum combustion, industry	18 440	Inhibits formation of phospholipids and S-containing amino acids
Manganese	Industry, coal combustion	16 230	Fever, pneumonia
Arsenic	Industry	9 570	Dermatitis, melanosis, perforation of nasal septum, possible carcinogen
Nickel	Coal combustion, industry	6 625	Dermatitis, dizziness, headaches, nausea, and carcinogenesis [also $Ni(CO)_4$]
Asbestos	Industry	6 080	Scarring of lungs, lung cancer
Cadmium	Industry	1 962	Gastrointestinal disorder, respiratory tract disturbances, carcinogenic and mutagenic
Mercury	Coal combustion, commercial, industry	777[c]	Tremor, skin eruption, hallucinations
Beryllium	Coal combustion, industry	156	Lung damage, enlargement of lymph glands, emaciation
Selenium	Ore refining, sulfuric acid manufacture, coal combustion	—	Depression, jaundice, nosebleed, dizziness, headaches

[a] Data from U. S. Environmental Protection Agency, "Air Pollution Emission Factors," AP-42, 2nd ed. USEPA, Research Triangle Park, North Carolina, 1973.

[b] Data from G. L. Waldbott, "Health Effects of Environmental Pollutants." Mosby, St. Louis, Missouri, 1973.

[c] Emissions may be larger than this estimate; see C. E. Billings and W. R. Matson, *Science* **176**, 1232–1233 (1972).

animals deleteriously. Sources, yearly emissions, and health effects of a number of such air pollutants are listed in Table 9.1. Although not all-inclusive, it gives a general impression of the problems which will remain after the five major air pollutants have been adequatey controlled. The paragraphs which follow will examine a few representative examples from Table 9.1 in detail.

Fluorides

The main sources of atmospheric emissions of fluorides are (1) steelmaking, where fluorspar (CaF_2) is used to increase fluidity of slags and aid in removal of phosphorus and sulfur; (2) manufacture of brick, tile, and cement from clays containing small percentages of fluorine; (3) aluminum production, where cryolite (Na_3AlF_6) is used as a solvent for electrolysis; (4) manufacture of phosphate fertilizer and phosphoric acid from phosphate rock consisting mainly of fluorapatite [$Ca_{10}F_2(PO_4)_6$]; and (5) combustion of coal. (Several of these industrial processes will be discussed in Chapter 11.) In the case of coal only trace quantities of fluorine are present, but with the tremendous amounts burned significant emissions can still occur. Natural sources of atmospheric fluorides are volcanoes, which emit HF, NH_4F, SiF_4, $(NH_4)_2SiF_6$, Na_2SiF_6, and KBF_4, and dust storms in areas where soils are high in fluoride. In some cases volcanic eruptions may be predicted on the basis of increased fluoride emissions which precede them.

The main pathway for removal of gaseous fluorides from the atmosphere is hydrolysis to nonvolatile materials which then precipitate:

$$3\ SiF_4 + 2\ H_2O \longrightarrow SiO_2 + 2\ H_2SiF_6 \tag{9.17a}$$

or reaction with silicate minerals or glass:

$$4\ HF + SiO_2 \longrightarrow SiF_4 + 2\ H_2O \tag{9.17b}$$

Most of the fluorides emitted as particulates do not hydrolyze readily. Among these are $Ca_{10}F_2(PO_4)_6$, CaF_2, Na_3AlF_6, AlF_3, Na_2SiF_6, and NaF. Their environmental effects depend largely on water solubility, which governs their ability to be incorporated into plants and animals. Solubility increases in roughly the order of the formulas given in the preceding list.

Accumulation of atmospheric fluoride by plants can result in altered metabolism; lesions in foliage; and changes in growth, development, and yield. For 1-day exposures of highly susceptible plants a 3–4 $\mu g/m^3$ level can induce leaf markings, and for periods of a month or more the threshold is about 0.5 $\mu g/m^3$. Domestic animals most sensitive to fluorides are dairy cattle. Long-term ingestion of feed containing more than 40 ppm (by weight) produces lameness, bone and tooth damage, and reduction of growth and milk production. It has been estimated that this level could be reached in forage crops exposed to 0.33 to 1.3 $\mu g/m^3$ HF for 30 days.[14]

There is some evidence[15] that the Meuse Valley and Donora, Pennsylvania air pollution disasters mentioned earlier may have been caused by fluorides rather than London smog. At least 30 mg/m^3 fluoride was present in the former case.[16] Certainly the incidence of chronic fluorosis, in which joints become calcified, is much greater in the vicinity of iron and aluminum works as well as fertilizer factories,[17] and recently the EPA has proposed emissions standards for the latter. In the Soviet Union a standard

[14] Committee on Biologic Effects of Atmospheric Pollutants, "Fluorides," p. 237. Nat. Acad. Sci., Washington, D. C., 1971.

[15] An overview is given by G. L. Waldbott, "Health Effects of Environmental Pollutants," pp. 6–9. Mosby, St. Louis, Missouri, 1973.

[16] K. Roholm, *J. Ind. Hyg. Toxicol.* **19**, 126–137 (1937).

[17] G. L. Waldbott, "Health Effects of Environmental Pollutants," pp. 155–161. Mosby, St. Louis, Missouri, 1973.

of 0.01 mg/m³ 24-hour average with maximum allowable peak concentration of 0.03 mg/m³ has been established for ambient air, but elsewhere regulations are aimed primarily at controlling exposure of workers rather than the general public. Since soluble and hydrolyzable fluorides are most harmful, wet scrubbers are usually suitable for emission control. The EPA estimates that their use in phosphate fertilizer plants would not increase costs by more than 1%.

Metals

Poisonous metallic elements in the environment are often referred to as "heavy metals" because lead and mercury (which will be discussed in Chapters 10 and 16) were characteristic examples. The term is broad enough, however, to include substances which are used precisely because they are not "heavy" (or more correctly, of large density). One of these, beryllium, has been chosen here as an example of metallic air emissions.

Unique properties, such as its high strength-to-weight ratio, transparency to x rays, ability to serve as a moderator in fission reactors (Section 6.1), and favorable characteristics of its alloys, have led to industrial use of beryllium despite its relative scarcity, high price, and toxicity. Beryllium is an excellent additive for rocket fuels because of its low density and the large negative free energy of formation of its oxide. The latter's high melting point (2800 K), high heat conductance, and high electric resistivity have led to ceramic applications such as spark plug insulators and liners for high temperature furnaces. Thus a considerable amount of beryllium comes into the environment, especially from extraction plants which convert ores such as beryl ($3 \, BeO \cdot Al_2O_3 \cdot 6 \, SiO_2$) into beryllium powder. Other major sources are ceramic plants, foundries, machine shops, propellant plants and rocket motor test facilities, and incinerators or open burning sites for waste disposal. Combustion of coal, which contains 1–3 ppm beryllium by weight, also liberates significant quantities to the atmosphere.

Most beryllium emissions are in the form of metallic powder or BeO particulates. They can be controlled by means of dry cyclones, wet scrubbers, fabric filters, and high efficiency particulate air (HEPA) filters. In most cases the cost of such equipment does not exceed 10–15% of the cost of manufacturing equipment. Such controls are important, because, in addition to acute and chronic berylliosis in workers associated with beryllium production plants, there are a number of cases where those living nearby were apparently affected by atmospheric levels as low as 0.015 μg/m³.[18] Acute poisoning produces inflammation of the respiratory system followed by a pneumonia-like process. In about 6% of cases it is followed by chronic berylliosis, which begins with progressive shortness of breath,

---

[18] G. L. Waldbott, "Health Effects of Environmental Pollutants," pp. 109–111. Mosby, St. Louis, Missouri, 1973.

weight loss, and cough. Beryllium is believed to interfere with passage of dioxygen from the lungs' alveoli to arterial blood. Eventually inflammation and scarring of lung tissue and heart damage occur. Beryllium compounds have also been found to cause malignant tumors in laboratory animals,[19] although it has not been determined whether lung cancer might result from berylliosis in humans.

Asbestos

One case where inhalation of particulate matter has been shown to lead to lung cancer is that of asbestos, a term which refers to fibrous material composed of any of six different crystalline silicates. Fibers of the most common of these, chrysotile, are built up from many hollow cylindrical *fibrils,* each of which has an outside and inside diameter of 34 and 18 nm, respectively. Each fibril is composed of a layer of SiO_4 tetrahedra (see Figure 2.5d), to which has been added a close-packed layer of OH^- ions. Mg^{2+} ions in octahedral holes hold the latter to the side of the $(Si_4O_{10})_n$ layer on which unshared oxygens are located. Because of a mismatch in size between the sandwiched layers the overall structure is not planar but curls into a cylinder. Although each such cylinder may be quite long and normal absestos fibers contain many of them joined in parallel, milling and other manufacturing operations invariably produce large numbers of individual fibrils, some of which are shortened as well. Such very small diameter particles are trapped in the lungs with a high degree of efficiency. Once there they apparently remain indefinitely — in some cases health effects have appeared as much as 30 years after exposure had ended.

The first published mention of asbestosis was in 1907. The disease is characterized by shortness of breath which results from scarring of lung tissue and extensive thickening of the pleura (the lining surrounding the lungs). More recently asbestos has been implicated as a causative agent in mesothelioma, a rare tumor of the pleura and the peritoneum (lining of the abdominal wall). This cancer, which is almost invariably fatal, is associated specifically with the asbestos fiber crocidolite, $Na_2Fe_5Si_8O_{22}(OH)_2$, much of which is imported to the United States from South Africa. Other, more common, cancers which localize in the lower lobes of the lungs have been observed in asbestos workers, especially smokers. Nearly 50% of persons diagnosed as having asbestosis eventually die of lung cancer. There is also inconclusive evidence[20] linking gastrointestinal cancers with asbestos-contaminated food or swallowing of sputum containing inhaled asbestos particles. This has important implications for cities such as Duluth, Min-

---

[19] G. L. Waldbott, "Health Effects of Environmental Pollutants," p. 109. Mosby, St. Louis, Missouri, 1973.
[20] I. J. Selikoff *et al., Amer. J. Med.* **42,** 487–496 (1967).

nesota, where asbestos fibrils from taconite tailings have been discovered in the water supply.

Dispersion of asbestos fibrils into the atmosphere may be controlled by careful handling of dry materials, substituting wet processing for dry, enclosing dusty operations, fabric filtration of effluents, and minimizing the use of asbestos in applications which lead to generation of emissions. The widespread spraying of asbestos insulation on steel frameworks of high-rise buildings is a good example of the last category. Eventually demolition of such buildings also releases significant numbers of fibrils. A problem nearly unique to asbestos is that fibrils are extremely stable, almost never aggregate to form larger ones, and thus may be reintroduced into the air even after being filtered, settling out, or spending many years as a component of a building.

Different industries release a wide variety of atmospheric and other pollutants. Quite often, as in the case of those discussed in this section, exposure of workers gives the first indication of problems. As such industries grow, toxic substances released by manufacturing operations reach the general public as well. Usually standards for such involuntary, continuous exposure are set considerably below those for a workplace. Toxic substances and the workplace environment will be discussed further in Sections 17.6 and 17.7.

EXERCISES

(The first group of exercises is based primarily on material found in this text.)

1. Calculate the Stokes' law terminal velocity for each of the following particles (assuming them to be spheres of diameter d):

   a. Fe_3O_4, $d = 0.1$ μm
   b. C, $d = 10$ μm
   c. $(NH_4)_2SO_4$, $d = 1$ μm
   d. $(NH_4)_2SO_4$, $d = 10$ μm

   The viscosity of air at 273 K is 170 μg/cm s and its density is 0.0013 g/cm^3.

2. What is a synergistic effect? Describe clearly how synergism between SO_x and particulates increases health damage from London smog.

3. Approximately 550×10^6 tonnes of coal were burned in the United States in 1975. Assuming an average sulfur content of 2% and no stack emission controls, calculate the amount of SO_2 emitted and compare it with data in Table 8.3.

4. Assume that the sulfur dioxide produced in Exercise 3 is removed from stack gases by the wet limestone method in which it is reacted with calcium carbonate. Write a chemical equation for the process. Calculate the amount of cal-

cium sulfate that would be produced in one year if all SO_2 were removed by this process.

5. Suppose that catalytic oxidation to sulfuric acid were used instead of the wet limestone process for removal of all the SO_2 in Exercise 3. Calculate the number of tonnes of sulfuric acid that would be produced in one year. How does this compare with the annual production of H_2SO_4 (25 million tonnes) by the United States chemical industry?

6. What would probably happen to ammonia introduced into a body of air already polluted with sulfur dioxide?

7. The approximate diameters of solid particles suspended in the atmosphere influence the behavior of particulate air pollution. Give at least three properties of particulates that depend on their size, and explain how the size of the particles makes a difference.

8. Name at least two devices for the control of particulate air pollution and explain clearly how they work.

9. Why is a "third body" (such as M in equation 9.4a) required in some gas phase reactions? What characteristic of such reactions requires a third body?

10. Calculate the geometric and arithmetic means for the following 5-min average measurements of SO_x concentrations (in ppm): 0.008, 0.012, 0.024, 0.042, 0.068, 0.110, 0.145, 0.22, 0.26, 0.54. How many samples exceed the geometric mean? The arithmetic? Repeat the calculations including the following measurements: 0.005, 0.015, 0.019, 0.035, 0.054, 0.091, 0.150, 0.18, 0.32, 0.87.

11. If, as stated in Section 9.5, a 2:1 mole ratio of $CaCO_3:SO_2$ is required in the wet limestone stack gas cleaning process, how much $CaCO_3$ would be required by one tonne of coal containing 3% sulfur?

12. How might a hydrogen economy affect the sodium absorption techniques for SO_x removal from stack gases?

13. Electricity consumption in the United States in 1971 was 5.5×10^{18} J. Assuming the cost estimates for cleanup given in Section 9.5, compare the costs and benefits of stack gas SO_2 removal.

14. Why do particles between 0.1 and 1 μm have unusually large effects on visibility?

(Subsequent exercises may require more extensive research or thought.)

15. List as many ways as you can of alleviating air pollution from SO_x. Try to evaluate each on the basis of

   a. Technological feasibility
   b. Economic feasibility
   c. Ability to provide a long-term solution to the problem
   d. Unexpected adverse effects upon other segments of the environment

16. Assume that you are manager of a 1000-MW$_e$ coal-fired electric power plant burning coal containing 3% sulfur and 65% carbon. Using data from Figure 5.3 calculate the amount of air needed to burn the sulfur and carbon and the ppm SO_2 in exhaust gas from the plant. (Assume, incorrectly, that the remainder of the coal contributes no gas to the stack and requires no oxidation; see H. C. Perkins, "Air Pollution," p. 274. McGraw-Hill, New York, 1974.)

17. Assuming that coal is available at the following prices:

    1.0% sulfur $0.35/10^6$ Btu
    3.3% sulfur $0.27/10^6$ Btu

    calculate the additional cost per year to burn low-sulfur coal in a 1000-MW_e plant. Assuming a 30-year plant lifetime, compare total costs of low-sulfur coal versus 20×10^6 spent on stack gas cleanup (see H. C. Perkins, "Air Pollution," p. 276. McGraw-Hill, New York, 1974).

18. In an experiment designed to measure concentrations of "respirable dust," R. E. Pasceri [*Environ. Sci. Technol.* **7**(7), 623–627 (1973)] used a high volume sampler preceded by a cyclone to collect his sample. Explain why such a system might be a good approximation to the human respiratory system.

10

Air Pollution—
Transportation Related

Science is always wrong, It never solves a problem without creating ten more.

George Bernard Shaw

On the basis of tonnage emissions, transportation—mainly the Otto cycle internal combustion engine—is by far the largest source of air pollution listed in Table 8.3. When economic damages from the major automobile emissions (CO, hydrocarbons, and nitrogen oxides) are considered the picture changes somewhat since health effects associated with London smog are fairly well known while those resulting from CO are not. (The indirect nature of some of the latter will be developed in the next section.) A recent report[1] estimates that total benefits from cleaning up automobile emissions would be in the range of $2.5 to 7×10^9 per year. Leaving aside questions of accurate assessment of damages from CO and lead, the 1.1×10^9 annual economic cost of photochemical oxidants in Table 8.3 is reason enough to apply considerable effort to abatement of automotive emissions.

10.1 CARBON MONOXIDE

The principal sources of carbon monoxide in the environment are incomplete combustion, combination of hydroxyl radicals with methane derived from decomposition of living matter, the oceans, and the growth

---

[1] Coordinating Committee on Air Quality Studies, "Air Quality and Automobile Emission Control," p. 121. Nat. Acad. Sci.—Nat. Acad. Eng., Washington, D. C., 1974.

and decomposition of plants containing chlorophyll. Chemical reactions related to the first source are shown in equations 10.1:

$$2\,C + O_2 \longrightarrow 2\,C\equiv O \qquad \Delta G^0_{298} = -221.15 - T(0.1796) = -274.67 \text{ kJ} \qquad (10.1a)$$

$$2C\equiv O + O_2 \xrightarrow[\text{slow}]{} 2\,CO_2 \qquad \Delta G^0_{298} = -566.25 - T(-0.1737) = -514.48 \text{ kJ} \qquad (10.1b)$$

$$CO_2 \xrightarrow[\text{T}]{\text{high}} C\equiv O + O \qquad (10.1c)$$

In any combustion process, carbon, either in elemental form or in a compound, is first oxidized to carbon monoxide. Because of the great strength of the triple bond in this molecule, most reactions of carbon monoxide have large activation energies, making the subsequent formation of carbon dioxide considerably slower. If there is insufficient dioxygen for the combustion reaction, conversion to CO_2 will be incomplete and large amounts of carbon monoxide will remain. If the combustion gases remain hot only briefly 10.1b cannot occur quickly enough to convert all CO to CO_2, even in the presence of excess O_2. At very high temperatures carbon dioxide dissociates to carbon monoxide and oxygen atoms, and upon rapid cooling CO cannot be reoxidized quickly enough to achieve an equilibrium concentration of CO_2. These reactions represent the principal sources of carbon monoxide in areas such as cities where there are high concentrations of human population.

On a worldwide scale there are much larger natural sources. Methane produced by decomposition of living matter can react with hydroxyl radicals to form methyl radicals and water:

$$CH_4 + \cdot OH \longrightarrow \cdot CH_3 + H_2O$$
$$\downarrow \text{(several steps)} \qquad (10.2)$$
$$C\equiv O$$

(The average concentration of hydroxyl radicals, which are produced by photolysis of water, is between 1 and 3×10^6 per cm^3 during daylight hours in the troposphere.) The methyl radicals can, in several additional reaction steps,[2] form carbon monoxide. This series of reactions produces about 80% of the carbon monoxide in nonurban atmospheres. The oceans apparently contribute CO to the air, although the source of aquatic CO is not precisely known. Together with decomposition of chlorophyll, they account for about 13% of global emissions.

$$CO_{(ocean)} \longrightarrow CO_{air} \qquad (10.3a)$$

$$C_{55}H_{70}MgN_4O_6 \xrightarrow[\text{decomposition}]{\text{growth}} C\equiv O + \text{other products} \qquad (10.3b)$$
(chlorophyll)

---

[2] J. C. McConnell, M. B. McElroy, and S. C. Wofsy, *Nature (London)* **233**, 187 (1971).

TABLE 10.1   Sources of Atmospheric Carbon Monoxide[a]

Source	Quantity (10^6 tonnes/year)	%
CH_4 oxidation	2800	80
Oceans	360	10
Man-made	245	7
Chlorophyll	90	3
Total	3495	100

[a] Data from *Chem. & Eng. News* July 3, p. 2 (1972).

Outside areas of high population density anthropogenic sources amount to only 7% of the total.

These estimates have been made on the basis of the isotopic composition of atmospheric CO in a manner similar to the method for radiocarbon-14 dating. Carbon monoxide from nonnatural origins is a result of the combustion of fossil fuels which have been stored beneath the surface of the earth for millions of years. The other sources make use of carbon atoms coming from organisms which until recently were living. Interaction of cosmic radiation with carbon dioxide in the upper atmosphere produces a small quantity of radioactive ^{14}C. Because living organisms continuously incorporate carbon dioxide through photosynthesis and other processes, this radiocarbon-14 reaches an equilibrium concentration in all living organisms. After an organism dies the amount of ^{14}C gradually decreases as the radioactive isotope decomposes; thus, the amount of ^{14}C is considerably smaller in fossil fuels than in living organisms. The quantity of ^{14}C in carbon monoxide produced by the incomplete combustion of fossil fuels is also smaller. By using a mass spectrometer to measure the $^{14}C : ^{12}C$ ratio in combination with rate constants for natural CO sources, the estimates shown in Table 10.1 were made.

CO Sinks

Because there are large natural sources for carbon monoxide, there must also be large natural sinks for it. Some of these are

$$C \equiv O + \tfrac{1}{2} O_2 \xrightarrow[\text{bacteria}]{\text{soil}} CO_2 \tag{10.4a}$$

$$C \equiv O + 3 H_2 \xrightarrow[\text{bacteria}]{\text{soil}} CH_4 + H_2O \tag{10.4b}$$

$$C \equiv O + \cdot OH \longrightarrow CO_2 + \cdot H \tag{10.4c}$$

Prior to 1970 both the natural sources and natural sinks were a mystery to most scientists, and even now little is known about equations 10.4a and 10.4b except that soil seems to be able to absorb large amounts of CO.[3]

---

[3] R. E. Inman, R. B. Ingersoll, and E. A. Levy, *Science* **172**, 1229–1231 (1971).

Since this capacity is destroyed if the soil is sterilized, it is theorized that soils contain certain bacteria which can make use of carbon monoxide in their metabolism, producing either carbon dioxide or methane. Soil organisms in the conterminous United States can handle approximately 560×10^6 tonnes of CO per year (more than four times the amount produced by automobiles). Extrapolation to the total land area of the earth (which is about 17 times that of the United States) is tenuous because of variations in climate and soil fertility; thus, it is not clear whether soils can provide a sink for all naturally produced carbon monoxide. Another possible sink is the reaction of CO with hydroxyl radicals to form carbon dioxide and hydrogen atoms. Based on the rate constant for 10.4c and estimates of hydroxyl radical concentrations,[4] this reaction may account for the removal of 50% of atmospheric carbon monoxide before it can come into contact with soils.

These recently discovered sinks have alleviated fears of a worldwide buildup of carbon monoxide in the atmosphere, but median annual concentrations in cities (3–13 ppm)[5] are far higher than those in natural surroundings (0.1 ppm from Table 8.2). Human activities have led to 8-hour averages as high as 40 ppm in commercial areas and 115 ppm in vehicles on congested city streets.[6] Transportation accounts for almost two-thirds of anthropogenic carbon monoxide, and there is a large correlation between observed concentrations and automobile traffic. Carbon monoxide levels are found to be unusually large at the edges of freeways, and they often show two peaks per day, one between 8:00 and 10:00 A.M. and the second between 4:00 and 6:00 P.M., correlating exactly with rush-hour traffic. It is important to note that within a city a major natural sink for carbon monoxide is greatly diminished because much of the soil has been covered by concrete and buildings. Approximately 98% of the carbon monoxide in the United States cities results from man-made sources.

Analysis of CO

The most commonly used method for determining atmospheric CO concentrations is nondispersive infrared (NDIR) analysis. This differs from the usual spectrophotometric ir measurements in that no grating or prism is used — all ir wavelengths pass through the sample and reference cells alternately (see Figure 10.1). The radiation is absorbed by a detector, raising the temperature of the latter in proportion to intensity. Use of gaseous carbon monoxide in the detector is an ingenious way to insure that it will have maximum sensitivity at wavelengths corresponding to CO's absorption

[4] B. Weinstock and H. Niki, *Science* **176**, 290–292 (1972).

[5] U. S. Environmental Protection Agency, "Air Quality Criteria for Carbon Monoxide," AP-62, p. 6–28. USEPA, Washington, D. C., 1970.

[6] U. S. Environmental Protection Agency, "Air Quality Criteria for Carbon Monoxide," AP-62, p. 6–22. USEPA, Washington, D. C., 1970.

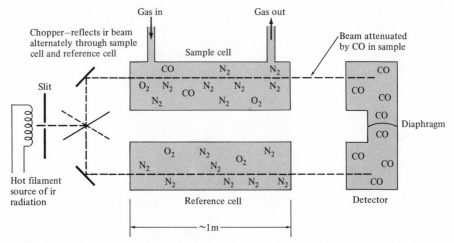

Figure 10.1 Nondispersive infrared analyzer for carbon monoxide. Since more ir radiation passes through the reference cell the temperature of the reference side of the detector is higher. Curvature of the diaphragm thus depends on CO concentration in sample cell.

spectrum. If the intensity of radiation passing through the sample cell is decreased because of the presence of CO, the sample side of the detector will not be as warm as the reference side. The resultant pressure differential is measured by distension of a flexible diaphragm. With proper calibration a direct meter or recorder readout of CO concentration is possible.

Commercially available instruments with cell path lengths of a meter or more can measure a range of 1–115 mg/m³ with response times of 1–5 min. This permits essentially continuous monitoring. Once the high initial investment is paid, a NDIR analyzer can be operated by inexperienced personnel, requires no wet chemicals, and is reasonably insensitive to changes in flow rate and ambient temperatures. Carbon dioxide and water vapor, which absorb strongly in the infrared (Figure 7.4) are the chief interferences, and their effects may be minimized by means of filter cells which remove the wavelengths which they absorb from sample and reference beams.

Effects of CO Air Pollution

Carbon monoxide has not been shown to produce detrimental effects on higher plant life for several weeks at exposure levels below 115 mg/m³ (100 ppm). It does inhibit nitrogen-fixing bacteria in clover roots when such levels are maintained continuously for a month or more.[7]

The principal effect of carbon monoxide on humans and other animals is

---

[7] U. S. Environmental Protection Agency, "Air Quality Criteria for Carbon Monoxide," AP-62, p. 7–2. USEPA, Washington, D. C., 1970.

its interference with the transfer of oxygen through the body. By forming a coordination complex with hemoglobin (Hb) in the red blood cells, carbon monoxide displaces oxygen and prevents the latter from being transported through the bloodstream:

$$Hb + O_2 \rightleftharpoons HbO_2 \qquad (10.5a)$$

$$HbO_2 + CO \underset{K \sim 210}{\rightleftharpoons} COHb + O_2 \qquad (10.5b)$$

Since the equilibrium constant for reaction 10.5b is approximately two hundred, small concentrations of carbon monoxide can compete successfully with ordinary concentrations of oxygen. The normal level of carboxyhemoglobin (COHb) in the blood cells is about 0.5%, most of which results from metabolism of heme groups. For concentrations of carbon monoxide below 100 ppm in the inhaled air the equilibrium level of COHb increases according to the approximate equation[8]:

$$\% \text{ COHb in blood} = 0.16 \times (\text{ppm CO}) + 0.5 \qquad (10.6)$$

However, the equilibrium is not established instantaneously and a period of hours is usually necessary before the maximum level of COHb is reached. Increased physical activity can shorten this time span significantly. The combined effects of ambient CO level, physical activity and length of exposure on % COHb in the bloodstream are shown in Figure 10.2.

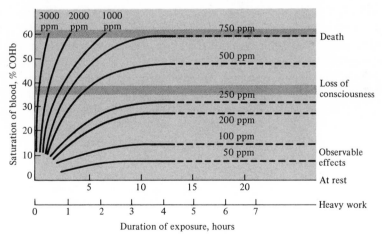

Figure 10.2 Dependence of levels of COHb in red blood cells on ambient air concentration, time of exposure, and activity level. Reprinted, with permission, from P. C. Wolf, *Environ. Sci. Technol.* **5**, 213 (1971). Copyright by the American Chemical Society.

[8] U. S. Environmental Protection Agency, "Air Quality Criteria for Carbon Monoxide," AP-62, p. 8–10. USEPA, Washington, D. C., 1970.

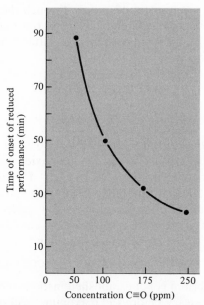

Figure 10.3 Impairment of time discrimination by carbon monoxide. Reduction by two standard deviations from average performance is taken as evidence of reduced performance. Source: U. S. Environmental Protection Agency, "Air Quality Criteria for Carbon Monoxide," AP-62, p. 8–18. US Govt. Printing Office, Washington, D. C., 1970.

Increased COHb in the blood has the effect of depriving various organs, especially the brain, of needed oxygen. This results in the impairment of corresponding physical abilities. One example is shown graphically in Figure 10.3, where time discrimination ability is related inversely to the concentration of carbon monoxide in the air breathed by human subjects. Such impairment could result in an accident should the subject be driving an automobile, but the number of such occurrences which could be attributed to CO air pollution in a year is almost impossible to estimate. Another complication in evaluating economic damages is the fact that cigarette smoking can raise the level of carbon monoxide in the lungs to between 400 and 500 ppm. From Figure 10.2 it can be seen that at these very high levels considerable COHb forms after only a short period of exposure. Smokers have been observed to have blood levels as much as 14 times normal.[9] Therefore, the effects of air pollution and smoking are difficult to separate.

Carbon monoxide in the atmosphere may also cause chronic health effects. For instance, a study done in Los Angeles[10] showed a strong correlation between carbon monoxide concentration and the mortality rate.

---

[9] U. S. Environmental Protection Agency, "Air Quality Criteria for Carbon Monoxide," AP-62, p. 9–8. USEPA, Washington, D. C., 1970.

[10] A. C. Hexter and J. R. Goldsmith, *Science* **172**, 265–267 (1971).

Twenty parts per million CO could account for 11 extra deaths per day in the Los Angeles area when compared with the mean yearly concentration of about 7 ppm. Attempts to find a similar correlation with the level of oxidants from photochemical smog were unsuccessful. Interpretation of such statistical studies is, of course, complicated by the fact that variables which correlated with CO level but were not considered may be the actual cause of deleterious effects.

Since abatement of carbon monoxide pollution depends on controls which must be imposed primarily upon the automobile, we shall postpone discussion of this topic until the effects of other types of pollution generated by the automobile have been enumerated (see Section 10.5).

10.2 NITROGEN OXIDES, NO$_x$

There are several natural sources of nitrogen-containing compounds (other than N$_2$) in the atmosphere.[11] Ammonia, nitrous oxide, and nitric oxide are all released by microbial processes. Each remains relatively independent of the others, however, because most atmospheric reactions involving changes in oxidation number of nitrogen proceed rather slowly. Ammonia returns to the soil or water environment by gaseous deposition (primarily because of its large solubility) or by precipitation of aerosols involving ammonium compounds. Sulfate and nitrate salts are important in this respect. Because of their water solubility they are often found in rainfall, and in such cases ammonia helps to neutralize acid rain. Nitrous oxide is also quite stable in the troposphere and is present in much higher concentrations than other nitrogen compounds (Table 10.2). Most of it is removed by biological sinks, but a small quantity diffuses to the strato-

TABLE 10.2 Estimated Annual Global Emissions and Background Concentration of Nitrogen Compounds (1965)[a]

Compound	Source	Emissions as N (10^6 tonnes/year)	Background concentration (ppm)
NH$_3$	Biological action	868	0.006
N$_2$O	Biological action	343	0.25
NO[b]	Biological action	212	0.0002–0.002
NO$_2$[b]	Combustion processes	14.6	0.0005–0.004

[a] Data from E. Robinson and R. C. Robbins, *J. Air Pollut. Contr. Ass.* **20**(5), 303–306 (1970).

[b] Most NO, whether from natural or anthropogenic sources, is oxidized to NO$_2$ within a few days; thus, the distinction made here is somewhat arbitrary, although conversion is more rapid and complete in urban atmospheres.

---

[11] E. Robinson and R. C. Robbins, *J. Air. Pollut. Cont. Ass.* **20**(5), 303–306 (1970).

sphere where its reaction with atomic oxygen produces NO (Section 8.4).

Reaction pathways involving NO and NO_2 are somewhat more complex and intertwined. Since methods of analysis often measure their combined total concentration these two compounds are usually treated together and labeled NO_x. By far the majority of NO_x comes from biological processes. The NO formed initially is oxidized by O or O_3 over a period of 4–6 days

$$NO + O + M \longrightarrow NO_2 + M \qquad (10.7a)$$

$$NO + O_3 \longrightarrow NO_2 + O_2 \qquad (10.7b)$$

maintaining a ratio $NO_2 : NO$ of about 2.5. Anthropogenic NO is formed by high temperature combination of dinitrogen and dioxygen (equation 8.8) followed by rapid cooling. This slows the reverse reaction so that equilibrium is not maintained. When NO is produced in urban atmospheres its conversion to NO_2 is more rapid and more complete than in rural areas, but the exact mechanisms which account for this are not well understood.[12] Approximately one-fifteenth of the total NO emitted to the atmosphere is a result of human activities.

A nitrogen dioxide molecule remains in the atmosphere for only 3 days on the average. Removal is approximately equally divided between direct gaseous deposition and reaction to form nitric acid:

$$2\ NO_2 + H_2O \rightleftharpoons HNO_3 + HNO_2 \qquad (10.8a)$$

$$\underline{3\ HNO_2 \longrightarrow HNO_3 + 2\ NO + H_2O \qquad (10.8b)}$$

$$3\ NO_2 + H_2O \longrightarrow 2\ HNO_3 + NO \qquad (10.8)$$

Thus, nitrogen as well as sulfur oxides contribute to acid rain.

Analysis of NO_x

Three techniques are commonly used for determination of NO_x. The *chemiluminescent* methods have been devised within the past decade and involve the reaction of O_3 with NO to produce NO_2 in an excited state (Section 8.3). When the latter decays to the ground state the luminescence emitted is proportional to the original concentration of NO (provided that O_3 was in excess). The method can be adapted to determine the sum of $NO + NO_2$ by treating the incoming sample with O, thus converting NO_2 to $NO + O_2$.

The *Griess-Saltzman technique* is based on reaction of NO_2 with sulfanilic acid to form a diazonium salt. The latter then couples with *N*-(1-naphthyl)-ethylenediamine dihydrochloride to form an azo dye whose absorption at 550 nm may be measured spectrophotometrically. Analysis must be completed within an hour of sample collection. The method can be

---

[12] R. S. Berry and P. A. Lehman, *Annu. Rev. Phys. Chem.* **22**, 57–58 (1971).

used to determine atmospheric concentrations of NO_2 from 40 to 1500 $\mu g/m^3$ (0.02–0.75 ppm). High concentrations of SO_2 may bleach the reagent, giving low results, and the stoichiometry of the process is not completely clear,[13] but it can be calibrated using permeation tubes and is easily modified for automated analysis.

The *Jacobs-Hochheiser method* has been adopted as standard by the EPA. It permits up to 2 weeks between sample collection and analysis, and is therefore more amenable to the National Air Surveillance Network (NASN). Polluted air is passed through aqueous NaOH, in which NO_2 is converted to sodium nitrite and some sodium nitrate. (From equation 10.8a and simple redox considerations one might expect 50% conversion of NO_2 to NO_2^-. However, a figure of 63% has been arrived at by experiment,[14] and NASN assumes all NO_2 is converted to NO_2^-.) The collected sodium nitrite is returned to a laboratory, treated with H_2O_2 to remove SO_2 interference, acidified, and used to diazotize sulfanilamide. The latter is coupled to *N*-(1-naphthyl)-ethylenediamine dihydrochloride as ın the Griess-Saltzman analysis.

In addition to questions about the stoichiometry of the Jacobs-Hochheiser method, there are problems involving efficiency of collection of NO_2. It varies nonlinearly with NO_2 concentration (Figure 10.4), and posi-

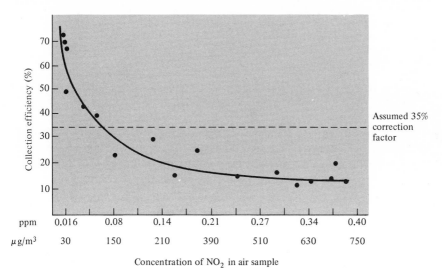

Figure 10.4 Variable collection efficiency of NO_2 in Jacobs-Hochheiser analysis. Reprinted, with permission, from T. R. Hauser and C. M. Shy, *Environ. Sci. Technol.* **6**(10), 892 (1972). Copyright by the American Chemical Society.

---

[13] F. P. Scarengelli, E. Rosenberg, and K. A. Rehme, *Environ. Sci. Technol.* **4**(11), 924–929 (1970).

[14] G. B. Morgan, C. Golden, and E. C. Tabor, *J. Air Pollu. Contr. Ass.* **17**, 300–304 (1967).

TABLE 10.3   Direct Effects of Atmospheric Nitrogen Dioxide[a]

Effect	NO$_2$ concentration		Duration
	ppm	μg/m³	
Increased incidence of acute respiratory disease in families	0.062–0.109	117–205	2–3 years
Increased incidence of acute bronchitis in infants and school children	0.063–0.083	118–156	2–3 years
Human olfactory threshold	0.12	225	—
Rabbits—structural changes in lung collagen	0.25	470	4 hours/day for 6 days
Navel orange—leaf abscission; decreased yield	0.25	470	8 months, continuously
Rats—morphological changes in lung mass cells characterized by degranulation	0.5 / 1.0	940 / 1880	4 hours / 1 hour
Mice—pneumonitis; alveolar distension; increased susceptibility to respiratory infection	0.5 / 0.5	940 / 940	6–24 hours/day for 3–12 months / 6–24 hours/day up to 12 months
Navel orange—leaf abscission, chlorosis	0.5	940	35 days, continuously
Rabbits—structural changes in lung collagen	1.0	1880	1 hour
Sensitive plants—visible leaf damage	1.0	1880	21–48 hours
Man—increase in airway resistance	5	9400	10 min
Monkeys—tissue changes in lungs, heart, liver, and kidneys	15–50	28 200–94 000	2 hours

[a] Source: U. S. Environmental Protection Agency, "Air Quality Criteria for Nitrogen Oxides," AP-84, p. 11–5. USEPA, Washington, D. C., 1971.

tive interference is caused by the presence of NO. The latter is probably not crucial since most of the NO would be oxidized to NO$_2$ anyway in an urban atmosphere. (One method of analysis for NO is to bubble it through KMnO$_4$ or some other strong oxidizing agent and determine it as NO$_2$.) The former difficulty has caused major turmoil at EPA, however, since those concentrations most important for determining compliance with the primary air quality standard of 100 μg/m³ may have been overestimated by as much as a factor of two.[15] Correction of the analytical method reduced the number of air quality control regions exceeding primary standards from 45 out of 247 to only two.

Most of the studies on which the 100 μg/m³ standard was based were in the range (> 100 μg/m³) where underestimation of actual NO$_2$ concentration was the rule. Hence, the standard itself was not affected by the error. A summary of representative effects of NO$_2$ which form the basis for the

---

[15] T. R. Hauser and C. M. Shy, *Environ. Sci. Technol.* **6**(10), 890–894 (1972).

standard is found in Table 10.3. In addition to the direct effects therein, NO_2 is the initiator of photochemical smog. Nitric oxide, at the concentrations observed in the atmosphere, exhibits no adverse effects except rapid oxidation to NO_2.

10.3 PHOTOCHEMICAL SMOG

The problem known today as photochemical smog was first observed in Los Angeles in the early 1940's, and a chemist, Arie J. Haagen-Smit, identified the role of sunlight and photochemical reactions in its formation during the 1950's. Los Angeles smog follows a cyclic pattern over a 24-hour period, the main steps of which are

1. Emission of NO_x and hydrocarbons to the atmosphere. The former absorb solar energy required to drive reactions of the latter.
2. Absorption of sunlight causing photodissociation of NO_2.
3. Consumption of NO_x and simultaneous buildup of "oxidant." The latter term includes O_3, O, peroxides, and excited states of O_2.
4. Oxidation of hydrocarbons to produce a wide variety of products. These include aerosols and eye irritants.
5. Dispersal of the pollutants.

Obviously, a temperature inversion or other meteorological or topographic factor which prevents dispersion of an air mass while steps 2, 3, and 4 are occurring adds considerably to the severity of the problem.

The hour-by-hour unfolding of a typical day of photochemical smog is shown in Figure 10.5. Hydrocarbons and NO_x emitted before sunrise accumulate in the atmosphere, and NO is oxidized to NO_2. As soon as solar radiation becomes available the reaction

$$NO_2 \xrightarrow{h\nu} NO + O \tag{10.9}$$

begins to occur rapidly, producing highly reactive atomic oxygen, which initiates a number of important reactions:

$$NO + O + M \longrightarrow NO_2 + M \tag{10.10a}$$

$$O + O_2 + M \longrightarrow O_3 + M \tag{10.10b}$$

$$O + Hc \text{ (olefin or aromatic)} \longrightarrow R\cdot + RCO\cdot \tag{10.10c}$$

(In these and equations to follow Hc represents any hydrocarbon of the type specified and R·, etc., represent *free radicals,* which have one or more unpaired electrons.) Ozone formed in 10.10b can also react with hydrocarbons yielding free radicals:

$$O_3 + Hc \longrightarrow RCO_2\cdot + \begin{array}{l} RCHO \text{ (aldehyde)} \\ \text{or} \\ R_2CO \text{ (ketone)} \end{array} \tag{10.11}$$

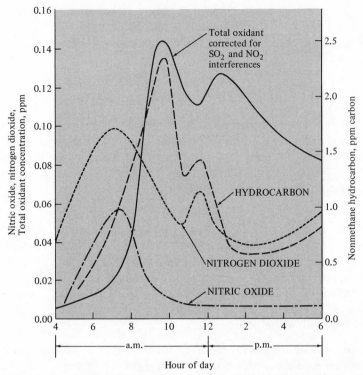

Figure 10.5 Hourly variation of selected pollutants in Philadelphia on Tuesday, July 18, 1967. Source: U. S. Environmental Protection Agency, "Air Quality Criteria for Photochemical Oxidants," AP-63, p. 2–17. US Govt. Printing Office, Washington, D. C., 1970.

some of which speed up production of NO_2 from NO:

$$RCO_2 \cdot + NO \longrightarrow NO_2 + RCO \cdot \qquad (10.12a)$$

Others may react with dioxygen to form peroxide radicals:

$$RCO \cdot + O_2 \longrightarrow RCO_3 \cdot \qquad (10.12b)$$

Note that reaction 10.11 can produce a variety of aldehydes and ketones, depending on the structures of the initial hydrocarbons. These eventually condense to form aerosols which limit visibility.

Another major class of compounds in photochemical smog is the eye irritants (lachrymators) such as peroxyacetylnitrate (PAN). These are produced by reaction of NO_2 with peroxide radicals:

$$CH_3C\begin{matrix} O \\ \diagup\diagdown \\ O-O\cdot \end{matrix} + NO_2 \longrightarrow CH_3C\begin{matrix} O \\ \diagup\diagdown \\ O-ONO_2 \end{matrix} \qquad (10.13a)$$

PAN

Both PAN and the related peroxybenzoylnitrate (PBzN) are very strong

PBzN

eye irritants, extremely toxic to plant life and contribute to the overall level of oxidant. A second reaction which helps to remove NO_2 from the region of photochemical smog is formation of acylnitrates:

$$RCO_2 \cdot + NO_2 \longrightarrow RC \overset{\displaystyle O}{\underset{\displaystyle ONO_2}{\big<}} \qquad (10.13b)$$

These compounds lack the peroxy linkage and hence are weaker oxidants.

The above equations (10.9 to 10.13) are intended to give representative examples of the types of processes which occur during stages 2, 3, and 4 of photochemical smog. By no means should they be taken as a complete list — even with respect to important reactions. Adequate representation of photooxidation kinetics of a single hydrocarbon (propylene) in contact with NO_x polluted air may require 81 elementary processes, and even a generalized mechanism containing only the most important steps includes 15.[16] The above discussion has concentrated on atomic oxygen and ozone to the detriment of mechanisms involving excited dioxygen (the $^1\Delta_g$ state), hydroxyl radicals, or chlorine atoms. The latter are produced by photodissociation of PbBrCl, the major form of lead emitted from automobile exhausts (Section 5.6).[17] Sulfur dioxide is also formed by — and interacts with — photochemical smog, and no notice of this complicating factor has been taken here. Nevertheless, the basic daily cycle of Los Angeles smog — emission of primary pollutants in the morning followed by their photochemical conversion to oxidants which subsequently produce a wide variety of products from hydrocarbons — should be evident.

Detection of Atmospheric Oxidants and Hydrocarbons

The standard method of analysis for photochemical smog is determination of *total oxidant* level. Polluted air is bubbled through a solution of KI, producing I_2 (or triiodide ions) which may be measured spectrophotometrically at 352 nm:

$$O_3 + 3 I^- + 2 H^+ \longrightarrow I_3^- + O_2 + H_2O \qquad (10.14)$$

---

[16] T. A. Hecht and J. H. Seinfeld, *Environ. Sci. Technol.* **6**(1), 47–57 (1972).

[17] For a general treatment of these mechanisms, see R. S. Berry and P. A. Lehman, *Annu. Rev. Phys. Chem.* **22**, 64–77 (1971).

This method is general for most oxidizing agents strong enough to react with I^-, not just O_3. Since many reducing agents (SO_2, for example) can react with I_3^-, the method gives a net oxidant level (total oxidant minus total reductant) unless they are removed. One method is scrubbing with CrO_3, but this also oxidizes NO to NO_2, necessitating a correction factor. If it is desired to measure O_3 alone a chemiluminescent method may be used. Reaction of O_3 with substances such as Rhodamine B or ethene

Rhodamine B

produces products in excited electronic states. Their decay to the ground state is accompanied by emission of characteristic visible radiation, and measurement of the level of luminescence by means of a photomultiplier determines the amount of reaction and hence of ozone present. The chemiluminescent method described earlier for NO can also be adapted to determine O_3.

Specific methods are available for other components of photochemical smog. Peroxyacetylnitrate was discovered by means of infrared spectroscopy using absorption bands at 5.75 and 8.62 μm. This method is complicated by the very low PAN concentration which required multiple reflections of ir radiation through the sample to give an effective path length of 500 m. Peroxyacetylnitrate, aldehydes, and ketones as well as hydrocarbons may be analyzed by means of gas chromatography using electron capture or flame ionization detectors. The latter type of detector gives a response proportional to number of carbon atoms in a compound and is used in the EPA standard reference method for total hydrocarbons. Recently, gas chromatographs using mass spectrometers as detectors have become available, making analysis of specific compounds in the complex smog mixture more feasible.

Effects of Photochemical Smog

Oxidants are estimated to have done approximately $\$1.1 \times 10^9$ worth of damage in the United States in 1970 (Table 8.3). Of this, the majority was attributed to effects on materials and the rest to those on vegetation. An important example of the former is deterioration of rubber because of oxidative scission of double bonds by ozone:

$$
\left[\begin{array}{cccc} H & CH_3 & H & H \\ | & | & | & | \\ C-C & =C & -C \\ | & & | \\ H & & H \end{array}\right]_n + O_3 \longrightarrow \begin{array}{c} H_3C \quad O \quad H \\ | \quad \diagup \quad \diagdown \quad | \\ R-C \qquad C-R' \\ \diagdown \quad \diagup \\ O-O \end{array} \xrightarrow{H_2O} \begin{array}{c} CH_3 \\ | \\ RC=O \end{array} + R'C \begin{array}{c} \diagup O \\ \diagdown OH \end{array} \quad (10.15)
$$

Natural rubber

This effect can be reduced through use of antiozonants, waxes, or ozone-resistant rubbers such as polychloroprene (in which the methyl groups of normal rubber are replaced by chloro). The annual cost to United States consumers for such changes is estimated at $160 to 180×10^6, and premature aging of unprotected rubber accounts for an additional 225×10^6. Extra labor for early replacement of damaged rubber products has been conservatively estimated at 75×10^6.[18] Attack of ozone on double bonds also reduces the strength of fabrics and causes bleaching of dyes, especially in cotton-polyester permanent press fabrics.

Oxidants can also inflict large-scale injury on plants, and this was one of the earliest observed manifestations of photochemical smog. Effects may range from complete collapse of cells to subtle changes in growth patterns, crop yields, and the quality of plant products. Injury to the most sensitive species can occur after exposure to 60 $\mu g/m^3$ (0.03 ppm) ozone for 8 hours. Shorter times are required if low levels of SO_2 are also present. The injury produced by PAN is characterized by under-surface bronzing or glazing of leaves. This may occur in sensitive species after exposure to 50 $\mu g/m^3$ (0.01 ppm) PAN for 5 hours. Not all such injury produces economic damage, but the value of crops and ornamental plants damaged in 1970 is estimated at 200×10^6.

In animals, ozone behaves similarly to ionizing radiation in that the possibility of free radical production exists and there probably is no threshold dose below which some response does not occur. It may contribute to accelerated aging and chromosomal injury. Ozone has been shown to reduce resistance to disease[19] in laboratory animals. In humans, eye irritation, an increased number of asthmatic attacks, and poorer performance of student athletes have all been attributed to photochemical oxidant levels around 200 $\mu g/m^3$ (0.1 ppm). Estimates of economic damage from such injury have not been made.

As in the case of CO and NO_x discussion of means of controlling photochemical oxidant pollution will be deferred to Section 10.5.

---

[18] T. E. Waddell, "The Economic Damages of Air Pollution," p. 80. U. S. Environ. Protect. Ag., Washington, D. C., 1974.

[19] U. S. Environmental Protection Agency, "Air Quality Criteria for Photochemical Oxidants," AP-63, pp. 8–12 to 8–15. US Govt. Printing Office, Washington, D. C., 1970.

10.4   AIRBORNE LEAD

Natural atmospheric concentrations of lead are in the range 0.0005–0.0006 $\mu g/m^3$. They result mainly from mobilization of silicate dusts from soils, which contain 16 ppm lead on average. Since most natural atmospheric lead is in the form of particulates the main sink is agglomeration and precipitation, sometimes alone but sometimes with rain. Human activities have contributed a large amount of lead to the atmosphere from the time the metal was first isolated in 1500 B.C. For example, lead analyses of annual ice layers in Greenland show an increase from 0.0005 $\mu g/kg$ in 800 B.C. to 0.21 $\mu g/kg$ in 1965.[20] Levels in Antarctic snow have also increased, but only to 0.02 $\mu g/kg$, perhaps because most lead smelting and consumption occurs in the northern hemisphere. The majority of man-made lead emissions are from automobile exhausts (Section 5.6), and they account for approximately 10% of urban particulates. In general, atmospheric lead concentrations are higher in cities than in surrounding rural areas, are larger during the day than at night, and decrease steadily as one moves farther from a street or highway. Table 10.4 summarizes the type of data available relating atmospheric lead levels to geographic areas. Lead aerosol concentrations in United States cities have been increasing steadily during the last few decades,[21] paralleling increased consumption of tetraethyllead.

TABLE 10.4   Atmospheric Lead Concentrations at Different Sites[a]

Location	Type of site	Lead concn. ($\mu g/m^3$)
North Central Pacific Ocean		0.0010
Greenland		0.005
California	White and Laguna Mountains	0.008
California	Remote mountains	0.12
Berlin	Quiet streets	0.4–0.5
Philadelphia		1.6
Berlin	Busy street	3.8
New York	2–75 m from traffic	4.1
Los Angeles	Central city	4.3–6.6
Detroit	5–150 m from traffic	4.8
Los Angeles	4-20 m from traffic	7.6

[a] Source: H. A. Waldron and D. Stöfen, "Sub-clinical Lead Poisoning," Table 4, pp. 10–11. Academic Press, New York, 1974.

[20] M. Murozumi, T. J. Chow, and C. Patterson, *Geochim. Cosmochim. Acta* **33**, 1247–1294 (1969).
[21] T. J. Chow and J. L. Earl, *Science* **169**, 577–580 (1970).

Analysis for Lead

The standard analysis for atmospheric lead[22] is based on the *dithizone method.* The sample is collected on a 0.45-μm membrane filter. If it is desired to trap gaseous (usually organic) lead compounds a tube filled with I_2 crystals may be used after filtration. (A recent report[23] indicates that more complete trapping of molecular lead may be accomplished by charcoal adsorption. The same authors indicate that organic lead may comprise 50% or more of the total, invalidating results obtained using only filtration for particulates.) The filtered sample is digested with HNO_3, H_2SO_4, and $HClO_4$ to solubilize the lead. Iodine crystals are dissolved in acid KI and reduced with sulfite:

$$I_3^- + H_2SO_3 + H_2O \longrightarrow 3\ I^- + SO_4^{2-} + 4\ H^+ \tag{10.16}$$

Again the lead is solubilized. Acidic Pb^{2+} from either source is made basic with NH_3 and extracted with a chloroform solution of dithizone:

$$Pb^{2+}_{(aq)} + 2\ S{=}C \rightleftharpoons Pb\,(dithizone)_2 + 2\ H^+_{(aq)} \tag{10.17}$$

(in $CHCl_3$)

Dithizone (in $CHCl_3$)

Following careful removal of excess dithizone (which has a green color) from the chloroform phase, the lead–dithizone complex may be determined spectrophotometrically at 510 nm.

Mass spectrometry is also important in lead analysis because it can be used to determine the source of the element. Lead has four stable isotopes: ^{204}Pb, ^{206}Pb, ^{207}Pb, and ^{208}Pb; and a given ore contains unique proportions of each. By determining the isotope ratios $^{206}Pb:^{204}Pb$, $^{206}Pb:^{207}Pb$, and $^{206}Pb:^{208}Pb$ from the mass spectrum of a sample its origin may be inferred. On the basis of such ratios, which are nearly identical for tetraethyllead in gasoline and atmospheric aerosols, the overwhelming contribution of the former to urban particulates has been determined. Atomic absorption analysis is also an important tool in the study of lead concentrations.

Effects of Lead

Acute lead poisoning has been a problem since antiquity. Its symptoms are abdominal pain, nausea, tingling and numbness of the hands and feet,

---

[22] National Academy of Sciences, "Airborne Lead in Perspective," Appendix B. Nat. Acad. Sci., Washington, D. C., 1972.

[23] J. W. Robinson and D. K. Wolcott, *Environ. Lett.* **6**(4), 321–333 (1974).

and muscle cramps. Fortunately, its occurrence has decreased rapidly during the twentieth century as sources of lead in the human diet have been discovered and eliminated. Acute poisoning may be treated by ingestion of calcium EDTA (ethylenediaminetetraacetic acid), which chelates Pb^{2+} and aids in its removal from the body. Lead is still the most common cause of poisoning among domestic animals, being obtained both from concentrated sources such as discarded batteries or from chronic ingestion of plants on whose leaf surfaces atmospheric lead has deposited. (Fortunately, most lead can be washed from leaf surfaces or human intake would be much larger.) The number of birds which die from ingestion of lead shot may be as high as one million per year, and shellfish have a remarkable facility for concentrating lead from their environments.

Chronic effects of lead in man are currently of much greater concern than acute. Certainly blood lead levels have been raised well above the natural concentration (Table 10.5). Under ordinary circumstances most of this lead is obtained from food rather than water or air, although absorption through the lungs is highly efficient and may account for increases in blood lead when atmospheric concentrations exceed 2 $\mu g/m^3$. Approximately 10% of blood lead in semirural areas is attributable to inhalation,[24] and proportionately larger fractions would be expected in cities, where atmospheric concentrations are higher. Higher blood levels are consistently found among persons exposed to large quantities of automobile exhaust. Lead intake is common among children with *pica*—the craving for unusual food. This often includes chips of lead-based paint, but recent evidence in-

TABLE 10.5   Lead Content of Human Blood[a]

Lead (ppm)	Significance
0.01	"Natural" blood lead level before man began using lead
0.10	Lower limit of "normal" blood level in the United States
0.25	Mean blood lead level in the United States
0.25	Suggested "danger" blood lead level for children
0.30	Mean blood lead level in Glasgow children
0.30	Lowest lead level found in industrially exposed adults having mild symptoms of lead poisoning
0.31	Mean blood lead level in Manchester children
0.40	Upper limit of "normal" blood lead level in the United States
0.40	Lower average blood lead level in children showing lead poisoning symptoms
0.40	Lowest approximate level found in industrially exposed adults having severe symptoms of lead poisoning
0.70	European "danger" threshold for occupational poisoning
0.80	United States "danger" threshold for occupational poisoning

[a] Source: T. J. Chow, *Chem. Brit.* **9** (6), 260 (1973).

[24] B. B. Ewing and J. E. Pearson, *Advan. Environ. Sci. Technol.* **3**, 75 (1974).

dicates that consumption of dust and dirt composed of lead fallout is a major source of unusual ingestion.[25] Urban dust has been found to have lead concentrations as high as 10 000 ppm by weight.

Large, soft cations such as Pb^{2+} or Hg^{2+} are especially prone to react with sulfhydryl (—SH) groups in proteins. Lead is known to inhibit at least two enzymes whose catalytic function is essential for biosynthesis of heme. Because this impairs the production of hemoglobin, oxygen starvation and anemia may result from elevated blood lead levels. For one enzyme, δ-aminolevulinic acid dehydrase, recent evidence indicates that current levels of environmental lead contamination are sufficient to produce inhibition and that no threshold exists below which blood lead has no effect.[26] It has also been suggested that blood lead levels may be associated with biochemical abnormalities in children's brains, since there is a direct relationship between the activities of blood and brain enzymes.[27] This may account for increased hyperactive and delinquent behavior among children, adults, and animals with high lead levels.[28] Thus, adverse *social* consequences may result from environmental lead contamination.

Many of the effects noted above have not been proved beyond a doubt. It is especially difficult to assess the impact of lead emitted from automobiles, since this is one, highly variable component in total lead exposure. In February 1975 the U. S. Court of Appeals for the District of Columbia set aside EPA regulations which provided for phasing out of leaded gasoline by 1979. The basis for this action was the above-mentioned lack of proof. However, an earlier EPA ruling that most service stations must sell at least one grade of unleaded gasoline was let stand because automobiles equipped with catalytic converters for exhaust emission control must burn unleaded fuel to avoid poisoning the catalysts. Although the controversy over health effects associated with atmospheric lead seems likely to continue for a number of years, the total lead consumption in motor fuel will probably decrease.

10.5 CONTROL OF AUTOMOBILE EMISSIONS

Automobiles emit hydrocarbons, lead, NO_x, and CO to the atmosphere. Nearly 100% of the latter three pollutants is found in engine exhaust, but about 15% of the hydrocarbons are lost by evaporation from the fuel tank and carburetor. Another 20% escape by leaking around the piston rings of

[25] T. M. Roberts, T. C. Hutchinson, J. Paciga, A. Cattopadhayay, R. E. Jervis, J. Van Loon, and D. K. Parkinson, *Science* **186**, 1120–1123 (1974); J. P. Day, M. Hart, and M. S. Robinson, *Nature (London)* **253**, 343–345 (1975).

[26] H. A. Waldron and D. Stoffen, "Sub-clinical Lead Poisoning," p. 122. Academic Press, New York, 1974.

[27] B. B. Ewing and J. E. Pearson, *Advan. Environ. Sci. Technol.* **3**, 84 (1974).

[28] *Chem. Brit.* **10**(6), 205 (1974).

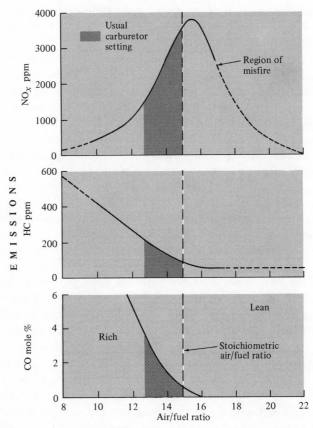

Figure 10.6 Effect of air–fuel weight ratio on exhaust emissions. Redrawn from A. J. Haagen-Smit. *Chem. Technol.* June, p. 335 (1972).

an Otto cycle engine (see Figure 5.12) to the crankcase, from which they evaporate. Fuel tank and carburetor emissions are currently controlled by adsorption on activated charcoal in a cannister through which escaping gases must pass. Positive crankcase ventilation (PCV), first added to automobiles in 1963, recycles vapors from crankcase to carburetor. Thus instead of escaping to the atmosphere these hydrocarbon emissions are burned.

Exhaust emissions produce conflicting control problems. Prior to regulation of emissions, carburetors were usually set to produce a "rich" mixture of gasoline and air—slightly more gasoline than the exact quantity which could be oxidized by the O_2 present was fed to the cylinders (Figure 10.6). This gave a smooth-running engine but did not allow complete combustion of CO and Hc. Adjustment of a "leaner" mixture containing the stoichiometric or a somewhat larger quantity of dioxygen greatly reduces CO and

Hc emissions, but produces *more* NO_x. This is because the former pollutants must be oxidized to remove them from exhaust gases while the latter is produced by oxidation of atmospheric dinitrogen.

Emission Standards

A chronology of regulations intended to reduce photochemical smog is presented in Table 10.6. You will note from the table that, in recognition of the aforementioned difficulty of simultaneously controlling NO_x as well as Hc and CO, regulation of the former did not begin until 1971 in California and 1973 nationwide. During the approximately 5 years between regulation of Hc and NO_x, emissions of the latter actually increased *more rapidly* than they would have in the absence of any controls. Such a reduction in one of the two chemical components needed to initiate photochemical smog might seem to be an adequate control measure, but in practice it is

TABLE 10.6   Chronology of Standards for Air Emissions from Motor Vehicles

| Year | Standard | Emission level (g/mile unless specified otherwise) | | |
		Hc	CO	NO_x
Prior to controls	—	11 (850 ppm)	80 (3.4%)	4–6
1966	California	275 ppm	1.5%	None
1968[a]	California and United States	410 ppm	2.3%	None
		350 ppm	2.0%	None
		275 ppm	1.5%	None
1970	California and United States	2.2	23	None
1971	California	2.2	23	4.0
1972	United States[b]	3.4	39	None
	California— old	1.5	23	3.0
	or new[b]	3.2	39	3.2
1973	United States	3.4	39	3.0
	California	3.2	39	3.0
1974	California	3.2	39	2.0
1975	United States	1.5	15	3.0
	California	0.9	9.0	2.0
1977	United States[c]	0.41	3.4	0.4

[a] 1968–1969 standards varied with engine size. The figures given are for 50–100, 101–140, and 140 cubic inch displacement, respectively.

[b] Based on a new test driving cycle used continuously after 1972.

[c] This represents a 90% reduction from 1970 levels for Hc and CO and the California 1971 standard for NO_x. However, in March 1975, EPA recommended freezing 1975 United States standards on Hc and CO through 1979, adopting 1975 California standards nationwide in 1980 and 1981, and postponing 1977 United States standards to 1982. EPA also recommended a nationwide standard of 2.0 g/mile NO_x for 1977.

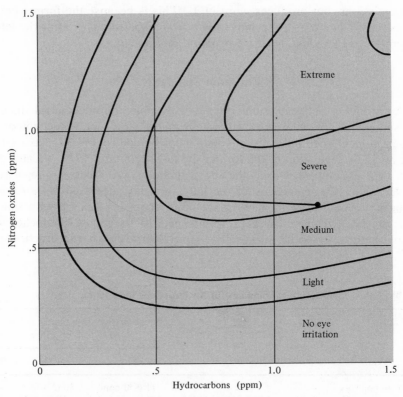

Figure 10.7 Effects of variations of atmospheric Hc and NO_x levels on development of eye irritation in photochemical smog. The straight line shows the effect of reducing Hc by 50% and simultaneously increasing NO_x by 10%. Data from A. J. Haagen-Smit, *Advan. Chem. Ser.* **113,** 169–185 (1972).

not. This is shown in Figure 10.7 where the concentrations of NO_x and Hc early in the morning of a smog day are plotted. The graph is divided by iso-irritation lines into areas of a similar degree of eye irritation. A 50% reduction of Hc (say from 1.2 to 0.6 ppm) accompanied by a 10% increase in NO_x (from 0.64 to 0.70 ppm) might actually *increase* eye irritation once smog developed. Obviously, unless a very large percentage (>95%) of Hc can be controlled, regulation of NO_x is also required to improve air quality.

The emission standards in Table 10.6 are the result of considerable negotiation between EPA, the auto industry, environmentalists, and, more recently, the Federal Energy Administration. As footnote *c* of Table 10.6 indicates, those projected beyond 1975 will probably have been modified by the time this is read. Test procedures have been changed by EPA several times since 1970 in an attempt to find a reproducible driving cycle representative of average automobile use in the United States. Thus, although California emission levels were less than 80% of 1970 in 1972, a new testing procedure required that gram per mile regulations be higher in the latter year. For the past 5 years applicable federal standards have changed on the average of every 6 months, making for a highly confusing situation.

In addition to adjusting the air–fuel ratio, retarding the spark and reducing the compression ratio of an internal combustion engine can effect small reductions in both Hc and NO_x emissions. Injection of air into exhaust manifolds can oxidize Hc and CO more completely and is in use by a number of manufacturers. Recirculation of exhaust gases to the cylinders reduces flame temperatures and hence NO formation from N_2 and O_2, and careful matching of the air–fuel ratio to power demands by means of fuel injection can allow an engine to run lean without misfiring. However, except for the last these changes give poorer fuel economy and driveability, and they are barely adequate in controlling pollution to 1975 requirements.

Both NO_x and Hc-CO standards can be met by using an *exhaust manifold thermal reactor* (EMTR). This device works on the same principle as air injection, but the exhaust manifold is insulated so that CO and Hc can be oxidized at a high temperature. If the air–fuel mixture is made very rich, NO_x emissions are small (Figure 10.6) and the reactor can take care of excess CO and Hc. A price must be paid in terms of fuel economy, however.

The Catalytic Converter

The exhaust control device which seems most compatible with fuel economy is the *catalytic converter* that has been installed on the majority of 1975 automobiles. Hot exhaust gases into which air has been injected are passed over a catalyst of platinum and palladium on the surface of silica or alumina pellets or a honeycomb of supporting metal. The catalyst bed is contained in a specially aluminized steel reactor that is about the size and shape of a muffler. It allows thermodynamically favored reactions such as oxidation of Hc and CO to take place rapidly at lower temperatures. Thus, the air–fuel ratio and spark timing in the engine may be set for maximum fuel economy and the catalytic converter will remove any increased amounts of pollution. As in the case of many catalysts, lead is a "poison"—small quantities can completely remove catalytic activity. (There is also some speculation that bromine from $C_2H_4Br_2$ added to exhaust lead from the cylinders may poison noble metal catalysts.) Since catalytic exhaust converters are required by EPA to last for 50 000 miles of service, cars so equipped must never burn leaded fuel—more than one tankful would destroy the catalytic activity and emissions would be nearly as great as for uncontrolled 1965 model vehicles.

Even with catalytic converters removal of NO_x is not simple, although at first it might appear to be. Platinum or palladium surfaces will speed most thermodynamically favored reactions, so NO can be converted to NH_3 or

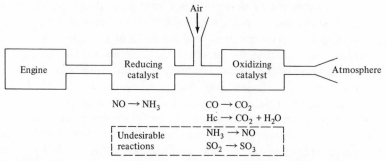

Figure 10.8 Dual catalyst exhaust gas treatment for removal of NO_x, Hc, and CO.

N_2. Most proposals for meeting 1977 emission standards involve a dual catalyst system like the schematic example in Figure 10.8. If there is a slight excess of fuel in the mixture entering the engine there will be an excess of Hc and CO over O_2 and NO in the exhaust gas entering the first catalyst. Apparently, NO is reduced by dihydrogen left on the catalyst surface following decomposition of Hc:

$$5\ H_2 + 2\ NO \longrightarrow 2\ NH_3 + 2\ H_2O \qquad (10.18)$$

Once NO has been reduced excess dioxygen is added and CO and Hc are removed by the second catalyst bed. Obviously, NO must be reduced first because addition of O_2 removes all reducing agents.

Problems with Catalytic Converters

Unfortunately, the NH_3 produced in equation 10.18 does not stay reduced. It will be recalled that the Ostwald process for converting NH_3 to HNO_3 begins with oxidation of ammonia to nitric oxide on a platinum catalyst. Thus, the second catalyst undoes much of the work of the first. This problem would not occur if NO could be reduced to N_2 instead of NH_3, because with a triple bond dinitrogen has a much larger activation energy for reoxidation. Production of N_2 would require that two NO molecules be adsorbed *close together* on the catalyst surface (Figure 10.9) so that their nitrogen atoms could bond together. Since NO concentrations in exhaust gases are quite small this is unlikely. The chief mechanism for formation of N_2 is apparently reaction of NO with NH_3:

$$6\ NO + 4\ NH_3 \longrightarrow 5\ N_2 + 6\ H_2O \qquad (10.19)$$

but this leaves substantial quantities of NH_3 to be reoxidized.

Two other problems with catalytic converters are the high temperatures at which they operate and the fact that they rapidly oxidize any sulfur in motor vehicle fuel to SO_3. There have been reports of grass fires started by converter-equipped cars whose exhaust systems were hotter than normal

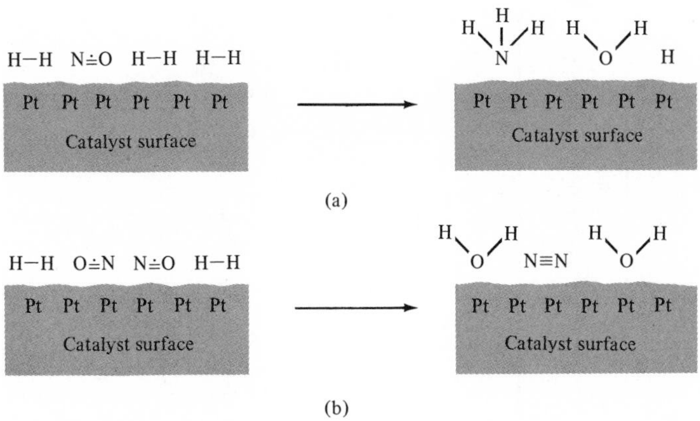

Figure 10.9 Possible mechanisms for reduction of NO on a platinum catalyst surface: (a) reduction to NH_3 and (b) reduction to N_2.

as a result of exothermic reactions on the catalyst bed. Of greater concern is increased emissions of sulfuric acid and sulfate aerosols by converter-equipped cars. Although all of the SO_2 emitted by noncatalytic exhausts is eventually converted to sulfate (Section 9.3), the catalytic system insures that oxidation will occur at ground level, producing sulfates exactly where a maximum number of persons will be affected. It was because of this problem that EPA proposed the postponement of enforcement of emissions standards detailed in footnote *c* of Table 10.6. George Bernard Shaw would probably be delighted if he could witness the new problems created by this scientific "solution" to automotive air pollution.

10.6 ALTERNATIVES TO THE OTTO CYCLE ENGINE

It is obviously not a simple matter to meet strict emissions standards (originally set for 1975 by the 1970 Clean Air Act, but currently postponed to 1977) by adding control devices or making adjustments on the standard automobile engine. Furthermore, the cost of meeting the current 1977 standards, which adequately control all three major automobile emissions, is expected to be from \$300 to \$500 per vehicle. Therefore, various alternative power plants have been proposed. The characteristics of a number of these are presented in Table 10.7.

The *Wankel rotary* engine, used in the Japanese Mazda automobile, produces less NO_x than the standard internal combustion engine. Carbon monoxide and hydrocarbon emissions are higher because of the large surface area of the Wankel combustion chamber, but these may be removed by means of an EMTR. However, complete control of NO_x is difficult, and efficiency of the Wankel at its current state of development is not good enough to make it a strong contender to replace the Otto cycle engine.

TABLE 10.7 Comparison of Some Alternatives to the Otto Cycle Engine[a]

Power system	Emissions (g/mile)					Maximum efficiency (%)	Ten-year cost (relative to Otto)
	HC	CO	NOx	SOx	Particulate		
Otto cycle, no control[b]	11	80	4–6	0.27	0.36	19	1.0
Otto cycle, 1975 U. S.	1.5	15	3.0	~0.27	0.1	19	1.06
Stratified charge (Honda CVCC)	0.2	2.0	0.8	~0.3	—	~22	~1
Wankel (with EMTR)	1.3	20	~2	~0.3	—	~15	~1
Diesel	3.5	5	4–12	—	c	21	0.82
Gas turbine	0.2–0.9	2–8	1.0–1.6	0.3	d	17	1.16
Rankine cycle (steam or organic working fluid)	0.013–0.4	0.35–2.0	0.25–0.6	0.2–0.3	d	14	1.04
Battery, electric (present; coal-fired plant, 2.5% S)	Negligible	Negligible	2	10	2.5	30	1.18
Hybrid (ICE-electric)[e]	0.4	4	0.4	—	—	20	1.15

[a] Source: R. U. Ayres and R. P. McKenna, "Alternatives to the Internal Combustion Engine." Johns Hopkins Univ. Press, Baltimore, Maryland, 1972.

[b] Values can be considerably higher for some uncontrolled engines.

[c] May be large if engine not properly tuned and maintained.

[d] Published data not available. Smokeless combustion should be readily attainable.

[e] Information supplied to the U. S. Department of Transportation by Minicars, Inc., Goleta, California.

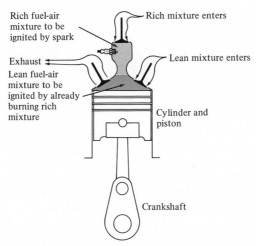

Figure 10.10 Stratified charge principle in the Honda CVCC internal combustion engine.

A very promising modification of the Otto cycle is the *stratified charge* principle. Running an engine with a large excess of air over fuel can reduce all three major pollutants, but too lean a mixture is almost impossible to ignite by means of a spark plug. In a stratified charge engine a small amount of a rich fuel–air mixture is provided at the spark plug. Once ignited it causes a much larger quantity of lean mixture to burn. In the Honda CVCC engine this is accomplished by means of two combustion chambers (Figure 10.10). Because combustion is smoother, slower, and occurs at lower maximum temperatures, all three major emissions are reduced well below 1975 standards. Only NO_x exceeds the 1977 (1982?) level, and no exhaust controls are needed. The stratified charge principle has produced similar results in a large American V-8 engine,[29] and little or no penalty in fuel economy seems to be required. The approach of modifying combustion to reduce emissions certainly gets closer to the basic problem than add-on exhaust devices and seems very promising.

Prior to the energy crisis the *diesel engine* was largely discounted as a replacement for the Otto cycle, even though it produces less CO and Hc emissions. Its higher efficiency and the availability of diesel fuel when gasoline was in short supply have revived interest more recently; but the diesel produces much NO_x and, if not properly maintained, considerable particulate emissions. The latter are not regulated now, but certainly would have to be if the majority of new cars were to be equipped with diesel engines. Diesels, despite higher initial costs, are more economical in the long run than other types of engines, accounting for their use in trucks and buses where long service is expected.

---

[29] *Environ. Sci. Technol.* **9**(1), 10 (1975).

Engines in which fuel is not burned in an enclosed cylinder fall into the category of *external combustion*. Since fuel need not be ignited by a spark but instead burns continuously, such engines permit greater variation in air–fuel ratio and hence are inherently less polluting, even in the absence of exhaust controls. The *gas turbine* has been under evaluation as an automotive power plant for more than a decade, and followers of the sports scene will recall that gas turbine-powered cars nearly ran off with the Indianapolis 500 mile race 2 years in succession. However, because exotic materials are required to withstand the high temperatures at which turbine blades must operate, such engines are costly. They also are somewhat slow to respond to driver demands for acceleration.

A second example of external combustion is the *steam* or *Rankine cycle* engine. Here combustion temperatures usually are lower than for the gas turbine, and emissions of all three major pollutants are extremely low. A number of makes of steam-powered cars were popular in the United States early in the twentieth century, but they eventually were displaced by internal combustion engines. Most of these early vehicles had large boilers which required considerable time between burner ignition and buildup of adequate pressure to drive the car. There also was a danger of explosion should pressure get too high, the range of cars was limited by their need to take on more water, and the water reservoir might freeze in winter. Most of these problems are overcome in modern designs by using a flash-boiler and a condenser to recycle water. In a flash-boiler only a small quantity of liquid is injected. It vaporizes rapidly, providing pressure as needed; thus, a large quantity of energy need not be stored in a boiler. Some modern Rankine engines are even based on working fluids such as freons, benzene, or other organic substances. In these latter cases one would have to evaluate carefully the environmental impact of production and possible loss of organic materials for tens of millions of vehicles. Rankine cycle engines are very attractive at first glance, but a number of potential manufacturers who were highly optimistic 5 years ago do not appear to have made more than superficial progress since then.

Two fundamental problems plague attempts to develop *electric powered automobiles*. The first has already been discussed in Section 7.2: The batteries of galvanic cells which are now available are far too heavy and expensive and they require lengthy charging cycles. Several United States firms now manufacture electric cars – usually based on lead-acid cells – but these are mainly aimed at short-trip applications such as commuting to work and shopping. Driving ranges of such vehicles are on the order of 80–160 km (50–100 miles), and even a rapid recharge would require 20–30 min (and waste energy as well). This might be partially overcome by means of a hybrid vehicle having both an internal or external combustion engine as well as electric motors.

Second, although electric cars are not expected to emit much pollution themselves, the problem would merely be shifted to electric power plants, increasing emissions of particulates and sulfur oxides and requiring storage of more radioactive wastes. Furthermore, it is highly unlikely that electric power production could be expanded rapidly enough to permit a changeover to electric vehicles in less than 20 years. [Remember that the net energy available, especially from nuclear plants, depends on how many new plants must be constructed each year (see Section 7.3).]

Evaluation of Alternatives

With the exception of the Wankel each of the alternative power plants described above achieves low emissions without add-on devices such as PCV or catalytic converters. This is advantageous because it eliminates temptation of the vehicle owner to disconnect or otherwise nullify the efficacy of the emission control system. Despite the certification of catalytic converters by means of 50 000 mile driving tests, there is no guarantee that similar systems in use by the general public will all maintain their integrity unless a major program of regular inspection and replacement is instituted. At present some states do not even inspect safety features such as brakes or lights except by random sampling at highway checkpoints.

For the near term the stratified charge Otto cycle engine appears to be an excellent choice for reducing automotive emissions. Looking somewhat farther into the future several of the alternatives mentioned may develop as serious competitors. The electric vehicle might be especially favored should a hydrogen economy develop. In any case energy conservation should not be ignored. Given enough time and high prices for fuel, living styles and travel habits may change significantly. Persons may live closer to workplaces, travel more often via mass transit systems, and de-emphasize the personal automobile, perhaps replacing many of its functions with nonpolluting machines such as bicycles. Social and behavioral changes such as these also afford tremendous opportunities for reduction of air pollution.

EXERCISES

(The first group of exercises is based primarily on material found in this text.)

1. Give equations for the reactions which produce the major portion of atmospheric CO in cities and in rural areas. Cite evidence that supports your choice.
2. Give a description (and equations where applicable) of the most common method of analysis for atmospheric:

   a. CO   b. NO_x c. Photochemical oxidant
   d. Ozone   e. Lead

3. Describe clearly a day in the life of a photochemical smog. What types of reactions would be occurring at each stage of the process?

4. Most proposals for meeting the 1977 pollution control requirements for automobiles involve the use of two separate catalytic reactors. Why two? What types of catalysts are to be used? Why would it be preferable to reduce NO to N_2 instead of NH_3 in a catalytic reactor? Why is is difficult to obtain N_2 by means of a catalytic process, especially if the concentration of NO is low?

5. Note that when dioxygen reacts with a free radical (equation 10.12b) another free radical is produced, but when NO_2 reacts (equation 10.13a) no new radical is formed. Explain. How many free radicals are formed when NO reacts with $RCO_2 \cdot$ (equation 10.12a)?

6. Referring to equation 8.8, what fundamental change in combustion is required in order to minimize production of NO? Explain why recirculating exhaust gases to the cylinders of an Otto cycle engine accomplishes this change.

(Subsequent exercises may require more extensive research or thought.)

7. What are the principal drawbacks of the Jacobs-Hochheiser analysis for NO_2? Explain how these problems have affected:

   a. The EPA primary air quality standard of 100 $\mu g/m^3$ NO_x.
   b. Regulations controlling emissions of NO_x from stationary sources.
   c. The 1970 Clean Air Act provision that automobile emissions of NO_x must be reduced by 90% in 1976.

   [See T. R. Hauser and C. M. Shy, *Environ. Sci. Technol.* **6**(10), 890–894 (1972).]

8. Look up the half-life of ^{14}C in a table of nuclides. Assuming that coal contains 65% carbon and that the current rate of production of ^{14}C has not changed over the geological time scale (about 10^{-12} of all carbon atoms currently are ^{14}C), calculate the number of ^{14}C nuclei that would remain in 100 tonnes of coal that was laid down 10^6 years ago.

9. It has been found that the presence of NO increases collection efficiency of NO_2 in the Jacobs-Hochheiser analysis. Why might this be so? (Equation 10.8a may be helpful.)

10. Why is carbon monoxide so much better at bonding to hemoglobin than O_2? (Consult a biochemistry text if you are not familiar with the active site of Hb to which O_2 binds.)

11. Computer simulations allow one to experiment with the environment without actually fouling things up. A good simulation of photochemical smog is available [B. J. Huebert, *J. Chem. Educ.* **51**(10), 644 (1974)]. Obtain the program and use it to determine the effect of varying intensity of sunlight, concentration of primary pollutants, etc., on oxidant level.

Suggested Readings

1. A. W. Adamson, "A Textbook of Physical Chemistry." Academic Press, New York, 1973. Chapter 18 is devoted to spectroscopy and photochemistry. Other physical chemistry texts should give similar discussions.
2. American Chemical Society, "Cleaning Our Environment." ACS, Washington, D. C., 1969; Supplement, 1971. Contains much useful information. The supplement is primarily a call for more research.
3. R. U. Ayres and R. P. McKenna, "Alternatives to the Internal Combustion Engine." Johns Hopkins Univ. Press, Baltimore, Maryland, 1972. A comprehensive analysis of existing and potential power plants for automotive vehicles.
4. W. Bach, "Atmospheric Pollution." McGraw-Hill, New York, 1972. This book, written by geographer, contains good discussions of meteorology, health effects, economic aspects, technology and control, and public action relating to air pollution.
5. R. S. Berry and P. A. Lehman, Aerochemistry of air pollution. *Annu. Rev. Phys. Chem.* **22**, 47–84 (1971). A comprehensive review biased (in the authors' words) "toward the application of aerochemical information to air resource management, rather than toward the fundamental chemistry." Includes an evaluation of various atmospheric models.
6. R. D. Cadle, "Particles in the Lower Atmosphere." Van Nostrand-Reinhold, Princeton, New Jersey, 1966. Concerned with all sources of particulates, not just air pollution.
7. R. D. Cadle and E. R. Allen, Atmospheric photochemistry. *Science* **167**, 243–249 (1970). A good general treatment of chemistry in the atmosphere.
8. E. N. Cantwell, "A Total Exhaust Emission Control System." E. I. DuPont de Nemours & Co., Wilmington, Delaware, 1970. Complete description of the EMTR in a reprint of a paper presented at a meeting in Quebec, 1970.
9. T. J. Chow, Our daily lead. *Chem. Brit.* **9**(6), 258–263 (1973). Facts and figures on environmental distribution of lead.
10. Committee on Biologic Effects of Atmospheric Pollutants, "Asbestos." Nat. Acad. Sci., Washington, D. C., 1971. Concludes that "it would be imprudent to permit additional contamination of the public environment with asbestos."
11. Committee on Biologic Effects of Atmospheric Pollutants, "Fluorides." Nat. Acad. Sci., Washington, D. C., 1971. This thorough report covers sources, sinks, analysis, effects on living systems, and population hazards of fluorides in nearly 300 pages.
12. Committee on Biologic Effects of Atmospheric Pollutants, "Lead: Airborne Lead in Perspective." Nat. Acad. Sci., Washington, D. C., 1972. A thorough report by a panel which some have accused of being biased toward industry [*Science* **174**, 800–802 (1971)].
13. Committee on Biologic Effects of Atmospheric Pollutants, "Particulate Polycyclic Organic Matter." Nat. Acad. Sci., Washington, D. C., 1972. Carcinogenesis and other effects of fused aromatic hydrocarbon pollutants are considered in detail.

14. Committee on Effects of Atmospheric Contaminants, "Effects of Chronic Exposure to Low Levels of Carbon Monoxide on Human Health, Behavior and Performance." Nat. Acad. Sci.—Nat. Acad. Eng., Washington, D. C., 1969.

15. R. A. Duce, G. L. Hoffman, and W. H. Zoller, *Science* **187**, 59–61 (1975); R. A. Carr and P. E. Wilkniss, *ibid.* **181**, 843–844 (1973); H. V. Weiss, M. Koide, and E. D. Goldberg, *ibid.,* **172**, 261–263 (1971). These papers and other references within them are part of a series of analyses of trace substances in remote areas aimed at determining anthropogenic contributions to global biogeochemical cycles.

16. J. R. Goldsmith and S. A. Landaw, Carbon monoxide and human health. *Science* **162**, 1352–1359 (1968).

17. R. F. Gould, ed., "Photochemical Smog and Ozone Reactions," *Adv. Chem. Ser. No. 113*. Amer. Chem. Soc., Washington, D. C., 1972. Specialist papers from two symposia.

18. K. Griggs, Toxic metal fumes from mantle-type camp lanterns. *Science* **181**, 842–843 (1973). Emission of beryllium from thorium oxide mantles of gas lanterns during initial use may be hazardous in enclosed spaces.

19. A. J. Haagen-Smit, The control of air pollution. *Sci. Amer.* **210**(1), 3–9 (1964). An early, readable discription of photochemical smog, its causes and control.

20. A. L. Hammond and T. H. Maugh, II, Stratospheric pollution: Multiple threats to earth's ozone. *Science* **186**, 335–338 (1974). The effects of hydrogen bombs, cosmic rays, freons, and SST's on stratospheric O_3 are considered. For specific details, see references footnoted in Chapter 8 of this book.

21. L. Hodges, "Environmental Pollution." Holt, New York, 1973. A physicist's view of air pollution is found in Chapters 3–6.

22. P. A. Leighton, "Photochemistry of Air Pollution." Academic Press, New York, 1961. The classic work on photochemical smog.

23. J. P. Lodge, Jr., ed., "The Smoake of London." Maxwell Reprint Co., Elmsford, New York, 1969. This reprinting of Evelyn's "Fumifugium" and R. Barr's "The Doom of London" gives a good account of the history of air pollution.

24. J. H. McFarland and C. S. Benton, The oxides of nitrogen and their detection in automotive exhaust. *J. Chem. Educ.* **49**(1), 21–23 (1972). Descriptions of methods and instruments for analysis of NO_x.

25. S. E. Manahan, "Environmental Chemistry." Willard Grant Press, Boston, Massachusetts, 1972. Chapters 11 through 15 are relevant.

26. D. F. S. Natusch, J. R. Wallace, and C. A. Evans, Jr., Toxic trace elements: Preferential concentration in respirable particles. *Science* **183**, 202–204 (1974).

27. H. B. Palmer and D. J. Seery, Chemistry of pollutant formation in flames. *Annu. Rev. Phys. Chem.* **24**, 235–262 (1963).

28. H. C. Perkins, "Air Pollution." McGraw-Hill, New York, 1974. The readable, up-to-date view of an engineer.

29. D. Shapley, Auto pollution: EPA worrying that the catalyst may backfire. *Science* **182**, 368–371 (1973). An early warning of sulfuric acid emissions from catalytic converter-equipped cars.

30. A. C. Stern, ed., "Air Pollution," 2nd ed., Vols. I–III. Academic Press, New York, 1968. The definitive reference work on the subject.

31. A. C. Stern, H. C. Wohlers, R. W Boubel, and W. P. Lowry, "Fundamentals of Air Pollution." Academic Press, New York, 1973. Treats air pollution from the perspective of a chemist, a meteorologist, and two mechanical engineers.

32. H. S. Stoker and S. L. Seager, "Environmental Chemistry: Air and Water Pollution". Scott, Foresman & Co., Glenview, Illinois, 1972. Chapters 1–7 are applicable to air pollution.

33. U. S. Environmental Protection Agency, "Air Pollution Emission Factors," AP-42, 2nd ed., USEPA, Research Triangle Park, North Carolina, 1973. Emissions data per unit of

output goods for a variety of industries. Must be combined with knowledge of industrial production to give total emissions.

34. U. S. Environmental Protection Agency, "Air Quality Criteria for Carbon Monoxide," AP-62; "Air Quality Criteria for Nitrogen Oxides," AP-84; "Air Quality Criteria for Photochemical Oxidants," AP-63. US Govt. Printing Office, Washington, D. C., 1970. A complete summary of sources, characteristics, and effects of these pollutants.

35. U. S. Environmental Protection Agency, "Air Quality Criteria for Particulate Matter," AP-49; "Air Quality Criteria for Sulfur Oxides," AP-50. US Govt. Printing Office, Washington, D. C., 1970. Thorough, well-written, and understandable descriptions of properties, sources, and toxicology of air pollutants.

36. U. S. Environmental Protection Agency, "Control Techniques for Asbestos Air Pollutants," AP-117; "Control Techniques for Beryllium Air Pollutants," AP-116. USEPA, Research Triangle Park, North Carolina, 1973.

37. U. S. Environmental Protection Agency, "EPA's Position on the Health Effects of Airborne Lead," USEPA, Washington, D. C., 1972.

38. U. S. Environmental Protection Agency, "Nationwide Air Pollutant Emission Trends, 1940–1970." USEPA, Research Triangle Park, North Carolina, 1973. Many useful data neatly summarized in tables.

39. T. E. Waddell, "The Economic Damages of Air Pollution," EPA-600/5-74-012. USEPA, Washington, D. C., 1974. A thorough estimate of the dollar costs of air pollution in 1970. A briefer treatment may be found in L. B. Barrett and T. E. Waddell, "Cost of Air Pollution Damage: A Status Report," AP-85. USEPA, Washington, D. C., 1973.

40. G. L. Waldbott, "Health Effects of Environmental Pollutants." Mosby, St. Louis, Missouri, 1973. A wealth of medical and toxicological information on nearly every possible air pollutant in language understandable to the nonspecialist.

41. H. A. Waldron and D. Stöfen, "Sub-clinical Lead Poisoning." Academic Press, New York, 1974. The authors have tried to assess information from all available sources. They conclude that lead can increase morbidity and mortality even when present below levels necessary to produce overt symptoms of lead poisoning.

IV
Earth

The optimist proclaims that we live in the best of all possible worlds; and the pessimist fears this is true.

James Branch Cabell

A MAJOR FACTOR IN THE RAPID GROWTH OF THE United States from a small collection of colonies into a world power during the past two centuries has been its rich endowment of natural resources. Excellent soils and extensive deposits of minerals containing important metals and nonmetals permitted both agriculture and industry to develop with maximum speed. But a recent study by the United States Geological Survey[1] indicates that known deposits of mineral resources are seriously depleted: This country is self-sufficient in only ten of the thirty-six important industrial raw materials and must import all or part of its requirements for the rest; to meet demands to the end of this century will require recycling, conservation, and new techniques for development of mineral resources, as well as the use of deposits currently classed as subeconomic or yet to be discovered.

On a worldwide basis annual consumption of some elements is uncomfortably close to estimated reserves, and there is little doubt that this will soon be true of many others. The discussion of ore formation in Chapter 2 has made it clear that minerals are nonrenewable resources—at least on any practical time scale—unless they are recycled. As quantities of solid wastes grow and reserves of high grade ores decline, concentrations of many of the elements will become larger in wastes than in the ores. Considering the difficulty and expense involved in disposing of solid wastes, this leads to the inescapable conclusion that techniques for separating, concentrating, and recycling these materials must be developed and applied.

Chapter 11 will consider the processes currently used for the extraction and purification of both metallic and nonmetallic minerals, together with associated pollution problems. Fertile soils are an extremely important resource, and their formation and conservation will be discussed in Chapter 12. Finally, the recent accelerated growth in production of solid wastes, current means for their disposal, and methods for recycling such "urban ores" will be treated in Chapter 13. The origins, properties, and reserves of energy-related minerals have already been discussed in Chapter 5. The fact that these materials are also in short supply dictates that the "entropy ethic" must be applied to mineral recovery or recycling, and note will be taken of the free energy cost of virgin and recycled materials whenever possible.

---

[1] U. S. Geological Survey, "United States Mineral Resources." US Govt. Printing Office, Washington, D. C., 1973.

11

Mineral Resources

There is many a rich stone laid up in the bowels of the earth,
many a fair pearl laid up in the bosom of the sea, that never
was seen, nor never shall be.

Joseph Hall (Bishop of Norwich)

. . . the cost and availability of the required energy are prob-
ably the single most important factors that will ultimately
determine whether or not a particular mineral deposit can be
worked economically.

United States Geological Survey

11.1 ESTIMATING RESERVES OF MINERAL RESOURCES

Accurate determination of the quantities of mineral resources in the
earth's crust is a difficult but important problem. More complete exploration, advances in technology, or increases in price may have major effects
on estimated exploitable reserves of any mineral (Section 2.6); thus, any
prediction, optimistic or pessimistic, must contain a large margin of error.

In attempting to predict the quantity of a given mineral which is available
for human exploitation, two approaches have been employed. The first ex-
trapolates, into the future, data on the rate of past industrial activity using
current levels of technology. For example, to estimate reserves of crude oil
(Section 5.4) the number of barrels discovered per meter of exploratory
drilling might be projected. Since this method does not allow for technological breakthroughs it may underestimate supplies of a mineral.

The second approach is to relate reserves of an element to its relative
abundance in the crust of the earth. It has been estimated,[1] for example,
that the number of tonnes of an element available in recoverable deposits
within the United States lies between 10^6 and 10^9 times its percent abun-

---

[1] V. E. McKelvey, *Amer. Sci.* **60**(1), 32–40 (1972).

TABLE 11.1 Estimated Annual World Consumption and Resource Reserves of the Elements

Element	Z	Principal sources	Quantity consumed[a] Tonnes/year	Rank	Resource reserves[b] (tonnes)
H	1	Water, methane	1.8×10^6	20	140×10^6
He	2	Natural gas	3.2×10^3	43	—
Li	3	Petalite, lepidolite, spodumene, lake brines	1.36×10^3	47	2×10^6
Be	4	Beryl	272	50	280×10^3
B	5	Na and Ca borates, brines	362×10^3 (B_2O_3)	24	1×10^6
C	6	Diamond	5.4 (industrial)	61	
		Graphite	272×10^3	26	
		Coal	1.6×10^9	1	
		Petroleum	816×10^6	2	
		Natural gas	363×10^6	3	
		Total C	2.8×10^9		$20 \times 10^{6\,c}$
N	7	Air	15×10^6	9	
		Soda niter	907×10^3 ($NaNO_3$)		2×10^6
O	8	Air	18.1×10^6	7	46.6×10^9
F	9	Fluorite	997×10^3	22	62.5×10^6
Ne	10	Air			
Na	11	Halite	90.7×10^6 (NaCl)	5	2.83×10^9
Mg	12	Seawater, magnesite	136×10^3 (Mg) 8.16×10^6 ($MgCO_3$)	11	2.09×10^9
Al	13	Bauxite	5.5×10^6	13	8.13×10^9
Si	14	Quartz	635×10^3	23	27.7×10^9
P	15	Apatite (phosphorite)	6.3×10^6	12	105×10^6
S	16	Sulfur, pyrite, natural gas	18.1×10^6	7	26×10^6
Cl	17	Halite	4.5×10^6	16	13×10^6
Ar	18	Air	18.1×10^3	35	—
K	19	Sylvite, carnallite	9.06×10^6	10	2.59×10^9
Ca	20	Calcite	54.4×10^6 (CaO)	6	3.63×10^9
Sc	21	Thortveitite	0.05 (Sc_2O_3)	68	2.2×10^6
Ti	22	Ilmenite, rutile	9.1×10^3 (Ti) 907×10^3 (TiO_2)	22	440×10^6
V	23	U, Pb vanadates	6.35×10^3	41	13.5×10^6
Cr	24	Chromite	1.27×10^6	21	10×10^6
Mn	25	Pyrolusite, psilomelane	5.44×10^6	14	95×10^6
Fe	26	Hematite, magnetite	281×10^6	4	5×10^9
Co	27	Co sulfides, arsenides	11.8×10^3	36	2.5×10^6
Ni	28	Pentlandite, garnierite	362×10^3	24	7.5×10^6
Cu	29	Chalcopyrite, chalcocite	4.8×10^6	15	5.5×10^6
Zn	30	Sphalerite	3.4×10^6	17	7.0×10^6
Ga	31	Bauxite	9.06	59	1.5×10^6
Ge	32	Germanite	90.66	53	150×10^3
As	33	Arsenopyrite, enargite	36.3×10^3	32	180×10^3
Se	34	By-product of Cu smelting	907	49	5×10^3
Br	35	Seawater, brines	99.7×10^3	29	250×10^3
Kr	36	Air			
Rb	37	Pollucite, salt deposits			9×10^6
Sr	38	Celestite	7.25×10^3	39	37.5×10^6
Y	39	Monazite, euxenite	4.5 (Y_2O_3)	68	3.3×10^6

TABLE 11.1 (Continued)

Element	Z	Principal sources	Quantity consumed[a] Tonnes/year	Rank	Resource reserves[b] (tonnes)
Zr	40	Zircon	907 (Zr) 181×10^3 (zircon)	27	16.5×10^6
Nb	41	Columbite, pyrochlore	1.18×10^3	48	2×10^6
Mo	42	Molybdenite	40.8×10^3	31	150×10^3
Ru	44	Platinum ores	0.12	66	1×10^3
Rh	45	Platinum ores	3	62	500
Pd	46	Platinum ores	21.8	58	1×10^3
Ag	47	Silver sulfides	7.25×10^3	39	7×10^3
Cd	48	Sphalerite	11.79×10^3	36	20×10^3
In	49	By-product of Zn and Pb smelting	9.07	59	10×10^3
Sn	50	Cassiterite	172×10^3	28	200×10^3
Sb	51	Stibnite	54.5×10^3	30	20×10^3
Te	52	By-product of Zn and Pb smelting	181	52	1×10^3
I	53	Brines, by-product of Chilean nitrate	3.63×10^3	42	50×10^3
Xe	54	Air			
Cs	55	Pollucite	1	64	300×10^3
Ba	56	Barite	2.9×10^6 (barite)	18	42.5×10^6
La-Lu		Monazite, bastnasite	1.81×10^3 (oxides)	45	15.1×10^6
Hf	72	Zircon	45.3	54	300×10^3
Ta	73	Tantalite	272	50	200×10^3
W	74	Scheelite, wolframite	27.2×10^3	33	150×10^3
Re	75	Molybdenite	1	64	100
Os	76	Platinum ores	0.06	67	500
Ir	77	Platinum ores	3	62	100
Pt	78	Platinum ores	27.2	56	1×10^3
Au	79	Gold, gold tellurides	1.45×10^3	46	400^d
Hg	80	Cinnabar	8.16×10^3	38	8×10^3
Tl	81	By-product of Zn and Pb smelting	27.2	56	50×10^3
Pb	82	Galena	2.5×10^6	19	1.3×10^6
Bi	83	By-product of Pb smelting	2.7×10^3	44	20×10^3
Th	90	Monazite, by-product of U extraction	45.3	54	720×10^3
U	92	Uranite	27.2×10^3	33	180×10^3

[a] Data from B. Mason, "Principles of Geochemistry," 3rd ed., Appendix III. Wiley, New York, 1966.

[b] Estimated as $10^9 \times$ percent crustal abundance given by B. Mason, "Principles of Geochemistry," 3rd ed., Table 3.3, p. 45. Wiley, New York, 1966. See text for discussion of methods of estimating resources.

[c] The estimated reserves are obviously too small by more than a factor of 10^3. This may be related to the fact that carbon, unlike the metals used to calibrate this method of estimating resources, was concentrated by biological rather than entirely geological processes. On the other hand, it may indicate that $10^9 \times$ percent crustal abundance is an underestimate of total resource reserves. A factor of 10^{10} or 10^{11} may be more reasonable.

[d] Gold is another example that belies uncritical reliance on this method of estimating reserves. Its great value and chemical inertness have probably contributed to the large reserves that have been found.

dance in the earth's crust. This type of estimate can only give order of magnitude results, of course, and ignores the problem of techniques for recovery from lower grade ores. This latter problem is illustrated by comparing iron with aluminum. Although iron is less abundant, reserves are 100 times those of aluminum, because no feasible techniques are available for recovery of the latter from many of its minerals.

Since the United States occupies about 7% of the land area of the world, *total reserves* of the elements can be roughly approximated by multiplying crustal percent abundances by 10^7 to 10^{10}. These estimates can be compared with the data on annual world *consumption* in Table 11.1. In a number of cases (lead and mercury, for example) reserves and consumption are uncomfortably close, and there is no doubt that for many minerals finite reserves are being approached by rapidly growing levels of consumption.

11.2 EXTRACTION OF METALS – GENERAL PRINCIPLES

A glance back to Table 1.1 or study of Table 11.1 will remind you that oxygen is by far the most abundant element on our planet. The crustal abundances of silicon, aluminum, iron, calcium, magnesium, sodium, and potassium follow in roughly the order given. Consequently, many ore-bearing minerals are oxides or silicates and also contain as impurities most of the metals mentioned above. The second largest class of ores consists of sulfides of the chalcophiles, and other ores contain sulfates, hydrated oxides, or carbonates formed by reactions with oxygen, water, or CO_2. In only a few cases (platinum, gold, and copper) are metals found in "native" (unoxidized) form.

There is a twofold problem in extracting a metal from its ore. First, the metal must be *chemically reduced* to elemental form. Such a reduction (equation 11.1) is usually accompanied by a positive free-energy change:

$$MX \longrightarrow M + X \tag{11.1}$$

The greater the affinity of the metal for X, the larger the value of ΔG. Reaction 11.1 can be made to occur by supplying free energy in either of two forms. A reducing agent, R, can release free energy upon combination with X, forcing the reaction to the right,

$$MX + R \longrightarrow M + RX \tag{11.2}$$

or electrical energy may be employed to force reaction 11.1 to the right. An example of each of these types of reduction will be treated a little later.

A second aspect of metallurgical extraction is the *removal of impurities* which may remain dissolved in the metal of interest after reduction. Such refining often involves the use of *slags* — mixtures of oxides which can dissolve unwanted constituents from the metal. The composition of a slag is carefully designed so that at the high temperatures of reactions like 11.2 it is immiscible with the metal and sufficiently fluid to be poured or skimmed from the metal surface. (Most oxides or silicates are much less dense than most metals; thus, slags collect at the top of a reaction vessel.) Because the melting point of a pure substance is decreased by a solute, mixtures of oxides are often used in slags, allowing them to remain liquid at lower temperatures, and minimizing the heat requirements of the process. Acidic and basic oxides may be employed to dissolve specific impurities and hence a slag is often designated as "acidic" or "basic." Finally, the mass of slag used should be minimal to avoid trapping metal, reduce the quantity of heat needed for melting, and minimize disposal problems.

Since silicon is the second most abundant element of the earth's crust, it is the commonest impurity in ores, and most slags contain primarily SiO_2, with smaller amounts of the oxides of aluminum, iron, calcium, magnesium, sodium, and potassium. Because it is more electronegative than most metals silicon forms partly covalent bonds with oxygen, and in SiO_2 the atoms are connected in a three-dimensional network (see Figure 2.5e.4). When the solid melts only a few of the Si–O bonds dissociate and the melt contains polymeric giant ions with fractional amounts of positive or negative charge per atom. Within each such ion part of the framework structure of silica is maintained, as shown in Figure 11.1. Because such large ions have difficulty diffusing around one another a silica melt tends to be highly viscous.

The addition of other metal oxides to molten silica has the effect of reducing viscosity and making the melt more basic. Oxides having the general formula RO (where R = Ca, Mg, Fe, Na_2, K_2) are basic. They react with water to give hydroxide compounds and in molten salts they liberate O^{2-}, which is a strong Lewis base. On the other hand, the polymeric ions produced when SiO_2 melts can accept pairs of electrons from oxide ions, thus behaving as Lewis acids. Part of the acidity of SiO_2 is thereby neutralized. At the same time the degree of polymerization of the giant ions in the silica melt is decreased, reducing the viscosity. This is shown in Figure 11.1b.

Since 2 mol of O^{2-} are required to saturate the acidic sites of SiO_2:

$$SiO_2 + 2\ O^{2-} \rightleftharpoons SiO_4^{4-} \tag{11.3}$$

the composition of a slag may be considered to be neutral when exactly 2 mol of RO have been added for each mole of SiO_2. (Other acidic oxides can neutralize different numbers of oxide ions. For example, Al_2O_3, P_2O_5, and SO_2 require, respectively, three, three, and one oxide ion for conversion to AlO_3^{3-}, PO_4^{3-}, and SO_3^{2-}.) When 2 mol of RO per mol of SiO_2 have been added the orthosilicate composition, $(RO)_2SiO_2$ or R_2SiO_4, results. An acid slag, which contains more acidic oxide than this composition, is useful for removing impurities such as calcium, magnesium, sodium,

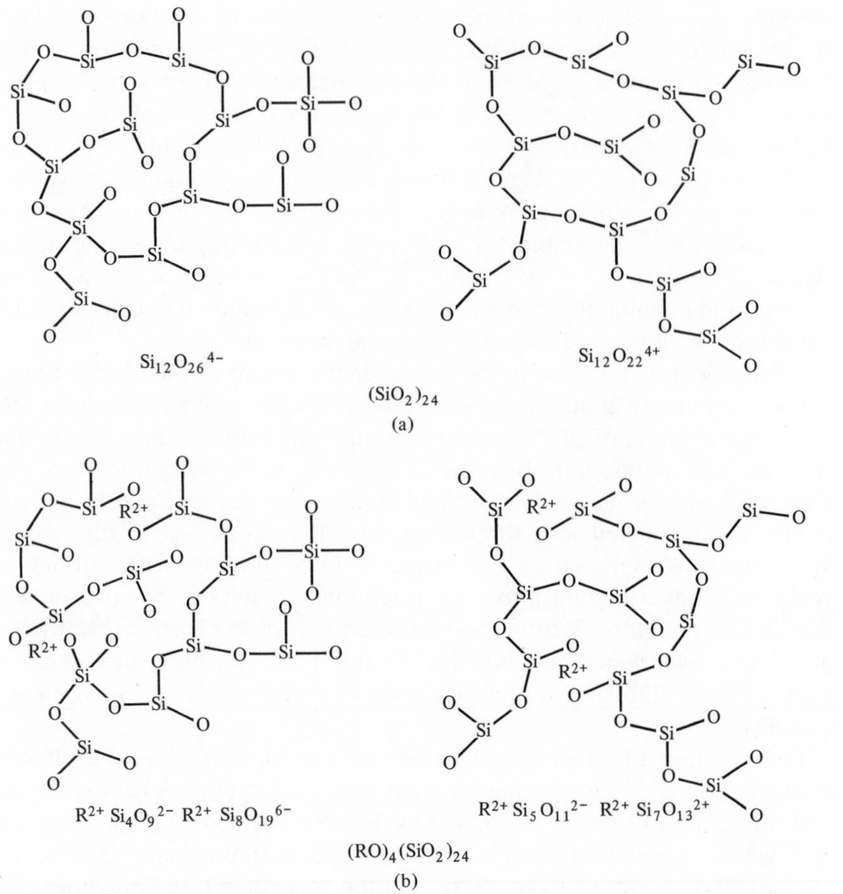

$Si_{12}O_{26}{}^{4-}$

$Si_{12}O_{22}{}^{4+}$

$(SiO_2)_{24}$

(a)

$R^{2+} Si_4O_9{}^{2-}$ $R^{2+} Si_8O_{19}{}^{6-}$

$R^{2+} Si_5O_{11}{}^{2-}$ $R^{2+} Si_7O_{13}{}^{2+}$

$(RO)_4(SiO_2)_{24}$

(b)

Figure 11.1 Structure within a silica melt: (a) giant ions in molten SiO_2 (two-dimensional representation) and (b) breaking of Si–O–Si bonds by RO (R = Ca, Mg, Fe, Na_2, or K_2).

potassium, or iron (which form basic oxides) from molten metals. Conversely, basic slags containing excess alkali or alkaline earth oxides can be used to remove the acidic oxides of silicon, phosphorus, or sulfur during ore refining processes.

Since many ores are oxides and some sulfides are converted to oxides before reduction, it is useful to consider reaction 11.2 when X is oxygen:

$$MO \longrightarrow M + \tfrac{1}{2} O_2 \qquad \Delta G_a \qquad (11.4a)$$

$$\tfrac{1}{2} O_2 + R \longrightarrow RO \qquad \Delta G_b \qquad (11.4b)$$

$$\overline{MO + R \longrightarrow M + RO} \qquad \Delta G = \Delta G_a + \Delta G_b \qquad (11.4)$$

This set of equations is similar to the half-cell reactions encountered in electrochemistry except that oxygen is being transferred instead of elec-

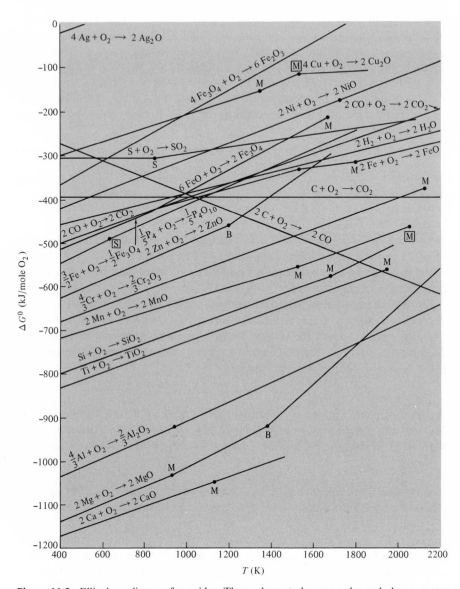

Figure 11.2 Ellingham diagram for oxides. Those elements lower on the scale have greater tendencies to form oxides at a given temperature. Temperatures of phase transitions for reactants and products, respectively, are indicated by M, ☐M☐ for melting point; B, ☐B☐ for boiling point; and S, ☐S☐ for sublimation point.

trons. Not surprisingly, metals and other substances may be ranked according to standard free energies of formation of their oxides. An *Ellingham diagram,* like the electrochemical series, indicates their tendency to be oxidized or reduced, but it presents this information as a function of temperature (see Figure 11.2). The lower it is found in the Ellingham diagram the more stable an oxide, and the greater the difficulty of recovering the metal. Thus, for example, at 1373 K magnesium will replace aluminum from its oxide:

$$2 \text{ Mg} + \tfrac{2}{3} \text{ Al}_2\text{O}_3 \rightleftharpoons 2 \text{ MgO} + \tfrac{4}{3} \text{ Al} \tag{11.5}$$

but at 2073 K $\Delta G^0 = +70$ kJ for equation 11.5 and the equilibrium constant is less than one.

Because ΔG^0 is plotted as a function of T in the Ellingham diagram, the slopes of the lines are given by $-\Delta S^0 [\Delta G^0 = \Delta H^0 + (-\Delta S^0)T]$. It was mentioned in Section 4.2 that ΔH^0 and ΔS^0 vary little with temperature unless a phase change occurs and this is borne out by Figure 11.2. Except at the melting or boiling points of reactants or products the slopes of the lines remain nearly constant.

Formation of metal oxides by reactions such as 11.4b usually involves a decrease in entropy since a solid is formed from solid or liquid metal plus gaseous oxygen. Thus the slopes $(-\Delta S^0)$ of most lines in the Ellingham diagram are positive. There are two exceptions to this rule. Reaction 11.6 has 1 mol of gaseous substance on each side

$$\text{C(s)} + \text{O}_2\text{(g)} \longrightarrow \text{CO}_2\text{(g)} \tag{11.6}$$

and the entropy change is therefore quite small, 0.8 J/K. Consequently, ΔG^0 for this reaction is nearly independent of temperature and its line is horizontal in Figure 11.2. The second exception also involves carbon:

$$2 \text{ C(s)} + \text{O}_2\text{(g)} \longrightarrow 2 \text{ CO (g)} \tag{11.7}$$

In this case there are 2 mol of gaseous product but only 1 mol of gaseous reactant and ΔS^0 is about $+167$ J/K, causing the curve in the Ellingham diagram to have a negative slope. As the temperature is raised CO becomes more stable and therefore carbon will reduce more metal oxides. Iron(II) oxide can be reduced above about 1000 K, zinc(II) oxide above 1275 K (this is above the boiling point of elemental zinc, which must be condensed from the vapor after carbon reduction), and aluminum(II) oxide above 2300 K. This latter reaction requires such a high temperature that it is impractical to construct a smelter to withstand the heat; thus, aluminum is recovered by an electrolytic process instead of carbon reduction.

The thermodynamic treatment of ore reduction just presented does not consider that reactants or products may be at activities (or concentrations) other than unity. It also ignores rates of the reactions involved. [In the carbon reduction of zinc(II) oxide, for example, a temperature of about 1375 K is necessary in order for zinc to be produced at a significant rate, although 1275 K is sufficient to produce an equilibrium constant greater than one.] In spite of these complicating factors the Ellingham diagram presents the basic information necessary for a general understanding of ore reduction in a simple, compact form.

11.3 IRON

As an example of the carbon reduction method of recovering metals from ores, consider iron- and steelmaking. The processes involved and major sources of pollution are shown schematically in Figure 11.3. Hematite (Fe_2O_3), magnetite (Fe_3O_4), limonite (hydrated Fe_2O_3), and siderite ($FeCO_3$) are the principal iron ores. Since most of these contain relatively high concentrations of the metal, little or no preliminary treatment or mineral dressing is required, and reduction by carbon in a blast furnace is the first major step. The modern blast furnace is based on techniques discovered in the fourteenth century, and except for vast increases in scale and some substitutions of materials (coke for charcoal, for example) it has not changed greatly.

Iron-making is a *countercurrent process.* The "charge," consisting of iron ore, coke, and limestone, enters the top of the furnace through an airlock-type pair of valves (the upper and lower bells) and preheated (900–1000 K) air is simultaneously forced through blowpipes (tuyères) at the bottom of the furnace. This blast of air passes up the shaft, reacting with some of the carbon in the coke to form CO and CO_2 and thereby heating the charge. The exhaust gases, which contain a high concentration of CO, can be trapped above the furnace and after cleaning may be burned to run the air compressors and preheaters. Liquid iron and a layer of slag which floats on top of it collect on the hearth at the bottom of the furnace. The iron is tapped about four times a day and allowed to run into a storage furnace where it is kept molten until transfer to a steelmaking furnace. Each tonne of iron requires 2–3 tonnes of ore, about 1 tonne of carbon, and nearly 6 tonnes of air. About 93% of the iron content of the ore is recovered as molten iron and another 4–5% can be trapped in the gas-cleaning operation, leaving no more than 3% unaccounted for.

Most ore in the lower portion of the furnace shaft is reduced by direct reaction with carbon. Since carbon monoxide is more stable than carbon dioxide at the temperature (> 1000 K) of this part of the furance the reaction is

$$FeO + C \longrightarrow Fe + CO \qquad (11.8)$$

Higher in the shaft, where the charge is cooler, indirect reduction (by CO instead of C) is apparently the principal reaction:

$$FeO + CO \longrightarrow Fe + CO_2 \qquad (11.9)$$

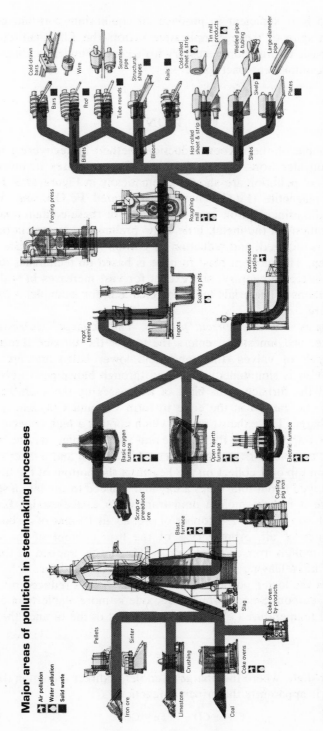

Figure 11.3 Major areas of pollution in steelmaking processes. Reprinted with permission from *Environ. Sci. Technol.* **5**(10), 1006–1007 (1971). Copyright by the American Chemical Society.

The excess oxygen in higher oxides such as Fe_2O_3 or Fe_3O_4 is also removed by reaction with C or CO, but at lower temperatures than equations 11.8 and 11.9 because of the lesser stability of iron(III) toward reduction.

At the high temperature of the furnace the limestone in the charge is decomposed (calcined) to lime,

$$CaCO_3 \longrightarrow CaO + CO_2 \qquad (11.10)$$

which serves as a *flux* to reduce the melting points of the silicon and aluminum oxide impurities. This reduces consumption of fuel (coke). The liquid slag flows down through the hottest part (> 2000 K) of the furnace rapidly enough so that silicon remains as the oxide even though it is thermodynamically unstable to carbon reduction. The basic character of CaO serves to neutralize the acid slag of SiO_2 and Al_2O_3, and, more importantly, to remove sulfur from the iron:

$$FeS + CaO + C \Longleftrightarrow Fe + CaS + CO \qquad (11.11)$$

This reaction tends to reverse when large concentrations of CaS build up in the slag, however; thus low-sulfur coal must be used for coking if sulfur content of the iron is to be minimized. Excess $CaCO_3$ is usually not added to the charge in order to force equation 11.11 to the right; because of its high melting point too much CaO increases the "free-running temperature" (melting point) of the slag, the temperature and thermal losses of the furnace, and the quantity of coke required as fuel. Most low-sulfur coal available in the eastern United States has been allocated to metallurgical use; hence, the incentive to develop western mines for electric power plant supply (Section 5.1).

The chief pollution problems associated with blast furnaces are solid wastes (slag) and particles suspended in the air blast. Some older blast furnaces were extremely bad air polluters, and numerous regions of Appalachia are still dotted with old slag heaps which support little or no plant life. Production of coke also evolves H_2S and some SO_2. Slags are sometimes salable as aggregate for cement, and air pollution can be controlled by bag houses, wet scrubbers, and electrostatic precipitators. Approximately 63×10^3 dm^3 of water per tonne of iron are required to quench coke and slag and for scrubbing of particulates; thus, water pollution is a secondary problem unless effluents are treated in sedimentation ponds.

11.4 STEEL

The pig iron produced by a blast furnace is brittle and suitable only for casting because of its high carbon content. The usual analysis of impurities shows 4% carbon, 2% phosphorus, 2.5% silicon, 2.5% manganese, and 0.1% sulfur by weight. Pig iron is converted to steel by oxidation of these impurities in preference to iron. Several techniques for such preferential oxidation are now in use. The *open hearth* and *electric furnaces* require external heating sources and use air at the surface of the molten mass as an oxidizing agent; the *oxygen furnace,* a modern variant of the now-obsolete Bessemer converter, uses dioxygen as the oxidizing agent. It is the most efficient steelmaking method, and is rapidly replacing the others, having increased from 2% of the United States production in 1959 to 65% in 1974.

Open Hearth

Open hearth steel accounted for 23% of United States production in 1974. Although rapidly being replaced by basic oxygen furnaces, some older open hearths have no facilities for air pollution control and are thus major contributors of dust and gases in the steel industry. The furnace is fired with power gas or natural gas preheated by passage through large chambers filled with an open lattice of firebricks. The direction of flow is reversed at suitable intervals so that the chambers which have been warmed by gases exhausting from the furnace can transfer heat to the incoming fuel gas and air. Because of this regenerative preheating system the high temperatures (~ 1900 K) necessary to melt scrap iron and slag can be achieved.

The usual charge to an open hearth furnace contains molten iron from a blast furnace and scrap iron or steel in roughly the proportions shown in Table 11.2. Enough limestone is added to produce the basic slag needed to aid removal of P_2O_5 and SiO_2 from the steel, and Fe_2O_3 and Fe_3O_4 are provided to help oxidize impurities. Atmospheric dioxygen at the surface of the melt also converts carbon, silicon, and manganese to oxides:

$$2 \, C + O_2 \longrightarrow 2 \, CO \qquad \Delta G^0_{1900} = -222 - T(0.180) = -565 \text{ kJ} \qquad (11.12a)$$

$$Si + O_2 \longrightarrow SiO_2 \qquad \Delta G^0_{1900} = -793 - T(-0.142) = -523 \text{ kJ} \qquad (11.12b)$$

$$2 \, Mn + O_2 \longrightarrow 2 \, MnO \qquad \Delta G^0_{1900} = -490 \text{ kJ} \qquad (11.12c)$$

$$\tfrac{4}{5} \, P + O_2 \longrightarrow \tfrac{1}{5} \, P_4O_{10} \qquad \Delta G^0_{1900} = -1214 - T(-0.518) = -230 \text{ kJ} \qquad (11.12d)$$

$$2 \, CO + O_2 \longrightarrow 2 \, CO_2 \qquad \Delta G^0_{1900} = -280 - T(-0.022) = -239 \text{ kJ} \qquad (11.12e)$$

TABLE 11.2 **Proportions of Scrap and Pig Iron Charged to Furnaces in the United States (1967)**[a]

Type of furnace	% Scrap	% Pig iron
Open hearth	41.6	58.4
Basic oxygen	29.3	70.7
Electric	97.9	2.1

[a] Data from U. S. Bureau of Mines, "Mineral Facts and Problems," p. 300. U. S. Department of the Interior, Washington, D. C., 1970.

Because of the presence of Fe_2O_3 and Fe_3O_4 little iron is oxidized by the air, and phosphorus can be converted to P_4O_{10} even though its line in the Ellingham diagram is above that of FeO. This mechanism for phosphorus removal, combined with the 12-hour length of an individual melt (allowing time for periodic chemical analyses of the impurities remaining), permits greater control over final composition than in other steelmaking processes. Because gas combustion provides heat (in addition to that of the exothermic oxidations 11.12) to melt iron, the open hearth can recycle as much as 50% scrap. Part of the decrease in percentage of iron and steel products recycled in recent years is the result of the shift from open hearth to basic oxygen steel production.

For each tonne of open hearth steel, 3–6 kg of dust is produced. About two-thirds of this is Fe_2O_3 in the form of particulates as fine as 0.001 μm in diameter. None of the other elements present (silicon, aluminum, manganese, calcium, magnesium, and phosphorus) accounts for more than 5% of the total particulates. The control of these emissions is especially difficult because of the high temperatures of the exhaust gases (up to 1500 K). Since most bag filters and cyclones are not designed to withstand temperatures above 500 K, electrostatic precipitators and venturi scrubbers are the most effective means of control. As in the case of the blast furnace, improper disposal of precipitator washings and scrubber liquids may contribute to water pollution.

Oxygen Steelmaking

Because of the recent availability of large quantities of low-cost dioxygen from distillation of liquid air, the *basic oxygen steelmaking process* is rapidly replacing all others throughout the world. This is fortunate from the standpoint of energy use because it is more than twice as efficient as the open hearth. For example, in April, 1963, when 100% of their steel production was by means of open hearth furnaces, the Lackawanna Plant of the Bethlehem Steel Corporation reported net energy requirements of 3.55×10^9 J/tonne.[2] In May, 1967, when the same plant had converted more than 80% of its production to basic oxygen furnaces, energy consumption had dropped to 1.03×10^9 J/tonne. Even when the energy required to obtain O_2 (0.24×10^9 J/tonne iron) is included this is a tremendous saving, although it becomes somewhat less significant when compared with the total energy (including blast furnace and finishing processes) for steel production.

There are several designs for oxygen steelmaking reactors. The *Kaldo process*, developed in Sweden, uses an oxidizing slag similar to that of the

[2] Rand Corporation, "The Growing Demand for Energy," Interim Report. Rand Corporation, Santa Monica, California, 1971.

open hearth, except that Fe_2O_3 and Fe_3O_4 are produced by a stream of dioxygen striking the surface of the molten pig iron. The sequence of reactions is the same as that of the open hearth but takes only about 35 min and requires no external fuel supply. The negative enthalpy changes of reactions 11.12 provide enough heat so that up to 40% of the charge can consist of steel scrap. This percentage is higher than for the *Rotor* or *L.D.* *processes* because the latter employ higher pressures and the stream of O_2 penetrates the surface of the molten metal, sweeping out CO without oxidizing it to CO_2. The heat of reaction 11.12e is thus lost and less scrap can be melted. The now obsolete Bessemer converter bubbled air through the metal, but was much less efficient than oxygen processes because much heat was carried off by dinitrogen. Also, nitrogen incorporated in the steels made them brittle and difficult to press or form. The L.D. process is most common for oxygen steelmaking in the United States.

All oxygen furnaces produce much the same type of air emissions — particles containing about 50% FeO and 25% Fe_3O_4 which are swept out by the stream of O_2. Hoods must be designed to fit closely around the mouth of the crucible containing molten steel in order to reduce the volume of gas delivered to venturi scrubbers or electrostatic precipitators. Because the gaseous effluent contains combustible CO a flame trap is necessary to prevent explosions which might be initiated by spark discharges in an electrostatic precipitator. Usually a vent is available which releases emissions directly to the atmosphere, should pressure build up in the pollution control equipment; thus, some pollution is still produced. Also, water used for gas cooling and scrubbing is contaminated by heat and particulates. The best water treatment is settling followed by recirculation via cooling towers, avoiding discharge to streams. Nearly all oxygen furnaces have integrated air pollution controls since they have been installed during the past two decades. Conversion from open hearth to oxygen steel production has therefore produced a reduction in air pollution as well as an increase in efficiency. From an environmental standpoint the only drawback is the smaller amount of scrap which can be recycled.

Electric Arc Furnace

The *electric arc furnace* is used primarily for making high grade steels from steel scrap, especially alloys containing nickel, chromium, manganese, vanadium, molybdenum, tungsten, niobium, and titanium. Many of these metals are more easily oxidized than iron and would be removed during other steelmaking processes. An electric arc struck between carbon electrodes and the metal surface provides temperatures above 2000 K as well as a slightly reducing atmosphere. When temperature is raised rapidly to the point where CO is more stable than most metal oxides, little of the alloying substance is lost from the scrap by oxidation. This is especially important because the composition of special materials can be maintained during recycling. Such furnaces will almost certainly play a major role in the recycling of iron and steel products even though their reliance on electric power makes their efficiencies rather low.

The newest electric arc furnaces are constructed with roofs which can be raised in order to facilitate loading of scrap metal, which usually tends to be bulky and of low density. Usually the electrodes can be raised and swung out of the way as well. These features complicate air pollution controls because effective hooding for collection of effluent gases is hard to design. Often large volumes of air must be handled and difficulties are encountered in the operation of collection equipment. At one installation a 15-year period of improvements in the original design was required before adequate control was achieved.

Pickling

Several operations peripheral to iron and steel production also produce various forms of pollution. Perhaps the most important of these is *pickling*. As steel is cooled and formed into sheets, rolls, or other salable products, a scale of iron(II) oxide builds up on its surface. This must be removed before a final finish can be applied, usually by pickling with sulfuric or hydrochloric acid:

$$\underline{\text{FeO}} + 2\,\text{H}^+ + \begin{array}{c} \text{SO}_4^{2-} \\ \text{or} \\ 2\,\text{Cl}^- \end{array} \longrightarrow \text{Fe}^{2+} + \text{H}_2\text{O} + \begin{array}{c} \text{SO}_4^{2-} \\ \text{or} \\ 2\,\text{Cl}^- \end{array} \qquad (11.13)$$

Hydrochloric acid is slightly more expensive but removes scale more rapidly, with less attack on the metal, and at lower acid concentrations. It is becoming more and more popular. When the acid concentration drops and the iron concentration increases to the point where scale removal becomes inefficient, spent pickle liquor is discarded or recovered. If pumped into a lake or stream a twofold problem ensues — the effluent is strongly acidic and the high concentration of iron may turn water brown, deposit slime, and exert a large oxygen demand (Section 14.2).

A good deal of pickle liquor is injected into wells from 1000 to 4000 m deep, but successful disposal depends on geological conditions. In some cases ground water supplies can be contaminated by vertical migration of injected wastes resulting from leakage of the well itself or through older boreholes through the formation into which the wastes are being pumped. Disposal in the vicinity of old faults may also initiate earthquake activity.

A number of processes have been devised to regenerate and recycle spent pickle liquor. Cooling sulfuric acid liquors precipitates $\text{FeSO}_4 \cdot$

7 H_2O, and allows the remaining acid to be recycled. Iron(II) sulfate can be used as a *flocculent* in municipal sewage treatment (Section 15.2) or roasted to iron(III) oxide and sulfuric acid.

$$4\ FeSO_4 \cdot 7\ H_2O + O_2 \xrightarrow{\text{roast}} 2\ Fe_2O_3 + 4\ H_2SO_4 + 24\ H_2O \qquad (11.14)$$

Another method of regeneration is controlled neutralization of either acid with lime followed by air oxidation to produce magnetite (Fe_3O_4) and calcium sulfate or calcium chloride. The iron oxide can be separated magnetically and recycled to a blast furnace. Calcium sulfate (gypsum) might find a market as wallboard, and $CaCl_2$ can be converted to reusable HCl and $CaSO_4$ by treatment with sulfuric acid. A similar process in which H_2SO_4 is added directly to hydrochloric acid pickle liquor giving HCl for reuse and solid $FeSO_4 \cdot 7\ H_2O$ is under development.

Two other processes are based on the relative volatility of HCl, which permits its expulsion from solution or solid chloride salts by heating. Sulfuric acid liquors may be treated with gaseous HCl followed by separation of $FeCl_2$ and roasting to regenerate iron(II) oxide and hydrogen chloride vapor:

$$Fe^{2+}(aq) + SO_4^{2-}(aq) + 2\ HCl(g) \longrightarrow FeCl_2(s) + 2\ H^+(aq) + SO_4^{2-}(aq) \qquad (11.15a)$$

$$FeCl_2(s) + H_2O \xrightarrow{\text{roast}} 2\ HCl(g) + FeO(s) \qquad (11.15b)$$

Direct roasting of hydrochloric acid liquors produces gaseous HCl and iron oxides by reactions similar to 11.15b. With all these regeneration techniques available it is somewhat surprising that more plants have not adopted them. Apparently the difficulty lies in lack of legislative restrictions on simpler, cheaper, but less effective means of disposal rather than technological problems with the regeneration techniques themselves.

11.5 ALUMINUM

A number of metals form such stable oxides that carbon reduction becomes unfeasible because of the extremely high temperatures required. The outstanding example is aluminum. It is exceeded in crustal abundance by only silicon and oxygen but is difficult to separate from aluminosilicate minerals, a fact which significantly limits economically recoverable reserves. The workable ore is bauxite, a hydrated aluminum oxide which is usually associated with iron, titanium, and silicon oxides; it contains only 30–70% Al_2O_3 and must be concentrated prior to electrolytic separation of aluminum. The usual method is the *Bayer process,* in which the impure mineral is extracted with 30% NaOH at 430 K and 800 kPa:

$$Al_2O_3(s) + 2\ OH^-(aq) \underset{\text{800 kPa}}{\overset{\text{430 K}}{\rightleftharpoons}} 2\ AlO_2^-(aq) + H_2O \qquad (11.16)$$

The oxides of iron, titanium, and silicon are not amphoteric (do not dissolve in excess base), and they are separated by sedimentation. After the solution has cooled, alumina is precipitated by adding a seed crystal of $Al_2O_3 \cdot 3\ H_2O$. Since the electrolysis is carried out in a molten salt containing fluoride ions, any water present would react to form HF, and must be removed by calcining at 1500 K:

$$Al_2O_3 \cdot 3\ H_2O(s) \xrightarrow{\ 1500\ K\ } \alpha\text{-}Al_2O_3(s) + 3\ H_2O \qquad (11.17)$$

The aluminum oxide is now in the crystal structure designated as α-alumina.

Since water is much more readily reduced to dihydrogen than aluminum(III) is reduced to the metal, electrolysis must be carried out in a nonaqueous medium which is a good solvent for Al_2O_3, does not react appreciably with any of the substances involved, and is not decomposed at a lower potential than alumina. In 1886, Hall in the United States and Héroult in France discovered that molten cryolite, Na_3AlF_6, met these criteria admirably. A modern *Hall-Héroult reduction cell* (diagramed in Figure 11.4) normally operates at a potential of 4.5 V, a current of 100 000 A, and a temperature of 1240 K. To produce one tonne of aluminum it consumes 1890 kg Al_2O_3, 450 kg anodic carbon, 70 kg cryolite, and 56×10^9 J electrical energy. Assuming the average 33% efficiency of electric power plants, this means that 168×10^9 J of primary energy is required.[3]

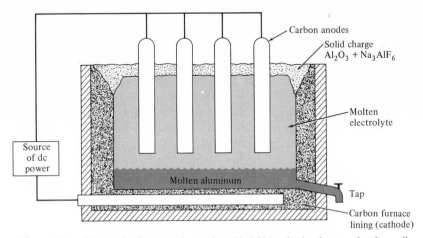

Figure 11.4 Schematic diagram of a modern Hall-Héroult aluminum reduction cell.

[3] J. C. Bravard *et al.*, "Energy Expenditures Associated with the Production and Recycle of Metals," ORNL-NSF-EP-24. Oak Ridge Nat. Lab., Oak Ridge, Tennessee, 1972.

An interesting feature of the Hall-Héroult process is that dioxygen produced by the oxidation half-reaction reacts immediately with the carbon anode to form CO_2:

$$\tfrac{2}{3} Al_2O_3 \xrightarrow{\text{electrolysis}} \tfrac{4}{3} Al + O_2 \qquad \Delta G^0_{1240} = 837 \text{ kJ} \qquad (11.18a)$$

$$\begin{array}{c} C + O_2 \longrightarrow CO_2 \\ \text{(anode)} \end{array} \qquad \Delta G^0_{1240} = -402 \text{ kJ} \qquad (11.18b)$$

$$\tfrac{2}{3} Al_2O_3 + C \longrightarrow \tfrac{4}{3} Al + CO_2 \qquad \Delta G^0_{1240} = 435 \text{ kJ} \qquad (11.18)$$

In effect the carbon — as well as the flow of electrons — serves as a reducing agent, and the anodes are slowly burned away as the electrolysis continues. In the absence of this effect the electric energy supply would have to be almost twice as great. Since only 14×10^9 J are required to produce enough carbon electrode for one tonne of aluminum, this indirect carbon reduction is more efficient than electrolysis.

The Bayer process also requires some energy input — about 44×10^9 J primary energy per tonne — and calcination of hydrated aluminum oxide to α-alumina requires about 9×10^9 J. Totaling all energy inputs gives 235×10^9 J/tonne aluminum.[4] Since the overall process is

$$Al_2O_3 \cdot 3 \, H_2O + \tfrac{3}{2} C \longrightarrow 2 \, Al + 3 \, H_2O + \tfrac{3}{2} CO_2 \qquad \Delta G^0_{298} = 986 \text{ kJ} \quad (11.19)$$

an overall efficiency of 7.7% can be calculated. Thus, a tremendous quantity of free energy (217×10^9 J) is wasted for every tonne of aluminum produced. The energy requirements of a number of other metallurgical processes are tabulated in Table 11.3.

Air and water emissions from electrolytic processes such as aluminum recovery are generally much less than in iron- and steelmaking. The calcining of hydrated alumina produces dust, but this is usually recovered by means of cyclone separators and electrostatic precipitators for recycle to the kiln. Because the Hall-Héroult cells cannot be completely enclosed there are usually emissions of carbon, alumina, and fluoride particulates and some gaseous fluorides. Effects resulting from the latter two substances were discussed in Section 9.6. These materials can be trapped by proper hooding arrangements, although some plants release large enough quantities of fluorides to affect the surrounding community and require the installation of scrubbers on their roof ventilators. The Bayer process must dispose of "red mud tailings" (a complex sodium aluminosilicate). About 5.0×10^6 tonnes of this material, which remains after aluminate ion is dissolved, are generated annually by eight plants along the United States Gulf Coast; it is usually impounded in large sedimentation ponds, but some has been dumped in the Mississippi River.

---

[4] This number differs from the one in Table 11.3 because it assumes a higher grade of bauxite.

TABLE 11.3 **Summary of the Energy Requirements for the Production and Recycle of Metals[a]**

Metal	Source	Energy requirements (GJ/tonne) in equivalent primary energy[b]		
		Energy	Recycle energy[c]	Gibbs free energy[d]
Iron	Magnetic taconite	18.77		−3.74
	$3\ Fe_2O_3 + 9\ C + \frac{9}{2}\ O_2 \longrightarrow 6\ Fe + 9\ CO_2$			
Raw steel (includes energy to produce iron)	Open hearth	~23.57		
	Basic oxygen	20.11	25–27%	
	Electric furnace	27.60		
Finished steel (includes energy to produce iron)	Average	55.19 ± 8.28[e]	45–50%	
Aluminum	Bauxite, 30% Al_2O_3	289.22		18.26
	$Al_2O_3 \cdot 3\ H_2O + \frac{3}{2}\ C \longrightarrow 2\ Al + 3\ H_2O + \frac{3}{2}\ CO_2$			
	Al scrap recycle (only ~3.3% recycled from outside the manufacturing plant)		8.88	
Copper	Sulfide ore (average of 1% and 0.3% ore)	89.25		−1.68
	$Cu_2S + O_2 \longrightarrow 2\ Cu + SO_2$			
	98% scrap recycle		3.39	
	Impure Cu scrap recycle		6.86	
Titanium	High grade rutile ore	614.62		11.45
	$TiO_2 + 2\ C \longrightarrow Ti + 2\ CO$			
	Lower grade ore	754.39		
	Ti scrap metal recycle		154.85	
Magnesium	Seawater (0.13% magnesium—source for ~80% of U. S. industry)	408.96	5.54	4.80
	$CaCO_3 + MgCl_2 + \frac{1}{2}\ CH_4 + \frac{1}{2}\ O_2 \longrightarrow \frac{3}{2}\ CO_2 + Mg + H_2O + CaCl_2$			

[a] Source: Most data used have been taken from J. C. Bravard, H. B. Flora, and C. Portal, "Energy Expenditures Associated with the Production and Recycle of Metals," ORNL-NSF-EP-24. Oak Ridge Nat. Lab., Oak Ridge, Tennessee, 1972 (available from National Technical Information Service, Springfield, Virginia).

[b] All electric energy use has been converted to the coal equivalent, burned in a power plant at 29.8% efficiency.

[c] Does not include energy to sort and separate the scrap from a waste stream; this would not be a factor when "home scrap" (that collected and recycled within the manufacturing plant) is used.

[d] Calculated for the reaction indicated at 298 K.

[e] Only estimates are available, ranging from 48.86 to 63.53; the estimates are also based on 1.516 tonnes raw steel per tonne of finished steel.

11.6 COPPER

Although aqueous electrolysis consumes less power than fused salt electrolysis, other techniques are usually even more economical for extraction of metal ions which are more easily reduced than water. However, aqueous electrolysis is widely used in the final purification of a number of metals, for example, the refining of high purity copper for the electric industry.

Impure copper from a smelter (for simplicity assume that the only impurities are silver and iron) is cast into a large plate and used as the anode in an electrolytic cell. A thin sheet of high purity copper is the cathode. During electrolysis the possible half cell reactions are as follows:

Anode	*Potential required for oxidation*
$Fe \longrightarrow Fe^{2+} + 2\,e^-$	$E^0 = +0.44$ V
$Cu \longrightarrow Cu^{2+} + 2\,e^-$	$E^0 = -0.34$ V
$Ag \longrightarrow Ag^+ + e^-$	$E^0 = -0.80$ V

Cathode	*Potential required for reduction*
$Fe^{2+} + 2\,e^- \longrightarrow Fe$	$E^0 = -0.44$ V
$Cu^{2+} + 2\,e^- \longrightarrow Cu$	$E^0 = +0.34$ V
$Ag^+ + e^- \longrightarrow Ag$	$E^0 = +0.80$ V

If the anode is maintained at a potential of -0.34 V, only iron and copper will dissolve; slightly contaminated metallic silver precipitates as "anode mud," from which it may be recovered by a subsequent process. (Relatively inert elements such as gold, silver, palladium, and platinum are important by-products of most electrolytic refining.) Since no silver goes into solution only Cu^{2+}, the most readily reduced ion in the electrolyte, will plate out as long as the cathode potential is more positive than -0.44 V, and Fe^{2+} will be left in solution. The overall reaction is simply transport of copper metal from anode to cathode and free energy is required only to overcome polarization and resistance effects. The current efficiency of copper refining is $\sim 90\%$.

Iron and aluminum are the only metals with greater world consumption than copper, and its extraction from ore illustrates several other important metallurgical techniques. Since it is a *chalcophile* copper is usually found as the sulfide, Cu_2S, often in minerals containing excess FeS and FeS_2. Although underground deposits of the zerovalent metal are known, techniques for concentrating ores containing as little as 0.3% copper have made open pit mining more profitable.

Flotation is an important example of these techniques. The first step is to grind the ore in the presence of water to produce a thin slurry. Flotation agents are then added and the slurry vigorously agitated either by fine streams of air bubbles or by saturation with air followed by partial evacuation (which causes air bubbles to form throughout the liquid). The suspended ore particles are swept to the surface where they "float" in a layer of foam. They are prevented from reentering the bulk liquid by the large surface tension of water. In order to float the particles must be hydrophobic — not wetted by water — and the purpose of flotation agents is to provide this hydrophobic character.

Xanthates and *aerofloats* are specific flotation agents used for the con-

$$R-O-C\underset{S^-M^+}{\overset{S}{\diagup}}$$

Xanthates

$$\underset{R-O}{\overset{R-O}{\diagdown}}P\underset{S^-M^+}{\overset{S}{\diagup}}$$

Aerofloats

$$R = C_2H_5, C_5H_{11}, \text{etc.}$$
$$M^+ = Na^+, K^+$$

centration of copper and other sulfide-containing ores. They have a polar or ionic molecular structure at one end and a nonpolar, covalent group at the other, and thus are similar to the surfactants in detergents (Section 14.4). The polar portion can serve as a soft Lewis base and is ideally suited to interact with soft (chalcophile) metal ions found in sulfide minerals. The hard (lithophile) metal ions found in oxides and silicates do not interact as strongly. Once molecules of the flotation agent have attached themselves to a suspended particle it is readily trapped by bubbles of air and carried to the surface. Since the chalcophilic particles attract greater numbers of xanthate or aerofloat molecules, they are concentrated in the froth. The process is indicated schematically in Figure 11.5. By varying the molecular structures of flotation agents processes can be designed which concentrate specific compounds.

Smelting

Following beneficiation by flotation the copper ore is further separated from oxide impurities by matte smelting, which utilizes the fact that molten sulfides are immiscible with molten oxides. When the mixture of oxides and sulfides is heated to about 1600 K a molten matte forms below the oxide slag. Copper concentrates there as a result of reaction 11.20:

$$FeS + Cu_2O \xrightarrow{1600 \text{ K}} FeO + Cu_2S \qquad (11.20)$$

When smelting is complete almost all of the copper has entered the matte, which contains $40-45\%$ copper. The remaining iron is removed by transferring the matte to a converter in which air is blown through the liquid. The temperature is raised to 1500 K by the exothermic reactions 11.21. The

$$2\,FeS + 3\,O_2 \longrightarrow 2\,FeO + 2\,SO_2 \qquad \Delta H^0 = -921 \text{ kJ} \qquad (11.21a)$$

$$Cu_2S + O_2 \longrightarrow 2\,Cu + SO_2 \qquad \Delta H^0 = -209 \text{ kJ} \qquad (11.21b)$$

first of these produces a slag which is poured off before the second is carried out. The final product is molten copper of 98% purity and a large

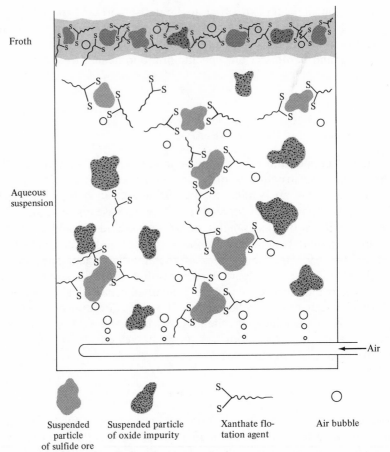

Froth

Aqueous
suspension

Air

| Suspended particle of sulfide ore | Suspended particle of oxide impurity | Xanthate flotation agent | Air bubble |

Figure 11.5 Flotation of sulfide ores. Air bubbles carry to the surface those suspended particles which interact most strongly with the flotation agent.

quantity of sulfur dioxide air pollution. Untreated gases emitted by the roasting and conversion processes may contain up to 8% SO_2, and many copper smelters are notorious for their effects on surrounding vegetation. (See Section 9.4 for methods of controlling these emissions.)

Hydrometallurgy

About one-tenth of the United States supply of copper is obtained by a *hydrometallurgical process* which involves leaching of the metal from low-grade ores:

$$Cu_2S(s) + \tfrac{5}{2} O_2(g) + 2 H^+(aq) \longrightarrow 2 Cu^{2+}(aq) + SO_4^{2-}(aq) + H_2O \qquad (11.22a)$$

$$Cu_2S(s) + \tfrac{5}{2} O_2(g) + 8 NH_3(aq) + H_2O \longrightarrow$$
$$2[Cu(NH_3)_4]^{2+}(aq) + SO_4^{2-}(aq) + 2 OH^-(aq) \qquad (11.22b)$$

If the ore contains basic oxides such as lime, an ammoniacal solution aids in dissolving the ore by forming the stable tetraamminecopper(II) complex ion. Once in solution the copper is precipitated as the metal by *cementation*. Light gauge shredded iron is added to the solution and because of its more negative reduction potential the iron replaces the copper. The cementation process has become more widely adopted as installation of air pollu-

$$Cu^{2+}(aq) + Fe(s) \longrightarrow Cu(s) + Fe^{2+}(aq) \qquad (11.23)$$

tion controls has increased the cost of smelting. Perhaps even more important is the fact that low-grade ores are not concentrated as effectively by flotation as by leaching. In some cases leaching may be carried out within the open pit mine itself.

Most of the iron used in copper cementation comes from scrap tin cans which are shredded and detinned prior to use at the rate of about three pounds per pound of copper. Cans are ideal for cementation because the steel is thin enough so that most of it can go into solution before the surface is entirely covered by precipitating copper. In the cities of Chicago, Houston, and Oakland tin cans are magnetically separated from refuse either at a landfill or after incineration. The 0.8% tin, along with labels and coatings or lacquers, is then removed from the cans by heat treatment at 900 K, and the remaining steel shredded. In 1972 over 700×10^6 scrap cans were "recycled" into copper by this technique.

11.7 OTHER METALS

One more general method for extraction of metals that is worthy of mention is the use of halide salts. These may be formed by direct reaction of ores with dichlorine, leaching with hydrochloric acid, treatment with gaseous hydrogen chloride, or chlorination accompanied by carbon reduction. Metal halides are usually more easily reduced than oxides, have lower melting and boiling points, and are almost always soluble in water. Furthermore, halides form low viscosity, highly conducting melts (because the monovalent anions tend not to form polymeric species), making them ideal for fused salt electrolysis—cryolite as the solvent in the Hall-Héroult aluminum reduction is an example.

Titanium

The *Kroll process* for recovery of titanium from ores containing rutile (TiO_2) uses halides. Gaseous $TiCl_4$ is purified by fractional distillation prior to reduction by molten magnesium:

$$TiO_2(s) + 2\ Cl_2\ (g) + C \xrightarrow{1100\ K} TiCl_4(g) + 2\ CO(g) \qquad (11.24a)$$

$$TiCl_4(g) + 2\ Mg(l) \xrightarrow{900\ K} Ti(s) + 2\ MgCl_2(l) \qquad (11.24b)$$

Titanium dioxide (a solid at the temperatures involved) would be more difficult to purify than TiCl$_4$ and the rate of reaction of TiO$_2$ with magnesium (which would take place at a solid-liquid interface) would be much less than that of 11.24b. The relative volatility of MgCl$_2$ is also an advantage because both the metal and its chloride can be distilled from the spongy titanium which is the final product of the process.

A new process for aluminum reduction also uses the chloride salt. After separation from bauxite alumina is chlorinated and AlCl$_3$ then undergoes electrolysis in completely closed cells, eliminating nearly all air emissions:

$$Al_2O_3 + 3\ Cl_2 \longrightarrow 2\ AlCl_3 + \tfrac{3}{2}\ O_2 \qquad (11.25a)$$

$$2\ AlCl_3(l) \xrightarrow{\text{electrolysis}} 2\ Al + 3\ Cl_2 \qquad (11.25b)$$

Dichlorine produced in the electrolysis cell is recycled to the chlorination reactor. Although this process is now only in the developmental stage, it appears to produce less air pollution and to require $\sim 30\%$ less electric power than the usual Hall-Héroult process.

Manganese and Cobalt from Nodules

A final example of a halide process is the recovery of metals such as manganese and cobalt from sea-floor *nodules*. These potato-size lumps of rock are found in deep ocean regions (3000–6000 m) and may represent a significant resource for a number of metals. They may be retrieved either by sinking barges to be loaded and refloated, by suction devices, or with dragline buckets. A hydrometallurgical process to recover manganese, nickel, copper, and cobalt in 95% yields has reached the stage of a pilot plant which can process one ton per day. After crushing and drying, the nodules are treated with gaseous hydrogen chloride at 400 K, which converts metals to chloride salts and reduces manganese(IV) to manganese(II):

$$MO(s) + 2\ HCl(g) \longrightarrow MCl_2(s) + H_2O(l) \qquad M = Ni, Co, Cu \qquad (11.26a)$$

$$MnO_2 + 4\ HCl \longrightarrow MnCl_2 + 2\ H_2O + Cl_2 \qquad (11.26b)$$

The latter reaction is a standard textbook method for preparation of Cl$_2$, which may be sold as a by-product. The water-soluble chlorides are easily separated from insoluble silicates, sulfates and oxides, but final separation of manganese, nickel, cobalt, and copper requires a liquid ion exchange process. (See Section 15.3 for a description of the principles of ion exchange.) Kerosene solutions of ion exchange materials which react selectively with Cu^{2+}, Ni^{2+}, and Co^{2+} extract the metals successively. The structures of the ion exchange reagents are proprietary information at present, but they probably form coordination complexes whose stability depends on the metal ion and/or pH. Metallic cobalt, nickel, and copper are deposited by electrolysis after their separation. Manganese(II) chloride is precipitated, dried, and converted to the metal by heating with a more active metal in much the same way that titanium is produced. This type of seafloor mining might significantly extend world supplies of a number of metals, but at this stage of the developing technology it is not clear what kinds of environmental damage may be done if and when the ocean floor is stripmined. It is clear that problems of international law will result as various countries try to lay claim to the richer portions of the sea.

11.8 CALCIUM AND MAGNESIUM MINERALS

A number of metals are of interest to industry primarily because of the properties of their compounds. A glance at Table 11.1 will confirm that many of the alkali and alkaline earth elements fall in this category. Magnesium and especially calcium minerals are often capable of forming extremely insoluble hydrates when treated with water. A simple example of this behavior is the conversion of gypsum into plaster of paris, a substance which hardens if it is mixed with water:

$$\underset{\text{Gypsum}}{CaSO_4 \cdot 2\ H_2O} \rightleftharpoons \underset{\text{Plaster of paris}}{CaSO_4 \cdot \tfrac{1}{2} H_2O} + \tfrac{3}{2} H_2O \qquad \Delta H^0_{298} = 16.5\ \text{kcal} \qquad (11.27)$$

This reaction can be forced to the right quantitatively at temperatures around 373 K (higher temperatures produce anhydrite, completely dehydrated calcium sulfate). The plaster may be hardened at room temperature by adding water. This spontaneously reverses reaction 11.27, producing insoluble gypsum (hardened plaster).

Portland cement which, when mixed with appropriate amounts of crushed stone, gravel, slag, or fly ash aggregates, is used to make concrete, is a somewhat more complicated mixture of compounds than plaster, but the principles of its hardening are the same. It consists of calcium and magnesium silicates and aluminates. The raw materials for cement production are clays or shales containing approximately correct proportions of oxides, and limestone to adjust the calcium content. (In some cases blast furnace slag may be used.) These materials are crushed, screened, and sometimes dried before passing into cylindrical kilns as large as 8 m in diameter and up to 200 m long. The kiln rotates around its long axis approximately once per minute, and is slightly inclined so that the raw materials pass slowly along it. At the low end the kiln is heated to 1800 K and hot gases move counter to the flow of solids in order to carry out the dehydration reactions. Upon exiting from the kiln the cement is ground to a fine powder and packaged for shipment. Energy and raw materials consumption for the manufacture of 1 tonne of cement are shown in Table 11.4.

TABLE 11.4   **Energy and Raw Materials Consumption in United States Portland Cement Production**[a]

Item	Quantity consumed per tonne produced
Fuel	7.82×10^9 J
Electric power	4.91×10^8 J (electrical), or 1.49×10^9 J (primary)
Water	1.11×10^3 dm^3
Labor	4.5 person hours
Limestone	1.32 tonnes
Shale	0.33 tonne
Gypsum	0.04 tonne
Dust emitted	0.08 tonne

[a] Data from R. N. Shreve, "Chemical Process Industries," 3rd ed., p. 171. McGraw-Hill, New York, 1967. Some European processes (vertical kiln) require only about half as much fuel per tonne of cement.

The principal environmental problems of cement production are the unsightliness of surface mining of raw materials and air emissions from kilns. A typical operation consumes on the order of 1.6 tonnes of raw materials and 0.3 tonne of coal to produce 1 tonne of cement. In the process 8 tonnes of gas passes through the kiln, picking up large quantities of dust. Collection of the dust particles, most of which are < 10 μm in diameter, requires a significant portion of the capital expenditure in a cement plant in the form of cyclone separators, fabric and gravel bed filters, and electrostatic precipitators. Because of the tremendous volume of particulates, extremely high collection efficiencies are required. For example, in the Lehigh Valley of Pennsylvania there are 16 kilns equipped with 99$^+$% efficient collectors, yet dustfall rates of over 12 tonnes/km^2 month have been recorded. If the recovered dust has a high content of alkali the amount which can be recycled to the kiln is limited. Some can be substituted for agricultural lime, but much is disposed of in abandoned quarries or storage piles. If not covered, enclosed, or sprayed with water to produce a crust, these may also contribute to air pollution.

11.9   SILICATES—SAND, STONE, AND GRAVEL

Sand and gravel consist primarily of silica but may also contain varying amounts of iron oxides, micas, and feldspars. Stone may range from cut building stone such as marble, granite, or limestone to various forms of crushed stone. These three materials far exceed in volume and value all other nonmetallic minerals, but their value per unit mass is extremely low. This means that transportation costs represent a large fraction of their cost, shipping distances must be small, and most industry operations are located near consumers—in areas of rapid urban growth. Thus objectionable environmental aspects such as noise from blasting, dust from crushing and screening operations, contamination of streams by sediment, solid wastes, and scarring of the land by open pit mining all occur on a large scale where they are readily observed by the general public.

The largest consumer of these commodities is the heavy construction industry — highways, large buildings, dams, and other public works projects. Industries such as glassmaking, iron and steel foundries, and metal extraction also require sizable supplies of sand, stone, or gravel. Although in heavily populated areas urban sprawl and public objection to environmental degradation have combined to limit readily available supplies, direct depletion is not a problem — the main detrimental effect is probably the energy required for processing. The glass industry is a case in point.

Glassmaking

A glass is a rigid, supercooled liquid having no definite melting point and a viscosity sufficiently high to prevent the ionic or molecular movement necessary to produce the highly ordered lattice of a crystal. Glasses may be formed from many chemical compounds or mixtures, but most commercial glasses are composed chiefly of fused oxides of sodium, calcium, and silicon with smaller amounts of other metal oxides. The material is not a pure chemical compound, but a general equation for the reaction might be written as

$$CaO + Na_2O + 5\ SiO_2 \xrightarrow{1700\ K} CaO \cdot Na_2O \cdot 5\ SiO_2 \qquad (11.28)$$

Figure 11.1 has already illustrated the molecular basis for the high viscosity of silica melts. By adding small amounts of alkali or alkaline earth oxides the melting point of the silica is lowered but its viscosity is only slightly decreased. Since viscosity increases rapidly with decreasing temperature, the polymeric ions of the cooling melt become immobilized well above the melting point, forming a glass.

The typical charge to a furnace might contain 50% (by weight) sand, 15% soda ash, 5% Na_2SO_4, 8% lime, and 25% cullet. The sand must be almost pure quartz since iron content exceeding 0.05% produces undesirable color. (An important means of making colored glasses is the incorporation of transition metal ions which absorb visible wavelengths.) Sodium is usually supplied as soda ash, sodium bicarbonate, salt cake (Na_2SO_4), or sodium nitrate, all of which decompose to Na_2O at elevated temperatures. Limestone, the principal source of calcium, also calcines to the oxide in the glass furnace. Cullet — crushed waste glass from imperfect or broken articles or from a recycling plant — may constitute from 10 to 80% of the charge. Although there is little difference between cullet and fresh silica sand insofar as glassmaking procedures or energy requirements are concerned, problems in separating glass from other wastes and colored glasses from each other, as well as the energy required to collect and transport the glass to a furnace, militate against recycling outside scrap. It is often more expensive in both free energy and money to recycle glass bottles than to manufacture new ones from raw materials. A detailed energy analysis of bottle recycling will be given in Section 13.4.

Because glassmaking furnaces operate at extremely high temperatures (1700 K or more), and since molten glass is similar in structure to the ceramic and refractory materials of the furnace itself, attrition of walls, ceilings, and floors by solution in the glass is a major problem. Attempts to improve furnace energy efficiency by better insulation are usually unsuccessful because higher wall temperatures allow such dissolution to proceed much more rapidly. In some cases a cooling water system is necessary. Thus, efficiencies of fossil-fueled glass furnaces range from 10 to 30%. Since molten glass is an electrolyte some European plants use electric heating to achieve thermal efficiencies as high as 70%, but when electric power generation losses are allowed for this is reduced to the same range as directly fired installations. Approximately 2×10^9 J are required to produce a tonne of glass in either case.[5]

Thus the relatively inefficient glassmaking process consumes large amounts of the intangible raw material, free energy. The same is true, although to a lesser extent on a per tonne basis, of the mining, grinding, screening, and transportation of other minerals in the sand-stone-gravel category.

11.10 SULFUR

The sulfur resources of the United States are listed in Table 11.5. Methods for obtaining sulfur from several of these sources have been mentioned previously: coal gasification (Section 5.3), petroleum refining (Section 5.6), and stack gas cleanup in electric power plants or ore smelters (Section 9.5). Use of the Frasch process to mine elemental sulfur has been declining in recent years, but it still produces about half of the United States supply. A sulfur well is drilled in much the same way as an oil well and then a set of three concentric pipes is slipped down the casing to a depth of from 150 to 800 m to reach elemental sulfur which is usually trapped between impervious layers of calcite ($CaCO_3$) and anhydrite ($CaSO_4$). Hot water (450 K) is forced down the well and enters the sulfur-bearing calcite layer where it melts the sulfur; hot air at 250–400 kPa is then forced down the inner pipe and its buoyancy used to lift the molten sulfur to the surface.

Because there are few areas other than the United States Gulf Coast where large deposits of elemental sulfur occur, the Frasch process is considerably less important on a worldwide basis. Both Japan and the

[5] H. F. Mark, J. J. McKetta, Jr., and D. F. Othmer, eds., "The Encyclopedia of Chemical Technology," p. 555. Wiley (Interscience), New York, 1963.

TABLE 11.5 **Estimated Sulfur Resources of the United States**[a]

Source	Identified resources (10^6 tonnes) available at			Hypothetical resources (10^6 tonnes)	Speculative resources (10^6 tonnes)
	<$25/tonne	$25–$35/tonne	>$35/tonne		
Elemental S (Frasch)	148	39	10	98	148
Sour natural gas	10	5	—	182	—
Petroleum sulfur	10	25	217	980	—
Pyrite	10	30	59	20	20
Sulfide ores (smelters)	20	39	39	98	196
Gypsum	—	148	6 900	1 770	—
Tar sands	—	—	10	—	—
Coal	—	—	21 000	19 300	—

[a] Data from A. J. Bodenlos, Sulfur. *In* "United States Mineral Resources" (D. A. Brobst and W. P. Pratt, eds.), Geol. Surv. Prof. Pap. No. 820, p. 613. US Govt. Printing Office, Washington, D. C., 1973.

U.S.S.R. obtain the majority of their sulfuric acid from SO_2 which is generated by oxidation of pyrites. Canada and France recover most of their sulfur from "sour" natural gas and petroleum. So far no country has begun to tap the vast resources of sulfur available in coal which is routinely burned as an energy source. The cost figures in Table 11.5 tell the story—sulfur from stack gas cleanup is at least 50% more expensive than Frasch process sulfur. By the year 2000, however, the price of sulfur is expected to rise considerably, making recovery from coal much more attractive.

It would be advantageous to recover sulfur from stack gases not only because of the reduction in air pollution levels but also as a means of conserving sulfur reserves. Total production of sulfur in this country was 8.9×10^6 tonnes in 1968 and is projected to reach 27×10^6 tonnes by 2000. The former figure is 1.8% and the latter 5.6% of the total identified resources of sulfur available below $35 per tonne (Table 11.5). Such reserves, therefore, cannot be expected to last for more than 50 years or so.

Sulfuric Acid

Most sulfur is converted into sulfuric acid, a compound which has a wide variety of end uses (Figure 11.6). It serves as a strong, inexpensive acid, a good dehydrating agent, and sometimes as an oxidizing agent. There are two major means for production of H_2SO_4. The *contact process* employs the same reactions and catalyst as the removal of SO_2 from stack gases (Section 9.5). The *lead chamber process* employs oxides of nitrogen and a

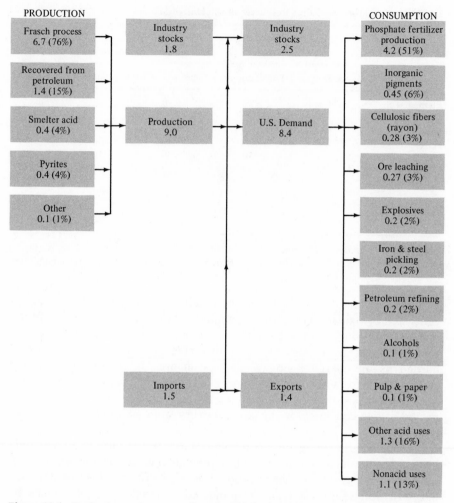

Figure 11.6 Production and consumption of sulfur in the United States, 1968 (10^6 tonnes).

lead surface to catalyze oxidation of SO_2 and SO_3 through the complicated series of reactions shown in Table 11.6. Since the chamber process has now been superseded by the contact process these reactions are of interest only because they represent a possible mechanism for formation of sulfuric acid aerosol from SO_2 emitted into an atmosphere containing NO_x and lead particulates. So far studies have neither confirmed nor denied the contribution of this reaction sequence to the production of H_2SO_4 in polluted air.

By comparison with fossil fuel combustion and ore smelting, manufacturing operations involving sulfur or sulfuric acid contribute little direct air pollution; but the low-cost material they provide indirectly prevents stack gas cleanup from becoming economically feasible. In the same indirect way

TABLE 11.6   Reactions in the Lead Chamber Process for Synthesis of Sulfuric Acid[a]

Type of reaction	Equation
Gaseous phase (homogeneous)	$2\ NO + O_2 \longrightarrow 2\ NO_2$
Gas–liquid interface	
Wetted Pb walls (heterogenous)	$SO_2 + H_2O \longrightarrow H_2SO_3$
	$NO + NO_2 + H_2O \longrightarrow 2\ HNO_2$
Gas–liquid interface (heterogenous)	$NO + NO_2 + 2\ H_2SO_4 \longrightarrow 2\ ONOSO_2 + H_2O$
Oxidation (liquid)	$H_2SO_3 + 2\ HNO_2 \longrightarrow H_2SO_4 + 2\ NO + H_2O$
Hydrolysis (liquid)	$ONOSO_2OH + H_2O \longrightarrow H_2SO_4 + HNO_2$
Overall reaction	$SO_2(g) + \frac{1}{2} O_2(g) + H_2O(l) \longrightarrow H_2SO_4(l)$

[a] Source: R. N. Shreve, "Chemical Process Industries," 3rd ed., p. 339. McGraw-Hill, New York, 1967.

much acid mine drainage (Section 5.2) could be attributed to the low cost of Frasch process sulfur. Most other sulfur-related water pollution results from the dumping of wastes by end-use industries (Figure 11.6), and these problems will be noted at the time each of those industries is discussed. The energy required for manufacture of sulfuric acid has been estimated to be on the order of 14×10^9 J/tonne.[6]

11.11   PHOSPHORUS

Fertilizers consumed 76% of the phosphorus production in the United States in 1968, with soap and detergent builders accounting for another 7%. A variety of other uses—animal feeds, matches, flame retardants, gasoline additives, and insecticides among them—consumed the remaining 21%, but none of these used more than 3% of the total. Phosphate rock, consisting primarily of sedimentary deposits of extremely insoluble fluorapatite, $CaF_2 \cdot 3\ Ca_3(PO_4)_2$, is the main source of phosphorus. Its treatment to make fertilizer is aimed primarily at increasing the solubility of the phosphate portion of the mineral, especially in the aqueous solutions of plant roots and stems. For use in detergents the orthophosphate in rock is usually converted to metaphosphate or shorter chain polyphosphates after solubilization.

Fertilizers

Several types of phosphate fertilizers are currently in use, but the initial step in their production is almost invariably treatment of phosphate rock

---

[6] C. E. Steinhart and J. S. Steinhart, "Energy," p. 335. Duxbury Press, North Scituate, Massachusetts, 1974.

with acid. Superphosphate is made by sulfuric acid treatment of fluorapatite; triple superphosphate by adding phosphoric acid:

$CaF_2 \cdot 3\ Ca_3(PO_4)_2 + 7\ H_2SO_4 + 3\ H_2O \longrightarrow$
(Fluorapatite)

$$3\ Ca(H_2PO_4)_2 \cdot H_2O + 2\ HF\uparrow + 7\ CaSO_4 \qquad (11.29a)$$
(Superphosphate)

$$CaF_2 \cdot 3\ Ca_3(PO_4)_2 + 14\ H_3PO_4 \longrightarrow 10\ Ca(H_2PO_4)_2 + 2\ HF\uparrow \qquad (11.29b)$$
(Fluorapatite)   (Triple superphosphate)

Phosphoric acid is obtained from phosphate rock or by oxidation of elemental phosphorus in the presence of water:

$CaF_2 \cdot 3\ Ca_3(PO_4)_2 + 10\ H_2SO_4 + 20\ H_2O \longrightarrow$

$$10\ CaSO_4 \cdot 2\ H_2O + 2\ HF\uparrow + 6\ H_3PO_4 \qquad (11.30a)$$

$$P_4 + 5\ O_2 \longrightarrow P_4O_{10} \qquad (11.30b)$$

$$P_4O_{10} + 6\ H_2O \longrightarrow 4\ H_3PO_4 \qquad (11.30c)$$

The source of the phosphorus is also phosphate rock, which is reduced by treatment with SiO_2 and coke in an electric furnace:

$$CaF_2 \cdot 3\ Ca_3(PO_4)_2 + 9\ SiO_2 + 15\ C \xrightarrow[\text{furnace}]{\text{electric}} CaF_2 + 9\ CaSiO_3 + \tfrac{3}{2}\ P_4 + 15\ CO \qquad (11.31a)$$

$$SiO_2 + 2\ CaF_2 \xrightarrow[20\%]{\text{about}} SiF_4 + 2\ CaO \qquad (11.31b)$$

The silica serves as an acidic flux which aids the removal of basic CaO from the phosphorus. Approximately 500×10^9 J (electric) or $15\ 600 \times 10^9$ J (primary) energy are required to produce one tonne P_4 in such a furnace. In addition, by whatever method it is manufactured, phosphate fertilizer produces gaseous fluoride emissions (see Section 9.6 for abatement techniques).

11.12   NITROGEN

Nitrogen, an atmophile in the geochemical classification, is the only element in large-scale industrial production which is obtained primarily from the atmosphere. The majority of natural *nitrogen fixation* is done by blue-green algae, the Azotobacteraceae, and symbiotic microorganisms which are associated with leguminous plants. All of these organisms contain an enzyme called *nitrogenase* which catalyzes fission of the triple bond in dinitrogen. Although it is estimated that only a few kilograms of nitrogenase are present in the biosphere, approximately 44×10^6 tonnes of nitrogen are fixed each year by microorganisms. Human processes are only beginning to rival biological nitrogen fixation—in 1960 $\sim 6 \times 10^6$ tonnes of synthetic ammonia were produced worldwide, but by 1970 plant capacity had grown to 60×10^6 tonnes. This represents a significant effect on the biogeochemical nitrogen cycle.

Most industrial nitrogen fixation is carried out by means of the *Haber process*. Dihydrogen and dinitrogen are combined directly at temperatures ranging from 675 to 900 K and pressures 100 to 150 times that of the normal atmosphere:

$$3 \ H_2 + N_2 \underset{800 \ K}{\overset{Fe}{\rightleftharpoons}} 2 \ NH_3 \qquad (11.32)$$

An iron catalyst containing 1% K_2O and Al_2O_3 permits appreciable reaction rates under these conditions. The appropriate mixture of N_2 and H_2 is prepared by high pressure catalytic reforming of natural gas:

$$\underset{\text{Natural gas}}{2 \ CH_4} + \underset{\text{Air}}{2 \ N_2} + \tfrac{1}{2} O_2 + H_2O \xrightarrow{\underset{1150 \ K}{Ni}} 2 \ CO + 5 \ H_2 + 2 \ N_2 \qquad (11.33)$$

or from coke oven gas. The latter is produced by reactions similar to those in coal gasification (Section 5.3). Thus in addition to free energy required to heat and pressurize the reactor, fossil fuels are raw materials for ammonia synthesis. Between 50 and 100×10^9 J primary energy are required to produce one tonne NH_3.

The majority of ammonia production from Haber process plants is used in nitrogen fertilizers. Anhydrous ammonia injected just below the soil surface is now the most common in the United States. Because less specialized equipment is needed for its application, ammonium nitrate, made by reacting ammonia with nitric acid, is also popular. The nitric acid is obtained via the *Ostwald process:*

$$4 \ NH_3 + 5 \ O_2 \xrightarrow[1200 \ K]{Pt\text{-}Rh} 4 \ NO + 6 \ H_2O \qquad (11.34a)$$

$$2 \ NO + O_2 \longrightarrow 2 \ NO_2 \qquad (11.34b)$$

$$3 \ NO_2 + H_2O \longrightarrow 2 \ HNO_3 + NO \qquad (11.34c)$$

Ammonium sulfate and urea are other important fertilizers derived from NH_3, and recently diammonium hydrogen phosphate, produced by reaction of NH_3 with phosphoric acid, has become a major component of the fertilizer market. It has the advantage that two of the three primary nutrients required for fertile soil — nitrogen, phosphorus, and potassium — are present in large amounts. Fertilizers are usually mixtures of several chemical compounds, and the standard method of representing their nutrient value is to give the weight percentage of N, P_2O_5, and K_2O. Diammonium phosphate is thus 16–48–0 fertilizer, containing 16% N and 48% P_2O_5.[7]

---

[7] The origin of this type of designation antedates the discovery that phosphorus(V) oxide has the molecular formula P_4O_{10}; thus, the percentage of P_2O_5 equivalent to the amount of phosphorus present is reported. The inertia of the fertilizer industry seems to be the chief reason why the simpler convention of reporting the percentage of each element instead of its oxide has not been adopted.

Some NO_x is emitted from nitric acid plants and control technology is usually required, but this is only a fraction of the contribution of stationary combustion sources or automobiles. Besides their high free-energy cost the major effect of nitrogen fertilizers appears to be their broad distribution in the environment. For instance, in some areas the nitrate concentration of groundwater has been significantly increased by fertilizer runoff. Under such circumstances well water may become dangerous for very young children whose stomachs are not sufficiently acidic to prevent reduction of nitrates to nitrites. The latter are capable of oxidizing Fe(II) to Fe(III) in hemoglobin, producing a condition known as methemoglobinemia. Nitrites also dilate capillary blood vessels, further decreasing the supply of dioxygen to the tissues. Fertilizer runoff (both phosphates and nitrates) may also contribute to accelerated eutrophication of lakes (Section 14.3). Since NH_3 is less readily leached from soils than NO_3^-, the switch to anhydrous ammonia fertilizer may help to alleviate this type of problem.

EXERCISES

(The first group of exercises is based primarily on material found in this text.)

1. What is a slag? Explain how the addition of CaO to a silica melt reduces the viscosity and increases the basicity of the slag. How many moles of CaO would have to be added to 1 mol of SiO_2 to produce a neutral slag?

2. What is an Ellingham diagram? How is the slope of a line in such a diagram related to the entropy change for formation of an oxide? For each of the reactions below give the sign of the slope of the line in an Ellingham diagram and explain clearly why the slope has that sign:

$$C + O_2 \longrightarrow CO_2$$
$$2\ C + O_2 \longrightarrow 2\ CO$$
$$3\ Fe + 2\ O_2 \longrightarrow Fe_3O_4$$

3. What substances are charged to an iron-making blast furnace? Give equations for the reactions that occur in the blast furnace.

4. What are the three most important techniques for converting pig iron to steel? Contrast them with regard to (a) energy consumption, (b) capacity for recycling scrap iron, and (c) amount of pollution produced.

5. What is meant by the term "pickling?" Write an equation for the chemical reaction involved. How is most pickle liquor disposed of? Describe and give equations for at least one method for recycling spent pickle liquor.

6. Describe and give equations for the Bayer and Hall processes for the recovery of aluminum from bauxite ore.

7. Explain clearly the flotation process for concentration of low-grade copper ores. What environmental problems are associated with such techniques for obtaining metals from deposits where the concentration is low?

8. Describe the smelting of copper ores. What type of air pollution is produced?

What other techniques have begun to be used as control of air pollution raises the cost of smelting? Give equations for reactions whenever possible.

9. What are some of the advantages of using chloride salts instead of oxides in the recovery of metals from ores? Describe and give equations for at least one such process.

10. Explain the relationship between the chamber process for synthesis of H_2SO_4 and a possible mechanism for production of sulfuric acid aerosol from coal-fired power plants.

11. What is the principal problem to be overcome in the conversion of phosphate rock to fertilizer? Give equations for the production of superphosphate and triple superphosphate fertilizer.

12. How is glass manufactured? Explain how the incorporation of CaO and Na_2O in silica promotes the formation of a glassy instead of a crystalline solid.

13. Explain why plaster of paris hardens when water is added.

14. Assuming that the coal burned contains 2.5% sulfur, calculate the emission factor (kg SO_2/tonne cement) for sulfur dioxide from a typical cement plant.

15. What are some of the characteristics of H_2SO_4 which account for its being produced in greater quantity than any other chemical commodity?

16. Using data given in Section 11.5 and equation 4.12, verify that only 17% of the free energy in coal and dioxygen is converted into aluminum by the Hall process. (Assume that reactants and products are at a temperature of 298 K.)

(Subsequent exercises may require more extensive research or thought.)

17. Is the open hearth steelmaking furnace really less energy-efficient than the oxygen furnace when the energy cost of O_2 and the energy required for blast furnace production of iron are included? (For the question assume the open hearth charge consists of 50% and the oxygen charge 30% scrap.)

18. Using a handbook compare the densities of the following substances:

Fe   Cu   Al Ca_2SiO_4 Mg_2SiO_4

How do these densities affect the use of slags in metal recovery?

19. What effect (if any) would you expect on the steel industry if a hydrogen economy (Section 7.1) were adopted in the United States?

20. The problem of fertilizers and crop yields has been with us for some time, and fixation of atmospheric dinitrogen has an interesting history. A good account is given in E. G. Rochow, G. Fleck, and T. R. Blackburn, "Chemistry: Molecules That Matter," Chapter 1. Holt, New York, 1974. Read it!

12

Clay Minerals and Soil

We live on the rooftops of a hidden world. Beneath the soil surface lies a land of fascination, and also of mysteries, for much of man's wonder about life itself has been connected with the soil. It is populated by strange creatures who have found ways to survive in a world without sunlight, an empire whose boundaries are fixed by earthen walls.

Peter Farb (from *Living Earth*, Harper & Brothers, 1959, p. 97.)

A thin layer of soil, usually no more than a meter in depth and consisting of a mixture of mineral solids, organic matter, water and air, is the basis for nearly all forms of life on the solid earth. The proportions of the four soil components may vary depending on climate, geographic location, and the composition of its parent material (usually igneous or sedimentary rock), but a ratio of about 50% mineral and organic matter and 50% fluid matter (air and water) is usually maintained. Soil is evolutionary in that the physical, chemical, and biological processes of *weathering* are constantly acting to change its composition. At the same time sedimentation on the sea floors deposits material which may eventually evolve into new soils. However, these processes require geological time scales — soil is not a renewable resource but must rather be conserved.

Because the soil contains a variety of life-forms at interfaces of all three phases of the biosphere (earth, air, and water), it has a unique ability to carry out the chemical and biochemical transformations which provide essential nutrients for living organisms and which cleanse the biosphere of unwanted or toxic compounds. Although it is impossible to discuss one component of soil without mentioning the effects of the others, it is useful to concentrate on mineral solids and organic matter separately before considering their combined interaction with soil air and water. We shall begin with minerals since the life history of a soil generally begins (and ends) with these materials.

12.1 WEATHERING OF SILICATES

Most of the substances and plant nutrients of soil were originally derived by weathering from the minerals of the earth's crust. The quantity of material affected makes the reactions involved the largest scale chemical changes which have occurred since the formation of the earth. The initial changes usually result from physical processes. When minerals formed under high pressures deep in the earth are forced to the surface the decrease in pressure permits expansion and cracking. Crystals of ice or of salts deposited by evaporation expand and enlarge such cracks, and rapid changes in temperature may also break up rock in the same way they can crack glass. Roots of plants established in the upper portion of a soil apply stress to unbroken rock at lower levels. All of these processes decrease the particle size of the soil minerals, bringing greater surface areas into contact with air, water, and the compounds dissolved in the soil solution. Particle size determines the rate of weathering to a certain extent, but chemical bonding within the crystalline minerals is the most important factor. Since water, dioxygen, and carbon dioxide are the most common reactive chemicals in contact with such compounds, rates of hydrolysis, oxidation, and carbonation will determine the rate of weathering. These rates will in turn be controlled by the availability of appropriate ions at the surface of the mineral crystal lattice.

Primary Mineral Weathering

Consider the weathering of olivine, $(Mg, Fe)SiO_4$, in which discrete SiO_4^{4-} tetrahedra are bound in the lattice by their mutual attraction for Mg^{2+} or Fe^{2+} ions. At the crystal surface some Mg^{2+}, Fe^{2+}, and SiO_4^{4-} ions will be partially exposed to contact with water, air, or carbon dioxide:

$$(Mg, Fe)SiO_4(s) + 4\,H^+(aq) \longrightarrow Mg^{2+}(aq) + Fe^{2+}(aq) + H_4SiO_4(aq) \qquad (12.1a)$$

$$2\,(Mg, Fe)SiO_4(s) + 4\,H_2O \longrightarrow$$
$$2\,Mg^{2+}(aq) + 2\,OH^-(aq) + Fe_2SiO_4(s) + H_4SiO_4(aq) \quad (12.1b)$$

$$2\,(Mg, Fe)SiO_4(s) + \tfrac{1}{2}\,O_2(g) + 5\,H_2O \longrightarrow$$
$$Fe_2O_3 \cdot 3\,H_2O(s) + Mg_2SiO_4(s) + H_4SiO_4(aq) \quad (12.1c)$$

Because they are not linked covalently to the rest of the structure, the SiO_4^{4-} tetrahedra are relatively easily dissolved. These weathering reactions release Fe^{2+} and Mg^{2+} for assimilation by plants, although oxidation of iron often results in formation of a new mineral, hydrated iron(III) oxide. The SiO_4^{4-} ions may condense to form polymeric silicate structures or they may precipitate with other cations of the soil solution, producing new minerals. The more soluble species (Mg^{2+} in this case) can be removed completely from the soil by leaching, eventually being transported to the ocean via streams and rivers. (Recall that magnesium for industrial use is recovered from the oceans — Table 11.3.)

The resistance of minerals to weathering is directly related to their crystal structures. Although olivine is readily dissolved, as the degree of linkage of SiO_4 tetrahedra through Si–O–Si bonds increases in pyroxene, amphibole, and biotite to its maximun in quartz (see Figure 2.5), the rate of weathering decreases markedly. Therefore as a soil develops, the minerals which crystallized first from magma are removed most rapidly, leaving behind those which formed last from the melt. The degree of weathering correlates with the types of minerals found in soil as indicated in Table 12.1.

Equations 12.1 indicate the importance of hydrogen ions in displacing other cations from SiO_4 groups in silicate minerals. These are usually supplied by water itself, by acid groups in the organic matter of the soil, or by dissolution from the air of acidic oxides such as CO_2 and SO_2. (The latter

TABLE 12.1   Representative Minerals and Soils Associated with Weathering Stages[a]

Weathering stage	Representative minerals	Typical soil groups
	Early weathering stages	
1	Gypsum (also halite, sodium nitrate)	Soils dominated by these minerals in
2	Calcite (also dolomite, apatite)	the fine silt and clay fractions are the
3	Olivine-hornblende (also pyroxenes)	youthful soils all over the world, but
4	Biotite (also glauconite, nontronite)	mainly soils of the desert regions
5	Albite (also anorthite, microcline, orthoclase)	where limited water keeps chemical weathering to a minimum.
	Intermediate weathering stages	
6	Quartz	Soils dominated by these minerals in
7	Muscovite (also illite)	the fine silt and clay fractions are
8	2:1 layer silicates (including vermiculite, expanded hydrous mica)	mainly those of temperate regions developed under grass or trees. In-
9	Montmorillonite	cludes the major soils of the wheat and corn belts of the world.
	Advanced weathering stages	
10	Kaolinite	Many intensely weathered soils of the
11	Gibbsite	warm and humid equatorial regions
12	Hematite (also goethite, limonite)	have clay fractions dominated by
13	Anatase (also rutile, zircon)	these minerals; they are frequently characterized by their infertility.

[a] Source: H. D. Foth and L. M. Turk, "Fundamentals of Soil Science," 5th ed., p. 158. Wiley, New York, 1972.

may be derived from air pollution; see Section 9.3.) Large quantities of water are required as reactants in equations such as 12.1, but water also performs an important function in carrying dissolved ions away from the weathering site and shifting the reactions to the right. Water is also readily adsorbed on minerals, causing expansion and flaking of surface layers and facilitating access by hydrogen ions. Soils in advanced weathering stages are usually found in regions of high rainfall and often are deficient in plant nutrients (see Table 12.1).

12.2 SECONDARY CLAY MINERALS

The majority of the ions dissolved by hydrolytic or oxidative weathering are neither incorporated in plants nor completely leached from soils but precipitate to form *secondary minerals,* the silicate and oxide clays (see Figure 12.1). Because they crystallize from solutions in which many nucleating sites are available, individual particles do not grow to large diameters. The large surface area of such colloidal particles combines with the crystal structures of the silicate clays to make them uniquely suitable as sites for soil–chemical reactions. The exact conditions for formation of clays from the parent feldspars and micas are not known, but weathering is characterized by depletion of alkali and alkaline earth metals, replacement of silicon by aluminum, and the appearance of Al^{3+} in octahedral rather than tetrahedral coordination.

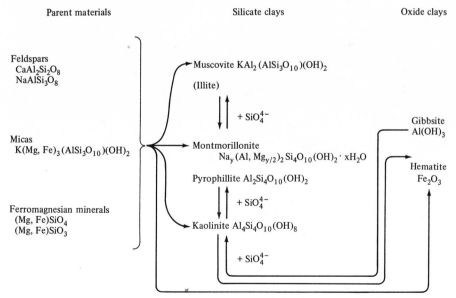

Figure 12.1 Formation of silicate and oxide clays. Note the successive depletion of alkali and alkaline earth metals followed by removal of silicon as weathering proceeds.

Examples of the three types of structures characteristic of the silicate clays are provided by muscovite, pyrophyllite, and kaolinite (Figure 12.2).

Muscovite and pyrophillite are examples of *triple layer structures*. Flat layers may be formed by SiO_4 tetrahedra (Figure 2.5d), all of which share three oxygens with other silicons. Usually the fourth (unshared) oxygen of each tetrahedron points toward the same side of the layer. Above this two-

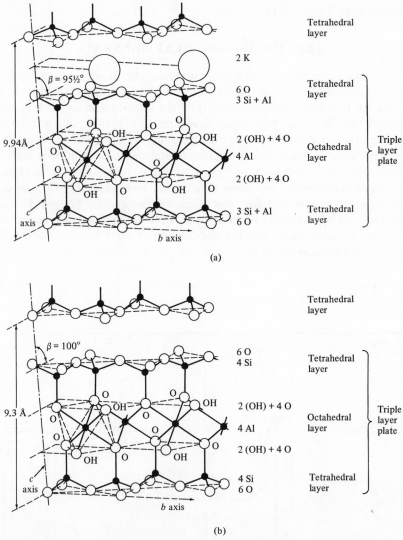

Figure 12.2 Structures of silicate clays: (a) structure of muscovite, $KAl_2(AlSi_3O_{10})(OH)_2$: (b) structure of pyrophyllite, $Al_2Si_4O_{10}(OH)_2$; and (c) structure of kaolinite, $Al_4Si_4O_{10}(OH)_8$. From: B. Mason and L. G. Berry, "Elements of Mineralogy," pp. 437–439. Freeman. San Francisco, California, Copyright © 1968.

dimensional array of SiO_4 tetrahedra is a layer of aluminum ions in octahedral sites. Some of the oxygen atoms in this second layer are shared with the SiO_4 tetrahedra of the first layer, others are provided by OH^- ions, and still others are shared with SiO_4 groups in the uppermost layer, which is simply an inverted version of the first one. Figures 12.2a and 12.2b show only a small portion of a cross section of one such triple layer or *plate*. It would extend ~ 2 μm in each of two dimensions (the length and breadth of the crystal or *soil platelet*). Each plate would thus be $\sim 20\ 000$ oxygen atoms long by $\sim 20\ 000$ oxygen atoms wide. The third dimension of the lattice is provided by stacking ~ 200 plates, one on top of another, making the platelet ~ 0.2 μm high. The lower portion of a second plate is shown at the top of each figure.

Muscovite differs from pyrophillite in that the former has one-fourth of the Si^{4+} ions in each of the two tetrahedral layers replaced by Al^{3+} ions, giving each of the triple layers one unit of excess negative charge for each substituted aluminum ion. These excess negative charges are balanced by K^+ ions which occupy sites between the triple layers of muscovite, binding them somewhat more tightly than in pyrophillite. Vermiculite is another clay having this illite type of structure, but it differs from muscovite in that many of the Al^{3+} ions in octahedral positions are substituted by Mg^{2+}, Fe^{3+}, and other ions. Substitution of Al^{3+} by divalent ions increases the negative charge on each triple layer or plate, allowing hydrated calcium or magnesium ions to occupy interlayer sites. Since other cations and water molecules are also attracted by the negative charges of the plates, vermiculite can adsorb large quantities of nutrients which otherwise would leach away. Thus, it is often used as a soil conditioner.

Montmorillonite differs from pyrophillite in the same way that vermiculite is derived from muscovite — about one-sixth of the Al^{3+} ions in octahedral holes undergo isomorphous substitution by Mg^{2+} ions. Again

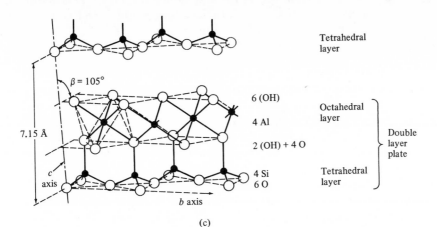

(c)

the negative charge developed on the plates allows a variety of cations to be held tightly enough to avoid leaching but loosely enough to be available to plants. Because there is little force of attraction between adjacent plates in the montmorillonite- or vermiculite-type structures, the clay particles expand readily when water is added. This has the effect of holding large quantities of water and making the interior as well as the exterior surfaces of the particles available for the adsorption of nutrient ions. Removal of water causes contraction of the clay platelets, accounting for the extensive cracking observed during dry periods and for settling of buildings in soils containing large fractions of montmorillonite clays.

The structure of kaolinite consists of stacks of *double layers,* and a portion of a cross section of one such layer is shown in Figure 12.2c. As in the previous structures a layer of aluminum octahedra is located above one of silicon tetrahedra, but instead of having a third layer the sixfold coordination of aluminum(III) is completed by hydroxide ions. Since they are adjacent to the oxygens of the SiO_4 portion of the next double layer plate, the hydroxide ions are responsible for hydrogen bonding which holds successive plates together. This stronger bonding favors the formation of larger particles and prevents the intrusion of water or cations between the plates, decreasing expansion and contraction and the number of sites to which cations may be bound. In kaolinite only oxygen atoms which are not bonded to more than one silicon or which have lost H^+ by acid dissociation are capable of attracting cations strongly to the outer surfaces of the particles.

It should be emphasized that the *soil solution* does not begin immediately at the exterior surface of a colloidal clay particle. Since the concentration of cations adsorbed on the surface is much greater than in the bulk solution, much water is associated by hydration with various metal ions. Water molecules are also held by hydrogen bonding to oxide ions in SiO_4 tetrahedra. (These effects are illustrated in Figure 12.3.) This adsorbed water is retained by soils dried at ordinary temperatures and is not available to plants. It can only be removed by heating.

12.3 COMMERCIAL USES OF CLAYS

Besides their importance as a fundamental component of soil, the clay minerals have many other uses. Vermiculite is often added to garden or nursery soils to increase their porosity and to plastics and concrete as a filler. Because of its ability to absorb water, montmorillonite (bentonite) forms gelatinous suspensions which have extremely high viscosities when still, but low viscosity when stirred. One use of such *thixotropic agents* is to keep freshly applied paint from running before it has dried. The kaolin group of clays are used for coating glossy paper and fillers for rubber, but principally in ceramics—brick, pottery, china, terra cotta, refractories, porcelain, and enamels.

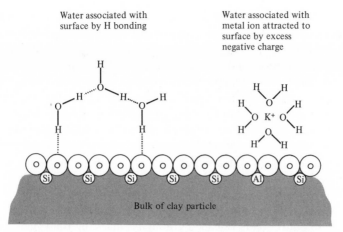

Figure 12.3 Association of water molecules at colloidal surfaces.

Although any ceramic contains many more ingredients than the clay itself, it is possible to represent the fundamental reactions involved in firing relatively simply. The initial process is removal of water, followed by conversion of silicon and aluminum oxides to the mineral mullite; at sufficiently high temperatures silica is converted to crystobalite. Depending on the quantity of alkali metal oxides present, crystals of mullite and crystobalite will be embedded in varying quantities of vitreous or glassy material, which acts as a bond, resists abrasion, and may impart translucency as in chinaware. More than 50 million tonnes of clay minerals are consumed annually by such industrial applications in the United States.

$$\underset{\text{Kaolinite}}{Al_4Si_4O_{10}(OH)_3} \xrightarrow{900\ K} \underset{\text{Alumina}}{2\ Al_2O_3} + \underset{\text{Tridymite}}{4\ SiO_2} + 4\ H_2O \qquad (12.2a)$$

$$\underset{\text{Alumina}}{3\ Al_2O_3} + \underset{\text{Tridymite}}{2\ SiO_2} \xrightarrow{1300\ K} \underset{\text{Mullite}}{3\ Al_2O_3 \cdot 2\ SiO_2} \qquad (12.2b)$$

$$\underset{\text{Tridymite}}{SiO_2} \xrightarrow{1750\ K} \underset{\text{Crystobalite}}{SiO_2} \qquad (12.2c)$$

12.4 ORGANIC MATTER

Once the physical and chemical weathering processes have broken down a portion of the parent minerals into colloidal clay, concurrently releasing nutrient ions, it becomes possible for plants and other organisms to begin to live in the developing soil. Photosynthetic plants manufacture organic matter — cellulose, hemicellulose, lignin, protein, fats, and waxes — which is decomposed by heterotrophic soil organisms when the plants die. Because of the varying reactivities and water solubilities of the organic residues, degradation occurs at different rates; recall that this is the first step in coal formation (Section 5.1). Cellulose and hemicellulose are rapidly decomposed, leaving lignin combined with protein. These latter constitute roughly 50 and 30% of the soil organic matter (humus). The degradation of cellulose, fats, and some protein yields H_2CO_3, HNO_3, and H_2SO_4 as well as carboxylic acids, all of which contribute to further weathering of the soil.

The relatively high concentration of protein is surprising since such compounds are normally readily decomposed, although much is probably incorporated into the cells of decomposing organisms. Soils with large fractions of clay have the greatest protein content, and it may be that either the protein molecules, the enzymes which degrade them, or both are so strongly adsorbed on the surfaces of clay particles that they are rendered unreactive. The nitrogen content of humus usually ranges between 3 and 6%, mainly in the form of peptide linkages in protein. Because of the inertness of humus this nitrogen is released gradually, allowing the organic matter to serve as a buffer for soluble compounds which might otherwise be leached away. Soils which have been depleted of humus will contain relatively small quantities of nitrogen and hence are amenable to the stimulation of plant growth by addition of nitrogen fertilizers.

The discussion of coal formation has already mentioned the stability of lignin, which requires the high temperatures and pressures of geological metamorphism for loss of carbon dioxide and water. However, dioxygen from soil air, mediated by enzymes in soil microorganisms, does induce a number of changes in lignin, the most important being oxidation of alcohol and methoxy groups to carboxylic acid functions, forming humic and fulvic acids. Hydrogen ions of these acid sites are readily exchanged for other cations, and the cation exchange capacity of humus is roughly twice as great as for montmorillonite. The exchange of potassium ions with hydrogen ions on a carboxylic acid site in humus is shown in equation 12.3:

$$\text{Humus}\left\{ -C\begin{array}{c} \nearrow O \\ \searrow OH \end{array} + K^+ \rightleftharpoons \text{humus}\left\{ -C\begin{array}{c} \nearrow O \\ \searrow O^-K^+ \end{array} + H^+ \right. \right. \qquad (12.3)$$

Humus is practically insoluble in neutral water, although some may go into colloidal suspension, but it does dissolve in basic solutions and some of its components may dissolve in acids. Like some of the clays, humus absorbs large quantities of water and exhibits swelling and shrinking. It is also an important factor in aggregation (structure formation) and helps to provide the aeration necessary for a healthy soil. With the development of an adequate quantity of humus together with a supply of clay minerals a soil reaches its maximum degree of *fertility:* The soil particles are able to store nutrients and supply them, when neeeded, in a form assimilable by plants.

12.5 SOIL NUTRIENTS

At least seventeen of the chemical elements have been shown to be essential to plant growth and plant nutrition. Table 12.2 lists these essential plant nutrients, the chemical forms in which they are most readily assimilated by plants, and their main sources in the soil. The division into macro- and micronutrient categories is made somewhat arbitrarily on the basis of

TABLE 12.2 Plant Nutrients[a]

Element	Source	Assimilable form	Function
		Macronutrient elements > 10 kg/hectare	
Carbon	Air, organic matter, mineral solids	CO_2, CO_3^{2-}, HCO_3^-	Major components of carbohydrates, proteins, fats and related compounds.
Hydrogen	Water-soil solution	H^+, OH^-	
Oxygen	Soil air	O_2	
Nitrogen	Organic matter	NH_4^+, NO_3^-, NO_2^-	Required for proteins, enzymes and chlorophyll.
Phosphorus	Organic matter, mineral solids	HPO_4^{2-}, $H_2PO_4^-$	Important role in energy transformation; assimilation of fats; larger proportions in seeds.
Potassium	Mineral solids	K^+	Essential for metabolic processes; remains in ionic form in plants.
Calcium	Mineral solids	Ca^{2+}	Part of cells walls as calcium pectate.
Magnesium	Mineral solids	Mg^{2+}	Active in enzyme systems, part of chlorophyll molecule, aids in translocation of P.
Sulfur	Organic matter, mineral solids	SO_3^{2-}, SO_4^{2-}	Important as a constituent of compounds including amino acids.
		Micronutrient elements	
Iron	Mineral solids	Fe^{2+}, Fe^{3+}	Plant enzyme systems; interrelated in plant utilization.
Manganese	Mineral solids	Mn^{2+}, Mn^{4+}	
Boron	Mineral solids	BO_3^{3-}, $B_4O_7^{2-}$	Calcium utilization.
Molybdenum	Mineral solids	MoO_4^{2-}	Reduction of nitrates; N fixation in legumes.
Copper	Mineral solids	Cu^+, Cu^{2+}	Components of enzymes; necessary for growth.
Zinc	Mineral solids	Zn^{2+}	
Chlorine	Mineral solids	Cl^-	Regulator of osmotic pressure and cation balance.

[a] In addition to the listed elements, others may also be essential to some or all plants. Among these are cobalt, vanadium, iodine, and strontium. Data based on H. D. Foth and L. M. Turk, "Fundamentals of Soil Science," 5th ed. Wiley, New York, 1972; R. G. Gymer, "Chemistry: An Ecological Approach." Harper, New York, 1973.

the quantity required for normal plant growth: macronutrients are generally required in quantities greater than ~ 10 kg/hectare.

Although carbon, oxygen, and hydrogen are available to plants from the atmosphere, water of the soil solution, or soil air, the majority of the macro- and micronutrients are obtained from mineral solids and organic matter of the soil. To be transported through the soil solution to appropriate sites for uptake by plant roots they must be in soluble form; hence, the large number of ions found in the second column of Table 12.2.

The exact mechanism by which various ionic forms of nutrients are absorbed by plant roots is not known, but two phenomena are well established. The process is selective—certain ions can be absorbed over a wide range of conditions while others are excluded—and it requires free energy to transport nutrient ions from a region of low concentration (the soil solution) to one of high concentration (the plant fluid). The free energy is supplied by oxidative metabolic reactions in plant roots; if dioxygen is kept from the roots net uptake of nutrients can be halted, even from concentrated soil solutions. Thus, soils which are completely saturated with water or whose very low porosity excludes air do not permit uptake of nutrients and greatly slow the growth of most plants.[1]

Root hairs are in intimate contact with soil solution, soil air, mineral and organic particles (see Figure 12.4), and root cells, like other plant and animal cells, are surrounded by membranes composed of lipids and pro-

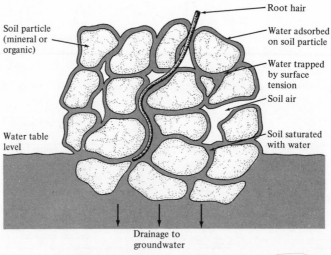

Figure 12.4   Structure of soil.

---

[1] Rice and some other plants which are capable of transporting O_2 from leaves and stems to roots are exceptions to this rule.

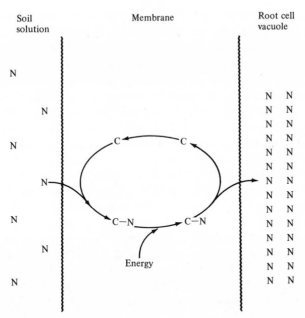

Figure 12.5 Transport of nutrient ions across a root cell membrane; N, nutrient molecule or ion and C, carrier molecule.

teins which are impermeable to most polar particles. Nutrient ions which have diffused to the surface of a root hair must be transported across these membranes to reach the plant fluids, as indicated schematically in Figure 12.5. In the diagram C represents a carrier molecule capable of forming a coordination complex with nutrient ions, N, neutralizing much of their charge and facilitating their transfer through the membrane.[2] In order for the nutrients to be transported against a concentration gradient, an input of energy must in some manner modify the carrier molecule making the transfer one way. The selectivity of nutrient absorption apparently depends on the ability of carrier molecules to bind to some ions but not others. Different carriers are required for Ca^{2+} and Mg^{2+}, but the same one can transport K^+ and Cs^+. The rules governing which carrier can react with which ions are probably similar to those for isomorphous substitutions in crystal lattices (Section 2.4).

12.6 ION EXCHANGE IN SOILS

Regardless of the difficulty in knowing exactly the structures and mechanism of action of carrier molecules, it should be clear that nutrients will not

---

[2] The mechanism of these carrier molecules is probably similar to that of the flotation agents described in Section 11.6 since they must make ions hydrophobic.

be available to a plant unless they are in chemical forms for which carriers exist and can diffuse to the root surface. Since insoluble compounds do not fit these requirements and soluble ones would tend to leach away without some storage mechanism, the presence of ion exchange sites on the mineral and organic soil particles is of utmost importance in plant nutrition.

As we have already seen, negatively charged sites exist in silicate clays because of isomorphous substitutions such as that of Al^{3+} by Mg^{2+}; anionic carboxylate groups in humus particles (micelles) are also negatively charged and will attract cations to their surfaces. The strength of this attraction depends on the relative concentrations of different ions in the soil solution, the magnitude of charge and radius of each ion, and the degree of stabilization resulting from hydration of the ion in the soil solution. The relative attraction of different ions for a colloidal micelle may be determined by measuring equilibrium constants for reactions such as 12.4. (Note that because of their greater charge the number of Ca^{2+} ions that can be accommodated by a soil micelle is only half as great as the number of K^+ ions. For this reason cation exchange capacities are reported in milliequivalents per 100 g of dry soil. One equivalent corresponds to one mole of positive charge or one-half mole of dipositive ions.)

$$
\begin{array}{c}
K^+ \quad K^+ \quad K^+ \\
K^+ \left(\begin{array}{c} - \text{Colloidal} - \\ - \text{micelle} - \end{array} \right) \begin{array}{c} K^+ \\ K^+ \end{array} \quad + \quad 5\,Ca^{2+}_{(1\,M,\ \text{soil solution})} \\
K^+ \quad K^+ \quad K^+
\end{array}
$$

$$
\begin{array}{c}
Ca^{2+} \quad Ca^{2+} \\
\left(\begin{array}{c} - \text{Colloidal} - \\ - \text{micelle} - \end{array} \right) \ Ca^{2+} \quad + \quad 10\,K^+_{(1\,M,\ \text{soil solution})} \\
Ca^{2+} \quad Ca^{2+}
\end{array}
$$

$$(12.4)$$

The usual order of cation attraction for micelles is $Al^{3+} > Ca^{2+} > Mg^{2+} > K^+ > Na^+$ (assuming equal concentrations). Hydrogen ion is difficult to place accurately, but is usually relatively weakly bound, as are the other singly charged ions. One might expect, on the basis of Coulomb's law, that for ions of equal charge the one having smaller radius would be more strongly bound to a negatively charged site, but the adsorbed ions are strongly hydrated (e.g., K^+ in Figure 12.3). For a given charge, the smaller the radius the greater the degree of association of water molecules with an ion. Thus, even through its ionic radius is smaller, the greater number of intervening water molecules prevents Na^+ from approaching the colloidal surface as closely as K^+, and Na^+ is more readily displaced by other cations.

TABLE 12.3   Cation Exchange Properties of a Typical Fertile Soil[a]

Depth (cm)	Cation exchange capacity (meq/100 g)	Exchangeable cations (meq/100 g)					% Base saturation	pH
		Ca^{2+}	Mg^{2+}	K^+	Na^+	H^+		
0–15	24.4	11.3	4.1	0.8	0.2	8.0	67	5.1
15–30	24.7	13.2	4.7	0.6	0.2	6.0	75	5.3
30–45	23.0	13.0	5.0	0.5	0.2	4.3	81	5.5
45–90	25.5	15.0	6.5	0.5	0.2	3.3	87	5.2
90–120	25.7	15.1	7.6	0.4	0.2	2.4	91	5.5
>120	22.1	13.4	7.2	0.3	0.2	1.0	95	5.8

[a] Data from G. D. Smith, W. H. Allaway, and F. F. Riecken, Prairie soils of the upper Mississippi valley. *Advan. Agron.* **2**, 157–205 (1950).

The capacity of soil to store nutrients by ion exchange is truly prodigious as indicated by the nutrients present in a typical prairie soil of the midwestern United States shown in Table 12.3. Assuming that the mass of the upper 15 cm layer of soil is of the order of 2×10^6 kg/hectare, the quantity of Ca^{2+} held in readiness for plant consumption is easily calculated as:

$$\frac{11.3 \text{ meq}}{100 \text{ g}} \times \frac{1000 \text{ g}}{1 \text{ kg}} \times \frac{2 \times 10^6 \text{ kg}}{1} \times \frac{1 \text{ eq}}{1000 \text{ meq}} \times \frac{20 \text{ g } Ca^{2+}}{1 \text{ eq}} = 4.52 \times 10^6 \text{ g } Ca^{2+}$$

This amounts to 4.5 tonnes of Ca^{2+} per hectare that can be exchanged with the soil solution.

Anions can be exchanged by soil particles, but only to a small extent; for example, some anions may replace OH^- ions at the surfaces of clay particles, especially kaolinite. And iron and aluminum oxide clays which are the product of extensive weathering (see Table 12.1) can be protonated to yield positively charged micelles. In gibbsite $[Al(OH)_3]$ each aluminum ion is surrounded by six hydroxide ions and each hydroxide shares two aluminums. Hydroxide ions at the surface can be protonated

$$(12.5)$$

providing positive sites to which anions may be bound. Unfortunately, by the time soil has weathered to the point where anion exchange sites develop, most of the nutrients have already been lost by leaching and drainage of the soil solution to groundwater.

In most fertile soils cations are held strongly by ion exchange, but anions remain in the soil solution where they are subject to leaching or can be removed by precipitation. The macronutrients most likely to be in short supply are those which occur as highly soluble anions (N in NO_3^-), those whose anions form very insoluble compounds (P as PO_4^{3-}), and those cations which are less tightly bound by negatively charged colloidal micelles (K^+). It is for this reason that nitrogen, phosphorus, and potassium are the most important components of fertilizers. Moreover, the form in which these elements are applied to soil is important. The conversion of phosphate rock to soluble form has already been discussed (Section 11.11). Nitrogen added to the soil as anhydrous ammonia, ammonium salts, or urea is mostly retained, at least until it is oxidized by soil bacteria to NO_2^- or NO_3^-. On the other hand, nitrate fertilizers often contribute far more to runoff because of their high solubility. Problems resulting from leaching of anionic nitrogen compounds into groundwater supplies were discussed in Section 11.12.

12.7 SOIL pH AND NUTRIENT AVAILABILITY

The acidity or alkalinity of a soil is determined by the types of cations adsorbed on its colloidal micelles, which in turn determine the pH of the soil solution. Calcium, magnesium, potassium, and sodium ions are termed *exchangeable bases* because when desorbed from soil micelles they increase pH:

$$\text{Micelle}\Big\{\text{Ca} + 2\,H_2O \rightleftharpoons \text{micelle}\Big\{{}^H_H + Ca^{2+}(aq) + 2\,OH^-(aq) \qquad (12.6)$$

$$\underset{\substack{\text{(Ca}^{2+}\text{ ion adsorbed on}\\ \text{soil micelle)}}}{} \qquad \underset{\substack{\text{(Two H}^+\text{ ions adsorbed}\\ \text{on soil micelle)}}}{}$$

The *percentage base saturation* of a soil is the percentage of the cation exchange sites which are occupied by these cations. Aluminum ions and, most importantly, hydrogen ions lower soil pH when they are released from micelle surfaces:

$$\text{Micelle}\Big\{H + H_2O \rightleftharpoons \text{micelle}\Big\{{}^- + H_3O^+(aq) \qquad (12.7a)$$

$$\text{Micelle}\Big\{Al + 5\,H_2O \rightleftharpoons \text{micelle}\Big\{{}^H_{\,H}_H + Al(OH)_4^-(aq) + H_3O^+(aq) \qquad (12.7b)$$

Since for all practical purposes an adsorbed aluminum ion produces the same effect as adsorbed hydrogen ions, the percentage of sites occupied by Al^{3+} or H^+ is referred to as the *percent hydrogen saturation*.

Either percent hydrogen saturation or percent base saturation may be related directly to pH for soils of similar minerology; for example, equation 12.8 has been found to hold for southern Michigan soils:

$$pH = \frac{187 - 0.3(CEC) - \%H \text{ saturation}}{24} \qquad (12.8)$$

If a cation exchange capacity (CEC) of 13 meq/100 g is assumed for an average soil, pH values calculated from equation 12.8 can range from 3.5 to 7.7, with pH = 5.6 corresponding to 50% hydrogen saturation. For other soils having different characteristics the equation would have to be modified, but most known soils fall in the range of pH from 4 to 10.

Soils of low pH usually occur in humid climates. Here excess water can force equilibria such as in equation 12.6 to the right so that nutrient ions leach away, leaving behind a soil with a high percent hydrogen saturation and low pH. In arid regions soluble cations such as Ca^{2+} and Na^+ are not carried away since they are released by weathering and may concentrate as carbonate salts. These compounds provide a reservoir of the cations which can saturate exchange sites:

$$CaCO_3(s) + micelle \begin{Bmatrix} H \\ H \end{Bmatrix} \rightleftharpoons micelle \begin{Bmatrix} Ca + H_2O + CO_2(g) \end{Bmatrix} \qquad (12.9)$$

Their hydrolysis

$$CaCO_3(s) + H_2O \rightleftharpoons Ca^{2+}(aq) + 2\ OH^-(aq) + CO_2(g) \qquad (12.10a)$$

$$Na_2CO_3 + H_2O \longrightarrow 2\ Na^+(aq) + 2\ OH^-(aq) + CO_2(g) \qquad (12.10b)$$

can result in soil pH values in the range 8–10—even greater than would be expected on the basis of 0% hydrogen saturation. When treated with HCl such calcareous soils are sometimes observed to bubble as a result of escaping CO_2.

The most important effect of variations in soil pH is their control of the availability of nutrients to plants; for example, as long as a soil has not reached 100% base saturation, an increase in pH reflects the presence of greater quantities of Ca^{2+}, Mg^{2+}, and K^+, both in the soil solution and adsorbed on micelles. In strongly basic soil solutions (pH > 9), the hydroxides and carbonates of calcium and magnesium begin to precipitate. Further increases in pH reduce the availability of these nutrients. Because its hydroxide is soluble, K^+ usually remains available in the soil solution at any pH corresponding to 50% base saturation or more. These and other relationships between pH and nutrient availability are indicated in Figure 12.6.

Iron, manganese, aluminum, and other ions whose hydroxides are extremely insoluble become less available as pH increases, but in some cases too low a pH releases such large quantities of these elements that

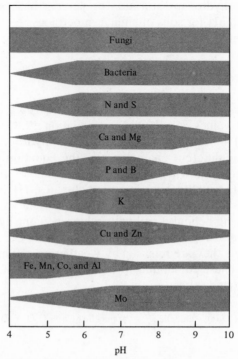

Figure 12.6 Relationship of nutrient availability (indicated by the breadth of the bars) and soil pH. Data from H. D. Foth and L. M. Turk, "Fundamentals of Soil Science," 5th ed. Wiley, New York, 1972; R. G. Gymer, "Chemistry: An Ecological Approach." Harper, New York, 1973.

toxic levels may be reached. Phosphorus is an especially interesting and important nutrient with regard to acid-base equilibria. Different carrier

$$H_3PO_4 \rightleftharpoons H^+ + H_2PO_4^- \qquad pK_1 = 2.13 \qquad (12.11a)$$
$$H_2PO_4^- \rightleftharpoons H^+ + HPO_4^{2-} \qquad pK_2 = 7.21 \qquad (12.11b)$$
$$HPO_4^{2-} \rightleftharpoons H^+ + PO_4^{3-} \qquad pK_3 = 12 \qquad (12.11c)$$

molecules are required for transport of $H_2PO_4^-$ and HPO_4^{2-} across root cell membranes, and the dihydrogen phosphate ion is the preferred form for assimilation. The pK values accompanying equations 12.11 indicate that $H_2PO_4^-$ will be the predominant form in solution over the range $2 < pH < 7$. Because of the insolubility of $FePO_4$ and $AlPO_4$ and the fact that large quantities of Fe^{3+} and Al^{3+} are released from their oxides at low pH, the maximum availability of phosphorus occurs between $pH = 6$ and $pH = 7$ (see Figure 12.6).

The effect of pH on soil organisms is also indicated in the figure, where it is evident that the capacity of fungi to survive at low pH is greater than for bacteria. Disease organisms can often be eliminated by high acidity, but useful species such as earthworms can be harmed as well. When pH drops below 5.5, nitrifying bacteria are affected and the rate of decomposition of humus is greatly slowed. This accounts for the low availability of nitrogen in acidic soils.

The acidity of a soil is usually adjusted by additions of sulfur (to make it more acidic) or lime (to make it more basic). The effect of sulfur is shown in equation 12.12:

$$S + \tfrac{3}{2} O_2 + H_2O \xrightarrow[\text{in soil}]{\text{bacteria}} 2\,H^+ + SO_4{}^{2-} \tag{12.12}$$

Lime is obtained from limestone or dolomite, which may be applied directly, as lime after calcining, or as slaked lime:

$$
\begin{array}{llll}
\text{Limestone} & CaCO_3 & & H_2O + CO_2 \\
\text{Lime} & CaO & + \text{micelle}\Big\} \begin{array}{c} H \\ \\ H \end{array} \longrightarrow \text{micelle}\Big\{ Ca & + H_2O \\
\text{Slaked lime} & Ca(OH)_2 & & 2\,H_2O
\end{array} \tag{12.13}
$$

In all three cases the effect is replacement of adsorbed hydrogen ions by calcium ions. Dolomite contains some $MgCO_3$ and its use results in adsorption of Mg^{2+} as well as Ca^{2+}.

The quantity of lime (or sulfur) required to raise (or lower) soil pH by a given amount depends on the total cation exchange capacity. Addition of lime to soil is similar to titration of a weak acid with a base—remember that many of the ion exchange sites of the humus are

$$-C{\overset{\displaystyle O}{\diagup}}{-}O^-\ \text{groups}$$

The soil acidity has an active component (H^+ ions dissolved in the soil solution) which is immediately neutralized, but the majority of the H^+ ions are held in reserve as a result of association with ion exchange sites. During the addition of lime, equilibria such as 12.7 shift to the right, replenishing H^+ in the soil solution. This produces a buffer action just as the dissociation of acetic acid does during its titration with sodium hydroxide. Therefore, the soil pH is increased very slowly by liming until 100% base saturation is approached, after which it shoots up rapidly and finally levels off. The quantities of lime or sulfur required to produce the same changes in pH are roughly proportional to the cation exchange capacities of various soils, and they can be calculated exactly by using equations such as 12.8.

12.8 SOIL MANAGEMENT

As a general rule soil can renew and conserve itself without outside intervention. Particles of organic matter, aggregation of mineral material by mutual attraction of negatively charged micelles for positive metal ions, and soil organisms all contribute to the formation of pore spaces which allow absorption of most rainfall, keeping water erosion to a minimum. Wind erosion is also slowed by aggregation of soil particles and the holding power of plant roots growing in a fertile soil. The many soil bacteria, fungi, and animals, together with catalytic surfaces afforded by colloidal clays, are capable of removing or detoxifying a variety of harmful substances which may be deposited in soil from the air or water. Air pollutants such as carbon monoxide, insecticides, herbicides, and plant or animal remains can all be assimilated. Despite these safeguards the excesses of man can contribute to degradation of soils in a variety of ways. To conserve and maintain the tremendous natural resource embodied in the soils of the world, it is necessary to understand the processes by which soils carry out their many functions.

Fertility of soil and the nutrients needed for plant growth have already been discussed. In a normal soil–plant system most of the material concentrated in plants is returned to the soil at the end of the growing season by their death and decomposition. This is most evident in grassland soils where almost the entire crop dies each year, but it also applies on a longer time scale to temperate zone forests in which the organic material concentrated in trees is eventually returned to the soil, but not on a yearly basis. Almost half of the organic matter in the forest soil is concentrated in its uppermost 15 cm, because the leaves and decomposing wood fall on the surface. This contrasts with the more even distribution and greater total amount of organic material in the grassland soil because of the annual death and decomposition of root systems.

Once cleared for cultivation the temperate zone grasslands are extremely fertile, considerably superior to forest soils from which much organic material is removed in the process of clearing. But continued cropping of either type of soil has the same effect: Organic material and other nutrients are rapidly depleted during the first few years of cultivation. A new steady state with a lower concentration of nutrients is reached after 50–100 years of continuous cropping. Fertilizers can be added to replace lost nutrients, but they cannot restore the soil structure and porosity afforded by decomposing organic matter. Proper conservation techniques therefore require the addition of at least enough crop residues to maintain organic matter at a level high enough to produce adequate crop yields and to prevent erosion. Quantities of organic matter in excess of this minimum apparently are even more beneficial in that they provide greater capacity for protecting plants against adverse weather conditions.

As mentioned earlier, organic matter is decomposed by a variety of microflora in the soil; when a specific compound becomes available, the bacteria or fungi which can metabolize it multiply rapidly until the supply of nutrients limits their growth. The quantity of carbon in most organic residues is roughly constant at 40–50%, but nitrogen content varies a good deal. If the ratio of carbon to nitrogen is too high, nitrogen may be the limiting factor in the growth of the decomposer organisms and thereby retard the rate of nutrient recycling; for example, barnyard manure or clover residues ($C:N = \sim 20:1$) often release excess nitrogen as NH_3 or NH_4^+ when they decompose. Straw ($C:N = 80:1$) usually requires an external source of nitrogen and, if incorporated in soil shortly before planting a crop, may cause microflora to compete with the desired crop for nitrogen supplies.

This problem can be overcome by applying nitrogen fertilizer at the time the organic material is returned to the soil. Most of the nitrogen will be incorporated in the microflora and will be released slowly as they die. On a smaller scale composting can be used to reduce wide $C:N$ ratios. Storing the organic matter in a compact pile under favorable conditions of moisture and aeration allows CO_2 and H_2O to escape while nitrogen is retained in the amino acids and proteins of the microflora. Providing additional nitrogen by adding fertilizer can greatly increase the population of decomposer organisms and therefore is often recommended for more rapid composting.

As noted earlier, temperate zone forests recycle less organic matter through the soil each year than do grasslands. This principle reaches its maximum in the rain forests of the tropics. Here nearly all of the nutrients are found in the growing plants, and because of the great rainfall the soils are usually in advanced stages of weathering. Few nutrients can be held in the soil in assimilable form and few remain to be released by further weathering. The luxurious jungle growth results from the efficiency and rapidity with which nutrients are recycled to the living portion of the ecosystem rather than the inherent fertility of the soil or the favorable climate. If such a jungle is cleared of its natural growth and cultivated for a few years, nearly all of the remaining nutrients are removed and the soil rapidly becomes unproductive. Addition of fertilizers to replenish the soil is not very helpful because there is no organic matter to retain nitrogen and the large amounts of aluminum and iron oxides of the highly weathered soils tend to fix phosphorus as $AlPO_4$ or $FePO_4$. Because of the alternating rainy and dry climate of many such regions, nutrients are rapidly leached away during part of the year, and the hydrous oxides that remain sometimes dry like bricks in an oven.

Conservation of such tropical soils requires different techniques from those used in temperate zones. Farming techniques which leave bare a minimum of the soil and maintain a high content of organic matter are necessary. Far too little research has been done in this area, and few improvements on the traditional methods of mixed cropping and shifting cultivation have been found. Shifting cultivation uses small plots which are worked until nutrient levels drop, at which time they are permitted to revert to jungle; after 10–20 years the plots may be cleared and worked again. Since such a system requires on the order of 20 hectares (50 acres) per person (far more area than in the temperate regions), the idea that problems of world food supply can be solved simply by bringing tropical lands under cultivation clearly rests on a shaky foundation.

While we are considering the conservation of soil organic matter, it may be well to point out that many of the exaggerated claims regarding nutritional quality of "organically grown" foods seem to have no basis in fact. Most comparisons of organic foods with those produced using "chemical" fertilizers have found no measurable difference. This makes sense from a molecular standpoint because a nitrate ion is a nitrate ion whether mined as saltpeter in Chile, produced by catalytic oxidation of Haber process ammonia, or synthesized by enzymatic oxidation of organic wastes in soil bacteria. Certainly, minimizing the use of synthetic pesticides can help to avoid contamination by trace residues, and composting is an excellent procedure for recycling what would otherwise be waste. As we have already seen most soils do need periodic replenishments of organic matter—perhaps more than is supplied by mechanized United States agriculture—but to claim that organic gardening is a cure-all is excessive.

12.9   DETOXIFICATION OF WASTES

The most telling argument in favor of large-scale application of organic matter such as manure or plant residues to agricultural land is the soil's unique ability to detoxify and absorb substances which if released to the air or water environments would be considered pollutants. The juxtaposition of a supply of dioxygen to act as an oxidizing agent, water to act as a solvent, extensive colloidal surfaces whose binding sites can reduce activation energies of otherwise slow reactions, and microorganisms whose enzymes can attack a variety of molecular structures provides what is probably the most efficient region in the biosphere for conversion of wastes into reusable forms. Synthetic materials—pesticides, plastics, and pollutants—can be decomposed as well as naturally occurring substances, as long as their structures are not too much different from those normally broken down by a given enzyme. Little is known about the exact mechanisms by which soils purify themselves and dispose of excess wastes, but some general principles are apparent.

There is usually a maximum rate at which a soil can carry out a particular chemical transformation. This is apparently the result of limitations on the population of bacteria or other microorganisms in the soil, and it obviously places an upper limit on the quantity of toxic material which can be assimilated by a given soil without killing off the decomposer organisms. However, this limit is far in excess of the amount of organic matter returned to the soil in a normal ecosystem. When a chemical substance which had not previously been in contact with a soil is incorporated in it there is often a time lag of days or perhaps even months before decomposition of the new substance becomes rapid. This may simply reflect the length of time for an adequate population of microorganisms to build up, but in some cases it appears that mutant forms of the decomposers, better adapted to attack the new substance, must develop before rapid decomposition can occur. In some cases, if toxic materials are initially added in small amounts and their concentrations built up gradually, soils can be "trained" to tolerate and decompose them.

The specificity of soils with respect to decomposition of foreign matter apparently depends on the structures of the active soil enzymes. This effect will be discussed in more detail in Section 18.4, when biodegradability and persistance of pesticides are considered, but a good example of soil specificity is the decomposition of urethanes. Methyl and propyl urethanes are not attacked but the ethyl compound is. Apparently the ability to bind to soil bacteria enzymes is strongly influenced by the length of the alkyl group.

| Methyl urethane | Ethyl urethane | Propyl urethane |

Some compounds are decomposed in soils without the intervention of living organisms. For example, studies of removal of air pollutants by soil have shown that sulfur dioxide concentrations can be reduced from 100 to 80 ppm by 15 min contact with soil, even though the soil had been sterilized previously. Similar results were found for oxides of nitrogen. Both of these pollutants were removed from the air more rapidly when unsterilized soils were used, however; and ethylene and carbon monoxide required the presence of soil microorganisms for their removal. In all cases, the reactions in soil are on the order of thousands of times faster than in the atmosphere.

EXERCISES

(The first group of exercises is based primarily on material found in this text.)

1. Describe the physical and chemical changes which occur as weathering converts rock into soil.
2. Explain clearly how nutrients and water are retained in soil as a result of the structures of soil minerals. What role does hydrogen bonding play in this phenomenon?
3. Explain how soil organic matter can serve as a buffer for soil nitrogen content and as a site for ion exchange processes. Why are such sites of great impor-

tance in plant nutrition? Why are cations more readily retained by soils than are anions?

4. How does the structure and composition of soil account for the fact that it is especially well suited to decompose and detoxify a large number of pollutants?

5. Why is the clearing and cultivation of tropical rain forests unlikely to result in as much of a contribution to world food supplies as the land area involved might imply? Explain why cleared forest land is less fertile than grassland.

6. Why is calcium less readily available to plants in acidic or basic soils than in those which are approximately neutral?

7. Explain why acid rainfall (Section 9.3) might be expected to affect soil fertility.

8. Assuming that farm manure corresponds approximately to $0.5 - 0.24 - 0.5$ fertilizer, calculate the weight percentages of nitrogen, phosphorus, and potassium in manure.

9. Calculate the number of kilograms of potential HF air pollution (Section 11.11) which could be avoided by spreading manure at the rate of 2.5 tonnes/hectare on a 100-hectare farm instead of a quantity of superphosphate fertilizer which would yield equivalent phosphorus.

(The subsequent exercise may require more extensive research or thought.)

10. What are some of the advantages and disadvantages of applying manure to agricultural land instead of fertilizer? Is there any benefit from using both at once?

13

Solid Wastes

Resources are not; they become.

E. S. Zimmerman (*World Resources and Industries*, 2nd ed., Harper, 1951, p. 15.)

. . . as chemical knowledge grows, we continue to discover that the useless rubbish of yesterday becomes the valuable resources of today.

Jacob Rosin and Max Eastman (*The Road to Abundance*, McGraw-Hill, 1953, p. 83.)

During the past two decades the quantity of solid waste collected in the United States has nearly doubled, and if predictions hold true it will double again over the next 20 years. The problem is not simply one of increasing population—per capita production of wastes has doubled in the United States since 1940. Table 13.1 indicates the sources of solid wastes and shows that industrial, mining, and agricultural activities produce much larger quantities than household or commercial sectors. The United States standard of living produces approximately 44 kg (97 lb) of solid waste every day for every person in the country. Of the solid waste that is collected for disposal, most is from household and commercial sources.

The technology used to handle these wastes lags well behind that used to control air and water pollution. As recently as 1966 well over half of the United States cities and towns with populations over 2500 were disposing of refuse in open dumps or by other unsanitary and nuisance-prone techniques. In the past there was little concern about the problem as long as disposal sites were far enough from public eyes (and noses) and in 1965 the federal government spent less than $300 000 per year for research on handling solid wastes. Prior to passage of the Solid Waste Disposal Act of 1965 it was the prevailing view that things no longer usable could simply be thrown away. The tremendous growth in production of municipal and

TABLE 13.1   Annual and Daily Production of Solid Wastes in the United States[a]

Type of waste	Total mass (10⁶ tonnes/year)		Per capita (kg/day)	
Household, commercial, municipal	226		3.1	
Collected		172		2.4
Uncollected		54		0.7
Industrial	100		1.4	
Mineral	1000		14	
Agricultural	1860		25	
Farm animal wastes		1360		19
Crop residues		500		7
Total (rounded)	3200		44	

[a] Data from R. J. Black *et al.,* "The National Solid Wastes Survey: An Interim Report."
U. S. Dept. of Health, Education and Welfare, Washington, D. C., 1968.

industrial refuse combined with increases in population has underscored
the fact (which should have been obvious to persons familiar with the law
of conservation of mass) that there is no longer an "away."

13.1   POPULATION × AFFLUENCE × TECHNOLOGY

The rapid growth in the quantity of solid wastes may be attributed to
several causes. Barry Commoner[1] has attempted to divide the environmental impact of any economic good into three major factors, any or all of
which may be important. The first is *population*—as the number of people
producing a pollutant increases, the amount of pollution can be expected to
increase as well. His second factor is *affluence*—production or consumption per capita. (Both of these factors affect the amount of refuse collected
since the total is climbing faster than per capita. The difference is attributable to population growth.) However, in most cases a third factor is
required to account for the entire increase in pollution levels. This is
technology—the amount of pollution produced per unit of economic good.
Total environmental impact is then given by the product of the three
factors.

$$\text{Environmental impact} = \text{population} \times \text{affluence} \times \text{technology}$$

$$= \text{population} \times \frac{\text{units produced}}{\text{population}} \times \frac{\text{pollution}}{\text{unit produced}} \quad (13.1)$$

Several aspects of the solid waste problem illustrate Commoner's technology factor quite well. Consider, for example, the economic good embod-

---

[1] B. Commoner, *Chem. Brit.* **8**(2), 52 (1972).

ied in the consumption of a glass of beer. Several competing technologies exist for transporting this good to the consumer: One is the use of large kegs which are available to the consumer through the services of a bartender in a centrally located taproom. A second is the use of glass containers which are returned by the consumer to the point of purchase for cleaning and refilling. The third and most rapidly growing technology is the nonreturnable bottle or can.

It is clear that the amount of pollution associated with these technologies increases in the order that they were given. A beer keg is sufficiently valuable to be reused many times. Each beer bottle, of whatever type, has an environmental impact due to the material and energy resources used for its production (Section 13.4) and when discarded it becomes part of the solid waste disposal problem. The environmental impact of beer consumption can be measured roughly, then, in terms of total bottles produced.

Table 13.2 gives the data needed for a comparison of impacts of beer consumption in 1950 and 1967. During that period the United States population increased by 30%, the affluence factor (consumption per capita) by only 5%, but the total number of bottles produced went up by 595%! The controlling factor in this increase is the more than 400% rise in the number of bottles per unit volume of beer consumed. The shift in technology toward the nonreturnable container is responsible for a tremendously augmented environmental impact.

There are many other forms of packaging which, like beer bottles, are being thrown into the solid waste heap by today's society. Self-service stores require that attractive packaging call a buyer's attention to a product; increased ease of shoplifting leads to the use of large packages for small items. Many of the most recently devised containers are composites of several materials (plastics laminated onto paper, glass bottles with metal rings from screw caps, and steel cans with aluminum pull tabs are some examples), greatly complicating separation of these materials prior to recycling. Changes in technology such as these have combined with increased population and affluence to raise the per capita annual consumption of packaging from 183 kg in 1958 to 238 kg in 1966 to 267 kg in 1971 and to

TABLE 13.2 Environmental Impact of Beer Consumption[a]

Factors	1950	1967	1967/1950	% Increase
a = Population	152×10^6	198×10^6	1.30	30
b = Affluence (dm^3/capita)	94.6	99.6	1.05	5
c = Technology (bottles/dm^3)	0.066	0.334	5.08	408
Total impact ($a \times b \times c$, bottles)	6.54×10^6	45.48×10^6	6.95	595

[a] Source: B. Commoner, The environmental cost of economic growth. *Chem. Brit.*, **8**(2), 62 (1972).

TABLE 13.3   Composition of Typical Municipal Refuse[a]

Physical (weight %)		Rough chemical (weight %)	
Cardboard	7	Moisture	28.0
Newspaper	14	Carbon	25.0
Miscellaneous paper	25	Hydrogen	3.3
Plastic film	2	Oxygen	21.1
Leather, molded plastics, rubber	2	Nitrogen	0.5
Garbage	12	Sulfur	0.1
Grass and dirt	10	Glass, ceramics, etc.	9.3
Textiles	3	Metals	7.2
Wood	7	Ash, other inerts	5.5
Glass, ceramics, stones	10		
Metallics	8	Total	100.0
Total	100		

[a] Data from E. R. Kaiser, Refuse reduction processes. *In* "Proceedings, the Surgeon General's Conference on Solid Waste Management for Metropolitan Washington," U. S. Pub. Health Serv. Publ. No. 1729. US Govt. Printing Office, Washington, D.C., 1967.

a projected 300 kg in 1976. In 1966 packaging cost the American public 25×10^9 — about 3.4% of the gross national product; $16 billion of this total was the value of raw materials, 90% of which were thrown away. In the United States packaging accounts for 47% of all paper production, 15% of aluminum, 75% of glass, 9% of steel, and 29% of plastics production.[2]

Even considering this tremendous increase in consumption of packaging, it accounts for only 20–30% of total household waste and 8% of commercial and industrial waste. The approximate composition of municipal refuse in terms of the major materials as well as the chemical elements present is indicated in Table 13.3. Because of this heterogeneity, the disposal of municipal wastes is considerably more difficult than for industrial waste, mine tailings, or agricultural debris.

In the industrial, mining, and agricultural sectors shifts to new technology have also exacerbated solid waste disposal problems. In the steel industry, as noted in Section 11.4, the shift to basic oxygen furnaces has reduced the amount of scrap recycled, although greater use of electric furnaces could reverse this trend. Automobiles have become more complicated, containing greater percentages of aluminum, copper, plastic, and other materials which make recycling more difficult. Techniques which permit lower grade ores (such as copper ores containing less than 1% copper, Section 11.6) to be mined naturally mean greater amounts of tailings. Obviously, the quantity of waste produced will be 99 times the copper production, or perhaps more if any flotation or other separation

[2] U. S. Environmental Protection Agency, press release, February 22, 1974.

agents are carried into the tailings. In agriculture the shift from pastures to feed lots and automated dairy barns concentrates animal wastes in such a small area that they cannot be handled naturally by the soil unless spread over a much wider region.

In a number of cases control of air or water pollution leads to new contributions of solid wastes. Fly ash from coal-fired power plants and particulates from iron and steel furnaces must be disposed of after they have been separated from effluent gases. The same is true of sludges from primary and secondary sewage (Section 15.2) and industrial wastewater treatment (Section 16.1). Great effort has been expended in order to find ways in which air and water pollutants can be trapped, but this is not a solution unless solid wastes are adequately disposed of.

13.2 MUNICIPAL WASTES

Industrial, mining, and agricultural wastes can usually be handled by their producers, often at the sites where they are produced. The technology required is usually not too complicated because of the homogeneity of the waste. In many cases the discarded material has (or may have at some future time) sufficient value to make its preservation worthwhile. This is especially true in mining where low-grade phosphate and copper ores (among others) are saved in the belief that developing technology will make their recovery feasible. The need for additional organic matter in soil leads to an obvious disposal method for agricultural wastes, provided they are not applied in such great quantities that groundwater supplies are contaminated.

Municipal wastes, on the other hand, can be simply thrown "away" by their producers. If the disposal technique contaminates the environment it does so at a great enough distance from most of the persons responsible that it can often be ignored. In 1966 the vast majority of solid waste was transported to land disposal sites, most of which, regardless of their name,

TABLE 13.4 Methods of Disposal of Municipal Wastes in the United States (1966)

Method	Total waste (%)		No. of sites
Land disposal	90		12 000
Sanitary landfill[a]		5.4	
Other		84.6	
Incineration	8		300
Air pollution controlled		2.4	
Other		5.6	
Miscellaneous (hog feeding, composting, ocean dumping, salvage operations)	2		—

[a] Daily cover, no open burning, no water pollution.

did not qualify as sanitary landfills (Table 13.4); nor did the majority of incinerators meet air pollution standards. In the remainder of this section we shall consider current methods for disposal of municipal refuse together with some of their associated problems.

Sanitary Landfills

In a sanitary landfill operation garbage and other wastes are spread in thin layers on the ground or in a specially excavated trench and then compacted by having a bulldozer drive over them. When the refuse reaches a depth of about 3 m a thin layer of soil is spread over the surface and compacted. Additional layers of trash may be added to achieve any desired depth, and when completed the fill is sealed with about one meter of compacted earth. For the landfill to be designated "sanitary," a thin layer of earth must be spread over the wastes at the end of each working day: this prevents the growth of large numbers of insects, rodents, and other disease-carrying pests and lessens the chance that burning initiated by spontaneous combustion or arson may contribute to air pollution. A sanitary landfill also must be sited to prevent surface or groundwater contamination, especially in rainy climates.

Much of the organic material in a landfill undergoes anaerobic decomposition, catalyzed by enzymes present in soil bacteria, producing CH_4, H_2S, H_2O, CO_2, H_2, and organic acids. The organic acids usually range from two to six carbon atoms in length, are capable of leaching carbonates (such as limestone) from surrounding soils, and may be the chief instigators of water pollution from landfills. Methane and dihydrogen produced in landfills usually are sufficiently dispersed in the atmosphere that no hazard is created, but 3–5 years are required before the quantities produced become negligible. If buildings are constructed too soon after completion of a landfill, they may trap the gases, creating an explosion hazard. On the other hand, the anaerobic conditions of decomposing municipal waste are ideal for filling abandoned coal mines since oxidation of pyrite to form acid mine drainage is effectively prevented (Section 5.2). In at least one case a project has been planned to trap and make use of combustible gases from a sanitary landfill as an alternative energy source.

Well-planned and carefully operated sanitary landfills need not be a blight on the neighborhoods in which they are located, but since the term *sanitary landfill* has been used to describe everything from open dumps to carefully operated sites, most persons are strongly opposed to construction of a landfill in their vicinity. The number of new sites available within metropolitan areas is thus quite limited, and old sites are rapidly being filled. In 1966 a survey of nearly 400 cities revealed that 48% of their landfills had fewer than 6 years of use remaining. Transportation and labor costs immediately begin to rise when it is necessary to use sites far from the concentrated population that is generating the refuse and this, combined with possibilities of water pollution, has convinced the Environmental Protection Agency that incineration is the best technique for municipal waste disposal in the near future.

Incinerators

Proper incineration of wastes requires temperatures between 1050 and 1250 K, so that complete combustion of oxidizable material can occur (leaving behind ash, glass, metal, and other materials amounting to about one-fourth the initial mass), but not so hot that slag formation can be a problem, complicating furnace operation and making separation of residues for recycling more difficult. The 300 municipal incinerators mentioned in Table 13.4 consume only about half of the incinerated waste in this country; most of the others are small, privately owned, and serve individual apartment or office buildings. These small-scale incinerators often are collectively responsible for large quantities of air pollution, especially particulates, and many have been forced to shut down or to upgrade their facilities as a result of strict air emission standards.

In the operation of a typical municipal incinerator (shown in Figure 13.1) refuse is first dumped from collection trucks into a large pit. From there it is transferred by crane to a hopper and slowly fed onto a grate where hot gases from the furnace dry and preheat it. As the material passes into the ignition chamber, preheated air is introduced from below the grate and combustion begins. The speed at which the grates conduct the refuse through each chamber can be adjusted to suit the composition of the waste and the temperature of the combustion chamber. A rotary kiln (similar to those in cement plants, Section 11.8) turns continuously, exposing all portions of the refuse to hot gases from the ignition grate. Any solid material which remains drops onto a conveyor that is submerged in water (region 6 in the diagram). The quenched solids are conveyed to trucks which transport them to landfill sites. Salvaging of steel or other noncombustibles is usually done by screening or magnetic separation just prior to dumping the residues in the trucks.

The hot gases from the drying grate meet those from the ignition grate and kiln in a mixing chamber where combustion is completed before they pass to a wet scrubber or other device for removal of particulates. In case of emergencies such as excessive temperature or loss of electric power gases can exit directly from the prequench chamber to the stack, but in normal operation they pass downward and then up through the wet scrubber.

In a great many European and a few American incinerators the objective is not merely to reduce the mass and volume of the wastes but to use the heat of combustion to generate steam. Some of it can be used within the

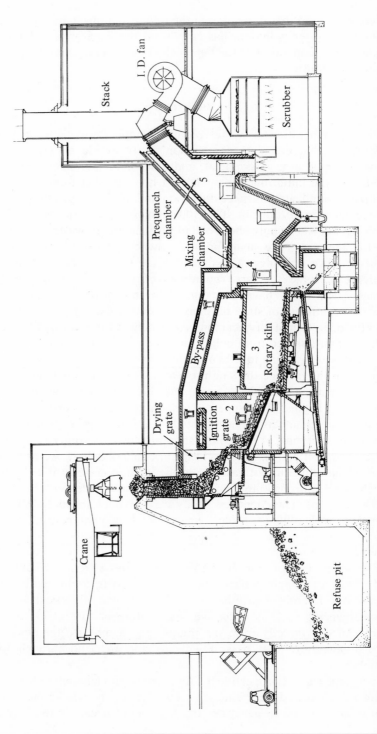

Figure 13.1 Design of typical municipal refuse incinerator. Source: A. J. Warner, C. H. Parker, and B. Baum, "Solid Waste Management of Plastics," p. A-117. Mfg. Chem. Ass., Washington, D. C., 1971.

incinerator plant itself and the rest sold to nearby industries or power plants. Two large-scale examples of steam-generating incinerators in the Western Hemisphere are the Des Carrieres plant in Montreal and the Chicago Northwest incinerator. Both differ from Figure 13.1 in that a third grate replaces the rotary kiln and the hot combustion gases pass through a boiler constructed of water-filled pipes. For this reason they are referred to as water-walled furnaces. (Each plant also has an auxiliary fossil fuel burner to ensure a continuous supply of steam should a strike interrupt input of rubbish.) The Montreal plant burns 99.5% of combustibles and its electrostatic precipitator allows less than 0.2 g dust/kg of gas to be emitted. It is estimated that if all municipal waste in the United States were burned in water-walled furnaces aproximately 10–15% of our electric power needs could be generated from the steam produced. The heating value of the average sample of municipal waste is about 10 GJ/tonne.

More recently the concept of using municipal refuse as part of the fuel supply for an existing electric power plant has found favor. Two 125-MW_e boilers at the Union Electric Company's Meramac plant in St. Louis have been successfully modified to burn gas, pulverized coal, and finely ground refuse. The latter is consumed at the rate of 300 tonnes/day over a 5-day week. This system appears to be attractive both economically and environmentally.[3]

Incineration of Plastics

Recently, there has been a good deal of concern about the products of combustion of plastics in incinerators. The quantity of plastics in packaging increased by more than 80% from 1965 to 1975 and the use of plastics in nonpackaging applications is also growing rapidly. The structures of some of the most common monomers and polymers used in synthetic plastics are shown in Table 13.5. The same monomer may be used in a variety of different materials and under many trade names, some of which are listed in the table. Most plastics consist primarily of carbon, oxygen, and hydrogen, as do the majority of combustibles in municipal refuse (see Table 13.3). The principal gases generated when they burn are carbon dioxide, water, and perhaps small quantities of carbon monoxide and nitric oxide. Because plastics have large heats of combustion they are valued as fuel in water-walled furnaces or on rainy days when more heat is needed to vaporize water from moist refuse.

A few plastics do create problems for air pollution control. Combustion

---

[3] U. S. Environmental Protection Agency, "Energy Recovery from Waste." US Govt. Printing Office, Washington, D. C., 1973.

TABLE 13.5 Structures of Monomers and Polymers Used in Common Plastics and Their Trade Names and Usages

Substance		Trade name	Uses	Total production (%)
Monomer	Polymer			
Ethylene $CH_2=CH_2$	Polyethylene $-CH_2-CH_2-CH_2-CH_2-CH_2-CH_2-$	—	Films and sheets for packaging and covers; toys; molded containers, tubing and piping; housewares; electric insulation; coatings on metal, paper, etc.; bottles; textiles; bristles.	54
Propylene $CH_2=CH-CH_3$	Polypropylene $-CH_2-CH-CH_2-CH-CH_2-CH-$ CH_3 CH_3 CH_3	—	—	—
Styrene $CH_2=CH$ (phenyl)	Polystyrene $-CH_2-CH-CH_2-CH-CH_2-CH-$ (phenyl groups)	Styrofoam	Refrigerators; insulation; packaging; dishes; glasses; bonding of adhesives; automotive instrument panels.	20
Vinyl chloride $CH_2=CHCl$	Polyvinyl chloride $-CH_2-CH-CH_2-CH-CH_2-CH-$ Cl Cl Cl	Vinyl, Saran (copolymer with vinylidene chloride)	Pipe and tubing; raincoats; building panels; curtains and draperies; upholstery; phonograph records; luggage; films.	11
Vinylidene chloride $CH_2=CCl_2$	Polyvinylidene chloride structure	—	—	—
Acrylonitrile $CH_2=CHCN$	Polyacrylonitrile $-CH_2-CH-CH_2-CH-CH_2-CH-$ $C\equiv N$ $C\equiv N$ $C\equiv N$	Orlon, Acrilan, Dynel	Textile fiber; rugs.	—

Phenol, formaldehyde	Phenolic	Bakelite, Formica	—	—
Urea, formaldehyde	Structure not definitely established	—	—	—
Dialcohol, diacid	Polyester	Dacron, Terylene, Mylar	Textile fibers, skis, fishing rods, films.	—
Cellulose, naturally occurring polymer		Rayon, Acetate	Textile and paper; packaging films; magnetic tapes; packaging.	—
Diacid, diamine $H_2N-(CH_2)_6-NH_2$	Polyamide	Nylon	—	—
Urethane $HO-(CH_2)_n-OH$, $O{=}C{=}N-(CH_2)_m-N{=}C{=}O$	Polyurethane	—	Foam padding for upholstery and linings for coats; cigarette and air filters; packaging; foam rubber applications.	—

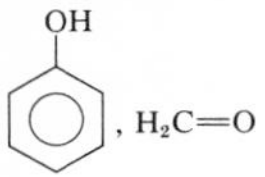

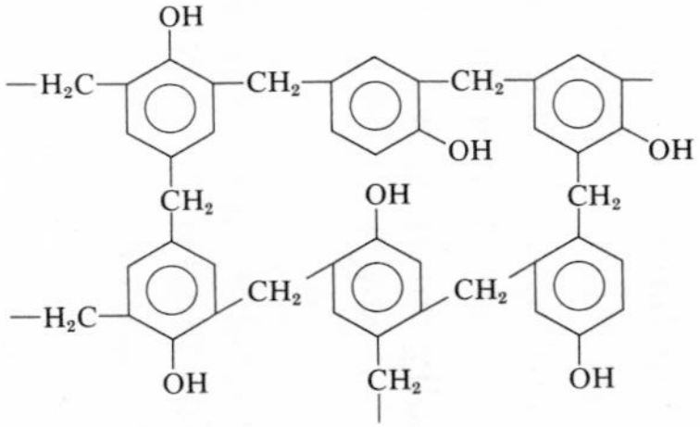

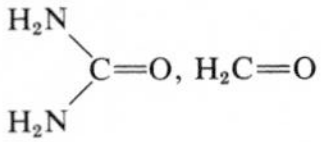

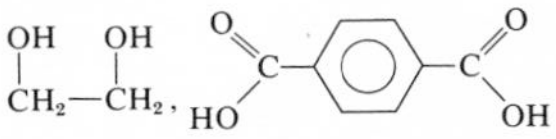

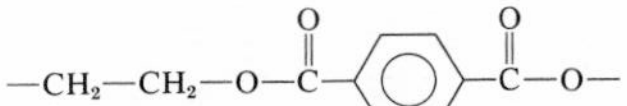

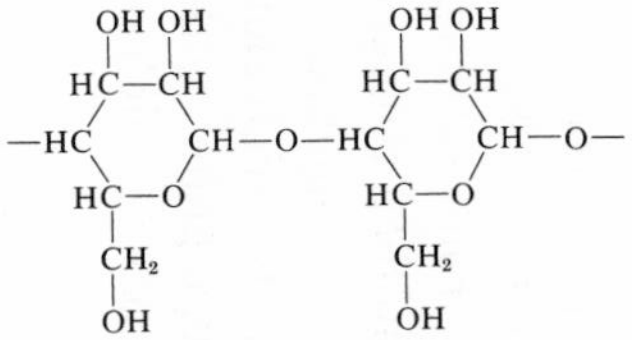

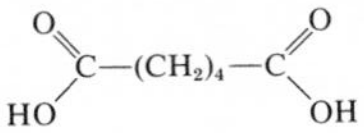

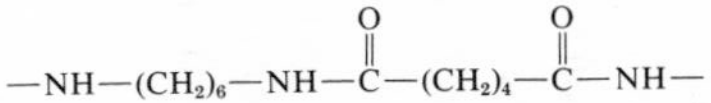

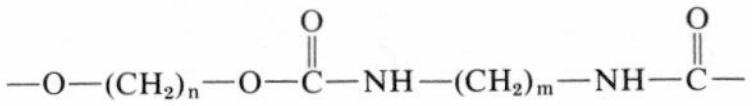

of polyvinyl chloride (PVC), polyvinylidene chloride, or copolymers of vinyl and vinylidene chlorides (such as Saran) yields hydrogen chloride:

$$-\left(CH_2-\underset{\underset{Cl}{|}}{CH}-CH_2-\underset{\underset{Cl}{|}}{\overset{\overset{Cl}{|}}{C}}-\right)_n + \tfrac{9}{2}n\,O_2 \longrightarrow n\,H_2O + 4n\,CO_2 + 3n\,HCl \quad (13.2)$$

Saran

Other chlorine-containing compounds in refuse will react in the same way. The HCl can increase the rate of corrosion within a incinerator, and if emitted to the atmosphere in large quantities can cause substantial damage to property, plants, and people. Public sentiment against chlorine-containing plastics has led some manufacturers to suggest alternative formulations containing high percentages of acrylonitrile, but this would appear to be a poor choice in view of the recent report[4] that polyacrylonitrile (Orlon) produces highly toxic HCN when burned.

Both HCN and HCl can be removed from incinerator stack gases by wet scrubbing. For HCl the removal efficiency is between 85 and 95%, but experiments with HCN have not yet been done. Some persons in the plastics industry have claimed that HCl does not complicate incinerator operation because wet scrubbers are necessary for control of particulates, but more than half of the incinerators now in operation have no air pollution controls at all, and some of the most modern ones (such as the Montreal and Chicago units) have only electrostatic precipitators because removal of gaseous emissions was not deemed to be necessary. Furthermore, the liquid effluent from scrubbers, which is usually neutralized with Na_2CO_3, will contain high concentrations of NaCN and NaCl which could cause water pollution problems.

Some other materials which cause problems in incinerators are wet grass (which can sift through grates without burning and can emit HCl and NO_2), and waterlogged garbage (which may be seared on the outside, leaving the inside to rot and cause odor problems). Auto tires and Styrofoam (both of which consist primarily of polymers) produce large amounts of smoke and soot. Molten aluminum can run down through grates, and magnesium reacts with iron oxide to form globs of steel. Ammunition and aerosol cans cause explosions which can damage equipment or sometimes injure workers. Large quantities of fats (lard) or plastics which have high heat value can melt steel components of the furnace. Cases have even been reported where sodium (from chemistry labs!) has ignited refuse prematurely and caused serious fires.

---

[4] J. W. Hill, *6th Amer. Chem. Soc. Great Lakes Reg. Meet., 1972* Abstract No. 27 (1972); L. M. Zabrowski, J. W. Hill, and M. W. Wehking, *Environ. Let.* **3**(4), 267–270 (1972).

Despite such problems incinerators show promise for recovery of the heat value of municipal waste, and they reduce the volume of waste by a large factor. Residues must still be transported to landfills, but the reduction of their volume significantly increases the usable lifetime of such sites. Moreover, these residues may constitute valuable "urban ores" which may be recycled. The average residue from a Washington, D. C. incinerator, for example, contains 28% iron—a figure not too much below that of ores which are currently being used, and most of the metal is already in reduced form.

Pyrolysis

Pyrolysis is defined as chemical change brought about by the action of heat alone. When applied to municipal waste it differs from incineration in that dioxygen is excluded and only enough combustion to supply heat for the process is permitted to occur. Destructive distillation of wood to produce methanol and coal gasification (Section 5.3) are other common pyrolytic processes. A major advantage of pyrolysis over incineration is that conditions may be adjusted so that low molecular weight hydrocarbons having significant commercial value, such as methanol and acetic acid, are produced.

The first commercial application of pyrolysis was the Destrugas process used in an 18-tonne/day plant at Kolding, Denmark. Waste is pyrolyzed between 1100 and 1300 K giving a gas consisting of 54% H_2, 23% CO_2, 10% CO, 10% CH_4, and small quantities of hydrocarbons and dinitrogen. Ammonia produced from protein apparently neutralizes HCl from PVC as long as the latter does not exceed 3% of the input. About 80% of the gas evolved is used to heat the plant and the remainder is sold. The mass of solid waste is reduced to about one-third in the char left behind. Little air pollution is produced because the reaction takes place in a closed system.

The flow diagram in Figure 13.2 outlines a pyrolysis scheme developed and tested in a 4-tonne/day pilot plant by Garrett Research and Development Company in the United States. In this process the wastes are first shredded to a size of 5 cm or less. Inorganic material is separated by a blast of air which carries the less dense, pyrolyzable organic material away. Reduction of inorganic content to less than 4% by weight is attained using 0.6 cm and fourteen mesh screens. The glass and metal of the inorganic fraction are separated from each other (using operations to be described in Section 13.3) and then sold. Prior to pyrolysis the organic fraction undergoes secondary shredding which reduces particle size to 24 mesh. Primary and secondary shredding each require about 1.5×10^8 J/tonne of refuse.

Pyrolysis takes place at 750 K and atmospheric pressure, under which conditions the liquid fuel fraction is maximized. The pyrolysis reaction is

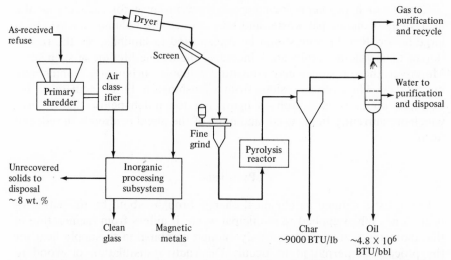

Figure 13.2 Flow sheet for refuse pyrolysis in recycling of solid wastes. Source: B. Baum and C. H. Parker, "Plastic Waste Disposal Practices," p. 28. Mfg. Chem. Ass., Washington, D. C., 1973.

slightly exothermic, minimizing the external heat requirements. About 40% of the organic material is converted to a heavy oil with about three-fourths as much heating value per unit volume as number six fuel oil (the most viscous and highest boiling oil used in industrial and electric power generation operations). The pyrolysis oil contains on the order of one-tenth the quantity of sulfur of most heating oils and can be burned in electric power plants which are equipped to handle it. Although oil is the principal pyrolysis product, fuel gas amounting to 27% of the input, a solid char accounting for 20%, and 13% water are also obtained.

The water is used in a scrubber to remove chloromethane (which is produced from PVC in the pyrolyzer instead of HCl) from the fuel gas. Since it is high in biochemical oxygen demand (Section 14.2), the water must undergo secondary sewage treatment before leaving the plant. After purification the fuel gas is burned to heat the pyrolyzer and the exhaust gases pass through a bag filter before venting to the atmosphere. The char, which has a heating value intermediate between lignite and bituminous coal and contains about 0.3% sulfur, could be burned in a power plant. However, pyrolysis introduces considerable void space into the char and there exists a possibility that it might be upgraded to activated charcoal for use in advanced wastewater treatment (Section 15.3). This would considerably improve the already favorable economics of the system.

The estimated value of the products of the pyrolysis system ranges from $6.75 to $21.80 per tonne refuse, which compares favorably with the operating cost of $6.00 per tonne. Even the minimum potential value of recoverable material would permit profits to be made. As reserves of petroleum and other sources of hydrocarbons diminish, the value of pyrolysis products may well increase, making the economics of the process more favorable. In any case pyrolysis can be pollution free, no external energy source is required, capital costs are about two-thirds those of incineration, and the solid residue is suitable for landfill and reduced in volume by more than 50%. These factors alone make the process attractive, even if salvagable products are used only for fuel.

Biodegradation

Because of complaints by many environmentalists that the stability of synthetic polymers makes them permanent wastes, there has been considerable effort recently to develop plastics which are biodegradable. There is no difficulty in making plastics that will degrade rapidly in the environment. Ultraviolet radiation and dioxygen are potent agents for decomposition of polymers, especially those containing unsaturated sites. The damage done to rubber by ozone has already been mentioned (Section 10.3). Much of the development of polymer chemistry has, in fact, resulted from efforts to develop antioxidants, stabilizers and plasticizers which retard degradation. Now it appears that this process is to be reversed.

For a variety of reasons biodegradability of plastics may turn out to be even more of a mixed blessing than biodegradability of detergent surfactants (see Section 14.4). Sunlight and air are ubiquitous. If containers degraded by these two agents are to protect their contents over an adequate shelf life some coating or stabilizer must be designed to protect the containers until they have been emptied. Those of frugal temperament might learn the hard way that plastic containers were not permanently stable. One has visions of pantry shelves covered with mixtures of preserves and decomposed polymer.

Perhaps a more reasonable approach is to design polymers which can satisfy the appetites of microorganisms in sanitary landfills, but here again problems arise. If incineration or pyrolysis were used prior to landfilling there would be no need for biodegradability. Both of these methods are capable of recovering at least the heat value of plastics, and pyrolysis shows promise of producing new raw materials as well, values which would be lost if biodegradable plastics were simply dumped. The nondegradability of plastics in landfills is thought to be an advantage by some because of reduced settling and no production of flammable gases. If at some time in the future additional supplies of reduced carbon compounds were needed, nondegraded plastics would still be minable from a landfill. All in all the creation of entropy pollution by development of biodegradable polymers seems a wasted exercise of chemists' talents.

13.3   RECLAMATION, RECYCLING, AND REUSE

Those familiar with the law of conservation of mass-energy will realize that it is impossible to "dispose" of solid wastes, but until the passage of the Resource Recovery Act of 1970 the U. S. Government had made no official recognition of this scientific law. Because wastes buried in a landfill or burned in an incinerator or dump were out of sight they were thought of as having been destroyed and no one gave them much further concern. The 1970 act stresses recovery by one of three techniques. *Reclamation* implies recovery of a component of waste for use in a manner different from its initial function. A good example is the use of tin cans for cementation of copper (Section 11.6). *Recycling* consists of recovering from waste the matter of which a product was made and reintroducing it into the production cycle for reproduction of the same item. Conversion of cullet (Section 11.9) back into glass bottles is a common example of recycling. Finally, *reuse* denotes multiple use of a given material or product and is typified by the returnable beverage bottle. Often no distinction is made among these three categories of resource recovery, but as we shall see they can be quite different.

Before considering the technology of scrap recovery it is useful to note the fractions of various materials which are currently drawn from wastes. Taking iron as an example, between 1951 and 1961 scrap recovery remained roughly constant at 36% of production, by 1969 it had dropped to 30% and by 1971 was at 27%. In the case of copper during the same two decades the use of scrap declined from 25 to 23%. Aluminum scrap accounted for only 3% of production in 1951, rose to 7% in 1961, and dropped to 5% by 1971, despite considerable publicity by various aluminum companies regarding their efforts in the field of recycling.

Much the same situation applies to nonmetallic minerals and forest products. About 35% of the United States paper fiber requirements were met by waste paper in 1946. By 1956 the figure had dropped to a little over 25% and reached 21% in 1966; projecting the trend to 1980 a figure of 17.5% is obtained. Although percentages of recycled paper have declined, the tremendous increase in consumption has permitted the actual tonnage recycled to increase from 7.3×10^6 in 1946 to 10.2×10^6 in 1966. This accounts for the claim that "We're recycling more paper now than we ever did." On the other hand, the amount thrown away has increased much more rapidly than what we have reclaimed. Estimates of the quantities of other materials which are potentially recoverable from scrap as well as the amounts that we now reclaim are shown in Table 13.6.

The principal reason for the decline in resource recovery is the development of new technology. Because there were profits to be made, great effort has gone into creation of more efficient and less expensive methods of harvesting, mining, and processing virgin materials, and at the same time

TABLE 13.6 Current and Potential Recycling of Resources in Trash (10⁶ tonnes, 1969)[a]

Resource	U. S. annual production		Estimated additional amounts available in urban solid wastes
	Total	From scrap[b]	
Inorganic			
Iron and steel	135	44.4	9–12.7[c]
Aluminum	3.4	0.18	0.9–1.08
Copper	2.0	0.54 ⎫	0.45
Zinc	1.4	0.09 ⎭	
Tin	0.07	0.021	0.03
Glass	14.5	1.8–4.5	9–10.8[d]
Construction aggregate	906	Negligible	10.8–13.5[d]
Organic			
Paper	53.5	10.8	32.6
Compost	Negligible	—	45–90.6
Protein	9.06[e]	—	4.5–9.06
Plastics	9.06	Negligible	3.6–6.3
Fuel	52 740 TJ[f]	Negligible	2109.6 TJ
Charcoal	1.8	— ⎫	9.06–45.3[g]
Coke	72.5	— ⎭	
Rubber	2.7[h]	0.18–0.9	1.8–2.5

[a] Data based on R. Grinstead, *Environment* **14**(3), 4 (1972).
[b] Production from scrap does not include in-plant recycling.
[c] Of which cans make up about two-thirds.
[d] Aggregate assumed to include glass and other mineral material present.
[e] Mainly for animal feed.
[f] Stationary energy sources.
[g] By pyrolysis.
[h] Of which about 2.5 is tires.

new products and new combinations of materials have made wastes far more difficult to separate. In many cases government-sponsored or government-approved subsidies also favor the use of virgin over reclaimed materials. For example, scrap iron requires an average of $4.54 per tonne in freight costs for return to a steel mill, while virgin iron ores costs ~ $1.80 per tonne for a trip of approximately equal distance. Such arbitrary rate differentials are routinely approved by the government. Perhaps the greatest subsidy of all is public acceptance (in the form of municipal waste collection) of the burden of handling packages and other superfluous materials. All of these circumstances have combined to squeeze resource recovery out of existence due to adverse economics, and to prevent adequate research into modern techniques for recycling.

Recently, a number of corporations have recognized that a profitable waste recovery system could be designed if all of the products that can be obtained from the wastes are kept in mind — especially if the system is to be operated by a municipality which otherwise pays fairly high costs for waste disposal. Air and water pollution controls, lack of landfill space, and public awareness of environmental considerations have also contributed to the feasibility of marketing resource recycling systems. While voluntary recycling programs are an excellent means of expressing the environmental awareness of small groups of individuals, the vast majority of solid wastes will go unreclaimed until institutions such as government and industry achieve a similar level of environmental consciousness.

Currently proposed resource recycling systems are composed of a variety of different technological attempts at separating all or most of the different components of municipal waste. Referring to Table 13.3, these components may be divided into the following categories in order of decreasing weight percent in refuse: paper, cardboard, and wood (53%); garbage and grass or leaves (22%); glass and ceramics (10%); metals (8%); and textiles, plastic, rubber, and other miscellaneous materials (7%). Methods of separating and recovering each of these categories of material will be discussed in the paragraphs which follow.

Paper and Cellulose Fiber

There are several factors which complicate the reclamation or recycling of cellulose fiber. Perhaps the most important is the mixture of various types of fiber, each of which has distinct properties, found in municipal refuse. About 90% of United States paper is made by chemical pulping (Section 16.5) in which lignin is dissolved from the cellulose, producing long, strong, easily bleached fiber. The other 10% of papermaking is done by mechanically grinding wood in the presence of water. Groundwood fiber, although less expensive to produce is generally shorter, weaker, and not as readily bleached since some lignin remains; its principal use is in newsprint, where permanence and whiteness are not very important. There is no way of separating higher quality fiber during recycling; thus, 100% recycled paper usually tends to be somewhat off-color and weaker than the virgin product.

Two other factors mitigating against recovery of cellulose fiber are the variety of synthetic materials incorporated into paper during its manufacture and the damage done to the fiber in the repulping process itself. Latex and clay coatings, wet-strength resins, impregnated wax, glues and adhesives, dyes, pigments, and permanent inks are among the chief contaminants incorporated in paper as a normal part of its use. Their number as well as the difficulty of removing them increases every year. Disintegrating paper in water for repulping damages hairlike fibrils attached to the cellulose fiber. This results in reduced strength, a softer "feel," and decreased dimensional stability in recycled paper products, properties which may be useful in some applications but are more often detrimental. As a result, although there is enough wastepaper to account for 60% of United States consumption, only one-third of it is recycled. On the other hand, in addition to the 20% of demand supplied by recycled paper, 25% of our total paper production comes from reclamation of waste wood from logging, sawmill, and demolition operations, leaving 55% to be supplied by trees cut specifically for pulpwood.

The figures in the preceding paragraph indicate that at least 85% of the United States demand could be supplied by complete recycling and reclamation of paper. One area in which considerable improvement could be made is recycle of newsprint. Since much of the low quality groundwood pulp is in this form, separation of newspapers prior to recovery of other cellulose fiber would improve its salability. A New Jersey company, Garden State Paper, chemically de-inks old newspapers and claims that the recycled product equals virgin material in quality for a slightly lower price. Home separation and voluntary collection of newspapers thus might contribute significantly to ease of recycling of other types of paper as well as reducing the consumption of virgin materials. However, as a number of voluntary recycling groups have found, there is little market for used newspapers at the present time. The small margin of profit, general resistance to change, publisher's financial interests in virgin pulping operations, and long-term contracts have prevented changeover to recycled newsprint. Currently no more than 5% is recycled.

Several techniques exist for separation of cellulose fiber from mixed refuse. Perhaps the most widely known is the Black Clawson process. Because cellulose fibers range from 20 to 50 μm in diameter by 4 to 8 mm in length, they can be separated from other wastes by slushing in water followed by coarse screening, centrifugal separation, and perhaps chemical treatment or flotation. Approximately 50% of the fiber can be removed in this way, the remainder is incinerated or pyrolyzed. Most of the cellulose recovered by the Black Clawson process is of low quality and is usually incorporated in asphalt roofing. Other companies use recovered cellulose fiber to make synthetic "lumber" by incorporation of resin binders. This latter type of reclamation has the advantage that materials which would ordinarily require virgin timber can be replaced.

Garbage, Grass, and Organic Matter

Garbage, grass, and organic matter are most readily recycled by composting—decomposition by aerobic bacteria. In a typical operation the municipal wastes are presorted to remove noncompostible materials and those which might have salvage value—paper, cardboard, rags, metals, and glass. Refuse is then shredded and stacked in long piles (windrows) where it degrades to humus much as it would in soil. Usually the decomposed material contains less than 1% of each of the three primary fertilizer nutrients; it may be enriched by addition of inorganic fertilizer if necessary. The final step is grinding and bagging for ultimate sale as soil conditioner.

The largest composting plant in the United States in 1969 was a mechanical system constructed in Houston in 1967. It was rated at 325 tonnes of refuse per day and was producing about 181 tonnes of compost per day in mid-1968, as well as salvaged paper, rags, and metal; but it ceased operating in 1971 for economic reasons. A major problem in composting often is lack of a ready market (at a profitable price) for the products. Composting is most common in the Netherlands where refuse contains a higher percentage of garbage and there is a good deal of luxury agriculture (e.g., bulb growing) to make use of the compost produced.

Textiles, Plastics, and Rubber

Textiles, plastics, and rubber are extremely difficult to recycle. Most polymers are delivered to manufacturer in high purity form, and even small variations in composition can adversely affect performance. Reclamation by pyrolysis or incineration in water-walled furnaces would appear to be the best means of handling them in mixed refuse. Since about 75% of the rubber produced in the United States appears in the form of tires, the proportion of this component of municipal refuse could probably be reduced by encouraging disposal through dealers and distributors. Substantial reduction should also be possible as a result of increased reuse, since many items appear to be discarded prematurely.

Glass and Ceramics

United States production of glass in 1970 was almost equally divided among three categories: one-third beverage bottles; one-third other containers; and one-third flat glass, glassware, and fiber glass. As we shall see a little later, the best means of recovery of the first two categories may well be reuse, but the last category would still be found in municipal waste and its separation remains important. This can be done fairly easily by means of the air classification and screening processes already mentioned, or by screening and/or magnetic separation of the solids left after pyrolysis or incineration. At this point two major problems intervene—the level of ceramic, stone, and metallic impurities must be low if the recovered glass is to be recycled; and, unless it is unimportant to the projected user, color sorting must be done.

The first problem can apparently be solved. The previously mentioned Garrett pyrolysis plant uses a flotation process to obtain sand-sized, mixed-color (green) glass of 99.7% purity in a 70% yield. The composition of this material is within 1% of that of typical virgin green glass with respect to the principal oxides (SiO_2, Na_2O, and CaO). It may be possible to remove colored glass from such a product by high intensity magnetic sorting. Recall that small quantities of transition metal ions are responsible for the color of most glasses. Since these ions are paramagnetic, glass containing them is attracted by a sufficiently powerful magnetic field, leaving behind colorless material containing only metal ions having inert gas electronic configurations. However, a great deal of electric power is required to achieve sufficient magnetic field intensity, and this this type of separation probably will not be feasible without further advances in technology.

Because of the problems inherent in recycling glass, a number of methods of reclamation have been developed. These applications are usually ones where color and purity are not very important — glass-wool insulating material, for example. Greater concern about conservation of free energy may increase its use. Others are substitution of ground glass for aggregate in cement and blacktop (glasphalt), manufacture of bricks or tiles, or use of the reclaimed material in place of talc or other mineral fillers for plastics. Certainly, as landfill sites become less available in metropolitan areas, these low-value outlets will increase in importance.

Metals

Iron is currently the only material recovered in significant quantities from municipal trash (about 2% of the available metal). Several circumstances have combined to make this possible. One is the fact that about two-thirds of the iron is in the form of tin cans which are readily reclaimed in cementation of low-grade copper ores (Section 11.6). A second is ferromagnetism, which makes iron-containing materials easily separable from other refuse. The fact that copper has a significantly higher value than iron also enters in because costs for detinning and for transportation from municipal dumps to mining sites are more easily accommodated. Recycling of this iron into tin cans or other steel products is much less favorable than reclamation because of high freight charges, impurities such as tin coatings and lead from the solder which holds the can together, and the lesser capacity of basic oxygen furnaces for melting scrap. Large quantities of steel from automobiles, used machinery, and in-plant scrap are recycled, but almost no municipal waste is included. The latter, which represents about 9% of steel production, could be a significant but not overwhelming portion of the input to steel manufacture.

The quantity of aluminum found in United States municipal waste is on the order of 10^6 tonnes/year — about 25% of total production. Because of the large amount of free energy consumed in its recovery from ore (Section 11.5), dumping or landfilling of aluminum represents a tremendous loss. Copper, zinc, tin, and lead are also present in amounts on the order of 10^5

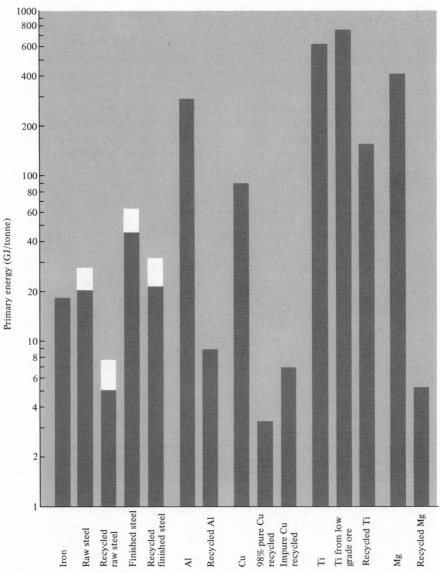

Figure 13.3 Energy required for production and recycle of metals. (Open bars represent ranges depending on different production techniques.) Data from Table 11.3.

tonnes/year. Markets for all of the nonferrous metals are readily available, but it is extremely difficult to separate them because of their low concentrations in trash. Recycling centers for all-aluminum cans are apparently economically and technologically feasible, but almost no attempts have been made to recover the other metals. Hydrometallurgical processes for separation of less active metals such as tin or lead might be a possibility, but if these were applied to aluminum the sizable quantity of free energy stored in the metal would be lost as soon as it was oxidized to Al^{3+}. The problem is to find a way of separating the metal without changing it chemically.

It is almost invariably true that in the case of metals reclamation, recycling, or reuse can provide significant free energy savings. This is shown on a logarithmic scale in Figure 13.3 which tabulates the energy required for production and recycle of iron, steel, copper, aluminum, magnesium, and titanium. In every case energy savings of a factor of two or more could be achieved by recycling, and in some cases more than tenfold savings could be made.

13.4   THE ENTROPY ETHIC AND SOLID WASTES

Much of the preceding discussion has been concerned with economic as well as technological feasibility of resource recovery. Economic factors are influenced by a great deal more than simply supply and demand in today's world, but most economic analysis tends to take a short-term (5–20 year) view, and many environmental problems have their basis in the fact that effects—"externalities" in the economist's jargon—not usually included in economic analysis are important. The question then arises: Can some easily quantifiable variable or variables be used to compare the impact on the environment of a variety of quite different processes? Is there some way to make rational decisions which take account of factors external as well as internal to conventional economic analysis? Specifically, in the context of this section, is there some scientific criterion by which the efficacy of different schemes for disposal, reclamation, recycling, or reuse of material resources can be judged?

One approach to these questions which gives considerable insight might be termed "thermodynamic economics" or the entropy ethic, and its basis has already been discussed in Sections 4.2 and 7.4. When the availability of resources on earth is considered on any time scale which even approaches the geochemical perspective of Part I, it becomes clear that the only resource of which there is a real shortage is free energy. Gold is an excellent example of a material resource that would be considered in short supply in an economic analysis. However, the quantity of gold in the oceans of the world is far greater than what has been mined from concentrated crustal deposits. Gold is not scarce but rather the free energy required to recover it from the extremely low concentration at which it occurs is the limiting factor. Thus the thermodynamic "cost" of any physical, chemical, or biological process is most appropriately measured in terms of the free energy lost as that process takes place.

To put it another way, if the earth is considered as a thermodynamic system, matter and energy are conserved. Only the *free* energy of the system (or the entropy of the universe) actually changes as a spontaneous process occurs. Therefore, either of these two quantities serves as a measure of the remaining capacity of the system to do useful work and the extent to which that capacity has been diminished by any specific process. Since the free-energy change is usually more easily computed it is the most convenient choice. It is usually calculated as $\Delta G = \Delta H - T \, \Delta S$, but since we will not be dealing with gases $\Delta H \simeq \Delta E$ and the change in internal energy may be substituted for the enthalpy term if desired. As a general rule entropy changes are not large and so the analysis can be confined to ΔE as a first approximation, but in some cases the $T \, \Delta S$ term does become important. The first of the two examples treated below ignores ΔS, but the second includes it.

Beverage Container Industry

Our first example involves a problem mentioned earlier — the consumption of a given quantity, in this case 1 dm³, of a beverage. In Table 13.7 the energy requirements are tabulated for each step in the life cycle of refillable and throwaway glass bottles, from acquisition and transportation of raw materials, through manufacture, filling, and transportation to the consumer, to disposal or recycle. The numbers in parentheses refer to the case where 30% of the bottles are recycled as opposed to those without parentheses where no bottles are recycled.

Two interesting conclusions may be drawn from the data in Table 13.7. First, with technology currently in the pilot plant stage of development, the energy cost for separation and color sorting of glass for the 30% of bottles to be recycled is *greater* than the energy cost of acquiring and transporting raw materials for 100% new bottles, whether returnable or throwaway. On thermodynamic grounds one would be better off *not* to recycle glass! This is a consequence of the ubiquity and low cost of the silica, lime, and soda ash required for glassmaking, so of course it does not apply to all trash components.

The second conclusion is even more striking. According to Table 13.7 it requires 18.2×10^6 J/dm³ to deliver a beverage in a throwaway bottle — more than twice the 7.585×10^6 J/dm³ required by the returnable bottle system. Hannon[5] has carried out similar analyses comparing throwaway containers of metal, paper, or plastic with returnable glass bottles and the energy ratios (ER = energy for throwaway/energy for returnable) range

---

[5] B. M. Hannon, *Environment* **14**(2), 11 (1972); and "System Energy and Recycling: A Study of the Beverage Industry," Doc. No. 23. Center for Advanced Computation, University of Illinois, Urbana, 1972 (revised March 17, 1973).

TABLE 13.7 **Comparison of Energy Consumption by Refillable and Throwaway Glass Beverage Bottles**[a]

Process	J/dm³ ($\times 10^{-3}$)	
	Refillable (8 fills)	Throwaway
Material acquisition		
100% new	251	1 310
30% recycle	(175)[b]	(926)[b]
Transportation of raw materials		
100% new	35	180
30% recycle	(25)	(127)
Container manufacture	2 155	11 320
Crown manufacture	539	540
Transportation to bottler	100	530
Bottling	1 505	1 500
Paper carrier manufacture	1 990	2 180
Transportation to outlet	985	510
Retailer and consumer	—	—
Collection and hauling	25	130
Separation, sorting (30% recycle)	(310)	(1 608)
Return for reprocessing (30% recycle)	(11)	(59)
Total:		
100% new	7 585	18 200
30% recycle	(7 820)	(19 420)

System energy ratios (ER = TA/REF)[c]

8 fills:	100% new	ER = 2.4
	30% recycle	ER = 2.5
15 fills:	100% new	ER = 3.3
	30% recycle	ER = 3.5

[a] Source: B. M. Hannon, "System Energy and Recycling: A Study of the Beverage Industry," Doc. No. 23, Table 2. Center for Advanced Computation, University of Illinois, Urbana, 1973.
[b] Figures in parentheses indicate recycle.
[c] Here, ER stands for energy ratio, TA for throwaway, and REF for refillable.

from 1.6 to 3.3 — it is always at least 50% more costly in terms of energy to use throwaways. To put this another way, reuse is favored over recycling or simply throwing bottles away, usually by a factor of two or more.

If the beverage industry had used nothing but returnable containers in 1970 its energy requirement would have been reduced by 40%. Since containers accounted for only 0.48% of United States energy consumption, this saving would have amounted to only 0.19%, a small fraction of the total, but it would have been sufficient to supply electricity to Pittsburgh, Washington, D. C., San Francisco, and Boston for about 5 months.

Automobile Industry

The automobile industry, which uses a far larger portion of material and energy resources, has been studied by Berry and Fels.[6] They defined the cycle of automobile manufacture and use as shown in Figure 13.4, with free energy increasing toward the top of the diagram. Their analysis includes both energy and entropy terms, and the results are summarized in Table 13.8.

The column labeled "Ideal" represents the calculated ΔG for the indicated process. For example, under mining and smelting, since iron is the

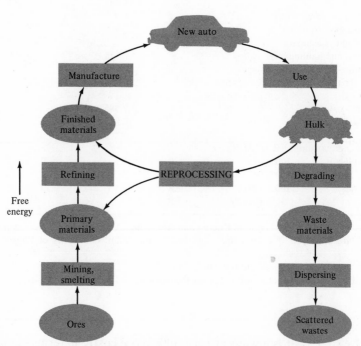

Figure 13.4 Materials cycle for automobile manufacture and use. Source: R. S. Berry and M. F. Fels, "The Production and Consumption of Automobiles," Report to Illinois Institute of Environmental Quality, Fig. 1. Department of Chemistry, University of Chicago, Chicago, Illinois, 1972.

---

[6] R. S. Berry and M. F. Fels, "The Production and Consumption of Automobiles," Report to Illinois Institute for Environmental Quality, 1972; R. S. Berry, Recycling, thermodynamics and environmental thrift. *Bull. Atom. Sci.* **27**(5), 22 (1971); also *in* "The Energy Crisis" (R. S. Lewis and B. I. Spinrad, eds.), p. 86. Educational Foundation for Nuclear Science, Chicago, Illinois, 1972.

TABLE 13.8   Real and Ideal Changes in Free Energy for Processing an Automobile[a,b]

Step	Ideal (10^6 J)	Real (10^6 J)
A. Mining and smelting	3700 to 10 800	-96 600
B. Manufacturing	Negligible	-37 600
C. Normal use	Negligible	Negligible
D. Recycling	Negligible	-51 100
E. Junking	-90	-90
F. Natural degradation	-6000	-6000

[a] Based on R. S. Berry, Recycling, thermodynamics and environmental thrift, *Bull. Atom. Sci.* **27**(5), 22 (1971); also appears *in* "The Energy Crisis" (R. S. Lewis and B. I. Spinrad, eds.), p. 86ff. Educational Foundation for Nuclear Science, Chicago, Illinois, 1972. More recent data have been obtained from Margaret F. Fels (private communication).
[b] In all the steps of this example, with the exception of step E (and the negligibly small changes in the ideal limits for steps B, C, and D), the overwhelming contribution to the change in free energy is given by ΔE, the energy change. Negative signs indicate losses or expenditures of energy.

principal component of the automobile, the ideal free energy change is dominated by that for equation 13.3,

$$Fe_2O_3 \longrightarrow 2\ Fe + \tfrac{3}{2} O_2 \tag{13.3}$$

although the mechanical energy required to raise the ore to the surface and ΔG for recovery of aluminum, copper, and other metals must be included as well. This ideal ΔG is the amount of energy required if all of the processes in the table could be carried out reversibly so that the minimum work would have to be expended. It should be noted that if this could be done free energy would be stored in the automobile (that is, ΔG would be positive for formation of a car). The work necessary to overcome this positive ΔG might be obtained from the system (that is, from the earth), in which case $\Delta G_{\text{system}} = 0$, or it might be obtained from solar energy. In the latter case the automobile would be a reservoir of stored solar energy, similar in many ways to a green plant or a sample of fossil fuel.

Another important aspect of Table 13.8 is the fact that in the ideal limit manufacture, use, and recycle of an automobile have free energy changes close to zero. (The use category excludes the burning of fuel during operation of the automobile since the problem being studied is the automobile cycle, not a fuel cycle.) This is because very little chemical processing occurs and there are few exo- or endothermic reactions involved. The major change is in the degree of order associated with the manner in which the various components of the vehicle are combined. The entropy changes for combination of the materials (manufacture), normal wear and tear (use), and recombination (recycling) into a new automobile are all small. However, when cars are junked at random (by parking them behind the barn, for instance) and allowed to decompose spontaneously there is a considerable increase in entropy and consequently a sizable loss of free energy (see steps E and F in Table 13.8). In the ideal case recycling would prevent this loss, and this is the reason that it seems so plausible to most scientists whose background includes thermodynamics.

Next, let us consider the real changes in free energy for the automobile cycle as listed in the table. First, note that the free energy consumed by recycling is about half of that released in mining and smelting. Thus the free energy saving which could be effected by recycling is approximately 45.6×10^9 J per automobile. This gives a maximum savings of over 30% of the total free energy value. (See the report by Berry and Fels for details.) Clearly the maximum possible amount of recycling is to be favored over simply letting cars decay wherever they stop running.

The largest free energy loss in the cycle, however, is the manufacturing step and its trend is toward increasing consumption. This means that extending vehicle life can effect greater savings than recycling. Berry and Fels estimated that by increasing the free energy of manufacture no more than 5%, the lifetime of an automobile could be tripled and the thermodynamic cost reduced by 62%. Thus, as in the case of the returnable bottle, the more durable, repairable automobile would be a considerable improvement over maximum recycling of the current product.

A third implication of the figures in Table 13.8 requires some discussion. The numbers in the column headed "Real" are all much larger than those in the "Ideal" case, except for rows E and F, which represent scattering of junk cars around the countryside and chemical degradation. In the processes devised by technological man 97% of the total free energy expended is wasted. Even with 100% recycling and extended lifetime vehicles, there would remain a great many highly inefficient steps in the cycle. In our current milieu of cheap sources of apparently boundless quantities of free energy there is little incentive for the development of more efficient technology. However, this era is coming to a close. A considerable effort on the part of scientists and technologists will have to be made in order to find new processes which can provide the same goods much more efficiently.

As a graphic illustration of the type of thinking that is required for such endeavors, consider Figure 13.5. The horizontal axis represents the entropy of various types of materials within the earth system, measured roughly by the concentrations of different elements from the average crustal abundance up to pure substances. On the vertical axis the free energy expended by various technologies for processing one material into another is represented by the maximum in the curves connecting different concentration points. Application of the entropy ethic requires that the curves be kept as low as possible—the minimum possible amount of energy should be used for a given transformation. Any suggested process, such as the one labeled "Proposed recycle," which far exceeds the energy requirements of current processes, is very likely an unrealistic one. Maximum

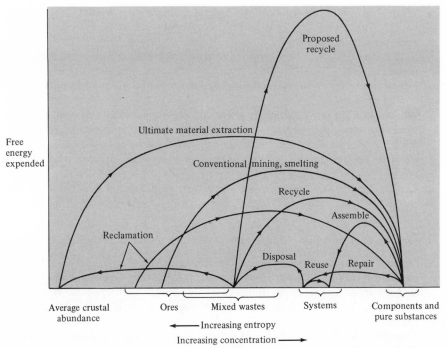

Figure 13.5 Energy and entropy effects in the processing of materials (neither scale is linear). Based on D. J. Rose, J. R. Gibbons, and W. Fulkerson, Physics looks at waste management. *Phys. Today* February, 32 (1972).

conservation of free energy—our most important resource—is represented by the lowest possible set of curves which can accomplish all the transformations that society feels are necessary. In general, this can be achieved by maximizing the efficiency of all steps in product cycles and by minimizing the distance of travel along the horizontal axis. At present the theoretical limits are well below actual practice, and a great deal of work and ingenuity will be required to bring the real values closer to ideality.

EXERCISES

(The first group of exercises is based primarily on material found in this text.)

1. What three major factors contribute to the environmental impact of a particular economic good? Explain how these three factors may be evaluated and combined in the case of a specific example.
2. What are the two principal techniques for handing municipal solid wastes? Explain the advantages and disadvantages of each.

3. What types of plastic produce HCl upon incineration? What types produce HCN? What technology is available for removal of these acids from the gases emitted from incinerators?

4. Explain the difference between pyrolysis and incineration as a method of solid waste disposal.

5. Define the terms *reclamation, recycling,* and *reuse.* Give an example of each type of process.

6. What metals are most commonly recycled? Based on energy and resource conservation, what metals are prime candidates for reclamation, recycling, or reuse?

7. Compare reclamation, recycling, and reuse in terms of the entropy ethic. Which process is most desirable?

8. Explain clearly why recycling glass beverage bottles is not feasible when the entropy ethic is considered. Is this generally true of materials other than glass? Why or why not?

(Subsequent exercises may require more extensive research or thought.)

9. Suppose you were assigned by the director of the EPA to do a thorough comparison of various transportation systems (automobile, bus, rapid rail transit, etc.) for intracity travel on the basis of free-energy consumption. What data would you have to include if the *entire* impact of each system were to be evaluated?

10. Recently, advertisements in national magazines have suggested incorporating a much larger fraction of aluminum in automobiles to reduce weight and save fuel. Suppose you had to prepare an environmental impact statement for such a change. What effects might it have?

11. Some states (Oregon was the first) have passed legislation requiring large deposits on *all* beverage containers and favoring reusable ones. What has been the effect of such laws on the states which adopted them? Does your state have similar legislation pending or in effect?

12. Air pollution controls for the automobile (Section 10.5) have been predicated largely on the assumption of annual model changes. Discuss the pros and cons of this approach both with respect to the rapidity of effecting air pollution control and the extent of automobile reuse.

Suggested Readings

Mineral Resources

1. J. C. Bravard, H. B. Flora, II, and C. Portal, "Energy Expenditures Associated with the Production and Recycle of Metals," ORNL-NSF-EP-24. Oak Ridge Nat. Lab., Oak Ridge, Tennessee, 1972. The energy required to produce and recycle magnesium, aluminum, iron, copper, and titanium is evaluated in detail. An excellent source of information.
2. A. H. Cottrell, "An Introduction to Metallurgy." St. Martin's Press, New York, 1967. Chapters seven through ten cover the basics of extraction of metals from their ores.
3. D. M. Evans and A. Bradford, *Environment* 11(8), 3–13 (1969). Details of the construction and possible problems of deep injection wells are given. The first author called public attention to the relationship between earthquake frequency and deep well injection of wastes.
4. National Academy of Sciences, Committee on Resources and Man, "Resources and Man." Freeman, San Francisco, California, 1969. Chapters 6 and 7 involve mineral resources from land and sea.
5. National Commission on Materials Policy, "Material Needs and the Environment, Today and Tomorrow." US Govt. Printing Office, Washington, D. C., 1973. This extensive compilation of data on the use of material resources in the United States makes twelve specific recommendations for changes in national materials policies. It also has collected a good deal of data from a variety of scattered sources, many of which also appear in this bibliography.
6. N. L. Nemerow, "Liquid Waste of Industry." Addison-Wesley, Reading, Massachusetts, 1971. Mainly useful for its catalog of emissions by a variety of industries and listing of abatement procedures. Contains many references.
7. R. H. Parker, "An Introduction to Chemical Metallurgy." Pergamon, Oxford, 1967. Presents the basic principles of thermodynamics, reaction kinetics, solution, and electrochemistry needed for an understanding of extraction metallurgy. Chapter 7, "Extraction and Refining of Metals," is organized according to general methods and basic principles rather than individual metals.
8. C. J. Ryan, "Materials, Energy and the Environment." National Commission on Materials Policy, Washington, D. C., 1973. This brief report, the result of a series of forums held on university campuses in 1972, urges conservation and recovery of scarce natural resources.
9. R. N. Shreve, "Chemical Process Industries," 3rd ed. McGraw-Hill, New York, 1967. An encyclopedia of chemical processes in nearly every industry except recovery of metals.
10. A. C. Stern, ed., "Air Pollution," 2nd ed., Vol. III. Academic Press, New York, 1968. Gives a detailed treatment of air emissions by nearly all industries and methods for abatement.

11. U. S. Bureau of Mines, "Mineral Facts and Problems." US Govt. Printing Office, Washington, D. C., 1970. Estimates of reserves, explanations of technologies used, discussions of environmental effects, and projections of demand to the year 2000 are included for almost every element and many compounds of commercial importance.

12. U. S. Geological Survey, "United States Mineral Resources." US Govt. Printing Office, Washington, D. C., 1973. The first comprehensive assessment by the Geological Survey of the nation's mineral resources in more than 20 years. Concludes that known deposits of raw materials are seriously depleted.

Soil

13. F. B. Abeles, L. E. Craker, L. E. Forrence, and G. R. Leather, Fate of air pollutants: Removal of ethylene, sulfur dioxide, and nitrogen dioxide by soil. *Science* **173,** 916 (1971).

14. H. L. Bohn and R. C. Cauthorn, Pollution: The problem of misplaced waste. *Amer. Sci.* **60**(5), 561 (1972). A perhaps overoptimistic view of the capacity of soil to absorb and detoxify pollutants.

15. H. D. Foth and L. M. Turk, "Fundamentals of Soil Science," 5th ed. Wiley, New York, 1972. An excellent reference on all aspects of soil chemistry. Photographs of models of the structures of soil minerals are especially helpful in explaining how structures affect chemical behavior.

16. R. G. Gymer, "Chemistry: An Ecological Approach." Harper, New York, 1973. The chapter on soil is useful and covers much the same material included here.

17. L. Hodges, "Environmental Pollution." Holt, New York, 1973. Chapters 11 and 13 are especially applicable to the material covered here. An extensive listing of sources of information is given at the end of the book.

18. S. E. Manahan, "Environmental Chemistry," Willard Grant Press, Boston, Massachusetts, 1972. Although the emphasis is primarily on water chemistry, the chapters on microorganisms, suspended solids, and soil are useful.

19. B. Mason and L. G. Berry, "Elements of Minerology." Freeman, San Francisco, California, 1968. An encyclopedia of common minerals with excellent figures showing their structures.

20. National Academy of Sciences, "Accumulation of Nitrate." NAS, Washington, D. C., 1972.

21. E. W. Russell and Sir E. J. Russell, "Soil Conditions and Plant Growth," 9th ed. Wiley, New York, 1961. An encyclopedic treatment of soils and soil chemistry.

22. R. Siever, The steady state of the earth's crust, atmosphere and oceans. *Sci. Amer.* **230**(6), 72–81 (1974). Describes weathering of soils and deposition of clay minerals on the ocean floor as part of a system of biogeochemical cycles.

Solid Wastes

23. American Chemical Society, "Cleaning Our Environment." ACS, Washington, D. C., 1969; Supplement, 1971. The section on solid wastes is quite useful, although some aspects of the problem have been omitted. The supplement consists primarily of recommendations for additional research rather than revised data.

24. R. S. Berry, Recycling, thermodynamics and environmental thrift. *In* "The Energy Crisis" (R. S. Lewis and B. I. Spinrad, eds.), pp. 86–93. Educational Foundation for Nuclear Science, Chicago, Illinois 1972; R. S. Berry and M. F. Fels, "The Production and Consumption of Automobiles," Report to the Illinois Institute for Environmental Quality. Department of Chemistry, University of Chicago, Chicago, Illinois, 1972. The latter reference

includes details which support the thesis of the former: The production and consumption of automobiles could be made more efficient by increased recycling, extended lifetime of the product, and especially by new, more efficient technology of manufacture.

25. D. H. M. Bowen, ed., "Solid Wastes." Amer. Chem. Soc., Washington, D. C., 1967-1971. This collection of 25 articles from *Environmental Science and Technology* provides a useful survey of the field.

26. B. Commoner, The environmental cost of economic growth. *Chem. Brit.* **8**(2), 52 (1972); "The Closing Circle." Knopf, New York, 1971. Both the article and the book state Commoner's thesis that growth of the wrong kinds of technology is a major factor in the recent environmental crisis in the United States.

27. Council on Environmental Quality, "Resource Recovery." US Govt. Printing Office, Washington, D. C., 1973. An assessment of technologies for resource recovery from municipal solid wastes.

28. E. M. Dickson, *Environment* **14**(6), 36 (1972). A description of a variety of ways in which the design of products hinders attempts at recycling.

29. J. M. Fowler and K. E. Mervine, "No Deposit, No Return." Environmental Resource Packet Project, Dept. of Physics and Astronomy, University of Maryland, College Park, 1973. This is one of a series of environmental resource packets and is especially useful because of the extensively annotated bibliography it contains.

30. R. Grinstead, *Environment* **12**(10), 2 (1970); **14**(3), 2 (1972); *ibid.* (4), 34–42 (1972). In this series of articles a research chemist describes current practices and available technology for resource recovery. Recycling of paper receives the greatest attention.

31. B. M. Hannon, *Environment* **14**(2), 11–21 (1972); "System Energy and Recycling: A Study of the Beverage Industry," Doc. No. 23, Center for Advanced Computation, University of Illinois, Urbana, 1973. These two references give details of the energy costs of various beverage container systems. The latter has been extensively revised to include aspects ignored by the former report. It definitely should be consulted since it provides the most up-to-date figures.

32. C. B. Kenahan, Solid waste. *Environ. Sci. Technol.* **5**(7), 594 (1971). A general description of a number of techniques developed for solid waste recovery by the U. S. Bureau of Mines.

33. C. E. Knapp, Can plastics be incinerated safely? *Environ. Sci. Technol.* **5**(8), 667 (1971). A general discussion of problems caused by plastics in incinerators and a description of a rather incomplete study of their effects.

34. R. A. Lowe, "Energy Recovery from Solid Waste." U. S. Environ. Protect. Ag., Washington, D. C., 1973. A rather detailed interim report on the St. Louis electric power plant which is partially fired by solid wastes.

35. G. M. Mallan and C. S. Finney. "New Techniques in the Pyrolysis of Solid Wastes." Garrett Research and Development Co., Inc., La Verne, California, 1972. A detailed description of pilot plant studies of an advanced system for pyrolysis of solid wastes.

36. National Center for Resource Recovery, Inc., "Resource Recovery from Municipal Solid Waste." Lexington Books, Heath, Lexington, Massachusetts, 1974. An excellent, detailed survey of this important topic.

37. H. Ness, Recycling as an industry. *Environ. Sci. Technol.* **6**(8), 700–704 (1972). A description of the operation of the recycling industry with special emphasis on recovery of copper and paper.

38. F. Rodriguez, The prospects for biodegradable plastics. *Chem. Technol.* **1**, 409–415 (1971). A good exposition of developments in a field of chemical research which seems unlikely to be of much use in solving solid waste problems.

39. D. J. Rose, J. H. Gibbons, and W. Fulkerson, *Phys. Today*, February, 32–39 (1972). A general approach to the problem of waste management with emphasis on physical processes, this paper gives some criteria by which the best solutions to solid waste problems may be identified.

40. G. Scott, Improving the environment: Chemistry and plastics waste. *Chem. Brit.* **9**(6), 267 (1973). A detailed excursion into the role of plasticizers, antioxidants, and activators in biodegradability of plastics.
41. A. J. Warner, C. H. Parker, and B. Baum, "Solid Waste Management of Plastics." Mfg. Chem. Ass., Washington, D. C., 1971; B. Baum and C. H. Parker, "Plastic Waste Disposal Practices." Mfg. Chem. Ass., Washington, D. C., 1973. These two industry-oriented reports contain a wealth of information regarding disposal practices for all solid wastes, with emphasis on plastics.

V
Water

Water is the only drink for a wise man.

Thoreau

You never miss the water till the well runs dry.

Rowland Howard

T HE TRUTH OF ROWLAND HOWARD'S WORDS IS borne out by the history of attempts to provide potable water supplies to large concentrations of human population. It is to be hoped that the success of such attempts will confirm Thoreau's words, but right now a number of wise persons are not completely convinced that drinking water supplies are adequately protected. Despite a long history of progress in water pollution control, problems of quantity and quality still remain. These were emphasized in 1975 at the first national meeting devoted to water resource management in the United States in 15 years,[1] which predicted a water crisis by 1985, especially if water-dependent technology for coal gasification and oil shale retorting (Sections 5.3 and 5.4) is developed in the western part of the country.

An argument against complacency in the water pollution control area is perhaps best made by considering the supply of pure drinking water. Prior to the concentration of populations in cities all that was necessary was to find a stream or spring or dig a well and be certain that one did not contaminate it with fecal wastes. In most cases natural processes would provide water free of bacteria and other contaminants. As more and more people came to use the same water supply, the problem of infections such as cholera, typhoid fever, dysentery, and hepatitis became more prevalent. By the middle of the nineteenth century sedimentation, clarification, and filtration of water supplies and underground sewers for removal of wastes were being introduced to prevent epidemics in large cities such as London. In 1888 a United States patent on chlorination of water was granted to Dr. Albert R. Leeds in Hoboken, New Jersey, and chlorination of its water supply was undertaken by the town of Adrian, Michigan in 1889. Chlorination or careful protection of reservoirs from contamination became common in the United States during the first third of the twentieth century, and people came to believe that drinking water was absolutely safe.

A number of these efforts to purify water had unexpected side effects, however. Sewers did remove feces and bacteria which might contribute to epidemics, but when dumped into a river untreated their oxygen demand was often sufficient to kill all fish. This led eventually to the construction of treatment plants which increased the rates of natural decomposition reactions, but even this was not enough. Increasing populations and industrial development resulted in much higher levels of plant nutrients such as phosphorus, nitrogen, and carbon in sewage effluents, and sped up the life cycle of lakes and estuaries.

Other industries released acids and heat (Section 5.2), radioactive substances (Section 6.2), heavy metals, and a variety of synthetic organic chemicals into natural waters; and agricultural development also produced inputs of fertilizer chemicals (Sections 11.11 and 11.12), pesticides (Sections 18.3–18.7), and sediments. Since a number of these substances were not readily decomposed or removed from rivers or lakes, they often remained in drinking water at low concentrations, but for the most part were ignored.

Within the past few years it has been discovered that such trace substances may have important consequences for human health.[2] A number of trace organic compounds are suspected carcinogens (Section 17.4), and there is apparently a significant relationship between their presence and total cancer mortality in the

---

[1] Sponsored by the Water Resources Council, Washington, D. C., April 22–24, 1975.

[2] J. Crossland and V. Brodine, *Environment* 15(3), 11–19 (1973); B. Dowty, D. Carlisle, J. L. Laseter, and J. Storer, *Science* 187, 75–77 (1975); *Chem. & Eng. News* November 18, 44–45 (1974).

New Orleans area. There is evidence that chloroform, bromodichloromethane, dibromochloromethane, and bromoform are produced from a common precursor *by chlorination.*[3] Thus the agent which removes pathogenic bacteria may at the same time introduce suspected carcinogens.

Passage of the Safe Drinking Water Act on December 17, 1974 and promulgation of strict EPA standards[4] made drinking water quality a key issue in 1975. However, interest and vigilance are almost certain to wane once it is believed that the problem has been "solved" again. It should be clear that for water pollution, as in the case of any environmental problem, there is no complete solution. New technologies, greater levels of affluence, and increasing population will continue to cause new difficulties and exacerbate old ones.

The chapters which follow make no attempt to survey every aspect of water pollution control. Rather some general principles are developed and applied to the specific problems of eutrophication, municipal sewage treatment, and industrial wastewater. In the latter case two specific industries (in addition to those mentioned in Chapter 11) have been chosen as examples. Numerous references to more highly specialized volumes are given for the benefit of readers who may wish to explore areas not covered in detail here.

---

[3] U. S. Environmental Protection Agency, Region V, "Joint Federal/State Survey of Organics in Drinking Water" (Preliminary Draft), p. 2. USEPA, Washington, D. C., 1975.
[4] *Federal Register* **40**(51), 11990–11998 (1975).

14

Water—General Principles

Fish say, they have their stream and pond;
But is there anything Beyond? . . .
One may not doubt that, somehow, good
Shall come of Water and of Mud;
And, sure, the reverent eye must see
A Purpose in Liquidity.

Rupert Brooke

14.1 UNIQUE PHYSICAL AND CHEMICAL PROPERTIES

Water is such a common, seemingly simple substance that we tend to forget the fact that its physical and chemical properties often lie at the extremes of the ranges established by the many substances chemists have synthesized. But in most cases it is just these unique properties which make water especially suitable or "fit" for the support of life as we know it. As we examine a number of its characteristics, it should become apparent why water plays so many and such important roles in the functioning of the biosphere.

Density

It is a well-known fact that the density of water, unlike that of most other liquids, reaches a maximum at 277 K and normal ice floats on the surface rather than sinking to the bottom of the liquid phase. The unusually low density of ice and of liquid water just above the melting point is attributed to the rather open structure necessary if both hydrogens and both oxygen lone pairs of each water molecule are to be involved in the formation of linear *hydrogen bonds*.[1] A number of important ecological and environmental consequences may be attributed to the expansion of water as it freezes. One—the initiation of rock weathering by ice formation in small cracks and fissures—has already been discussed (Section 12.1).

If water behaved as most liquids, lakes would freeze solid in regions of the world where temperatures drop below 273 K for extended periods. Fish and other aquatic organisms would have had to adapt to being frozen for much of their lives. Instead, a layer of water at 277 K collects at the bottom of most lakes in winter while only the surface layer freezes (Figure 14.1). In summer the opposite situation applies—a layer of water warmed by solar radiation (epilimnion) collects at the surface above colder, denser liquid at the bottom (hypolimnion). The boundary between the epilimnion and hypolimnion is known as the thermocline. In spring and fall, as the surface layer warms and cools, a point is reached at which the entire lake is nearly uniform in temperature so that vertical mixing (turnover) can occur. This distributes dioxygen and nutrients throughout the water.

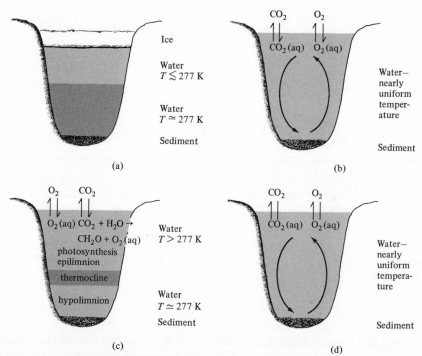

Figure 14.1 Thermal stratification of a lake through four seasons: (a) winter, (b) spring (turnover), (c) summer, and (d) fall (turnover).

---

[1] W. L. Masterton and E. J. Slowinski, "Chemical Principles," pp. 292–295. Saunders, Philadelphia, Pennsylvania, 1973; any general chemistry text will have a similar discussion.

In summer the hypolimnion serves an important function since certain fish and other organisms prefer the lower temperatures it provides. However, except for spring and fall turnovers, there is no mechanism by which O_2 dissolved in surface waters from air or produced by photosynthesis can reach the lower strata of a lake. If too much organic matter (which must be decomposed by oxidation) falls into the hypolimnion it may become anaerobic (dioxygen deficient) and species which live there may be unable to survive. One of the dangers of thermal pollution (Section 5.2) of a deep lake is that warm water discharged to the epilimnion will postpone (or even prevent) fall turnover, leaving the hypolimnion anaerobic.

Surface Tension

The water molecule's ability to form large numbers of strong hydrogen bonds is related to another unusual physical property, its large surface tension. Because water molecules strongly attract one another and have lesser attraction for other types of molecules, it is difficult to break through the surface of liquid water. As a result of this property, many insects are able to walk on the surfaces of lakes and streams. More importantly, large surface tension is related to the rise of a liquid in a tube of small diameter, a property known as *capillarity*. The movement of water in plant vessels of small diameter and the retention of water in small interstices between soil particles are facilitated by its large surface tension.

Polarity

Because of its bent structure and the sizable difference in electronegativity between hydrogen and oxygen, the water molecule is extremely polar. When ionic substances dissolve in water each ion is immediately surrounded (hydrated) by from four to eight dipolar water molecules. The resultant ion-dipole attractions stabilize the ions in solution. Since dipolar water molecules align with an electric field, water also has one of the highest *dielectric constants* of known substances. The complete statement of Coulomb's law:

$$F = q_1 q_2 / D r_{12}^2 \qquad (14.1)$$

where F is the force between two charges q_1, q_2 separated by a distance r_{12} in a medium of dielectric constant D, shows that for the same pair of particles at the same distance the force decreases with increasing D. Thus, ions dissolved in water are less likely to approach closely enough to pack into a regular crystal lattice. Its polarity and high dielectric constant make water an excellent solvent for many of the ionic compounds which constitute the earth's crust. Thus, it is ideal as a medium for transporting inorganic nutrients within living organisms.

Thermal Properties

When compared with those of other common liquids, the thermal properties of water are also unusual. Its large *heat capacity* of 75.5 J/mol K allows a given quantity of heat to be stored with minimal temperature change, and its large *thermal conductivity* allows for rapid transfer of heat to or from adjacent substances. Because a large number of hydrogen bonds must be broken upon melting or vaporizing water, its *heats* of *fusion* (6.00 kJ/mol) and *vaporization* (44.1 kJ/mol) are unusually large.

Organisms (such as *Homo sapiens*) that require a narrow range of temperature for their internal environments have used these properties to produce excellent thermoregulation systems. Heat produced by metabolic chemical reactions is transferred to water in the bloodstream and thereby distributed throughout the body. Should the temperature begin to rise, perspiration is allowed to vaporize and energy is transferred to the surroundings. The body temperatures of most warm-blooded animals lie in the range of 308–311 K, where the heat capacity of water passes through a minimum. Thus, attempts to increase or decrease this temperature meet with a small amount of negative feedback since slightly more than twice as much heat is required for a two-degree change than for a one-degree change.

The temperature of the entire biosphere is also moderated by the storage of solar energy in water or its phase changes. Temperatures near large bodies of water usually do not fall much below its freezing point because of the heat of fusion given off, and solar energy trapped in one geographic region may be transferred to another by means of atmospheric water vapor or warm ocean currents such as the Gulf Stream. Hence, winter temperatures in European coastal cities as much as 1500 km closer to the North Pole are comparable to those in Philadelphia or New York. Of the common liquids only water, by virtue of its unusual thermal properties, could accomplish this sort of climate modification.

Water and Plant Growth

The growth of terrestrial green plants is another excellent example of the fitness of water's unusual properties. Plant cells require some internal fluid to maintain the rigidity of cell walls and provide a medium where metabolic reactions can proceed at reasonably rapid rates. Besides its role as a reactant in photosynthesis (Section 6.6) water dissolves other necessary substances such as atmospheric CO_2 and soil nutrient ions. Because the leaves and chloroplasts of a plant must be in contact with the air to absorb CO_2 and emit O_2, much of whatever liquid is present will tend to be evaporated by solar heat. The unusually large heat of vaporization of water minimizes the amount of such *transpiration* for a given amount of solar energy input. Nevertheless, it still requires 100 kg of soil solution over the lifetime of a crop to produce 1 kg of plant material and nearly three-fourths of the latter consists of water which usually is removed by drying before the crop is used. Much greater quantities of any other liquid would be required in the soil to achieve the same yield.

The Water Cycle

By far the majority of water in the biosphere is found in the oceans (Figure 14.2) where its ability to perform some of the functions just described is limited because it contains high concentrations of salts dissolved during rock weathering processes. The supply of fresh water—especially that not tied up in polar ice caps or flowing underground—is much smaller, and it is this which is currently being used, reused, and reused again in many parts of the United States and the world. Therefore, the primary emphasis in the sections which follow will be on the pollution of freshwater supplies. However, it should not be forgotten that many freshwater rivers empty into estuaries which are the breeding grounds for a large portion of oceanic life. Many such estuaries have been heavily polluted by human and industrial wastes.[2] Remnants of plastic packaging wastes have been found in mid-ocean regions,[3] and organic debris from North American rivers has been

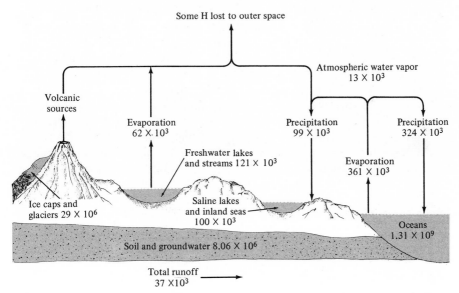

Figure 14.2 The water cycle. Volumes of water in each reservoir are given in km³. Flows from one reservoir to another are in km³/year. Data from L. Hodges, "Environmental Pollution," pp. 126–127. Holt, New York, 1973.

[2] See, for example, the description of Galveston Bay by L. J. Carter, *Science* **167**, 1102–1108 (1970).
[3] J. B. Colton, Jr., F. D. Knapp, and B. R. Burns, *Science* **185**, 491–497 (1974).

carried as far as 1600 km from shore by underwater currents. Ground-water has sometimes been contaminated by deep injection wells (Section 11.4) and trace quantities of DDT, lead, and other contaminants have been found in polar ice caps. Apparently no portion of the water cycle is entirely free of human influence.

14.2 CRITERIA OF WATER QUALITY

There are several general criteria relating to the quality of fresh water that depend on one characteristic which is shared by a variety of compounds. Measurement of that one characteristic is thus more important than analysis for each individual compound. The example most familiar to the reader is probably *pH*, which is defined as

$$\text{pH} = -\log a_{\text{H}^+} \simeq -\log [\text{H}^+] \tag{14.2}$$

where a_{H^+} is the activity of (hydrated) hydrogen ions in solution and the square brackets indicate concentration in mol/dm³. The pH may be measured using a glass electrode and pH meter or somewhat more rapidly and less accurately by means of color-calibrated indicator papers. Natural waters usually do not exceed the pH range from 3 to 10, and in most cases remain between pH = 6 and pH = 9. Closely related to pH (which is an *intensity* factor) is *alkalinity*—the *capacity* of natural water to neutralize acid. The difference between intensity and capacity factors is emphasized by the calculations in Exercise 1. The main contributors to alkalinity in natural waters are HCO_3^-, CO_3^{2-}, and OH^-, although phosphates, silicates, ammonia, and other bases must also be considered.

The pH of natural waters is extremely important because it determines the solubility and chemical form of most substances; for example, hydroxides of many metals are insoluble—the higher the pH the less metal is available in the water (see Table 14.1 for pH values at which common metals precipitate), unless of course hydroxo complexes are formed. In the

TABLE 14.1 Precipitation of Metal Hydroxides from Dilute Solutions as a Function of pH[a]

Metal	pH	Metal	pH
Fe^{3+}	2.0	Ni^{2+}	6.7
Al^{3+}	4.1	Cd^{2+}	6.7
Cr^{3+}	5.3	Zn^{2+}	6.7
Cu^{2+}	5.3	Co^{2+}	6.9
Fe^{2+}	5.5	Hg^{2+}	7.3
Pb^{2+}	6.0	Mn^{2+}	8.5

[a] Data from John G. Dean, Frank L. Bosqui, and Kenneth H. Lanouette, *Environ. Sci. Technol.* **6**(6), 518–22 (1972).

case of amphoteric metals such as aluminum, chromium, lead, and zinc, a sufficiently high pH will redissolve the hydroxide precipitate. Anions of weak acids undergo protonation equilibria (Exercise 10) which determine the predominant species at a given pH. These same effects regulate the availability of nutrients in the soil solution (Figure 12.6). The metabolism of aquatic organisms is also affected by pH since many of the necessary chemical reactions are catalyzed by acids or bases.

Suspended Solids

Another general indication of water quality is the amount of suspended solid material. This may be measured by filtering and weighing or according to the turbidity of a sample. Natural suspended solids are usually secondary soil minerals, and these colloidal particles adsorb metal ions and other positively charged species extensively (Sections 12.2, 12.6). Hence, even insoluble materials may be transported over considerable distances if suspended particulate material is present. Pesticides are often adsorbed as well. Anthropogenic sources of suspended solids include organic matter from municipal sewage and a variety of inorganic substances such as fly ash, cement dust, or iron oxides collected by wet scrubbers.

There are two somewhat different processes by which the rate of aggregation of colloidal solids may be increased so that they may be removed rapidly from water.[4] *Coagulation* involves reduction of electrostatic repulsions between the particles (recall that because of isomorphous substitutions most colloidal clay platelets are negatively charged, Section 12.2). Small quantities of salts added to the suspension can coagulate many colloids if a number of positive ions equal to platelet's negative charge become adsorbed. If too much salt is added, however, so many positive ions may become adsorbed that the platelets become positively instead of negatively charged. Hence an optimum salt concentration, above which restabilization of the colloid may occur, is usually observed for coagulation.

Flocculation depends on reagents which form chemically bonded bridges between colloidal platelets, linking them into much larger aggregations. An example is provided by the use of partially hydrolyzed polyacrylamide to flocculate municipal sewage solids (Figure 14.3). In the presence of positively charged metal ions this polyelectrolyte can aggregate a large number of platelets into a single particle which, according to Stokes' law (Section 9.1), should settle rapidly. Of course, both coagulation and flocculation are dependent on pH since hydrogen ions are capable of occupying negative sites on colloid platelet surfaces, reducing or increasing overall charge.

---

[4] V. K. LaMer and T. W. Healy, *Rev. Pure Appl. Chem.* **13,** 112 (1963); not all water chemists would agree with this classification.

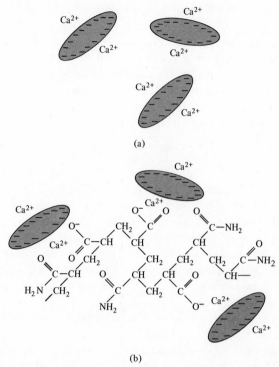

(a)

(b)

Figure 14.3 Flocculation of negatively charged colloidal particles by a polyelectrolyte. (a) Colloid particles with Ca^{2+} ions adsorbed. (b) Colloid particles linked by partially hydrolyzed polyacrylamide (schematic).

Redox Reactions

Another important criterion of water quality is the potential and capacity for oxidation or reduction. In natural waters contaminants are best removed by oxidation. The products of most such reactions are non-poisonous and in the case of macronutrients (Table 14.2) usually are inorganic forms readily assimilated by plants. When organic wastes are decomposed in the absence of dioxygen (anaerobic conditions) or other oxidizing agents, the products obtained (Table 14.2) are often poisonous and certainly are not in forms appropriate for plant or animal nutrition, although anaerobic decomposition can be beneficial in some cases (bottom muds for example).

The *reducing potential* (an intensity factor) of a water sample may be reported in terms of E, the potential of the sample relative to a normal hydrogen electrode, or pE, where

$$pE = \frac{EF}{2.303RT} = \frac{-\Delta G}{2.303nRT} = -\log a_{e-} \tag{14.3}$$

TABLE 14.2   Removal of Macronutrient Elements from Contaminated Water

Element	Principal product obtained under conditions of high pE (aerobic)	Principal product obtained under conditions of low pE (anaerobic)
Carbon	CO_2, CO_3^{2-} (depending on pH)	CH_4
Hydrogen	H_2O	—
Nitrogen	NO_3^-	NH_3, NH_4^+ (depending on pH)
Sulfur	SO_4^{2-}	H_2S, HS^- (depending on pH)
Phosphorus	HPO_4^{2-}, $H_2PO_4^-$ (depending on pH)	HPO_4^{2-}, $H_2PO_4^-$; sometimes PH_3

(R and T represent the gas constant and temperature, respectively; F is the Faraday constant; n is the number of moles of electrons transferred; and a_{e^-} is the activity of electrons in solution. Although even a hydrated electron cannot last more than a microsecond in solution, the concept of electron activity and pE, like a_{H^+} and pH, is a useful one.) The smaller the pE the greater the a_{e^-}, and the more difficult it is for substances to lose electrons (be oxidized). The range of pE in aqueous solution is limited by reduction and oxidation of water to 0 to 21 at pH $= 0$ and -14 to 7 at pH $= 14$. For neutral water in equilibrium with the atmosphere p$E \cong 13$, while under anaerobic conditions where methane and CO_2 are being produced by microorganisms pE may drop to -4.[5]

In addition to the potential for oxidation or reduction, one must also be concerned with a capacity factor which is related to pE in the same way that alkalinity is related to pH. Since nearly all of the oxidizing capability of natural waters is supplied by atmospheric or photosynthesized O_2, the *oxygen demand* of a sample is the preferred capacity factor. Two different methods are used to measure it. In the *biochemical oxygen demand* (*BOD*) test a water sample (diluted if oxygen demand is expected to be high) is saturated with dioxygen. It is then incubated at constant temperature (usually 293 K) for a known period (usually 5 days—designated by BOD_5). This allows time for microorganisms in the wastewater to mediate oxidation of organic matter. Then the remaining amount of dissolved O_2 is determined and BOD is obtained by subtraction. The precision of such a technique is not very high ($\sim \pm 20\%$), but it has the advantage of paralleling the microbial oxidations which would occur in natural waters.

In order to avoid the minimum of 5 days required for determination of BOD, to include those wastes which are only slowly oxidized by O_2 (some have 10-day half-lives), and to avoid problems of poisoning of microorganisms by toxic substances in the wastes, the *chemical oxygen demand* (*COD*) test has been devised. In this case the water sample is treated with a known quantity of a strong oxidizing agent (usually $K_2Cr_2O_7$) which rap-

---

[5] For a more complete description, see S. Manahan, "Environmental Chemistry," Chapter 3. Willard Grant Press, Boston, Massachusetts, 1972; W. Stumm and J. J. Morgan, "Aquatic Chemistry," Chapter 7. Wiley (Interscience), New York, 1970.

TABLE 14.3   Water Intake, BOD, and Suspended Solids Associated with Various Industries (1968)[a]

Industry	Water intake		Waste concentration (ppm) of process water			Total wastes (Gg)		
	Cooling (Tg)	Process (Tg)	BOD	COD	SS	BOD	COD	SS
Food processing	1 879	1 279	87	114	703	111	146	899
Textile mills	104	480	304	327	70	146	157	34
Paper mills	2 866	6 507	336	3 565	388	2 186	23 197	2 525
Chemicals	15 556	3 233	130	378	225	420	1 222	727
Petroleum and coal	5 417	417	52	210	76	22	88	32
Rubber	422	105	17	57	30	2	6	3
Primary metals	16 006	5 313	18	80	259	95	425	1 376
Domestic sewage[b]	—	—	170	—	210	3 313	—	3 994
Pure water	—	—	1–5	1–5	Variable	—	—	—

[a] Most data from U. S. Environmental Protection Agency, "The Economics of Clean Water," pp. 37–38. US Govt. Printing Office, Washington, D. C., 1972.
[b] Data for 1963 from "Cleaning Our Environment: The Chemical Basis for Action," p. 97. Amer. Chem. Soc., Washington, D. C., 1969.

idly oxidizes most of the organic matter. The remaining $Cr_2O_7^{2-}$ is then determined by back titration with a suitable reducing agent. Although COD can be determined in 2.5 hours and is reproducible, it may not give a complete picture of the amount of pollution either. For instance, dichromate oxidizes Cl^- to Cl_2 and thus an uncorrected COD includes the chloride content of a sample even though Cl^- would not be oxidized in natural waters and probably would not be considered a pollutant. Interference by chlorides can be eliminated by adding mercury(II) sulfate prior to oxidation.

It should be emphasized that BOD and COD were designed for and apply best to organic wastes. Benzene and other aromatics, pyridine and straight-chain aliphatic hydrocarbons are not determined by either test. Thus, a wide variety of potential pollutants may be missed if only BOD or COD is considered. Nevertheless, oxygen demand is a good general indication of industrial pollution as well as municipal sewage. Table 14.3 indicates the quantities and average concentrations of BOD and suspended solids (SS) associated with a number of types of industry.

Other Water Analyses

In addition to those already described, nonspecific analyses for color, odor, taste, turbidity, total dissolved solids, radioactivity, and methylene blue active substances are often applied to natural waters. (The last of these tested for the alkylbenzenesulfonate surfactants used in detergents until

1965.) A large number of analyses for specific substances have also been devised to detect the many compounds which may dissolve in water or become adsorbed on suspended particles. Criteria for permissible and desirable levels of many of these are listed in Table 14.4. Obviously, one would hate to be assigned the task of looking for synergistic effects among all of the substances which might be found in a water supply.

14.3 NATURAL WATER—EUTROPHICATION

Eutrophication is a natural process in the aging of a confined body of water. To see how it works we might refer to Figure 14.1c which shows a stratified lake in summer. As nutrients are supplied to the warm, aerobic epilimnion via stream drainage, organisms grow, reproduce, and die. Their remains settle into the colder, possibly anaerobic hypolimnion where they may not all decompose. Thus, a layer of bottom sediment builds up and the lake gradually becomes shallower. Eventually the depth is not great enough for stratification to occur, the entire lake is heated in summer and more rapid growth of algae and shoreline plants occurs. Finally, the lake is converted to a marsh, choked with plant life, and eventually it disappears altogether.

For most natural lakes the time scale for eutrophication is measured in tens of thousands of years, although it depends on the size of the lake, the area of the basin draining into it, and the quantities of nutrients in the drainage basin. The process can be accelerated tremendously, however, if quantities of nutrients far in excess of those normally leached from the drainage area are provided. As an example of accelerated eutrophication, some ecologists have estimated that Lake Erie has undergone changes that would normally have required ten thousand years during the 70-year period since 1900. Accelerated eutrophication often results in so-called "algal blooms" during which increased nutrient levels permit algae to multiply very rapidly until they have completely exhausted the supply of one or more elements. At this point all of the algae starve and die, causing a rapid depletion of dissolved dioxygen. Because most of the supply of dissolved dioxygen is consumed by decomposition of the dead algae, fish may be asphyxiated, and massive fish kills are often associated with algal blooms. The final result is deposition of large quantities of sediments, the remains of both algae and higher organisms.

Law of the Minimum

Most plans for the control of accelerated eutrophication are based on the *law of the minimum* which was first clearly stated by the chemist Justus von Liebig in 1840. Under steady-state conditions the essential material available in amounts most closely approaching the critical minimum needed

TABLE 14.4 Water Quality Criteria, Domestic Water Supplies[a,b]

Quality parameter	Permissible criteria (ppm unless otherwise indicated)	Desirable criteria (ppm unless otherwise indicated)
1. Color (Co-Pt scale)	75 units	<10 units
Odor	Virtually absent	Virtually absent
Taste	Virtually absent	Virtually absent
2. Turbidity	—	Virtually absent
3. Inorganic chemicals		
pH	6.0–8.5	6.0–8.5
Alkalinity ($CaCO_3$ units)	30–500 mg/l	30–500 mg/l
Ammonia	0.5	<0.01
Arsenic	0.05	Absent
Barium	1.0	—
Boron	1.0	
Cadmium	0.01	
Chlorides	250	<25
Chromium (hexavalent)	0.05	Absent
Copper	1.0	Virtually absent
Dissolved oxygen	>4.0	Air saturation
Fluorides	0.8–1.7 mg/l	1.0 mg/l
Iron (filtrable)	<0.3	Virtually absent
Lead	<0.05	Absent
Manganese (filtrable)	<0.05	Absent
Nitrates plus nitrites (as mg/1 N)	<10	Virtually absent
Phosphorus	10–50 μg/l	10 μg/l
Selenium	0.01	Absent
Silver	0.05	—
Sulfates	250	<50
Total dissolved solids	500	<200
Uranyl ion	5	Absent
Zinc	5	Virtually absent
4. Organic chemicals		
Carbon chloroform extract (CCE)	0.15	<0.04
Methylene blue active substances	0.5	Virtually absent
5. Pesticides		
Aldrin	0.017	—
Chlordane	0.003	—
DDT	0.042	—
Dieldrin	0.017	—
Endrin	0.001	—
Heptachlor	0.018	—
Heptachlor expoxide	0.018	—
Lindane	0.056	—
Methoxychlor	0.035	—
Organic phosphates plus carbamates	0.1	—
Toxaphene	0.005	—
Herbicides 2,4-D plus 2,4,5-T, plus 2,4,5-TP	0.1	—

TABLE 14.4 (Continued)

Quality parameter	Permissible criteria (ppm unless otherwise indicated)	Desirable criteria (ppm unless otherwise indicated)
6. Radioactivity		
Gross beta	1000 pCi/l	< 100 pCi/l
Radium-226	3 pCi/l	< 1 pCi/l
Strontium-90	10 pCi/l	< 2 pCi/l

[a] Source: "Report of the Committee on Water Quality Criteria." Federal Water Pollution Control Administration, U. S. Department of the Interior, US Govt. Printing Office, Washington, D. C., 1968.

[b] For a good discussion on the basis of these criteria, see J. A. Borchardt and G. Walton, Water quality. *In* American Water Works Association, Inc., "Water Quality and Treatment," 3rd ed., Chapter 1. McGraw-Hill, New York, 1971.

will be the limiting factor in growth and reproduction of an individual species of plant or animal. Of course, the "critical minimum" will be different for different organisms, and for the same organism the critical minimum of one essential material may be much different from that of another.

On an atomic scale the law of the minimum becomes very simple. Suppose we consider the synthesis of glycine, an essential amino acid. It requires

$$5 \text{ H} + 2 \text{ C} + 2 \text{ O} + \text{N} \longrightarrow \underset{\underset{\text{H}-\text{N}-\text{H}}{|}}{\overset{\overset{\text{H}}{|}}{\text{H}-\text{C}-\text{C}}} \overset{\text{O}}{\underset{\text{OH}}{\diagup}} \tag{14.4}$$

5 mol of hydrogen atoms, 2 mol of carbon atoms, 2 mol of oxygen atoms, and 1 mol of nitrogen atoms to produce 1 mol of glycine. If we were to start with 5 mol of hydrogen, 2 mol of carbon, 2 mol of oxygen, but only $\frac{1}{2}$ mol of nitrogen, we could only make $\frac{1}{2}$ mol of glycine. In fact, it does not matter how much hydrogen, carbon, and oxygen we have (as long as it is more than 2.5 mol, 1 mol, and 1 mol, respectively), the amount of glycine that can be synthesized is determined by the limiting element, in this case, nitrogen.

Of course, the requirements of a living organism, even the simplest, are far more complicated than the example above. An organism might require that its supply of nitrogen be in a special form that is easily metabolized; thus, the presence of a certain number of moles of nitrogen atoms might not be a sufficient condition for the organism's continued existence. Nevertheless, it would probably be a necessary condition, unless the organism had an excess of stored nitrogen or was able to utilize some other element in its place. (Certain mollusks, for example, are able to make use of strontium when their supplies of calcium are depleted or cut off.) Another problem in the application of the law of the minimum is that it applies only to steady-state conditions. When rapid changes or fluctuations are taking place more than one element may be limiting, and the identity of the limiting elements may change from one instant to the next.

Many attempts at applying the law of the minimum to the eutrophication problem have concentrated on finding the limiting element which controls the rate of growth of algae. Referring to Table 1.1, which lists the abundance of the elements in the human body (and, by rough extrapolation, in most organisms), we find that the elements present in greatest quantity are hydrogen, oxygen, carbon, nitrogen, calcium, and phosphorus. While these need not necessarily be limiting, they are required in larger concentration than the other elements, and most studies of the problem have focused on carbon, nitrogen, or phosphorus. Hydrogen and oxygen are present in an almost limitless supply, the hydrogen coming from water and the oxygen from air. (Although nitrogen is abundant in air, it is not readily available to living organisms because of the large $N\equiv N$ bond energy.) Calcium is generally available via soluble compounds.

Several problems have arisen in attempts to decide whether carbon, nitrogen or phosphorus is the culprit in accelerated eutrophication. First of all, it has been found that under certain circumstances algae may absorb more of an abundant nutrient than is needed for immediate growth. The excess may be stored for use at a later time. Second, different algae have different requirements for different nutrients, making experiments either difficult to reproduce or even perhaps irrelevant to the actual situation in a given body of water. Even worse, some nutrients may be supplied at essentially constant concentrations because they are replenished from solid or gaseous phases. For example, the insolubility of most phosphate salts insures that sediments will contain large quantities of phosphorus. Measuring the concentration of phosphorus in the liquid phase will not determine the total amount of phosphorus available because solubilization of phosphorus from sediments will constantly replenish the liquid phase as phosphorus is incorporated in organisms. The same is true of dissolved forms of carbon, which may be replenished by CO_2 from the air. Finally, it should be remembered that the law of the minimum can be exactly applied only in a steady-state situation. Algal blooms are probably very far removed from the steady-state; thus, the basic concept of a limiting reagent may be faulty.

These complicating factors have caused considerable argument among the proponents of one or another element as the limiting factor in algal blooms, but the consensus appears to be that regardless of what element is limiting, the solution to the problem is to find an element whose availability can be reduced to such an extent that it *will become* the limiting factor. More specifically, if man's activities are causing increased nutrient levels, algal blooms, and accelerated eutrophication, which of these activities may be most readily controlled to reduce the concentration of one element to the point where steady-state conditions will be attained, the law of the minimum will apply, and premature aging of a lake can be prevented? To answer this question it is probably more appropriate to look at the sources of carbon, nitrogen, and phosphorus in lakes and estuaries.

Carbon may enter natural waters in the form of CO_2 from the atmosphere, carbonate salts from soils or rocks, or organic components of human or animal wastes. Nitrogen and phosphorus are also present in sewage, although to a lesser extent than carbon. The major source of nitrogen is agricultural runoff. Both nitrates and ammonium compounds are usually water-soluble, and the negatively charged nitrate ion is only weakly bound by ion exchange sites in soils. Unless they are immediately utilized by plants, most nitrate fertilizers end up in streams or lakes. The amount of phosphorus fertilizers used is about half as great as nitrogen, and agricultural runoff accounts for much less of the phosphorus in natural waters because of the insolubility of phosphates. The largest source of phosphorus appears to be detergents, which account for about 30–40% of phosphorus entering the aquatic environment.[6] This fraction is even higher in streams near large concentrations of human population. The public effort to control phosphate-containing detergents has resulted because this element appears to be easiest to control.

14.4 DETERGENTS AND PHOSPHATES

Why are phosphates used in detergents? Let's take a moment to find out by considering the historical development of soap and detergent production. There are three main types of soiling materials which must be removed from articles to be washed. The first is *dirt,* usually defined as finely divided (often colloidal) inorganic particles. *Greases* and *oils* are typified by hydrocarbons and fats, and *stains* include a wide variety of intensely colored substances which may be present in low concentrations. Dirt, once its van der Waals or coulombic attraction for a fabric or other surface is broken, is generally water soluble, but greases, oils, and many stains are not. The latter must be emulsified by *surfactant* or *detergent molecules.*

Surfactants consist of molecules that have long, nonpolar hydrocarbon chains connected to charged or highly polar end groups (such as carboxylic acid anions or protonated amine groups). They are similar to the molecules that formed the precursors of biological cells (Section 3.5) in that when dissolved in water above a certain concentration they tend to form *micelles.* Nonpolar groups cluster together within a micelle, leaving polar end groups in contact with surrounding water. Obviously greases, oils, and organic stains prefer the nonpolar interior of a micelle and hence can be solubilized.

---

[6] R. D. Grundy, *Environ. Sci. Technol.* **5**(12), 1184–1190 (1971).

Soaps are produced by strongly basic hydrolysis of animal fats:

$$
\begin{array}{c}
\text{CH}_3(\text{CH}_2)_{16}\text{C} \underset{\text{O}-\text{CH}_2}{\overset{\text{O}}{\diagdown}} \\
\text{CH}_3(\text{CH}_2)_{16}\text{C} \underset{\text{O}-\text{CH}}{\overset{\text{O}}{\diagdown}} \quad + 3\text{ K}^+ + 3\text{ OH}^- \longrightarrow 3\text{ CH}_3(\text{CH}_2)_{16}\text{C} \underset{\text{O}^-\text{K}^+}{\overset{\text{O}}{\diagdown}} + \text{CH}_2-\text{CH}-\text{CH}_2 \\
\text{CH}_3(\text{CH}_2)_{16}\text{C} \underset{\text{O}-\text{CH}_2}{\overset{\text{O}}{\diagdown}} \qquad\qquad\qquad\qquad\qquad\qquad \overset{|}{\text{OH}}\;\;\overset{|}{\text{OH}}\;\;\overset{|}{\text{OH}}
\end{array}
$$

$$(14.5)$$

Fat   Soap
                                              (potassium stearate)

Because stearic acid is a relatively weak acid, solutions of its potassium salt are basic. If they are made neutral or acidic, stearate anions are made less polar as a result of protonation to form stearic acid, and detergent action is greatly lessened.

Most soaps do not work well in *hard water* where relatively high concentrations of Ca^{2+} and other cations are present. Anions such as stearate precipitate with hard water cations:

$$
Ca^{2+} + 2\text{ CH}_3(\text{CH}_2)_{16}\text{C} \underset{\text{O}^-\text{K}^+}{\overset{\text{O}}{\diagdown}} \longrightarrow Ca\left[\text{CH}_3(\text{CH}_2)_{16}\text{C} \underset{\text{O}}{\overset{\text{O}}{\diagdown}}\right]_2 + 2\text{ K}^+
$$

$$(14.6)$$

forming large quantities of a flocculent, scummy material. This forms "bathtub rings" and tends to be trapped in fabrics, especially if washing is done by machine. Moreover, a large fraction of the detergent molecules are removed from solution before they have a chance to emulsify anything. Calcium and other hard water ions can be removed and soap molecules can be maintained in anionic form by adding sodium carbonate, sodium metasilicate, or trisodium phosphate, but all of these raise the pH of the washing solution to 12 or more, making it somewhat caustic to human skin and sometimes accelerating wear of the articles being washed.

During the 1940's synthetic detergent molecules came into use, the most popular of which were alkylbenzenesulfonates (ABS):

$$
\begin{array}{c}
\text{CH}_3-\underset{\overset{|}{\text{CH}_3}}{\text{CH}}-\text{CH}_2-\underset{\overset{|}{\text{CH}_3}}{\text{CH}}-\text{CH}_2-\underset{\overset{|}{\text{CH}_3}}{\text{CH}}-\text{CH}_2-\text{CH}-\text{CH}_3 \\
\\
\underset{\text{SO}_3^-\ \text{Na}^+}{\bigcirc}
\end{array}
$$

ABS Surfactant

The sulfonate group ($-SO_3^-$) is less basic than the carboxylate group, allowing strong detergent action at lower pH and fewer problems with caustic reactions. The rate of ABS production grew especially rapidly when it was discovered that the synthetic detergents interacted synergistically with the category of compounds known as *builders*. The least expensive and most effective builders are tetrasodium pyrophosphate ($Na_4P_2O_7$) and pentasodium tripolyphosphate ($Na_5P_3O_{10}$). Since the synergistic effect permitted better cleaning with smaller quantities of synthetic surfactants, soaps were rapidly outpaced in the marketplace, and much larger quantities of phosphate builders began to be consumed. Currently soap production is less than one-fourth that of synthetic detergents.

The ABS detergents had one glaring fault — because of the branched hydrocarbon side chain the aquatic microorganisms whose enzymes had catalyzed decomposition of soaps (or their precursor fats) were unable to handle ABS. The synthetic detergents were so slowly biodegradable that relatively high concentrations built up. Numerous streams (and occasionally tap water) became covered with soap suds and in some cases sewage treatment facilities were interfered with. Although foam problems were mainly aesthetic, public objection was strong. ABS detergents were banned in some European countries and United States manufacturers agreed voluntarily to discontinue their use.

By 1965 linear alkylbenzenesulfonate (LAS) detergents had been developed. These molecules were relatively easily biodegradable because their linear side chains had been designed to mimic the nonpolar portions of the

$$CH_3-CH_2-CH_2-CH_2-CH_2-CH_2-CH_2-CH_2-CH_2-CH_2-CH-CH_3$$

$$SO_3^- \quad Na^+$$

LAS Surfactant

structures of soaps such as potassium stearate. However, like the ABS synthetic detergents, they still required large quantities of phosphate or other builders for maximum cleaning power at minimum price. Most clothes washing products contained well over 50% inorganic constituents, for example, and those for mechanical dishwashers contained less than 5% organic surfactant.[7]

Although their synergistic interaction makes it impossible entirely to separate the effect of a builder from that of a surfactant, the main way the former improves cleaning power appears to be its ability to *sequester* (form

[7] J. C. Harris and J. R. Van Wazer, "Detergent Building." *In* "Phosphorus and Its Compounds" (J. R. Van Wazer, ed.), Vol. 2, Chapter 27, p. 1740. Wiley (Interscience), New York, 1961.

soluble coordination complexes with) calcium and other hard water cations. This is illustrated in equation 14.7 for tripolyphosphate:

$$
Ca^{2+} + {}^-O-\overset{O}{\underset{O_-}{P}}-O-\overset{O^-}{\underset{O}{P}}-O-\overset{O}{\underset{O_-}{P}}-O^- \rightleftharpoons
$$ (14.7)

Phosphate builders also maintain pH at sufficiently high levels (~ 10) so that the sulfonate groups of surfactants are not protonated, and there is some evidence that they speed up the release of dirt particles from fabrics.

The large proportion of builder in most detergent formulations combined with the rapid growth in consumption of synthetic detergents (2.5×10^6 tonnes/year in 1970) releases large quantities of polyphosphates to sewage treatment plants. There they are hydrolyzed to orthophosphate:

$$
P_3O_{10}^{5-} + 2\ H_2O \longrightarrow 2\ HPO_4^{2-} + H_2PO_4^-
$$ (14.8)

which is subsequently released to natural waters. By 1970 the problem of accelerated eutrophication and the possibility of controlling it by reducing phosphorus levels had been publicized extensively by environmentalists. Both public officials and detergent manufacturers began to look for ways of effecting such reductions.

Three general approaches were employed. The first involved a crash program of research to develop inexpensive, new builders that could sequester hard water cations but would not contribute phosphorus to natural waters. For a time it appeared that nitrilotriacetic acid (NTA) would be an ideal choice. In basic solutions or when used as the trisodium salt NTA is an excellent sequestering agent for ions such as Mg^{2+} or Ca^{2+}. It forms coordinate covalent bonds using

three —C(=O)(O⁻) groups and one >N: group:

NTA   Mg(NTA)⁻

As in the case of polyphosphates most NTA is decomposed in sewage treatment plants and loses its ability to sequester metal ions.

Unfortunately, at about the same time that NTA was being introduced into detergents, research came to light which linked it to certain types of birth defects. They are apparently the result of NTA's ability to transfer heavy metal ions through the placenta and into an unborn fetus. The extremely large chelating ability of this ligand and the relatively nonpolar exterior of the complex formed make such action a definite possibility (see Section 17.4). Because there is no guarantee that all NTA would be degraded by sewage treatment and because of the potential it would have for permanent damage to individuals as yet unborn if it were introduced into water supplies, the Surgeon General of the United States banned all use of NTA in detergents. So far no other new builder has been devised.

The second approach to the phosphate problem has been generally adopted in states where phosphate content of detergents has been limited or banned. This is to use previously known but less effective builders involving carbonates, silicates, citrates, or borates, sometimes in conjunction with phosphates. However, because they are derived from weaker acids than tripolyphosphoric, the sodium salts of these builders hydrolyze to give more basic solutions. (Sodium metasilicate can result in pH values as high as 13.) Thus if they are accidently ingested by curious children, the detergents containing such builders are more caustic and produce more serious damage. At one point two arms of the United States government were at odds over the phosphate problem for just this reason. The Office of the Surgeon General favored banning sodium carbonate because of several poisoning incidents, while the Environmental Protection Agency favored banning phosphates.

The third approach to the problem is to remove phosphate at the sewage treatment plant. This usually requires improved secondary or tertiary (Sections 15.2 and 15.4) treatment and thus demands funds for the necessary construction. In 1971 the United States and Canada entered into a five-year agreement to construct such plants around Lake Erie, but much effort south of the border was stymied until a 1975 United States Supreme Court decision released construction funds appropriated by Congress which had been impounded by the Executive branch of the government.

It would appear that the best approach to the phosphate problem is some combination of the second and third approaches. Large amounts of phosphate and other builders are not required in soft water areas, and a reduction in phosphate content might reduce the load (and hence the cost) on sewage treatment plants in others. Some states have banned phosphates entirely, and many have restricted total phosphorus to 8.7% in clothes-washing detergents. So far there has not been a definitive evaluation of the efficacy of these procedures.

EXERCISES '

(The first group of exercises is based primarily on material found in this text.)

1. What concentration of NaOH is required to give a pH of 11.0? What concentration of NH_3 would produce the same value? $[K_b(NH_3) = 1.8 \times 10^{-5}]$ Calculate the amount of H_2SO_4 that could be neutralized by 1 cm³ of each solution. Although both have the same concentration (intensity factor) of H^+, their alkalinities (capacity factors) are very different. Explain.

2. List at least three unusual properties of water and explain how the unusual behavior makes water especially fit for the support of life.

3. Explain why different decomposition products are obtained for the macronutrient elements carbon, hydrogen, nitrogen, phosphorus, and sulfur under anaerobic and aerobic conditions. Why is this difference far more important with regard to water pollution than to air pollution?

4. Describe the BOD and COD tests. Write equations for the reactions involved in O_2 determination. Explain the differences between the two tests. Which is more reproducible?

5. State clearly Liebig's law of the minimum and explain how it applies to the problem of eutrophication. What factors complicate application of the law of the minimum?

6. How are soaps and detergents manufactured? What is the difference between ABS and LAS detergents? Why are builders such as phosphate used?

7. What role does hydrogen bonding play in making water unique among liquids?

8. Linear alkylsulfonate detergents are biodegradable whereas ABS detergents are not. What aspect of the structure of LAS detergents apparently accounts for this difference?

(Subsequent exercises may require more extensive research or thought.)

9. Use Stokes' law (equation 9.1) to calculate the settling rate in water of an Fe_2O_3 particle whose radius is 10 μm. (Density and viscosity values may be obtained from a handbook.)

10. Using the equilibrium constants

$$K_1 = \frac{[H^+][HCO_3^-]}{[CO_2]} = 2.2 \times 10^{-7} \qquad K_2 = \frac{[H^+][CO_3^{2-}]}{[HCO_3^-]} = 4.7 \times 10^{-11}$$

and assuming activities equal concentrations, calculate the fraction of total carbonate available as CO_2, HCO_3^-, and CO_3^{2-} at each of the following pH values: 4, 5, 6, 7, 8, 9, 10, 11, 12. Plot the fraction of each species versus pH (see S. Manahan, "Environmental Chemistry," pp. 25–26. Willard Grant Press, Boston, Massachusetts, 1972).

15

Municipal Wastewater Treatment

Do you know what these piles of ordure are, collected at the corners of streets, those carts of mud carried off at night from streets, the frightful barrels of the nightman, and the fetid streams of subterranean mud which the pavement conceals from you? All this is a flowering field, it is green grass, it is mint, thyme, and sage; it is game, it is cattle, it is the satisfying lowing of heavy kine; at night it is perfumed hay, it is gilded wheat, it is bread on your table, it is warm blood in your veins, it is health, it is joy, it is life.

Victor Hugo

Although wastes from a variety of industries flow into municipal sewage treatment plants in a great many cases, the main treatment problem is removal of the organic matter provided by human excrement. This is usually measured in terms of quantity of suspended solids and oxygen demand (usually BOD). Treatment processes are usually divided into two classes, each of which mimics and accelerates natural means for removing solids and BOD. *Primary treatment* involves physiochemical techniques such as screening, degritting, coagulation, flocculation, and sedimentation which remove suspended material. Since much of the latter is readily oxidizable, BOD is reduced as well. *Secondary treatment* is a biochemical process in which air oxidation mediated by microorganisms removes BOD remaining after primary treatment. Concurrently, additional solids are removed by filtration or sedimentation. Recently, it has become evident that primary and secondary sewage treatment are inadequate in many areas where water is reused extensively and so a number of *tertiary* or *advanced wastewater treatment* techniques are under development or being adapted from methods already in use for treatment of industrial wastes. Some of these are capable of removing as much as 99% of the waste from an effluent.

15.1 THE IMPORTANCE OF MICROORGANISMS IN WATER PURIFICATION

One of the primary aims of water purification and wastewater treatment has been the elimination of pathogenic bacteria, and more recently there has been concern that some viruses may survive conventional water treatment. On the other hand, the bacteria and fungi which break down organic matter in sewage are extremely useful to mankind as are the flora of vertebrate intestines which break down complicated components of food allowing them to be absorbed rapidly. Microorganisms are also important in the mobilization of toxic substances such as mercury (Section 16.3) in the aquatic environment. Since the metabolism of bacteria, fungi, and algae is quite versatile (more so than for higher organisms), it is usually true that if they do not decompose a particular compound it will remain in the environment for a long period. Thus, some information regarding microbes is essential for an understanding of water chemistry.

There are many ways of classifying the myriad different types of microorganisms. *Bacteria* consist of very small (0.3–30 μm diameter) rod-shaped, spherical, or spiral cells which are sometimes equipped with "tails" (flagella) to provide locomotion. Because of their small size and large surface area per unit mass, substances in the surrounding medium are readily accessible to bacteria and their enzymes readily catalyze a variety of reactions. Most bacteria depend on external organic matter for both their energy supply and the carbon from which most of their structural components are made, although a few carry out photosynthesis and thus can manufacture their own carbon compounds.

All *fungi* are *heterotrophic* (require an external source of organic material to supply energy and carbon; see Section 3.5) and most consist of long filaments some 5–10 μm in diameter. Fungi usually are *aerobic* (require dioxygen for their metabolism) and are tolerant of greater quantities of acids and heavy metals than are bacteria. Although many of the products of fungal decomposition enter natural waters, fungi do not grow well in water and will not be considered further here. *Algae* are usually *autotrophic,* producing organic matter from CO_2 and H_2O by photosynthesis (Section 6.6). However, in the absence of light algae also degrade stored organic matter to supply energy and thus consume rather than supply O_2. Rapid growth of blue-green algae "blooms" in eutrophic lakes has already been described (Section 14.3).

Bacteria

As a general rule aquatic bacteria obtain their supply of energy from oxidation–reduction, and any reaction which yields more than 20–30 kJ of free energy per mole of electrons transferred appears to be satisfactory in this regard. The specific reactions employed depend on the supply of dioxygen. If it is plentiful the half-reaction

$$\tfrac{1}{4} O_2(g) + H^+(10^{-7} M) + e^- = \tfrac{1}{2} H_2O(l) \qquad pE^0(10^{-7} M\ H^+) = 13.75^1 \qquad (15.1a)$$

serves as the electron acceptor. The principal energy supply of *aerobic bacteria* is provided by oxidation of carbohydrate (CH_2O):

$$\tfrac{1}{4} CH_2O + \tfrac{1}{4} H_2O = \tfrac{1}{4} CO_2(g) + H^+\ (10^{-7} M) + e^- \qquad pE^0(10^{-7} M\ H^+) = -8.20 \qquad (15.1b)$$

The overall reaction is thus

$$\tfrac{1}{4} O_2(g) + \tfrac{1}{4} CH_2O = \tfrac{1}{4} CO_2(g) + \tfrac{1}{4} H_2O \qquad \begin{aligned} pE^0(10^{-7} M\ H^+) &= 21.95 \\ \Delta G &= -125\ kJ \end{aligned} \qquad (15.1)$$

Other species such as NH_4^+ and HS^- are also readily oxidized by O_2, yielding the products shown in the middle column of Table 14.2.

Two types of bacteria can survive in the absence of dioxygen. The *obligate anaerobes* require O_2-free conditions exclusively, while the *facultative anaerobes* utilize dioxygen when it is available but turn to other electron acceptors when it is not. The methane-producing bacteria *Methanobacterium, Methanobacillus, Methanococcus,* and *Methanosarcina* are examples of the former category. They derive their energy from the disproportionation of carbohydrate:

$$\tfrac{1}{4} CH_2O = \tfrac{1}{8} CH_4 + \tfrac{1}{8} CO_2 \qquad \Delta G = -23\ kJ \qquad (15.2)$$

Because it is relatively insoluble in water, methane formed by this reaction leaves the aquatic environment rapidly, and since 1 mol of CH_4 requires 2 mol of O_2 for complete oxidation, this is an efficient means of reducing BOD. Despite the fact that reaction 15.2 liberates only one-fifth as much free energy per mole of electrons as 15.1, it is extremely important under the anaerobic conditions of stagnant swamp water and bottom sediments in streams or lakes. Similar *fermentation* processes are employed in the major sludge treatment method, but aside from local use in heating digestors, use of the methane as a fuel appears not to be economically feasible.

Strong oxidizing agents such as nitrate ion may also be utilized as electron acceptors by bacteria. Often NO_3^- is reduced only as far as NO_2^-, but in the special case of *denitrification* the reduction proceeds to N_2. To determine energy changes the half reaction:

$$\tfrac{1}{5} NO_3^- + \tfrac{6}{5} H^+(10^{-7} M) + e^- = \tfrac{1}{10} N_2(g) + \tfrac{3}{5} H_2O \qquad pE^0 = 12.65 \qquad (15.3)$$

---

[1] The superscript zero refers to reactants and products in standard states except those where other concentrations are indicated.

is coupled with oxidation of carbohydrate or other organic matter. The overall reaction

$$\tfrac{1}{5} NO_3^- + \tfrac{1}{4} CH_2O + \tfrac{1}{5} H^+ = \tfrac{1}{10} N_2(g) + \tfrac{1}{4} CO_2(g) + \tfrac{7}{20} H_2O \qquad \Delta G = -119 \text{ kJ} \quad (15.4)$$

is an important link in the biogeochemical nitrogen cycle since it restores atmospheric N_2 fixed by *Azotobacter* and *Rhizobium*. As we shall see in Section 15.4, similar reactions may be employed to remove nitrate from sewage treatment effluents.

Coliform Count

Since direct tests for the presence of pathogenic bacteria require large samples, expert analysis, and considerable time, the absence of these organisms is evaluated indirectly by determining the most probable number of coliform bacteria. From 100 to 400×10^9 of these benign organisms are excreted from the intestines of a single human every day. They are affected by the aquatic environment in a manner similar enough to the pathogenic bacteria which cause typhoid fever and chlorea so that their absence can be taken as an indication that water has not been recently contaminated by fecal material and is safe to drink.

A coliform count is performed by making appropriate dilutions of the sample and incubating it at 308 K for 48 hours in a lactose nutrient broth. If the medium becomes turbid the presence of bacteria is indicated and it can be spread over an agar gel plate. After an additional 24-hour incubation the number of bacterial colonies on the plate is counted under a microscope.[2] The test is based on the assumption that even a single living coliform present in the sample will divide and grow in the culture medium, eventually producing a colony. The commonly accepted standards for coliform counts are drinking water, $< 1/100 \text{ cm}^3$; normal, unpolluted inland lakes, $10–100/100 \text{ cm}^3$; suspicious range (mild sewage pollution), $1000–5000/100 \text{ cm}^3$; definite evidence of sewage pollution, $5000–10\ 000/100 \text{ cm}^3$; heavy sewage pollution, dangerous, $10\ 000–100\ 000/100 \text{ cm}^3$; and sewage, $> 100\ 000/100 \text{ cm}^3$.

15.2 PRIMARY AND SECONDARY TREATMENT

Primary and secondary sewage treatment have already been defined in terms of the physical, chemical, and biological techniques employed. From

---

[2] "Standard Methods for the Examination of Water and Wastewater," 13th ed., pp. 635–711. American Public Health Association, American Water Works Association, and Water Pollution Control Federation, 1971. The more recently accepted but equally accurate membrane filter technique is described in detail in this reference, as are modifications to determine fecal coliforms or fecal streptococci instead of total coliforms.

TABLE 15.1 **Performance of Conventional Treatment of Municipal Wastes**[a]

Measure	Primary	+Secondary
Removal of		
BOD	35%	90%
COD	30%	80%
Suspended solids	60%	90%
Refractory organics	20%	60%
Total N	20%	50%
Total P	10%	30%
Minerals	—	5%
Cost/m³	$0.01 to $0.015	$0.02 to $0.04
United States population (in millions) having type of treatment in		
1962	103.7 (88%)[b]	61.5 (52%)[b]
1968	129.7 (93%)[b]	85.6 (61%)[b]
1973	162.3 (98%)[b,c]	107.7 (65%)[b,c]

[a] Data from Office of Water Planning and Standards, "National Water Quality Inventory," Table XI-3. US Govt. Printing Office, Washington, D. C., 1974; American Chemical Society, "Cleaning Our Environment: The Chemical Basis for Action," p. 108. Amer. Chem. Soc., Washington, D. C., 1969.
[b] Percentage (of total population served by sewers) which has the indicated type of treatment.
[c] 1.6% of sewered population had tertiary treatment.

the standpoint of the EPA and other regulatory bodies, however, primary treatment is usually defined as that which removes 30–40% of total BOD. When followed by secondary treatment the removal efficiency increases to 85–93%. Performance with respect to other water quality parameters is indicated in Table 15.1. Note that with regard to soluble inorganic compounds (minerals), nitrogen, phosphorus, and stable (refractory) organic compounds removal efficiencies are much lower than for BOD and suspended solids. As the last part of the table indicates installation of primary and secondary sewage treatment facilities has been growing more rapidly than population for the last two decades in the United States.

A schematic flow chart of the major operations involved in most primary and secondary treatment plants is shown in Figure 15.1. The role of flocculents and coagulants in sedimentation has already been discussed (Section 14.2). Lime [CaO or Ca(OH)$_2$], alum [Al$_2$(SO$_4$)$_3$], and iron(III) chloride are the most common coagulants employed for sedimentation and sludge thickening. Two main techniques are employed for carrying out the oxidation required in secondary treatment. The trickle filter consists of a bed of slime-coated rocks over which the sewage is sprayed. Water trickling over the rocks is brought into contact with the air and with bacteria which oxidize organic wastes, accelerating the natural process by which this material would be removed. In the activated sludge process air or dioxy-

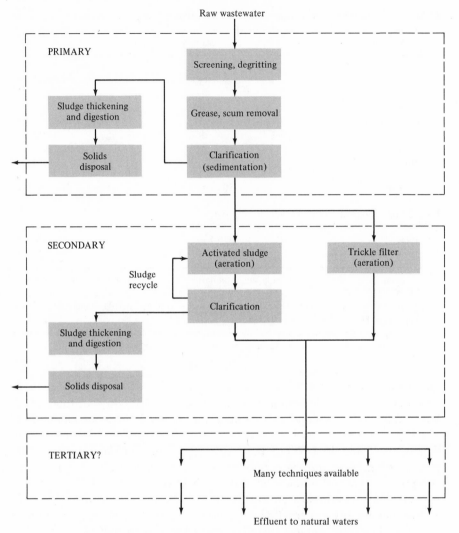

Figure 15.1 Flow diagram for municipal wastewater treatment.

gen is bubbled through a large tank containing the sewage. Bacteria collect on the sludges in this aeration tank and catalyze the oxidation of the waste. Following aeration the solids are allowed to settle out and are recirculated to the aeration tank to maintain the bacterial colony. The purified liquid is released to a lake or stream or treated by advanced techniques.

Sludge Handling

The most difficult problem facing sanitary engineers is what to do with the large quantities of sludge generated by both primary and secondary treatment. Sludges are semiliquids with solids content ranging from 0.5 to 5%. This corresponds roughly to the consistency of thin mud. The extent of the sludge problem may be seen by considering a typical city. The Chicago Sanitary District handles sewage from a population of 5.5 million people with an industrial contribution the equivalent of 3 million more. The total volume of waste is 6.6×10^6 m³/day, corresponding to 815 tonnes of solid material if all of the water could be removed. The city spent 14.5 million dollars in 1965 to dispose of these solids, approximately 46% of its total budget for sewage treatment.[3]

The four main objectives in sludge-handling processes are (1) to convert organic matter in the solids to a relatively stable form; (2) to reduce the volume and mass of the sludge by removing water; (3) to destroy or control harmful bacteria and other organisms; and (4) if possible, to obtain by-products whose sale can reduce the cost of sludge treatment. To accomplish these objectives the sludge is concentrated by sedimentation, coagulation, flocculation, and occasionally flotation. Some are digested in the presence of aerobic or anaerobic bacteria to obtain methane, which can be used to power other portions of the sewage treatment plant. Water may be removed from sludge by vacuum filtration or centrifugation or by heat drying prior to or during incineration. Final disposal methods include landfills, conversion to fertilizer, or dumping at sea.

None of these techniques is entirely satisfactory, either on technological or economic grounds. Sedimentation ponds and lagoons require valuable space and often produce undesirable odors. Incineration requires at least 25% solids if any usable heat is to be generated, and unless adequate air pollution controls are installed may simply convert one problem into another. Ocean dumping sites are already overused, and in the case of New York City there is evidence that sludges previously dumped almost 20 km offshore may be drifting back toward recreational beach areas. Some cities (notably Milwaukee) heat their sludge to kill bacteria and then package it as fertilizer, but since the nutrient level is low supplements must be added to maintain a consistent formula. The presence of metal ions (such as Fe^{3+} used to precipitate phosphates or Pb^{2+} washed into treatment plants by storm runoff from city streets contaminated by auto emissions) also mitigates against reclamation of the nutrient value of sludge. Much is currently disposed of in landfills since transportation to more remote sites such as farmland or coal mines is expensive.

15.3 TERTIARY OR ADVANCED WASTEWATER TREATMENT

As Table 15.1 indicates, not all of the impurities in a wastewater stream are removed by primary and secondary treatment. Ordinarily, the remaining quantities would be disposed of naturally in the river which re-

---

[3] V. W. Bacon and F. E. Dalton, *J. Water Pollut. Contr. Fed.* **40**, 1586 (1968).

ceived the effluent, but the growth of United States population and industry and its concentration along rivers require that most water be used many times before it reaches the ocean. In many cases the increments of pollution remaining after ordinary treatment are too large for natural removal prior to the next use. Thus, a number of *tertiary* or *advanced wastewater treatment* techniques are becoming important, although at present only a very small percentage of municipal systems use them (Table 15.1). Several of these methods are also applicable to industrial wastes where BOD and suspended solids may not be the most important indicators of water quality, and indeed a number of them find much greater use in such applications. In addition, they may be employed at the beginning (water purification) as well as the end of the human water use cycle.

Activated Carbon Adsorption

Refractory organic compounds, which are not oxidized by secondary treatment, have been implicated in environmental cancer problems (see the discussion in the introduction to Part V). These substances, whether dissolved or in colloidal suspension, may be removed by passing water over a bed filled with granulated, activated charcoal. This substance has an extremely large surface area and is capable of adsorbing large numbers of a variety of different molecules. When adsorption sites are saturated the carbon can be regenerated by heating to 1225 K in an atmosphere of air and steam, which oxidizes the adsorbed compounds. As much as 77% removal of organic matter and 90% removal of suspended solids can be effected at a cost of about $0.02/m³ wastewater.[4]

Activated carbon is also effective in removing odors and tastes (such as that of chlorine) from drinking water, and some water distribution companies use it routinely for this purpose. Several small units which can be attached to an individual's home faucet are also available in the United States.

Ultrafiltration

Although the term *reverse osmosis* is widely used, *ultrafiltration* is more descriptive of this process for removing molecular or ionic contaminants whose diameters or hydration shells are larger than that of a water molecule. The technique makes use of an osmotic membrane, usually made of cellulose acetate, whose pore sizes are quite small (Figure 15.2). Since the larger impurity molecules cannot fit through the pores, the membrane serves as an extremely effective filter, hence, the name of the process.

---

[4] American Chemical Society, "Cleaning Our Environment: The Chemical Basis for Action," p. 126. Amer. Chem. Soc., Washington, D. C., 1969.

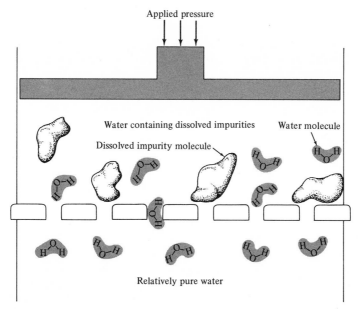

Figure 15.2 Schematic diagram of ultrafiltration. Since molecular size of dissolved impurities is large they cannot pass through the membrane pores: H—O—H indicates water molecules and shading indicates impurity molecules.

The term *reverse osmosis* applies because ultrafiltration is the opposite of a spontaneous process. Because of the greater disorder (entropy) when a concentrated solution of impurities is diluted, there is a strong tendency for water molecules to flow *up* through the membrane in Figure 15.2. The minimum pressure necessary to counteract this tendency is the osmotic pressure, Π, which may be calculated as

$$\Pi = \frac{nRT}{V} = MRT \qquad (15.5)$$

where M represents the number of moles of solutes per dm^3 of solution above the membrane and the other symbols have their usual meaning. Rather large pressures are required to reverse the normal osmotic flow, especially if the filtration is to be done rapidly, and the most difficult problem in ultrafiltration is supporting the rather fragile membrane so that it does not break while maintaining adequate space for water to flow through.

A pilot plant for ultrafiltration has operated for 2000 hours at pressures 25 times normal atmospheric in Pomona, California.[5] It was able to remove 88% of dissolved solids, 84% of COD, 98% of phosphate, and about 75%

[5] American Chemical Society, "Cleaning Our Environment: The Chemical Basis for Action," p. 130. Amer. Chem. Soc., Washington, D. C., 1969.

of nitrogen compounds. The technique is widely used in treating industrial wastes and can be employed for desalination of seawater.

Electrodialysis

Some of the advanced wastewater techniques are designed primarily for removal of dissolved ionic substances. These methods are also useful for desalination of seawater for irrigation and drinking purposes, and find numerous applications in treatment of industrial wastes which contain ionic impurities. The standard laboratory purification technique, distillation, usually requires too much energy to be competitive, although in areas of large solar flux it can be made feasible by using the sun's heat to evaporate the water. Pure water will also freeze out of a solution of salts, and in some areas less free energy is consumed by this means of desalination. Studies of the transport of icebergs from the Antarctic to the United States Southwest or other dry coastal areas have indicated that natural freezing may be a viable technique,[6] although it may entail some problems (Exercise 11).

Electrodialysis is a standard technique for removal of ions from solution which may be applied at room temperature. A schematic diagram of a typical cell is shown in Figure 15.3. A strong electric field produced by the anode and cathode at the right and left ends of the cell causes anions to migrate to the right and cations to migrate to the left. The cell is divided into smaller units by membranes, some of which are permeable only to anions and others of which are permeable only to cations. By alternating the membranes along the large cell some subcells will allow both anions and cations to escape, producing deionized water. Other subcells will allow both anions and cations to diffuse in, and therefore a greater concentration of ionic impurities will be produced. Every other subcell along the electrodialysis cell will produce pure water while the other half of the subcells produce concentrated brine which may be disposed of or treated in some way for the removal of the ionic compounds dissolved in it.

The cation and anion permeable membranes are the most important aspect of the electrodialysis process. They usually contain negatively or positively charged sites, respectively, near the small orifices through which the ions must pass. These sites can attract ions of opposite charge, permitting them to pass through the membrane, while repelling ions of like charge. A basic problem with electrodialysis occurs when large quantities of organic material are also present in the water sample to be purified. Since the size of many organic molecules is much greater than that of the inorganic ions, such molecules often become trapped in the passageways through the

---

[6] W. Weeks and W. Campbell, "Icebergs as a Fresh Water Source: An Appraisal," Research Report No. 200. U. S. Army Cold Regions Research and Engineering Laboratory, Hanover, New Hampshire, 1973.

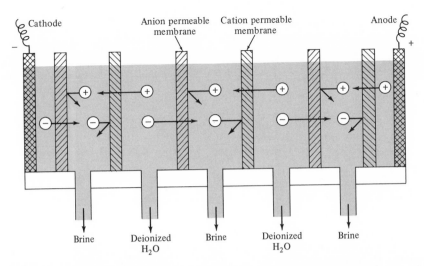

Figure 15.3 Schematic diagram of electrodialysis cell: ▨ anion permeable membrane and ▧ cation permeable membrane.

membranes, blocking them so that no ions can pass through. Colloidal particles can also produce the same effect. Therefore, electrodialysis units must shut down periodically so that fouling materials may be removed by reverse flushing of the membranes. The efficiency of the units also deteriorates markedly between these periodic cleanups. Nevertheless, electrodialysis can remove 40–50% of dissolved inorganic materials and this is often enough to get rid of the increment added by each water use.

Ion Exchange

Ion exchange is well known as a technique for water softening and may also be useful for purification of wastewater. It is often used in laboratories for water deionization instead of distillation. The basic principle governing the ion exchange process is illustrated in Figure 15.4. Ion exchange resins may be synthetic polymeric materials or naturally occurring zeolites (see Sections 2.4 and 12.6). In either case the resin material has positively or negatively charged sites to which anions or cations will be attracted. Depending on their concentrations in solution, ions will exchange between the sites on the resin and the dissolved state. A cation exchanging resin might first be "loaded" with hydrogen ions by passing strong acid over it. If, at a later time, a solution containing calcium ions were placed in contact with the resin, some of the hydrogen ions would go into solution, having been replaced by calcium ions.

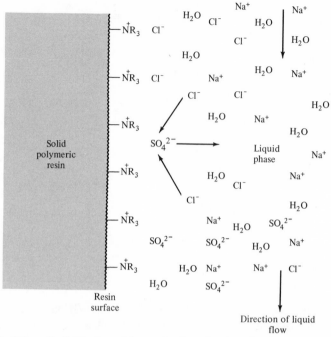

Figure 15.4 Schematic illustration of the mechanism of action of an ion exchange resin. Sulfate ions are replaced from the positively charged resin surface by a concentrated chloride ion solution. Overall reaction: Resin$\}$ SO_4^{2-} + 2 Cl$^-$(aq) $\longrightarrow$ 2 resin$\}$ Cl$^-$ + SO_4^{2-}(aq).

$$2 \text{ Resin} \Big\} \text{ H}^+ + \text{Ca}^{2+}(aq) \rightleftharpoons \text{resin} \Big\} \text{Ca}^{2+} + 2 \text{ H}^+(aq) \qquad (15.6a)$$

Similarly, hydroxyl ions can be displaced from an anion exchanger by some other type of anions:

$$2 \text{ Resin} \Big\} \text{OH}^- + \text{SO}_4^{2-}(aq) \rightleftharpoons \text{resin} \Big\} \text{SO}_4^{2-} + 2 \text{ OH}^-(aq) \qquad (15.6b)$$

If water is passed over two such ion exchange resins in succession, dissolved cations will be replaced by H$^+$ and dissolved anions by OH$^-$ which will then combine to form water:

$$2 \text{ H}^+(aq) + 2 \text{ OH}^-(aq) \longrightarrow 2 \text{ H}_2\text{O} \qquad (15.6c)$$

$$2 \text{ Resin} \Big\} \text{ H}^+ + \text{Ca}^{2+}(aq) + 2 \text{ resin} \Big\} \text{OH}^- + \text{SO}_4^{2-}(aq) \longrightarrow \text{resin} \Big\} \text{Ca}^{2+} + \text{resin} \Big\} \text{SO}_4^{2-} + 2 \text{ H}_2\text{O}$$

$$(15.6)$$

The overall effect (equation 15.6) is removal of ions from the sample.

After a certain period of use the ion exchange resins must be regenerated with the original cations and anions (in the above case H^+ and OH^-). This is done by passing a concentrated solution of the cations or anions over the resin, a process known as backflushing. It is not necessary that the original cations and anions be H^+ and OH^-, as long as they are not detrimental to the environment; for instance, the typical home water softener uses sodium ions and chloride ions, small increments of which are not harmful to most persons. (Those on low-sodium diets should not use water softened in this way.) When these ions have been exchanged for calcium ions, carbonate ions, and other constituents of hard water, the resin is back-flushed with a saturated brine solution.

Ion exchangers are extremely efficient water purifiers, but they are also expensive. As a general rule, resins are not subject to fouling as is the case with electrodialysis membranes, but they must be back-flushed often in order to remove the undesirable ions which have accumulated. Because ion exchange can remove nearly all of the ions in a given sample of water, some treatment programs have proposed treating only a part of the effluent and then using it to dilute the untreated material. This method would reduce the concentrations of undesirable inorganic materials enough so that there would not be an accumulation in the water supply but would be less expensive than treating all of the effluent for complete removal of ions.

Removal of Bacteria and Viruses

Although as many as 90% of all bacteria are removed by activated sludge treatment, the 10% that remain may cause problems when water is reused. Thus, some bactericidal treatment is required either upon release of treated wastewater or when it is purified prior to reuse. The preferred method in the United States is chlorination, usually with dichlorine, although ClO_2 is occasionally used because it is a stronger oxidizing agent. Approximately 7 g of Cl_2 will dissolve in 1 dm³ of pure water at 293 K and normal atmospheric pressure. It reacts within a few seconds to form hypochlorous acid:

$$Cl_2 + H_2O \rightleftharpoons HOCl + H^+ + Cl^- \qquad (15.7a)$$

which is weakly dissociated:

$$HOCl \rightleftharpoons H^+ + OCl^- \qquad (15.7b)$$

In dilute solutions above pH = 4 reaction 15.7a lies entirely to the right. Hypochlorous acid predominates in solution up to pH = 6.0, at which point 15.7b also begins to shift to the right. Above pH = 9 a dilute dichlorine solution contains > 95% OCl^-. Because these equilibria can be approached from either side, calcium and sodium hypochlorites can supply equivalent disinfectant power to Cl_2 (at the same pH). They are often used in small scale applications where handling of compressed Cl_2 gas is uneconomical.

When ammonia or amines are present *chloramines* are formed:

$$RNH_2 + HOCl \rightleftharpoons RNHCl + H_2O \tag{15.8}$$

(R might be H, an alkyl group, or even an amino acid residue.) Chlorine in combination with nitrogen compounds is still available to oxidize impurities (as is that in the OCl^- form), but the reduction potentials of these species are lower than for HOCl and hence their disinfecting power is less. Formation of chloramines does have the advantage of longer-lasting bactericidal effect, however; these compounds may persist throughout a water distribution system. *Breakpoint chlorination* refers to the process of adding a large excess ($> 2:1$ mole ratio $Cl_2 : NH_3$) of dichlorine. This completely oxidizes the amines, leaving residual HOCl which may persist for some time. Because of the dissociation of hypochlorous acid and the formation of chloramines, more than 150 times as much Cl_2 must be added to a given volume of water at pH = 10 than at pH = 5 to achieve equivalent bactericidal action.

Although chlorination does an excellent job of reducing coliform counts (and hence of removing the even more susceptible pathogenic bacteria), there are indications that it is not always successful in killing viruses.[7] The complicated bioassays necessary to detect these organisms are just beginning to be done on a large scale, so this problem may become more evident in the future. One method for removing both bacteria and viruses is *ozonation,* which uses O_3 as an oxidizing agent. Ozone generated by an electric discharge in dry air is bubbled in fine streams through the water to be treated, sometimes in conjunction with ultrasonic waves which break down larger waste particles, aiding oxidation. Air containing 2% O_3 is more effective than Cl_2 in removing tastes and odors and it is capable of oxidizing many refractory organic compounds. Costs are on the order of $0.02/m^3$, but since some states (Florida, for example) require chlorination of wastewater, ozonation has been slow to be adopted in the United States. It is far more popular in Canada and Europe, having been used to purify drinking water in Paris for over 65 years.[8]

15.4 REMOVAL OF NITROGEN AND PHOSPHORUS

Because of their special relationship to eutrophication considerable effort has been expended on techniques for removal of nitrogen and phosphorus

---

[7] O. J. Sproul, L. R. Larochelle, D. R. Wentworth, and R. T. Thorup, Water reuse. *Chem. Eng. Progr. Symp. Ser.,* p. 130. American Institute of Chemical Engineers, New York, 1967.

[8] Detailed evaluations of alternative disinfection processes are given by E. J. Laubisch, "Chlorination and Other Disinfection Processes." *In* "Water Quality and Treatment," 3rd ed., Chapter 5. American Water Works Association, Inc., McGraw-Hill, New York, 1971.

from municipal wastewater. In the case of phosphorus, several methods are in routine use, but nitrogen removal is still undergoing pilot plant tests. If nitrogen is present in the -3 oxidation state it can be removed from water by a process known as *ammonia stripping*. Since ammonium ions are in equilibrium with ammonia in aqueous solution

$$NH_4^+ + OH^- \rightleftharpoons NH_3 + H_2O \qquad (15.9)$$

increasing the pH of the solution will increase the amount of ammonia present. It can then be swept out of solution by a stream of air. At pH values above 11.25 the concentration of NH_3 is more than 100 times that of NH_4^+ (Exercise 3), and the process is theoretically possible. The pH is usually raised by addition of lime, which may already have served the dual purpose of acting as a coagulant.

Completely oxidized nitrogen may be removed by *denitrification* (Section 15.1). Since secondary treatment is an oxidative process, conditions which oxidize nitrogen compounds to NO_3^- can often be arranged. Denitrification bacteria are then added together with a nutrient (reducing agent). Although carbohydrate appears in equation 15.4, methanol is usually employed in sewage treatment:

$$6\ NO_3^- + 5\ CH_3OH \xrightarrow[\text{bacteria}]{\text{denitrification}} 3\ N_2 + 5\ CO_2 + 7\ H_2O + 6\ OH^- \qquad (15.10)$$

Pilot plant operations have shown that 82% of the nitrogen in a sewage sample can be converted to N_2 by means of such a nitrification–denitrification process. Ammonia removals of over 90% have been attained in pilot air stripping plants.[9]

The simplest method for *phosphate removal* is lime, alum, or iron coagulation:

$$5\ Ca^{2+} + 4\ OH^- + 3\ HPO_4^{2-} \longrightarrow \underline{Ca_5(PO_4)_3(OH)} + 3\ H_2O \qquad (15.11a)$$

$$Al_2(SO_4)_3 + 2\ PO_4^{3-} \longrightarrow \underline{2\ AlPO_4} + 3\ SO_4^{2-} \qquad (15.11b)$$

$$FeCl_3 + PO_4^{3-} \longrightarrow \underline{FePO_4} + 3\ Cl^- \qquad (15.11c)$$

Such processes can effect 90% removal of phosphorus for a little over one cent per cubic meter of waste.[10] Because the precipitates formed in reactions 15.11 are often colloidally stable, small amounts of an anionic polyelectrolyte flocculent are required if they are to settle rapidly.

Under conditions of high dissolved oxygen and pH and with hard water, phosphorus removal by the activated sludge process can be as high as

---

[9] American Chemical Society," Cleaning Our Environment: The Chemical Basis for Action," p. 133. Amer. Chem. Soc., Washington, D. C., 1969.

[10] American Chemical Society, "Cleaning Our Environment: The Chemical Basis for Action," p. 131. Amer. Chem. Soc., Washington, D. C., 1969.

60–90%. This phenomenon was originally attributed to "luxury phosphorus uptake" by aerobic bacteria in the sludge, i.e., the bacteria were assumed to be assimilating far more phosphorus than was necessary for their continued existence. However, more recent data[11] indicate that at high rates of aeration CO_2 is swept out of the activated sludge treatment tank, raising pH so that $H_2PO_4^-$ is converted to HPO_4^{2-}. The latter precipitates readily according to equation 15.11a. Thus the greater phosphorus content in the activated sludge is present as hydroxylapatite, not in the bacterial cells.

15.5 DISPOSAL OF TREATED WASTEWATER IN SOILS

Because the effluent from primary and secondary sewage treatment is high in nutrients, numerous schemes have been proposed to make use of it for growing crops in either the water or land environment. While not strictly classed as tertiary or advanced treatment, a number of these appear to show considerable promise.

In East Lansing, Michigan, a project has been designed to treat secondary effluent by passing it through a series of eutrophic, artificial lakes.[12] Plants grow rapidly in the lakes and are harvested several times a year and used as animal feed. After treatment is partially complete fish will be grown and the last lake in the series will be usable for recreation (other than swimming).

Perhaps even more promising is spray irrigation of crop and forest land with secondary effluent. Pathogenic organisms die rapidly in a soil environment, and most nutrients are rapidly assimilated by growing plants unless the soil is overloaded. Usually the capability to handle nitrogen is the limiting factor, and as much as 3 tonnes of phosphorus per hectare can be absorbed. Experiments at Pennsylvania State University[13] have shown that 99% removal of phosphorus and 60–85% removal of nitrogen can be effected by the "living filter" of soil. Both crop yields and forest growth were increased with minimal effects on parameters of soil quality such as exchangeable calcium or potassium. Recently, the EPA has publicly praised the success of land disposal in Muskegon, Michigan. This is the largest pilot project for the technique to date.

Agricultural Wastes

Like human wastes, agricultural wastes can contaminate waters. Since one head of cattle produces about 16 times as much waste as one human, the seriousness of the problem can immediately be seen. In 1965 the

[11] A. D. Menar and D. Jenkins, *Environ. Sci. Technol.* **4**, 1115 (1970).

[12] *Environ. Sci. Technol.* **5**(2), 112–113 (1971).

[13] *Environ. Sci. Technol.* **6**(10), 871–873 (1972).

production of livestock (cattle, hogs, sheep, poultry, and horses) yielded 1032 million tonnes of solid and 394 million tonnes of liquid waste.[14] The waste disposal problem is especially serious on specialized livestock farms and feedlots where from 10 to 50 thousand animals are concentrated in a small area. Wastes equivalent to those of a city are produced, and are often released without adequate treatment.

Such wastes would be disposed of naturally by absorption into soil, but the spreading of manure is complicated by its concentration in a small area and problems of adequate nutrient balance, objectionable odor, and the spread of pathogenic bacteria. Methods are now being developed in which specially designed plows inject the manure 10–20 cm below the soil surface. Such recycling of manure would help to provide soil organic matter and reduce the need for fertilizer, thus obviating some of the problems of agricultural runoff mentioned in Section 11.12. It is unfortunate that current low-cost fertilizers are often used in preference to manure, since this practice consumes energy and leads to water pollution.

EXERCISES

(The first group of exercises is based primarily on material found in this text.)

1. Given the pE^0 values in equations 15.1a, 15.1b, and 15.3, use equation 14.3 to calculate ΔG^0 for equations 15.1 and 15.4. Compare your values with those given in the text.

2. Given that pE^0 for complete oxidation of carbohydrate to CO_2 is -8.20 and pE^0 for reduction of CO_2 to CH_4 -4.13, write half-reactions and calculate pE^0 and ΔG^0 for equation 15.2.

3. Calculate the pH of an aqueous solution which contains 100 times more NH_3 than NH_4 ($K_b = 1.8 \times 10^{-5}$ for NH_3).

4. Describe the major goals of domestic sewage treatment and the processes which are employed to achieve these goals.

5. Explain some of the techniques employed in handling sludge from sewage treatment plants. What are some of the problems involved? Give equations for at least three reactions used to help coagulate sludges.

6. Give at least three examples of advanced wastewater treatment processes and explain the chemistry involved.

7. Explain the principle on which ion exchange resins operate. Diagram the surface of a cation exchange resin. How are ion exchangers used in water softening?

8. Suppose it is desired to purify a 1 M solution of sugar in water by ultrafiltration. What pressure will be required initially (at 298 K) to force the solvent

---

[14] H. D. Foth and L. M. Turk, "Fundamentals of Soil Science," 5th ed., p. 342. Wiley, New York, 1972.

through a semipermeable membrane? How will the pressure vary as the purification process proceeds?

9. Describe methods for the removal of phosphates during sewage treatment. Give equations where possible.

10. Name and give equations for at least two methods for removal of nitrogen compounds from secondary sewage treatment effluents.

(Subsequent exercises may require more extensive research or thought.)

11. A proposal has been made that fresh water be supplied to desert areas by towing icebergs from Antarctica [*Chem. & Eng. News* July 16, 32 (1973)]. Given that approximately 2×10^9 m^3 water is required to supply a population of one million for one year, calculate the size of a single iceberg 100 m thick which could supply the city of San Diego [population 720,000; see *Chem. & Eng. News* August 6, 40 (1973)].

12. Calculate the amount of energy required to melt the iceberg in Exercise 10. Using data on incoming solar energy in Section 6.4, calculate the area of flat plate collectors necessary to obtain the required heat.

13. Cattle in the United States excrete approximately 1.3×10^9 tonnes of solid and liquid waste per year. Using data from H. D. Foth and L. M. Turk ("Fundamentals of Soil Science," p. 332. Wiley, New York, 1972), how much 10–10–10 fertilizer would be equivalent to this quantity of manure? What would be the limiting nutrient in the manure?

14. The Technologically Oriented Instructional Games project (c/o Professors Fred Fink and Ron Rosenberg) at Michigan State University, East Lansing, Michigan 48823) has available a computer simulation of water quality management which allows the user to act as supervisor of a wastewater treatment plant. Obtain the WAQUAL game, get it to run on your university's computer, and try it!

16

Industrial Wastewater

Man is a rational animal who always loses his temper when he is called upon to act in accord with the dictates of reason.

Oscar Wilde

He that will not apply new remedies must expect new evils.

Francis Bacon

16.1 SOURCES OF INDUSTRIAL WATER POLLUTION

In addition to its function in carrying off sewage produced by workers, water plays two main roles in manufacturing industries: It may serve as a source or sink for heat or it may be directly involved in some chemical process as a reactant, product, or solvent. Table 14.3 has already made the distinction between cooling and process water use in industry. It is an important one because cooling water contains far fewer impurities in most cases, and usually is not treated. Process water will therefore be of primary concern in this chapter, although as we shall see presently, one minor source of mercury in the environment has resulted from the use of its compounds to control algae and slimes which built up in cooling water pipes.

Table 14.3 makes clear the fact that the quantities of BOD and suspended solids in industrial process water are comparable to those in municipal wastewater. However, this is only part of the story. One indication that more than just the usual organic matter of municipal sewage is to be found in industrial effluents is the fact that COD for paper mill and chemical plant wastes is much greater than BOD. This is the result of materials sufficiently refractory not to be oxidized in 5 days by atmospheric O_2 but susceptible to dichromate oxidation. An important aspect of water pollution from industry is that it cannot be characterized by one or two non-

393

specific criteria such as BOD, suspended solids, or coliform count. Depending on the type of industry a wide variety of hazardous substances may be dissolved, suspended, or adsorbed on suspended particles in effluents. This is especially true of the chemical industry because of the large number of different substances which are handled.

Many different types of industrial water pollution have already been described in this book. Among these are acid mine drainage and thermal pollution (Section 5.2), effluents from petroleum refining and oil spills (Section 5.7), radioactive materials (Section 6.2), solids and acid resulting from air pollution control (Chapter 9), spent pickle liquor and other effluents associated with steel production (Section 11.4), and contamination from other metals processing (Sections 11.6 and 11.7). Sulfuric acid is widely used in a great many chemical processes as well as in the manufacture of rubber and plastics. Spent solutions must be neutralized or recycled as in the case of pickle liquor. The same is true of sodium hydroxide (caustic) used in vegetable processing. A 10–20% NaOH solution is usually employed to peel potatoes prior to manufacture of french fries, for example. Food processing also releases high concentrations of BOD, often in the form of objectionable materials such as blood, entrails, grease and fat, or

TABLE 16.1 Occurrence of Metals or Their Compounds in Effluents from Various Industries[a]

Industry	Al	Ag	As	Au	Ba	Be	Bi	Cd	Co	Cr	Cu
Mining operations and ore processing	x		x					x			
Metallurgy and electroplating		x	x			x	x	x		x	x
Chemical industries	x		x		x			x		x	x
Dyes and pigments	x		x					x		x	x
Ink manufacturing									x		x
Pottery and porcelain			x							x	
Alloys						x					
Paint					x					x	
Photography		x		x				x		x	
Glass			x		x				x		
Paper mills	x									x	x
Leather tanning	x		x		x					x	x
Pharmaceuticals	x										x
Textiles	x		x		x			x			x
Nuclear technology						x		x			
Fertilizers	x		x					x		x	x
Chlor-alkali production	x		x					x		x	
Petroleum refining	x		x					x		x	x

[a] Data from U. S. Environmental Protection Agency, "Water Quality Criteria Databook," Vol. 2. US Govt. Printing Office, Washington, D.C., 1971; J. G. Dean, F. L. Bosqui, and K. H. Lanouette, *Environ. Sci. Technol.* **6**(6), 518–521 (1972).

hair, which degrade slowly. Processing of natural textile fibers (wool or cotton) requires alkaline solutions and may lead to emission of as much as 1.5 kg of impurities per kilogram of fiber.

From all of these examples of industrial water pollution two—heavy metals and pulp and paper—have been selected for detailed examination in the remainder of this chapter. Small (sometimes large) quantities of a variety of metals are released by almost every industry. Since many of them are toxic and relatively little is known about their cycling in the biosphere, they constitute a very real problem. The pulp and paper industry uses more process water than any other in the United States (Table 14.3), and its emissions often occur in otherwise unspoiled areas of natural beauty. Since its total load of wastes is the largest in Table 14.3, it too is a prime target of environmentalists' concern. It is hoped that the reader will be able to generalize from specific details on these two types of water pollution to others which have received less than complete treatment herein.

16.2 HEAVY METALS

The ubiquity of heavy metals in industrial processes is illustrated nicely by Table 16.1. Almost every type of process involves release of at least trace quantities of half a dozen or more metals in one form or another.

| | | | | | | | | | | Metals | | | | | | | | |
Fe	Ga	Hg	In	Mn	Mo	Os	Pb	Pd	Ni	Sb	Sn	Ta	Ti	Tl	U	V	W	Zn
		x		x	x		x								x	x		
		x	x				x		x								x	x
x	x	x				x	x				x	x	x		x			x
x							x			x			x	x				
x	x								x									
										x					x			
	x	x				x		x				x						
						x	x					x						x
			x				x								x			
									x				x			x		
		x				x	x				x	x				x		
x	x																	x
x	x	x				x						x						
x	x						x		x	x								
		x													x			
x	x		x				x		x									x
x	x						x				x							x
x							x		x									x

Increasing numbers of industrial plants can therefore increase concentrations of these metals in lakes or streams, and the phenomenon of *bioamplification* (Section 16.3) often multiplies such concentrations manyfold within living organisms.

Some general techniques are available for removal of heavy metals from industrial effluents. Since many of their hydroxides are insoluble, appropriate adjustments of pH can cause them to precipitate. Lime or limestone is often used for this purpose. The pH at which a number of common metal ions initially precipitate was given in Table 14.1. The two main difficulties with hydroxide precipitation are the problem of separating metals if recycling is desirable and the fact that many hydroxides are flocculent and extremely hydrophilic. This makes the complete removal of water rather difficult.

If the metal to be recovered is valuable, electrodeposition techniques may be used to obtain it in pure form. A dilute solution may be concentrated by evaporation and then placed in an electrolytic cell. Different metals will deposit on the cathode at different potentials, allowing recovery of a high purity sample. Anodes for such cells are made of graphite or other insoluble materials. (Similar techniques are used to obtain metals from their ores; see Section 11.6.) Another process which depends on the electromotive force series is cementation (Section 11.6). Scrap iron or shredded tin cans may be used to treat effluents containing metal ions such as Ni^{2+} and Cu^{2+} which are more easily reduced than Fe^{2+} or Fe^{3+}. Once Fe has displaced the less active metals, Fe^{2+} and Fe^{3+} may be removed from the effluent by lime coagulation.

Another important means of removing metals from fairly concentrated solutions is solvent extraction. This depends on the formation of *chelate compounds* between the metal ions and negatively charged ligands. When neutral compounds are formed they are usually soluble in nonpolar solvents which are immiscible with water. The metals may thus be sequestered into the organic phase and removed from the water. Some common sequestering agents are ethylenediaminetetraacetic acid (EDTA), 8-quinolinol (8-hydroxyquinoline or oxine), nitrilotriacetic acid (NTA), and acetylacetone. As an example of the solvent extraction process, the following equations describe the reactions involved in the sequestration of chromium(III) by 8-quinolinol:

$$\text{[structure]}-OH + CO_3^{2-} \rightleftharpoons \text{[structure]}-O^- + HCO_3^- \qquad (16.1a)$$

8-Quinolinol
(oxine)

$$Cr^{3+} + 3 \quad \rightleftharpoons \qquad\qquad\qquad\qquad (16.1b)$$

Oxinate ion

Uncharged, relatively
hydrophobic coordina-
tion complex, $Cr(oxine)_3$

$$Cr(oxine)_3(aq) \rightleftharpoons Cr(oxine)_3(\text{nonpolar solvent}) \qquad (16.1c)$$

In addition to the methods just enumerated, ultrafiltration, ion exchange, activated carbon adsorption, and other techniques described in Section 15.3 are often applied to industrial waste.

16.3 MERCURY

As an example of the problems involved in heavy metal pollution we shall consider mercury, a subject of much controversy in recent years. In addition to the weathering of igneous rock, which releases about 800 tonnes of mercury per year, mostly to fresh water and eventually to the oceans, a variety of human activities are responsible for dispersion of mercury in the environment. As Table 16.2 indicates, on a yearly basis anthropogenic sources far outweigh natural ones, but since the latter have been in effect over the geological time scale, mankind has not yet made significant contribution to the total amount of mercury in the oceans.[1] It is also interesting to note that the largest anthropogenic source of environmental mercury is vapor released from fossil-fueled power plants and other combustion processes. Much of this is eventually removed from the atmosphere by rain and finds it way into natural waters, and the observation has been made that fish taken from lakes in industrial areas contain more mercury than those from nonindustrial areas, despite the fact that there was no direct input of mercury to the water.[2]

Annual industrial consumption of mercury in the United States is summarized in Table 16.3. Mercury is used primarily in elemental form because it is the only unreactive metal that is a liquid at ordinary temperatures and hence can be used as a fluid conductor of electricity. Inorganic mercury compounds such as HgO ("red lead") are often used in paints

---

[1] J. Gavis and J. F. Ferguson, *Water Res.* **6**, 989–1008 (1972).
[2] J. Gavis and J. F. Ferguson, *Water Res.* **6**, 1005 (1972).

TABLE 16.2   Sources of Mercury in the Environment[a]

Source	Quantity of mercury (tonnes)	
	Production	Estimated releases
Total world production of Hg from ores (1900–1970)	361 000	120 000[b]
World production of Hg from ores (1970)	10 000	3 300[b]
Total world release from fossil fuels (1900–1970)	—	150 000
World release from fossil fuels (1970)	—	4 800
Total anthropogenic release (1900–1970)	—	270 000
Annual release by rock weathering	—	800
Total release by rock weathering (1900–1970)	—	5 700
Total quantity of Hg in oceans (approximately 60 000 year half-life before sedimentation)	—	45 000 000
Total Hg released by weathering during earth's lifetime	—	1 600 000 000

[a] Data from J. Gavis and J. F. Ferguson, *Water Res.* **6**, 989–1008 (1972).
[b] Estimated on the basis of one-third of production since this fraction was unaccounted for in the United States, 1945–1958.

because of their ability to poison barnacles and other organisms which might attach themselves to ships' hulls. As in the case of a number of categories in Table 16.3 it is difficult to estimate how much of the mercury consumed in paint actually finds its way into aquatic ecosystems.

The organomercurials, some of whose structures are shown in Figure 16.1, are often purposely dispersed in the environment in the form of fungicides, bactericides, and algicides. Acetatophenylmercury(II) (PMA) was until recently widely used in the pulp and paper industry to control algae and slime in water supplies. The methylmercurials are very effective in preserving seeds from growths of fungi and mold until they can be

TABLE 16.3   Industrial Consumption of Mercury in the United States (1969)

Industry	Consumption (tonnes)
Chlor-alkali	1575
Electrical apparatus	1417
Paints	739
Scientific instruments (thermometers, barometers, etc.)	531
Dentistry	232
Catalysts	225
Agricultural (seed dressings, etc.)	204
Laboratory use	155
Pharmaceuticals	55
Pulp and paper	42
Other	736
Total	5911

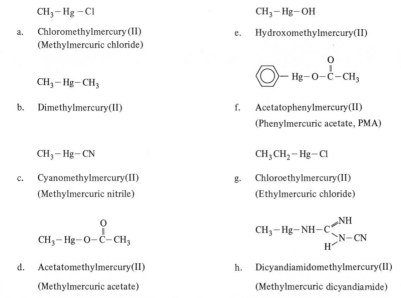

a. Chloromethylmercury(II)
(Methylmercuric chloride)

b. Dimethylmercury(II)

c. Cyanomethylmercury(II)
(Methylmercuric nitrile)

d. Acetatomethylmercury(II)
(Methylmercuric acetate)

e. Hydroxomethylmercury(II)

f. Acetatophenylmercury(II)
(Phenylmercuric acetate, PMA)

g. Chloroethylmercury(II)
(Ethylmercuric chloride)

h. Dicyandiamidomethylmercury(II)
(Methylmercuric dicyandiamide)

Figure 16.1 Structures of some organomercurial compounds released to the environment. Common names are given in parentheses.

planted. Such fungicides were responsible for poisoning of an entire New Mexico family which used low priced seed grain to feed a hog that they later slaughtered and ate,[3] and Sweden has banned their use since 1966 because of the deaths of birds which ate freshly planted seeds.

Minamata Disease

The most disastrous case of environmental mercury poisoning, and perhaps the most illustrative of the complicated ecological pathways of this heavy metal, began some 25 years ago in the Japanese fishing port of Minamata. The Chisso Chemical Company constructed a plastics plant which used organomercurials as catalysts, and each year a small quantity of chloromethylmercury(II) was discharged into the Minamata River and adjacent Yatsushiro Bay. Beginning in 1953 a strange illness affected the people of Minamata—they became tired and irritable, complained of headaches, numbness in arms and legs and difficulty in swallowing and began to experience blurring of vision, loss of hearing and muscular coordination, inflammation of gums, and a metallic taste in their mouths. Since the cause of these symptoms was not evident they came to be known as *Minamata disease*.

---

[3] G. L. Waldbott, "Health Effects of Environmental Pollutants," p. 147. Mosby, St. Louis, Missouri, 1973.

By 1963 it had been determined that the "disease" was caused by mercury (at levels between 27 and 102 ppm dry weight) in the staple food of the community—fish taken from the bay. Although Chisso eventually controlled its mercury releases, 43 persons had died from Minamata disease by 1960 and numerous babies with congenital defects had been born. By 1975 some Japanese doctors estimated that as many as 10 000 persons had been affected, 703 seriously and permanently maimed and more than 100 killed by Minimata disease. Tragically, other similar episodes have occurred more recently in the Japanese cities of Niigata and Ariake.

Bioamplification

An important aspect of Minamata disease is the extent to which mercury became concentrated in fish relative to the waters of the bay. Certain substances, of which chloromethylmercury(II) is an excellent example, tend to concentrate along *ecological food chains,* a process known as *bioamplification.* Any substance which is slightly more soluble in the tissues of simple organisms than in the surrounding water will be a more concentrated component of the diet of the more complex species which feed on the simple ones. Thus, at each higher *tropic level* in the food chain the substance becomes more concentrated (Figure 16.2). Since humans are usually at the tops of food chains, high concentrations of poisonous substances may occur in human diets, with tragic results such as Minamata disease.

Molecules which can be concentrated in biological organisms are usually hydrophobic, but not extremely so (see Section 17.1), a class into which chloromethylmercury(II) falls almost exactly. The common name methylmercuric chloride (and numerous statements in the environmental literature) might lead one to think that in aqueous solution the equilibrium

$$CH_3-Hg-Cl \rightleftharpoons CH_3-Hg^+ + Cl^- \tag{16.2}$$

lies to the right, but the equilibrium constant[4] is $10^{-5.45}$, indicating that, like $HgCl_2$, $CH_3-Hg-Cl$ is very weakly dissociated. The uncharged species $CH_3-Hg-Cl$ is apparently readily transported across body membranes and collects in fatty tissues, from which there are few mechanisms for expelling it. The absorption of mercury is also favored at low pH since the $CH_3-Hg-OH$ which forms when pH exceeds 5 is much less soluble in organic phases and more soluble in water than $CH_3-Hg-Cl$.[5] This, rather than the insolubility of $CH_3-Hg-CH_3$ formed at high pH, seems capable of accounting for the fact that fish in neutral or basic waters usually contain smaller quantities of mercury.

[4] R. B. Simpson, *J. Amer. Chem. Soc.* **83**, 4711–4717 (1961).
[5] W. A. Willford, Great Lakes Fishery Laboratory, Ann Arbor, Michigan (private communication).

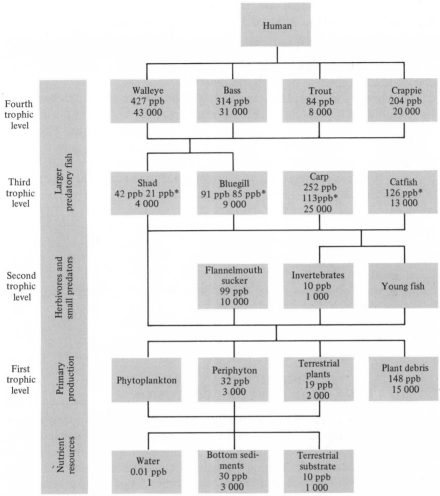

Figure 16.2 Bioamplification of mercury in Lake Powell, Utah and Arizona. The mean level of mercury in parts per billion (ppb) and the amplification factor above the surrounding water are given for each species. Thus the Walleye contains a concentration of mercury 43 000 times that of the water. Data from L. Potter, D. Kidd, and D. Standiford, *Environ. Sci. Technol.* **9**(1), 41–46 (1975). * Based on whole body analyses. Other values are for axial muscle.

The Mechanisms of Mercury Poisoning

Once in the human body organomercurials and divalent mercury both react with a large number of enzymes, usually inhibiting catalysis of essential metabolic reactions.[6] This apparently results from the strong bonds

---

[6] J. L. Webb, "Enzyme and Metabolic Inhibitors," Vol. 2, Chapter 7. Academic Press, New York, 1965.

formed between the large, soft Hg^{2+} ion and sulfhydryl (—SH) groups on the amino acid cysteine (Figure 3.2.A.6) in the enzymes:

$$CH_3—Hg—Cl + enzyme—SH \rightleftharpoons CH_3—Hg—S—enzyme + H^+ + Cl^- \quad (16.3)$$

The mercury-containing group can disrupt the enzyme's structure,[7] consequently impeding its catalytic function.

Methylmercuric chloride is especially poisonous because it can penetrate the membranes separating the bloodstream from the brain and those in the placenta through which nutrients must pass to an unborn fetus. The latter was apparently the cause of congenital birth defects in the Minamata disaster. Mercury poisoning is treated by having the victim ingest British antilewisite (BAL), calcium ethylenediaminetetraacetate (Ca_2EDTA), or some other strong chelating agent which can remove H^{2+} in a coordination complex.

$$
\begin{array}{c}
H_2C—CH—CH_2 \\
|\quad\ |\quad\ \ | \\
OH\ \ SH\ \ \ SH
\end{array}
$$

2,3-Mercaptopropanol
(BAL)

Anion of EDTA

This is similar to methods for treating lead poisoning (Section 10.4).

16.4   DETECTION AND ABATEMENT OF MERCURY POLLUTION

In the late 1960's analyses of mercury levels in fish—first in Canada and then in the United States—revealed that in many cases they exceeded the U. S. Public Health Service standard of 0.5 ppm for human foods. [It has been shown that cats fed nothing but fish which contained between 0.3 and 0.5 ppm of mercury developed mercury poisoning.[8] For a 70-kg human the minimum dosage at which symptoms of mercury poisoning appear is 0.3 mg/day. This corresponds to 600 g (1.3 lb) of fish containing 0.5 ppm of mercury each day—a quantity which might be approached by persons on high protein diets.] Most measurements of mercury levels employ either atomic absorption spectroscopy, neutron activation analysis, or gas chromatography.[9] The former has been selected by the U. S. Public Health Ser-

---

[7] See your general chemistry text or W. L. Masterton and E. J. Slowinski ("Chemical Principles," 3rd ed., Chapters 14 and 23. Saunders, Philadelphia, Pennsylvania, 1973), for a discussion of enzyme structure and its relationship to catalysis.

[8] *Environment* **15**(6) 23 (1973).

[9] For a brief description of these techniques, see D. A. Skoog and D. M. West, "Principles of Instrumental Analysis," Chapter 5, Atomic absorption; Chapter 24, Gas chromatography. Holt, New York, 1971. Also see B. Keisch, "The Atomic Fingerprint: Neutron Activation Analysis," United States Atomic Energy Commission Information Booklet WAS 303. USAEC, Washington, D. C., 1972.

vice as a standard technique, although it appears that some losses of volatile organomercurials are possible. Neutron activation analysis produces consistently higher mercury analyses and is apparently more accurate, but it requires that the sample be irradiated in a nuclear reactor. Neither method distinguishes between organomercurials and inorganic mercury, although the difference in their toxicities would make such a distinction useful. The latter distinction can be made by careful use of gas chromatography,[10] but the technique has not been applied as a standard method.

Chlor-Alkali Production

In the United States high concentrations of mercury in freshwater fish were almost invariably associated with chemical plants for the manufacture of dichlorine, sodium hydroxide, and dihydrogen. In 1968 approximately 25% of United States dichlorine (and much higher percentages in countries such as Germany and Italy) was produced in *mercury cells,* which operated as outlined schematically in Figure 16.3. In such cells nearly saturated brines were electrolyzed to produce gaseous Cl_2 and H_2, leaving NaOH behind in solution. In order to prevent Cl_2 and H_2 from recombining explosively the former gas was produced at a carbon anode in an *electrolyzer cell* which used liquid mercury as the cathode. Because *sodium amalgam* (a solution of sodium in mercury) is relatively stable it formed at the cathode instead of gaseous H_2:

$$\text{C anode:} \qquad Cl^-(aq) \longrightarrow \tfrac{1}{2} Cl_2(g) + e^- \qquad (16.4a)$$

$$\text{Hg cathode:} \quad Na^+(aq) + e^- + x\,Hg \longrightarrow \underset{\text{(Sodium amalgam)}}{Na(Hg)_x} \qquad (16.4b)$$

$$x\,Hg + Na^+(aq) + Cl^-(aq) \longrightarrow Na(Hg)_x + \tfrac{1}{2} Cl_2(g) \qquad (16.4)$$

The liquid amalgam was then allowed to flow into a *decomposer* where it was made the anode. The sodium was oxidized back to Na^+ and dihydrogen was released at a carbon cathode:

$$\text{Na(Hg) anode:} \quad Na(Hg)_x \longrightarrow Na^+(aq) + e^- + x\,Hg \qquad (16.5a)$$

$$\text{C cathode:} \quad H_2O + e^- \longrightarrow H_2(g) + OH^-(aq) \qquad (16.5b)$$

$$Na(Hg)_x + H_2O \longrightarrow Na^+(aq) + OH^-(aq) + H_2(g) + x\,Hg \quad (16.5)$$

The mercury (except for a small fraction) was then recycled to the electrolyzer and concentrated NaOH could be removed for sale.

---

[10] G. Westöö and M. Rydälv, *Var Foeda* **23**, 183 (1971); G. Westöö, *Acta Chem. Scand.* **22**, 2277 (1968).

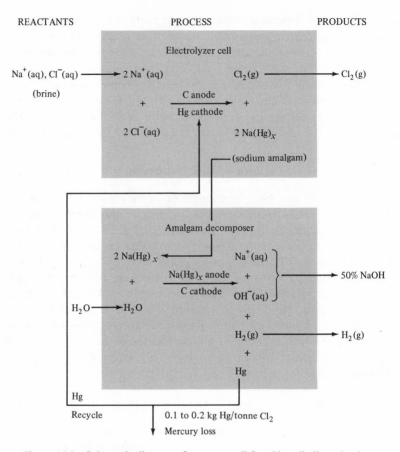

Figure 16.3   Schematic diagram of mercury cell for chlor-alkali production.

The 0.1–0.2 kg of mercury lost per tonne of Cl_2 seems a small quantity until one takes account of the fact that nearly 2500 tonnes of Cl_2 were produced in mercury cells each day during 1960. Thus, every 2–4 days a tonne of mercury was lost to the environment. Because elemental mercury is relatively inert and insoluble, most persons concerned with these losses assumed that it would settle rapidly to the bottoms of rivers and lakes into which it was discharged and remain there. In fact most of it did just that, but between 1965 and 1970 chemists in Sweden and then in the United States discovered that certain anaerobic bacteria (*Methanobacterium amelanskis*) were capable of methylating and oxidizing elemental mercury, converting approximately 1% of what was in the sediments to CH_3—Hg—Cl, CH_3—Hg—OH, and CH_3—Hg—CH_3. This permitted mercury to enter the freshwater food chain and accounted for the high concentrations observed in fish.

Abatement Procedures

Once the role of mercury cell chlor-alkali production in freshwater mercury pollution was recognized, steps to control discharges were taken rapidly. Between 1969 and 1972 there was an 85% reduction in the quantity of mercury released. Many existing mercury cells have been equipped with settling ponds or liquid cyclone collectors which separate the metal on the basis of its density. In some cases this additional recycling of mercury has proved economically as well as environmentally beneficial.

The most important abatement procedure, however, has been conversion from mercury cells to *diaphragm cells* which employ a porous asbestos diaphragm to separate anode and cathode. This allows sodium ions to diffuse toward the cathode but reduces migration of hydroxide ions back toward the anode where they would react with Cl_2 or be discharged:

$$Cl_2 + OH^- \longrightarrow HOCl + Cl^- \qquad (16.6a)$$

$$OH^- \longrightarrow \tfrac{1}{2} H_2O + \tfrac{1}{2} O_2 + e^- \qquad (16.6b)$$

Besides zero discharge of mercury the diaphragm cell has the advantage of requiring 10% less free energy (10.6×10^9 J/tonne Cl_2 versus 11.9×10^9 J/tonne) than a typical mercury cell.[11] The recent introduction of dimensionally stable anodes[12] in diaphragm cells has resulted in additional energy savings.

Unfortunately, the prevention of further mercury discharges is only a small portion of the abatement problem. Ninety-nine percent of all mercury discharged is still present in bottom sediments downstream from chlor-alkali plants. In some cases concentrations as high as 500–1000 ppm have been detected. If bacteria continue to convert these mercury deposits to methylmercurials, it may take several hundred years until they are depleted. In the meantime, fish will continue to accumulate large concentrations of mercury and humans will have to be careful of how much fish they consume. Attempting to dredge up the mercury-containing sediments apparently makes the problem worse instead of better because it stirs up the bottom of the stream and can actually increase the mercury concentrations, at least temporarily. When this method was tried in Minamata the mercury content of fish was found to increase.

Perhaps the best solution for the time being is to cover the concentrated mercury deposits to prevent their oxidation. A layer of fluorspar (CaF_2) from 2 to 3 cm thick can reduce production of methylmercurials by about 80%, provided that additional quantities of mercury are not deposited on top of it. Advanced wastewater treatment applied to domestic sewage can also contribute to the reduction of mercury pollution. Since nitrate and

---

[11] R. N. Shreve, "Chemical Process Industries," 3rd ed., pp. 231–240. McGraw-Hill, New York, 1967.
[12] *Chem. & Eng. News* Nov. 9, p. 32 (1970).

phosphate ions are needed to carry out anaerobic oxidation, reduction of their concentrations limits the populations of anaerobic bacteria and they cannot produce methylmercurials. Because of the high concentrations of mercury in areas adjacent to chlor-alkali plants, it may eventually be possible to "mine" such deposits. All that needs to be found is a method of preventing the mining process from producing a temporary increase in mercury levels in the surrounding waters.

16.5 PAPER MILLS

Since paper mills produce by far the largest total quantities of wastes listed in Table 14.3, it is essential that we consider the chemical processes which lead to such emissions. The basic raw material is cellulose – the most abundant organic substance in nature and a versatile, rapidly renewable source of reduced carbon. The chief object of *pulping* is to break down the solid structure of wood into a suspension of long fibers which can be refined and dried in the *papermaking* process. Pulping can be accomplished by a simple grinding process (*groundwood pulp*), but more than 90% requires some chemical treatment. In 1966 more than 60% of United States pulping was done by the *kraft* (alkaline sulfate) process and nearly 20% by the *acid* or *neutral sulfite* treatment.[13] In 1970 more than 250 kg of paper and paperboard products were consumed for each person in the United States.[14]

Kraft Pulping

The term *kraft* is derived from the German word meaning "strong," and the process was developed in that country in the late nineteenth century. It employs basic hydrolysis in the presence of sulfide ions to remove lignin (Section 5.1) as well as oils and resins from wood, and is especially suited to coniferous trees. Although precise equations cannot be written, the process may be summarized as

$$\text{Lignin} + \text{Na}_2\text{S} + \text{NaOH} \longrightarrow \begin{array}{l}\text{alcohol and acid hydrolysis products,}\\ \text{CH}_3\text{—S—CH}_3, \text{CH}_3\text{—SH, etc.}\\ +\\ \text{Na}_2\text{SO}_4\end{array} \qquad (16.7)$$

The cellulose fibers which remain are the longest of all pulps, accounting for the strength of the paper produced, but the pulp is brown and difficult to

---

[13] H. Gehm, "State-of-the-Art Review of Pulp and Paper Waste Treatment," p. 226. U. S. Environmental Protection Agency, US Govt. Printing Office, Washington, D. C., 1973.
[14] H. Gehm, "State-of-the-Art Review of Pulp and Paper Waste Treatment," p. 230. U. S. Environmental Protection Agency, US Govt. Printing Office, Washington, D. C., 1973.

bleach. The mercaptans and sulfides produced by reaction 16.7 cause air pollution problems because of their foul odor.

The result of 3–5 hours digestion is so-called *black liquor*. This contains hydrolyzed lignin and most of the chemicals originally added, and the desired cellulose fiber can be separated from it by screening. An important factor in the process is recovery and recycle of black liquor. This is done by concentrating it and burning in a furnace to remove organic compounds and regenerate Na_2S:

$$Na_2SO_4 + 2\,C \xrightarrow{\Delta} Na_2S + 2\,CO_2 \qquad (16.8a)$$

$$2\,NaOH + CO_2 \longrightarrow Na_2CO_3 + H_2O \qquad (16.8b)$$

Sodium hydroxide is then regenerated by addition of slaked lime:

$$Na_2CO_3 + Ca(OH)_2 \longrightarrow 2\,NaOH + \underline{CaCO_3} \qquad (16.9)$$

and the black liquor (which has now been converted to white liquor) is recycled to the digestor.

The cellulose pulp must be washed, an operation which may release diluted black liquor to the environment. It is usually bleached using chlorine dioxide produced from dichlorine and sodium chlorite:

$$NaClO_2 + \tfrac{1}{2}\,Cl_2 \longrightarrow NaCl + ClO_2 \qquad (16.10)$$

This oxidizes lignin, tannin, and other organic compounds responsible for the brown color. The pulp can then be converted into paperboard, brown bags and wrapping paper, building paper, and strong white papers.

Sulfite Pulping

Most high-quality writing and printing papers are produced by the acid sulfite process. Calcium and magnesium bisulfite, produced by combination of SO_2 with the respective carbonates, are used to sulfonate double bonds in lignin:

$$2\,RC{=}CR' + Ca(HSO_3)_2 \longrightarrow RCH{-}C\!\!\begin{smallmatrix} R' \\ \diagup \\ \diagdown \\ SO_3^- \end{smallmatrix}\ \ Ca^{2+}\ \ {}^-O_3S\!\!\begin{smallmatrix} R' \\ \diagdown \\ \diagup \end{smallmatrix}\!\!C{-}CHR \qquad (16.11)$$

The process requires 6–12 hours at temperatures from 400 to 475 K and nearly seven times normal atmospheric pressure. As in the kraft process pulp is filtered following digestion. The filtrate is evaporated to dryness and burned to recover sulfur dioxide, magnesium oxide, and heat which generates steam for use in the process:

$$Mg(HSO_3)_2 \longrightarrow MgO + H_2O + SO_2 \qquad (16.12a)$$

$$\underset{\substack{\text{(Organic} \\ \text{residues)}}}{C}\ \ + O_2 \longrightarrow CO_2 \qquad (16.12b)$$

Although MgO can be slaked and recycled:

$$MgO + H_2O \longrightarrow Mg(OH)_2 \qquad (16.13a)$$

$$Mg(OH)_2 + 2\ SO_2 \longrightarrow Mg(HSO_3)_2 \qquad (16.13b)$$

this process is more difficult with the less soluble calcium salts. Until stream pollution standards were imposed most calcium-containing liquors were simply dumped.

Papermaking

Once lignin and other material have been removed and the pulp has been bleached, it is beaten or refined and various fillers, sizings, coloring, and wet-strength resins are added to produce appropriate properties for a variety of end uses; for example, all papers except those intended to absorb water contain fillers such as talc, china clay, or titanium dioxide. Then papermaker's alum $[Al_2(SO_4)_3 \cdot 18\ H_2O]$ is added as a coagulant and the pulp is passed through a *Fourdrinier machine*, which forms it into a continuous sheet and removes water by suction, pressure, and heating. By the time this final product has been attained a large quantity of a variety of different chemicals (Table 16.4) has been consumed and many opportunities for their release into both air and water have occurred.

16.6   ABATEMENT OF PAPER MILL POLLUTION

A large portion of the wastewater problem in paper mills is associated with oxygen demand. Therefore the same techniques employed in primary and secondary municipal waste treatment are often used. Effluents are screened, their acid or base content is neutralized, and suspended solids may then be separated by sedimentation or flotation. In 1971 more than 63% of the 118 kraft mills in the United States were equipped with mechanical clarifiers and 18% more had settling basins. The remainder were not controlled. Of the 77 sulfite pulping operations only a little over half had clarifiers.

Biochemical oxygen demand can be removed almost completely by biological oxidation. This is often done in shallow, large-volume lagoons, but these are subject to odor problems. Experiments begun with activated sludge secondary treatment in 1950 had been scaled up successfully on only a dozen mills by 1970, although some others were artificially aerating holding lagoons. Nevertheless, biological treatment is quite successful in removing BOD and will certainly be installed on more plants in the future. The chief substances remaining after such secondary treatment are refractory organic compounds (those which account for the much higher COD than BOD for paper mills in Table 14.3). Some of these contribute to objectionable color in the effluent.

TABLE 16.4 Estimated Consumption of Chemicals and Free Energy in Pulp and Paper Manufacturing (1960)[a]

Chemical	Consumption (Gg)
Pulp manufacture (all types)	
Wood	46 800 to 468 000
Salt cake (Na_2SO_4)	839
Limestone ($CaCO_3$)	173
Lime (CaO)	404
Sulfur	432
Soda ash (Na_2CO_3)	354
Ammonia	56
Magnesium hydroxide	43
Pulp bleaching	
Chlorine	331
Caustic soda (NaOH)	172
Lime	134
Sodium and calcium hypochlorite	128
Sulfuric acid	51
Sodium chlorate	31
Papermaking	
China clay ($Al_2O_3 \cdot 2\ SiO_2 \cdot 2\ H_2O$)	850
Starches	525
Alum [$Al_2(SO_4)_3 \cdot 18\ H_2O$]	382
Rosin	136
Waxes	86
Titanium dioxide	48
Wet-strength polymer resins	19
Slimicides (Cl_2, mercurials, etc.)	9
Energy	
Electric power for kraft pulping (GJ/tonne)	1.0^b
Electric power for sulfite pulping (GJ/tonne)	$11-16^b$
Steam for kraft pulping (GJ/tonne)	12^c
Steam for sulfite pulping (GJ/tonne)	2^c

[a] Total production of paper and paperboard products in 1960 was 31.2 Tg. Data from R. N. Shreve, "Chemical Process Industries," p. 632. McGraw-Hill, New York, 1967.

[b] Electric—multiply by approximately 3 to obtain primary energy consumed.

[c] Much steam is generated by burning in-plant wastes; see text.

Alternate Pulping Process

In part because of incomplete cleanup of water effluents but largely because of air pollution regulations, the paper industry is currently considering alternate methods for delignification of wood and bleaching of pulp. One of the most promising approaches involves dioxygen, which may also be used for treatment of wastes and black liquor. One oxygen bleaching plant is currently operating in the United States,[15] and oxygen pulping processes are under development.[16] The latter eliminate odor problems, reduce water requirements for bleaching, and substantially increase pulp yields. They also require less power for refining into paper. Due to uncertainties regarding costs they have yet to be widely adopted, although a 450-tonne/day plant is under construction by Weyerhaeuser Corporation in Everett, Washington.

Several other sulfur-free pulping processes have been proposed,[17] but none has been developed as far as those involving dioxygen. Reduction in pulp mill pollution could also be accomplished by increased recycling of paper (Section 13.3), since de-inking and repulping require less chemical treatment and produce smaller quantities of waste. Most paper recycling plants currently send their wastes to municipal sewage treatment plants.

EXERCISES

(The first group of exercises is based primarily on material found in this text.)

1. Explain the chemistry involved in at least two techniques for removal of metal ions from industrial effluents.
2. If it was desired to recover a metal in high purity form from industrial waste, what method would probably be chosen? Why?
3. Describe the operation of a chlor-alkali plant using mercury cells. Give balanced equations for the reactions involved.
4. How can mercury losses from chlor-alkali plants be prevented? Why is this not a complete solution to mercury pollution problems?
5. Explain why chloromethylmercury(II) is more damaging to human health than other mercury compounds. How is it produced from elemental mercury in the environment?
6. Describe the two main processes for pulping wood to form paper. Include chemical equations where possible. What are some of the problems in recycling of cellulose fiber?
7. What techniques are usually used for abatement of paper mill discharges?
8. Based on the discussion in Section 16.3, account for each of the following facts:
   a. Organomercurials are incorporated into fish more readily in salt water than in fresh water.

---

[15] *Chem. & Eng. News* July 22, pp. 15–16 (1974).

[16] *Pulp & Pap.* October, 120–121 (1973); A. W. J. Dyck, *Amer. Pap. Ind.* October, 20–21 (1973).

[17] W. H. M. Schweers, *Chem. Technol.* 4(8), 490–493 (1974), and references therein.

b. Fish in water at pH = 7 or above are found to contain less mercury than in acidic waters.

(Subsequent exercises may require more extensive research or thought.)

9. Given the following dissociation constants:

$$CH_3\text{—}Hg\text{—}Cl \qquad K = 10^{-5.45}$$
$$CH_3\text{—}Hg\text{—}OH \qquad K = 10^{-9.5}$$

Calculate the pH at which equal concentrations of the chloro and hydroxo mercury species would exist. (Assume that the chloride ion concentration is 1 ppm. Then repeat the calculation assuming it is exactly at the permissible criterion of 250 ppm given in Table 14.4.)

10. What correlation might there be between use of NaCl and $CaCl_2$ for highway deicing and mercury content of fish? [see *Science* **172,** 1128–1131 (1971); **175,** 1142–1143 (1972); **176,** 288–290 (1972)].

Suggested Readings

1. H. E. Allen and James R. Kramer, eds., "Nutrients in Natural Waters." Wiley (Interscience), New York, 1972. A collection of chapters by different authors, most of which were presented at an ACS symposium in 1971, this book concentrates on nutrients and eutrophication, especially with respect to the Great Lakes.
2. American Public Health Association, American Water Works Association, Water Pollution Control Federation, "Standard Methods for the Examination of Water and Wastewater," 13th ed., American Public Health Association, Washington, D. C., 1971.
3. American Water Works Association, Inc., "Water Quality and Treatment," 3rd ed., McGraw-Hill, New York, 1971. The field is fairly completely covered by 28 authors in 19 chapters.
4. F. Bakir *et al.,* Methylmercury poisoning in Iraq. *Science* **181,** 230–241 (1973). A report by scientists from the Universities of Baghdad and Rochester on an epidemic of mercury poisoning which occurred in 1972.
5. G. C. Berg, "Water Pollution." Scientists' Institute for Public Information, New York, 1970. A brief, general survey.
6. L. J. Carter, Galveston Bay: Test case of an estuary in crisis. *Science* **167,** 1102–1108 (1970). A good example of problems which can occur in the biologically essential area where fresh and salt water meet.
7. *Chem. & Eng. News* June 22, pp. 36–37 (1970). An early description of mercury discharges and contamination of fish in the United States.
8. *Chem. & Eng. News,* Water pollution controls to cost a bundle. Oct. 15, p. 13 (1973); EPA sees no economic blocks to clean water. Feb. 4, p. 16 (1974). Reports on economic impacts of water pollution control.
9. G. V. Cox, Industrial waste effluent monitoring. *Amer. Lab.* July, pp. 36–40 (1974). This paper summarizes the methods of wastewater analysis which meet Federal standards. The data, presented in table form, contain references to the literature for each parameter.
10. J. Crossland and V. Brodine, Drinking water. *Environment* **15**(3), 11–19 (1973). A survey of problems relating to contamination of drinking water supplies.
11. R. L. Culp and G. L. Culp, "Advanced Wastewater Treatment." Van Nostrand-Reinhold, New York, 1971. An all-inclusive and fairly detailed treatment.
12. S. L. Daniels and D. G. Parker, Removing phosphorus from waste water. *Environ. Sci. Technol.* **7**(8), 690–694 (1973); C. E. Adams, Jr., Removing nitrogen from waste water. *ibid.* pp. 696–701. This pair of articles surveys methods for removal of the two elements most often implicated in accelerated eutrophication.
13. J. G. Dean, F. L. Bosqui, and K. H. Lanouette, Removing heavy metals from waste water. *Environ. Sci. Technol.* **6**(6), 518–522 (1972). Surveys techniques for treatment of water containing heavy metals.
14. P. R. Dugan, "Biochemical Ecology of Water Pollution." Plenum, New York, 1972. A

microbiologist's view of water pollution, in terms which can be understood by any scientist.

15. R. C. Eckert, H.-m. Chang, and W. P. Tucker, Oxidative degradation of phenolic lignin model compounds with oxygen and alkali. *Tappi* **56**(6), 134–138 (1973). This article in the journal of the *Technical Association of the Pulp and Paper Industry* describes mechanisms of degradation of model lignin structures in oxygen pulping.

16. *Environ. Sci. Technol.,* Are you drinking biorefractories too? **7**(1), 14–15 (1973). Problems with organic compounds in drinking water along the lower Mississippi River.

17. *Environ. Sci. Technol.,* Cleaning up: Paper industry's mess. **8**(1), 22–24 (1974). Concludes that some returns on investments in pollution control equipment are possible if proper plans are made.

18. *Environ. Sci. Technol.,* Ozonation seen coming of age. **8**(2), 108–109 (1974). Predicts a rosy future for ozonation in water and sewage treatment.

19. *Environ. Sci. Technol.* **8**(10), October 1974. This special issue devoted to water pollution contains much useful information.

20. *Environ. Sci. Technol.,* Changes are in store for pulping technology. **9**(1), 20–21 (1975). A qualitative description of nonsulfur pulping techniques.

21. G. M. Fair, J. C. Geyer, and D. A. Okun, "Elements of Water Supply and Wastewater Disposal," 2nd ed., Wiley, New York, 1971. A detailed treatment of all aspects of the subject.

22. J. Gavis and J. F. Ferguson, The cycling of mercury through the environment. *Water Res., 6,* 989–1008 (1972). A thorough review of what is known about cycling of mercury in local and global environments.

23. H. Gehm, "State-of-the-Art Review of Pulp and Paper Waste Treatment." U. S. Environmental Protection Agency, US Govt. Printing Office, Washington, D. C., 1973. A somewhat industry-oriented view of paper mill waste treatment, but contains much information.

24. L. J. Goldwater, Mercury in the environment. *Sci. Amer.* **224**(5), 15–21 (1971). Argues against a "panicky reaction" to episodes of mercury poisoning. Also presents an excellent photograph of chromosomes damaged by high blood levels of mercury.

25. R. D. Grundy, Strategies for control of man-made eutrophication. *Environ. Sci. Technol.* **5**(12), 1184–1190 (1971). Reports on the status of scientific and legislative attempts at control of eutrophication. Concludes that long-term solutions will have to go beyond control of detergents or even phosphates alone.

26. D. E. Gushee, Clean water: What is it? How will we achieve it? *Chem. Technol.* June, pp. 334–344 (1973). An analysis of scientific and political aspects of passage of the Federal Water Pollution Control Act Amendments of 1972, Public Law 92-500.

27. I. R. Higgins, Ion exchange: Its present and future use. *Environ. Sci. Technol.* **7**(13), 1110–1114 (1973). A description of uses of ion exchange in water purification.

28. L. Hodges, "Environmental Pollution." Holt, New York, 1973. Chapters 8–11 are concerned with water pollution.

29. G. P. Howells, T. J. Kneipe, and M. Eisenbud, Water quality in industrial areas: Profile of a river. *Environ. Sci. Technol.* **4**(1), 26–35 (1970). A thorough report on the biological status of a major industrial river—the Hudson—which has been carefully studied.

30. J. H. Hubschman, "Lake Erie: Pollution abatement, then what? *Science* **171**, 536–540 (1971). A proposal that excess productivity resulting from accelerated eutrophication be harvested for human use.

31. G. E. Hutchinson, Eutrophication. *Amer. Sci.* **61**(3), 269–279 (1973). A thorough treatment of the scientific background of the problem.

32. G. F. Lee, Role of phosphorus in eutrophication and diffuse source control. *Water Res.* **7**, 111–128 (1973). Argues that attempting to control phosphorus inputs to lakes is a sound approach to controlling accelerated eutrophication, but that a better understanding of the relationship of phosphorus input and plant growth is necessary.

33. L. B. Luttinger and G. Hoché, Reserve osmosis treatment with predicted water quality. *Environ. Sci. Technol.* **8**(7), 614–618 (1974). Describes applications of ultrafiltration in water purification.

34. J. McCaull and J. Crossland, "Water Pollution." Harcourt, New York, 1974. Written at a popular level by two members of the staff of *Environment*.

35. S. E. Manahan, "Environmental Chemistry." Willard Grant Press, Boston, Massachusetts, 1972. Manahan, an analytical chemist, discusses water chemistry in somewhat greater detail than is done here, but at a somewhat lower level than Stumm and Morgan.

36. National Academy of Sciences, "Eutrophication: Causes, Consequences, Correctives." Nat. Acad. Sci., Washington, D. C., 1969. These proceedings of a symposium held in 1967 contain 33 articles by different authors. The introduction, summary, and recommendations are quite useful.

37. Wayne A. Pettyjohn, "Water Quality in a Stressed Environment." Burgess, Minneapolis, Minnesota, 1972. A collection of readings dealing with ground water and surface water contamination.

38. F. G. Pohland, "Anaerobic Biological Treatment Processes," Advan. Chem. Ser. No. 105, Amer. Chem. Soc., Washington, D. C., 1971. Proceedings of a symposium.

39. R. N. Shreve, "Chemical Process Industries." McGraw-Hill, New York, 1967. Chapters 13 and 33 are devoted to chlor-alkali and pulp and paper industries, respectively.

40. L. G. Sillén, The ocean as a chemical system. *Science* **156**, 1189–1197 (1967). An expert on the equilibria of ionic solutions considers the largest example in the world.

41. H. S. Stoker and S. L. Seager, "Environmental Chemistry: Air and Water Pollution." Scott-Foresman, Glenview, Illinois, 1972. Chapters 8–14 are devoted to water pollution.

42. W. Stumm and J. J. Morgan, "Aquatic Chemistry." Wiley (Interscience), New York, 1970. Detailed descriptions of chemical equilibria in natural environments.

43. G. Sykes and F. A. Skinner, eds., "Microbial Aspects of Pollution." Soc. Appl. Bacteriol. Symp. Ser. No. 1. Academic Press, New York, 1971. A wide variety of papers relating to the role of microorganisms in water pollution.

44. U. S. Environmental Protection Agency, "Annual Report on Fish Kills by Pollution." US Govt. Printing Office, Washington, D. C. An indication (though not infallible) of the extent of pollution of natural waters.

45. U. S. Environmental Protection Agency, "Water Quality Criteria Data Book," Vol. 2, US Govt. Printing Office, Washington, D. C., 1971. Results of a review of more than 5000 publications on inorganic chemicals in water with regard to toxicity, carcinogenicity, mutagenicity, and teratogenicity.

46. U. S. Environmental Protection Agency, "The Economics of Clean Water," 3 vols. US Govt. Printing Office, Washington, D. C., 1972. This report details the investment levels needed to meet water quality objectives by both municipalities and industries.

47. U. S. Environmental Protection Agency, "Atmospheric Emissions from the Pulp and Paper Manufacturing Industry." USEPA, Research Triangle Park, North Carolina, 1973. Information on the nature and quantities of air emissions from chemical pulping, especially the kraft process.

48. U. S. Environmental Protection Agency, "National Water Quality Inventory," 2 vols. US Govt. Printing Office, Washington, D. C., 1974. The first systematic survey of United States water quality. Also sets goals for 1977–1983.

49. U. S. Environmental Protection Agency, Interim primary drinking water standards. *Fed. Regis.* **40**(51), 11990–11998 (1975).

50. G. L. Waldbott, "Health Effects of Environmental Pollutants." Mosby, St. Louis, Missouri, 1973. Chapter 12 deals with poisoning by lead and mercury.

51. R. A. Wallace, W. Fulkerson, W. D. Shults, and W. S. Lyon, "Mercury in the Environment: The Human Element," ORNL NSF-EP-1. Oak Ridge Nat. Lab., Oak Ridge, Tennessee, 1971. This report contains, in its own words, "All you may ever want to know about mercury," although a number of things have been learned since its publication.

52. J. L. Webb, "Enzyme and Metabolic Inhibitors," Vol. 2. Academic Press, New York, 1965. Chapter 7 gives very thorough coverage of the chemical properties and health effects of mercury compounds.

53. Walter J. Weber, Jr., "Physicochemical Processes for Water Quality Control." Wiley (Interscience), New York, 1972. Chemical processes are emphasized, whether they pertain to natural waters, water supplies, or wastewater. The author holds that artificial distinctions among these classes are becoming less relevant as multiple usage of water becomes commonplace.

54. J. M. Wood, Environmental pollution by mercury. *Advan. Environ. Sci. Technol.* **2,** 39–56 (1971). This general survey pays special attention to the methylation of inorganic mercury.

55. J. M. Wood, Biological cycles for toxic elements in the environment. *Science* **183,** 1049–1052 (1974). Considerable space devoted to methylation of mercury. Concludes that care must be taken in deciding what compounds of a toxic element are to be monitored if its full environmental impact is to be assessed.

VI
Life

A thing is right only when it tends to preserve the integrity of the community, and the community includes the soil, water, fauna, and flora, as well as the people.

Aldo Leopold

A NUMBER OF PARALLELS MAY BE DRAWN BEtween the biosphere as a whole and any particular organism (such as *Homo sapiens*) within it. Each has evolved over a relatively long period of time and is dependent on complex interactions among its constituent parts. Each is a highly organized, nonequilibrium system which requires a nearly continuous flow of free energy from its surroundings in order to maintain an improbable, low entropy state. Finally, each has natural mechanisms which attempt to cleanse it of substances which are present in excess of the necessary quantities and to synthesize those which are absent.

These parallels emphasize the fact that Leopold's statement is not based merely on altruism, but rather reflects enlightened self-interest on the part of the human race. Substances which tend to disrupt the natural environment are often capable of destroying *homeostasis* — the regulation and buffering of rapid, large changes — within the human body. Acute or chronic illness may then follow. The extinction of certain species from the biosphere is not simply to be regretted and subsequently forgotten. In many cases it serves as an indicator that something is wrong — something which requires human study to determine its cause and may necessitate curtailment of certain human activities. An excellent example is the observation of eggshell thinning in predatory birds, which led to the curtailment of use of persistent pesticides.

There is another, equally valid argument for the preservation of diversity of species implied in Leopold's words. The discussion in Section 4.9 indicated that a steady-state, nonequilibrium system should exhibit some maximum in the degree of order to be expected as a function of energy flow, and in Section 7.5 the thesis that the optimal energy flow for the earth was not too different from current solar input was supported. Attempts at increasing the order of the biosphere by decreasing the number of species present (as is being done on a large scale by modern, mechanized agriculture) require constant inputs of free energy. They could conceivably tip the system so far in one direction that the negative feedback mechanisms which normally maintain homeostasis may be overcome, resulting in instability and catastrophic change.

The chapters which follow will examine the mechanisms by which the human body protects itself from toxic substances as well as some of the ways in which sufficiently large quantities of such materials can affect the body. Many of the most dangerous ones are the products of chemists' ingenuity, some inadvertently and some by design. In either case social mechanisms by which their release can be controlled are necessary. In addition, the steps which seem to be required in order to feed a constantly growing human population will be examined. Some of them — the dispersal of large quantities of pesticides is a good example — may be detrimental to the environment, and thus, eventually, to humans. Finally, it will be shown that agriculture, which most persons think of as a means of trapping solar energy, has developed into a major consumer of fossil fuels.

17

Toxic Substances

Man is but a reed, the most feeble thing in nature, but he is a thinking reed. The entire universe need not arm itself to crush him. A vapor or a drop of water suffices to kill him. But if the universe were to crush him man would still be more noble than that which killed him, because he knows that he dies, and the advantage which the universe has over him: The universe knows nothing of this.

Pascal

In Chapter 3 the origin and evolution of the molecules which make up living systems were summarized and the reader was referred to other sources[1] for details regarding their structures and functions. With the exception of the lipids, which will be described in the next section, the assumption of familiarity with elementary principles of biochemistry will again be made in this chapter. If more detailed information on specific topics is desired it may be found in several of the suggested readings at the end of Part VI.

17.1 LIPIDS

Lipids are water-insoluble substances which may be extracted from cells by organic solvents such as chloroform, ether or benzene. They serve four general functions: (1) as structural components of membranes, (2) as intracellular storage sites for metabolic free energy, (3) as substances by which metabolic fuel may be transported, and (4) as protective components

---

[1] Any general chemistry textbook should give basic material regarding the structures and functions of carbohydrates; proteins and enzymes; nucleic acids; and fats and lipids. If yours does not, see W. L. Masterton and E. J. Slowinski, "Chemical Principles," 3rd ed., Chapter 23. Saunders, Philadelphia, Pennsylvania, 1973.

of the interface (cell walls and skin) between many organisms and the outside environment.

Two major classes of lipids may be identified. The *neutral fats* contain three fatty acid molecules condensed with the three hydroxyl groups of glycerol and are completely nonpolar or hydrophobic. They serve primarily to store free energy in the form of fat droplets in cells, especially the adipose cells of vertebrates. Waxes, steroids, terpenes, vitamins, and prostaglandins are other types of nonpolar lipids which constitute a smaller weight percentage than neutral fats in most organisms. The *amphipathic* or

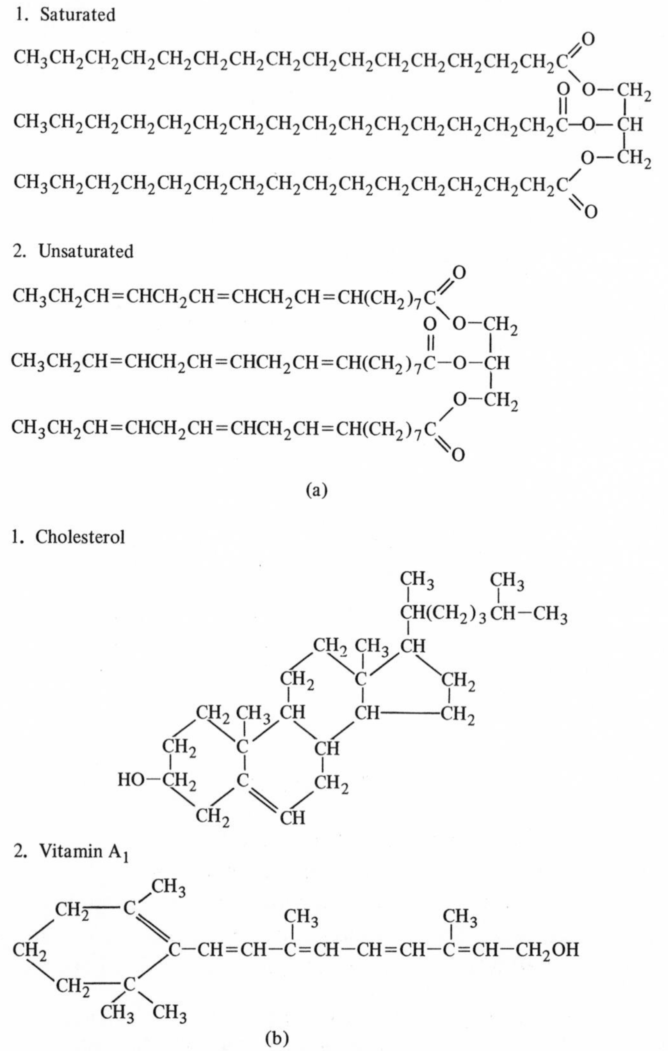

1. Saturated

2. Unsaturated

(a)

1. Cholesterol

2. Vitamin A_1

(b)

Figure 17.1 Some typical lipids. (a) neutral fats, (b) miscellaneous, (c) amphipathic (polar), and (d) general representation of amphipathic (polar) lipid.

polar lipids are similar to the soap or detergent molecules discussed in Chapter 14 in that they possess a highly polar, hydrophilic "head" and one or more nonpolar, hydrophobic "tails." These molecules are most often found as structural components of membranes. Examples of both major classes of lipids are shown in Figure 17.1.

The most important aspect of lipids in relation to environmental pollution is their nonpolar, hydrophobic nature. Compounds of low polarity such as DDT, chloromethylmercury(II), and other organic or organometallic substances tend to concentrate in the lipid portions of humans and other organisms, initiating the process of *bioamplification* (Section 16.3). Furthermore, hydrophobic compounds may be much more likely to be adsorbed by an organism because of the nature of the lipids in the membranes

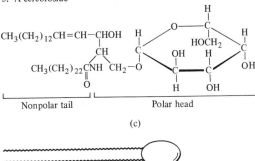

1. Phosphatidylcholine

Nonpolar tail   Polar head

2. Sphingomyelin

Nonpolar tail   Polar head

3. A cerebroside

Nonpolar tail   Polar head

(c)

Nonpolar tail   Polar head

(d)

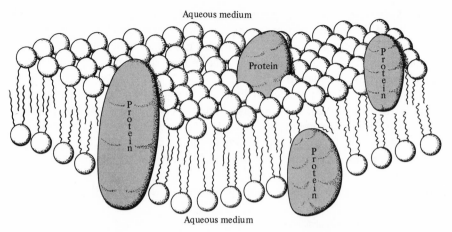

Figure 17.2 Cross section of a typical cell membrane. The nonpolar lipid "tails" form the interior of the membrane while polar "heads" extend into the aqueous medium. Globular protein molecules are embedded in the lipid bilayer.

through which they must pass. Biological membranes may consist of several layers of cells, a single layer of cells, or in the simplest case, a single cell wall. In any case a substance must pass through one or more membranes to enter and affect the organism or one of its cells. Cell walls usually consist of a mixture of about 60% protein and 40% lipid together with a small amount of carbohydrate. The lipids are mainly amphipathic and are arranged in a bilayer with their hydrophilic heads extending into the aqueous media on either side of the membrane and their hydrophobic tails mingling in the interior (See Figure 17.2). Protein molecules may be located on the membrane surface or may penetrate partially or fully through the lipid bilayer.

Because the interior of the membrane is extremely hydrophobic, ionic compounds have difficulty diffusing through it. There is a strong positive correlation between biological activity of any type and hydrophobic character of a molecule,[2] although activity reaches a maximum and then declines when hydrophobic character becomes great enough that the diffusing molecule begins to be held within the hydrophobic core of the bilayer. Thus as a general rule nonpolar molecules or un-ionized forms of ionizable molecules pass most readily across the membrane barriers.

In order to enter the human body any foreign substance must pass several such barriers.[3] Absorption from the external environment can occur through membranes in the mouth, the gastrointestinal tract, the lungs, and

[2] C. Hansch, *Accounts Chem. Res.* **2**, 232–239 (1969).

[3] The reader who wishes a more detailed treatment is referred to G. L. Waldbott, "Health Effects of Environmental Pollutants," Chapter 6. Mosby, St. Louis, Missouri, 1973.

/or the skin. Inside the body additional impediments such as the blood–brain barrier, cell membranes, or in the case of a pregnant woman, the placental barrier may be encountered. In each case metal ions or highly polar substances will be turned back unless they can be converted in some way to nonpolar forms. One instance of such a conversion, the role of the chelating agent nitrilotriacetic acid (NTA) in transferring heavy metal ions across the placental barrier, was mentioned at the close of Chapter 14.

17.2 DETOXIFICATION OF FOREIGN SUBSTANCES

All organisms have mechanisms by which undesirable substances can be excreted, allowing a constant, low concentration to be maintained. The principal routes of excretion from humans are via the urine and the bile, but sweat, expired air, expectorated mucus, saliva, milk, and secretions (other than bile) into the gastrointestinal tract are also important in the case of specific compounds. Sulfur compounds (such as H_2S) and their selenium and tellurium analogs are often excreted in sweat and exhaled air, halogens are removed via skin and saliva, and the contamination of cow's milk and human milk by pesticides, antibiotics, and other drugs is well known.

Aside from the respiratory system (whose defenses against foreign substances are discussed in Sections 9.1 and 9.4), the kidneys and the liver are the most important organs for elimination of toxic substances, and it is worthwhile to consider their roles in more detail. In the kidney (Figure 17.3) three distinct processes occur. The first, *glomerular filtration,* is an *ultrafiltration* in which blood pressure forces plasma from an artery through a membrane permeable to all but protein and very large lipid molecules. Thus, polar and nonpolar substances alike are forced into a kidney tubule which leads through a series of convolutions to a central collecting system and then to the bladder. However, many substances (such as Na^+, K^+, Cl^-, sugar, and water) required to maintain homeostasis are forced into the tubule, outside the body's membrane system. These must be *reabsorbed,* and a large number of capillary blood vessels contact the numerous convolutions of the tubule for this purpose. Reabsorption may be *active,* in which case free energy is required to drive "pumps" which move specific species back into the bloodstream, or *passive,* in which case diffusion depends on a concentration gradient and the lipid solubility of the substance being transported. The tubules and the capillaries associated with them are exposed to a wide variety of toxic substances and are especially susceptible to damage should those substances be held in the lipid bilayers of their membranes.

Nonpolar substances, which can diffuse back into the bloodstream from the tubules more readily, are more difficult to eliminate from the body. Usually they must be converted to polar form by some metabolic process of oxidation, reduction, or hydrolysis. Such reactions usually occur in the

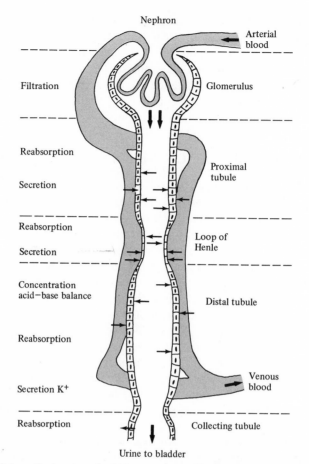

Figure 17.3 Schematic drawing of a kidney nephron. The human kidney contains approximately 10^6 such units. Source: D. C. Weber and Y. Nosé, *in* "Engineering Principles in Physiology" (J. H. U. Brown and D. S. Gann, eds.), p. 322. Academic Press, New York, 1973.

liver and are catalyzed by special enzymes. By introducing polar groups such as —OH or —NH$_2$ they form products which are less lipid-soluble and therefore less easily reabsorbed by the kidney.[4] Although the liver enzymes are not highly specific and a large number of different ones are available, man-made compounds for which there are no natural sources will be metabolized very slowly unless they are similar in structure to naturally occurring molecules toward which the liver enzymes are active. This is the case with DDT, for which several reaction steps and considerable time are required to produce a polar, non-lipid-soluble structure.[5] As a general rule

---

[4] In some cases, however, these metabolic processes may be disadvantageous, as in the case of methanol whose oxidation products, formaldehyde and formic acid, are probably the cause of the blindness associated with its chronic ingestion.

[5] R. D. O'Brien, "Insecticides: Action and Metabolism," pp. 125–129. Academic Press, New York, 1967.

pesticides and other synthetic organic compounds have a higher than normal probability of causing damage to health because they are likely to be retained in living organisms and the mechanisms by which they might be excreted may be inefficient or nonexistent.

17.3 ENZYME ACTIVATION AND INHIBITION

Acute toxicity of many substances (such as lead and mercury) results from their interactions with proteins, and especially with the enzymes, which are highly effective and specific catalysts for most essential metabolic reactions. Every enzyme contains an *active site* (or catalytic site) which binds the substrate (a reactant molecule or molecules) and aids in producing the desired products. Most enzyme-catalyzed reactions proceed 10^8 to 10^{11} times faster than the uncatalyzed reactions, and specificity is so great that in some cases only a single molecular structure and no other is affected.

The ability of enzymes and other proteins to perform their functions is intimately related to their three-dimensional structures. Such a structure is determined by the sequence of amino acids along the "backbone" of the polymer (see Figures 3.2 and 3.3), which in turn controls the manner in which the polypeptide folds back on itself to form a helical or globular structure. As early as 1894 Emil Fischer proposed the *"lock and key" hypothesis* which states that the substrate fits the active site as a key fits a lock. More recently, the lock and key hypothesis has been modified slightly to include the possibility that the active site may be *induced* to adopt a shape into which the substrate fits accurately. Such a change may result from the substrate (much as a hand influences the shape of a glove) or from some other molecule (an *"activator"*) which is bound to the active site or to some other portion of the protein in such a way that it can produce changes of about 0.1 nm or more in the dimensions of that site. The concept of enzyme flexibility also permits the reverse of activation—an *inhibitor* molecule may bind to or distort the active site so that the substrate cannot be bound or the cleavage or formation of bonds cannot be accomplished. Both activators and inhibitors are useful because they can regulate the functions of a cell by turning on and off the enzymes which catalyze the reactions necessary to accomplish those functions.

Heavy metals are quite effective in inhibiting the action of enzymes. Soft metal ions such as Hg^{2+}, Pb^{2+}, and Cd^{2+} are strongly bound by —SH and —SCH$_3$ groups in cysteine and methionine side groups. The former is especially important because —SH groups are associated with the active sites in a number of enzymes. In other cases one metal ion may replace another in an enzyme, reducing or eliminating its biological activity. Such substitutions usually follow the rules set forth in Section 2.4 for isomorphous substitutions in minerals, although in some cases a foreign metal may bind to a different site from the native one. An example of the latter,

TABLE 17.1   Examples of Enzymes Inhibited by Heavy Metals[a]

Metal ion	Enzyme inhibited
Hg^{2+}	Alkaline phosphatase, glucose-6-phosphatase, lactic dehydrogenase
Cd^{2+}	Adenosine triphosphatase, alcohol dehydrogenase, amylase, carbonic anhydrase, carboxypeptidase (peptidase activity only), glutamic-oxaloacetic transaminase
Pb^{2+}	Acetylcholinesterase, alkaline phosphatase, adenosine triphosphatase, carbonic anhydrase, cytochrome oxidase
As^{3+}, As^{5+}	Pyruvate dehydrogenase

[a] Source: *J. Chem. Educ.* **51** (4), 236 (1974); from *Annu. Rev. Biochem.* **41,** 91 (1972).

nonisomorphous substitution is the replacement of Zn^{2+} by Hg^{2+} in carbonic anhydrase. Even though Hg^{2+} may not coordinate to the same site left by Zn^{2+}, the structure of the protein apparently is modified sufficiently that it becomes inactive. Table 17.1 lists some examples of enzymes whose catalytic activity is impaired by metal ions. This type of inhibition is usually reversible.

An example of irreversible inhibition is provided by diisopropylphosphorofluoridate (DFP), a representative of a class of toxic organophosphorus compounds which includes nerve gases and some insecticides. This compound reacts with hydroxyl groups on serine residues to form a covalently bonded derivative as shown in Figure 17.4. Any enzyme which requires participation of a serine hydroxyl function at its active site will be inhibited DFP. Probably the most important example is acetylcholinesterase, an essential participant in the functioning of the central nervous system. This effect will be discussed further in Section 18.6.

The active sites of many enzymes contain metal ions, as do those of a number of transport proteins such as hemoglobin. The action of many toxic, nonmetallic, inorganic species results from coordinate covalent bonding or oxidation–reduction involving these metals. The effect of CO

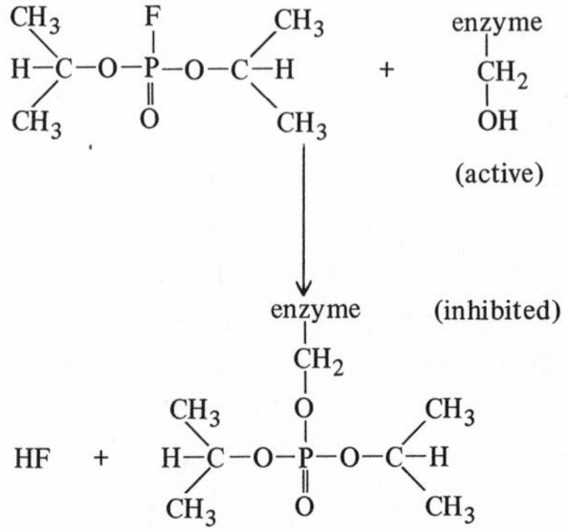

Figure 17.4 Irreversible enzyme inhibition by diisopropylphosphorofluoridate (DFP).

on hemoglobin has already been documented in Section 10.1. Carbon monoxide, cyanide ion, and hydrogen sulfide inhibit the complex enzyme cytochrome $a + a_3$ which controls an essential step in respiratory oxidation of carbohydrate and other energy storage materials. This is apparently a result of coordination of these ligands to iron (or possibly copper) atoms in the enzyme. Fluoride ions inhibit a variety of enzymes, often by forming complexes with Mg^{2+} ions or by forming fluorophosphato complexes with magnesium or other metal ions.[6]

When environmental pollution introduces a metal ion or some other foreign substance into a protein whose molecular weight is 6000 or more it may seem almost trivial, but by a slight disruption of three-dimensional structure such a species may completely destroy biological activity. The molecular architecture of a protein molecule is so finely tuned to perform a specific function exceedingly well that it is rather delicately susceptible to any external factor which may disrupt that function.

17.4 MOLECULAR MECHANISMS OF MUTAGENESIS, CARCINOGENESIS, AND TERATOGENESIS

The information necessary to specify the structures of the thousands of different proteins and enzymes in the average living cell is stored in the base sequence of *deoxyribonucleic acid* (DNA) molecules. This information is transcribed from DNA to messenger-RNA which carries it to the ribosomes where protein synthesis occurs. At these latter sites translation of the three-letter words of nucleic acid structure into the twenty-letter alphabet of amino acids required for protein sequences is accomplished by transfer-RNA. Since errors in specification of the amino acid sequence will often affect the structure of a protein or enzyme in a way that destroys its function, any interference with information transfer by environmental contamination can produce major effects on the health and welfare of the host organism. Because of the role of DNA in the transmission of information from parent cells to progeny, any alteration of its sequence of nitrogenous bases will continue to appear in subsequent generations, providing them as well as the parent cell with incorrect information. In highly developed, multicellular organisms this is most crucial in the case of the sex or germinal cells which combine to produce a new organism. Changes in the DNA of other (somatic) cells will be transmitted to the progeny of that specific type of cells, but will not persist for a long period because of the eventual death of the entire organism.

Changes in the base sequence of DNA are known as *mutations*, and substances which can produce them are called *mutagens*. The most frequently

[6] Committee on Biologic Effects of Atmospheric Pollutants, "Fluorides," p. 70. Nat. Acad. Sci., Washington, D. C., 1971.

observed mutations are "transitions" in which one purine (adenine or guanine) replaces another, or in which there is an exchange of pyrimidines (thymine or cytosine). (See Figure 3.2 for structures of these compounds.) "Transversions," in which a purine replaces a pyrimidine or vice versa are much rarer. Transitions and transversions usually produce a single incorrect amino acid along a protein backbone. Another more serious type of mutation results from insertion or deletion of a base along the DNA chain. Since the genetic code is read as triplets of bases, an insertion (or deletion) modifies all triplets and garbles all specifications of amino acids unless deletion (or insertion) restores the correct code at a later point along the chain.

Mutations such as those just described need not always produce deleterious effects on the organism. For example, if only one amino acid is incorrectly specified along the backbone of a protein, the three-dimensional structure may not be greatly affected and protein function may not be destroyed. This is usually the case when a hydrophilic amino acid is replaced by another hydrophilic residue or a hydrophobic–hydrophobic replacement occurs, because the hydrophobic interactions which determine three-dimensional protein structure are not affected much. On the other hand, substitution of glutamic acid (hydrophilic) for valine (hydrophobic) in a chain of approximately 140 amino acids in human hemoglobin produces abnormally shaped red blood cells which no longer transport oxygen efficiently. This condition results in an inherited disease known as sickle cell anemia. Occasionally, a mutation will occur which improves the ability of a cell or a species to survive, but in an organism as highly developed as a human the probability of a favorable mutation is far smaller than for a deleterious one.

Some mutagens can react directly with the nitrogenous bases of DNA. Nitrous acid, for example, converts amino groups to keto functions. Thus, cytosine can be transformed into uracil as shown in Figure 17.5. The latter pairs by hydrogen bonding with adenine rather than guanine, altering the sequence of any DNA for which such a strand would serve as a template.

Other mutagens are similar enough to one of the nitrogenous bases that they may be incorporated into DNA itself. Their mutagenic character is a result of minor structural differences which reduce the accuracy of the

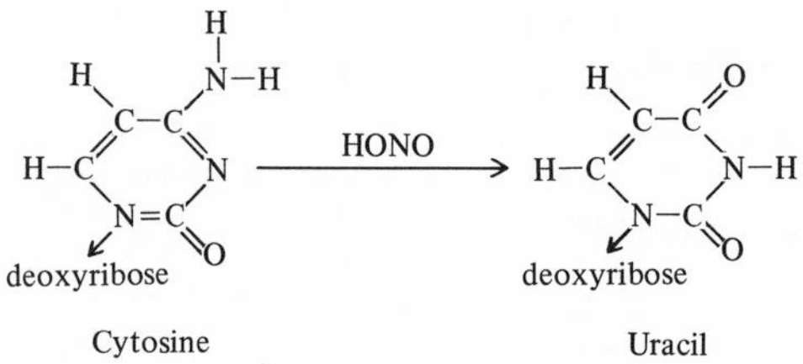

Figure 17.5 Conversion of cytosine to uracil by nitrous acid.

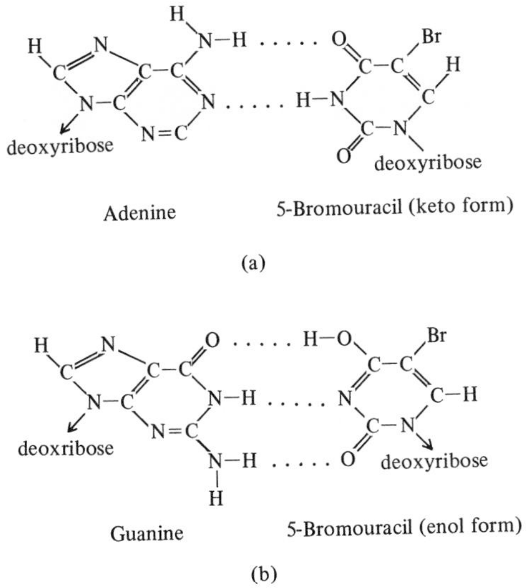

Figure 17.6 Mutagenic activity of 5-bromouracil: (a) hydrogen bonding of the normal keto form and (b) hydrogen bonding of the enol form.

replication of the double helix. For example, 5-bromouracil in its normal keto form hydrogen bonds with adenine (as would uracil or thymine, see Figure 17.6), but in its enol form it base pairs with guanine. Substitution of —Br for —CH_3, should 5-bromouracil replace thymine, partially stabilizes the enol form and makes errors in transmission of genetic information more likely.

Other damage to DNA may be of a grosser nature. For example, many planar aromatic hydrocarbons are thought to be able to position themselves (intercalate) between the flat layers of hydrogen-bonded base pairs in the interior of the double helix, forcing it partially to uncoil and causing errors in transmission of transcription of the genetic code (See Figure 17.7). Ionizing radiation and a variety of chemicals can actually break both strands of the double helix, resulting in damage to chromosomes which can be seen under the microscope. Broken DNA strands can be repaired by certain enzymes, but in many cases, especially when there are multiple breaks, two ends may be joined which were not formerly adjacent. Sometimes an entire gene (that portion of a DNA molecule which specifies a given polypeptide) may be lost. This means that the progeny of a cell with such a mutation will lack the capacity to synthesize some particular protein or enzyme. Should this material be needed for cell metabolism at some later time it will be unavailable and the cell may die.

Both *teratogenic* (producing birth defects) and *carcinogenic* (producing cancer) substances appear to be related to mutagens, leading to the hypothesis that they too may act by modifying DNA in some way. Active

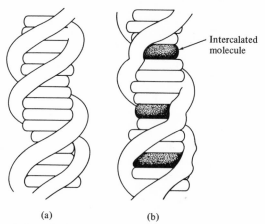

(a)   (b)

Figure 17.7 Intercalation of planar molecules between base pairs of DNA: (a) secondary structure of normal DNA and (b) DNA containing intercalated molecules. Source: L. Fishbein, W. G. Flamm, and H. L. Falk, "Chemical Mutagens," p. 30. Academic Press, New York, 1970.

carcinogens and mutagens often contain electrophilic (electron poor, Lewis acid) sites which are capable of reacting with nucleophilic (electron rich, Lewis base) sites present in DNA. As a rule nearly all carcinogens have been found to be mutagens, although not all mutagens induce the growth of tumors. One theory of the origin of cancer holds that as a result of modification of DNA by radiation, chemical mutagens, or virus-derived nucleic acids the synthesis of control molecules which slow cell division when the surroundings are crowded is prevented. Thus, the cancerous cells proliferate rapidly and are incapable of providing to their progeny the information necessary for synthesis of growth controls. In a great many cases such a change in genetic material may lie dormant for a long period, awaiting a second, initiating factor which triggers a cancerous growth. This sort of behavior complicates the assignment of causative factors in cancer research and greatly increases the difficulty of screening for carcinogenic effects the more than 500 new substances introduced commercially each year.

17.5 CLASSIFICATION OF TOXIC SUBSTANCES

Because of our lack of detailed knowledge of the mechanism (or more likely the mechanisms) by which mutagenic, teratogenic, and carcinogenic substances produce their effects on a molecular level, it is impossible at this time to make accurate and specific correlations of these effects with molecular structures. The variety of effects which other toxic agents can produce on metabolism and vital organs — both synergistically and antagonistically — complicates matters even further. Nevertheless, a number of classes of substances which are prime candidates for further investigation of toxic effects can be identified.[7] In many cases these or related materials have been shown to produce chromosome damage in test-tube experiments, and all may be considered as *potential* genetic hazards. The classes of materials are as follows:

1. *Physical agents.* Ultraviolet and ionizing radiation can provide sufficient energy to break a variety of chemical bonds, causing chromosome damage. Inhalation of small fibrils of asbestos (Section 9.6) can result in deterioration of lung tissue (asbestosis) and a rare cancer, mesothelioma, which has a 20-year latency period.

2. *Metals.* Nickel, cadmium, mercury, and lead, none of which has been shown to be essential for life, are introduced into the environment in large quantities by man's activities. Organometallic derivatives of these metals are often volatile, may be concentrated in lipid tissues, and sometimes cause chromosome damage,[8] making them especially dangerous.

3. *Nitrites and nitrates.* The proven mutagenicity of nitrous acid, the ability of nitrate to be reduced to nitrite, and the possibility of production of extremely carcinogenic and mutagenic nitrosamines suggest that nitrates and nitrites are prime candidates for further evaluation of health hazards.

4. *Organochlorine (or perhaps organohalogen) compounds.* A variety of solvents, such as those used in glues, nail polish removers, and cleaning fluids have been shown to produce chromosome damage, as has gasoline. Pesticides such as dieldrin, aldrin, and heptachlor induce the growth of tumors in test animals. Dichlorvos (DDVP, 2,2-dichlorovinyl-dimethylphosphate), the active ingredient in vaporizing resin strip pest killers, produces breaks in root tip chromosomes, although it has not been proved to have any effect on humans. Polychlorinated biphenyls (PCB's) have also been introduced into the environment in large quantities, and recently have been replaced in some applications (for which they were banned by EPA) by polybrominated biphenyls.

5. *Alkylating agents.* These include dialkyl sulfates, nitrosamines and nitrosamides, epoxides, alkane sulfonic esters, lactones, some aldehydes, and sulfur and nitrogen mustards as well as the aforementioned pesticide DDVP. Ethyleneimine (aziridine) has been shown to cause mutation in fruit flies (*Drosophila*), wheat, and barley, as well as chromosome aberrations in cultured human cells. Aziridine and its derivatives are used in textiles for crease proofing, in adhesives and coatings, in lubricant additives and jet fuels, as insecticides, chemosterilants, and soil conditioners and for a variety of other purposes. *N*-Nitrosonornicotine has recently[9] been

---

[7] Much of the information regarding this list of substances was obtained from J. Schubert, *Ambio* **1**(3), 81–89 (1972).

[8] S. Skerfving *et al., Environ. Res.* **7**(1), 83 (1974).

[9] D. Hoffman, S. S. Hecht, R. M. Ornaf, and E. L. Wynder, *Science* **186**, 265–267 (1974).

shown to be present in tobacco and may well be a major factor in the correlation between smoking and lung cancer. As a result of the limited amount of research done so far, more than 25 alkylating agents have been shown to be possible mutagens and/or carcinogens.

6. *Aromatic hydrocarbons.* Benzene and 3,4-benzpyrene are well-known examples of this class which have carcinogenic properties. A number of dyestuffs are derived from aromatic compounds and some are carcinogenic. Many of these dyes are finding increasing markets in household products such as tissues and towels which come in close contact with mucous membranes. After use large quantities of these compounds are discarded into the environment.

7. *Anesthetics.* Spontaneous abortions and premature deliveries among female anesthetists have been observed at 2–4½ times the normal rate. This effect has been correlated with the presence of halothane (bromochlorotrifluoroethane) at 90 ppm and nitrous oxide at 7000 ppm. One cannot help noting the similarity of halothane to the freons (which also have anesthetic properties) used as propellants in aerosol sprays. It is possible (although so far no research has been done on the question) that similar effects might result from excessive inhalation of spray mists. It is well established that freons affect heart rhythms by a quite different mechanism.

8. *Air pollutants.* In addition to 3,4-benzpyrene formed by coal combustion a variety of substances formed in photochemical smog are potential mutagens and carcinogens. These include ozone, nitrogen dioxide, sulfur dioxide, aldehydes, ketones, and peroxides.

9. *Food additives.* Substances on the GRAS (Generally Recognized as Safe) list of the Food and Drug Administration have achieved that status largely because they have been in use for a long time during which no ill effects have been discovered. However, the difficulties of detecting carcinogenic and mutagenic effects are great enough that if these compounds fall into any of the categories listed above they probably should be examined more carefully. Some examples are nitrites; coal tar dyes; sugar substitutes; preservatives such as sodium sulfite, benzoic acid, butylated hydroxyanisole, and butylated hydroxytoluene; and a number of emulsifiers, sequestrants, and stabilizers. Recently, it has been shown that the widely used beverage preservative diethyl pyrocarbonate reacts with ammonia to form the well-known mutagen, urethane.[10]

10. *Naturally occurring toxicants.* Numerous substances such as estrogens, fungal and seafood toxins, nitrates and other inorganic compounds, and excess amino acids occur naturally in foods and may be toxic, mutagenic, or carcinogenic. There is evidence that foods preserved by smoking contain 3,4-benzpyrene, for example. While in the past mutations

---

[10] G. Löfroth and T. Gejvall, *Science* **174,** 1248–1250 (1971).

caused by such agents probably contributed to the development and evolution of species, one of the attributes of modern civilization is that all human life is valued and, insofar as possible, natural selection is prevented from destroying those possessed of unfavorable genes. Thus, it may well be advantageous to reduce the rate of mutation resulting from naturally occurring substances below what it was in the past. Excessive amounts of these materials, as well as those introduced by man's activities, probably ought to be avoided in foods.

17.6 THE VINYL CHLORIDE EPISODE

The length of the above catalog of potentially toxic substances might lead one to the conclusion that all chemical compounds are carcinogenic (or mutagenic, etc.) if taken in large enough quantities over a long enough period, but this appears not to be the case. Only a small fraction of all substances have been found to cause cancer or mutations. The problem is to sift them out from the great majority of compounds, and to prevent their introduction into the environment. Some of the factors which make this problem extremely difficult become evident when we consider the history of a recently discovered chemical carcinogen, vinyl chloride.

Polymers of this colorless gas were introduced into commerce during the early 1940's as a substitute for natural rubber. Because of the many uses for polyvinyl chloride (PVC) and the development of other applications (as an aerosol propellant, for example), production of vinyl chloride grew to 2.43×10^9 kg in the United States in 1973. Polyvinyl chloride plastics now account for about 1% of the United States gross national product. In January of 1974 B. F. Goodrich Company voluntarily revealed to federal and state officials that since 1971 three workers associated with PVC polymerization operations at its resins plant at Louisville, Kentucky, had died of angiosarcoma of the liver. Because this form of cancer is extremely rare (no more than 25 cases are usually reported per year in the United States), this disclosure caused an immediate reaction among regulatory officials.

At the request of the U. S. Department of Health, Education and Welfare, Britain's Imperial Chemical Industries published the results of toxicological studies by Professor Caesare Maltoni of the University of Bologna, Italy, a few weeks later. In August 1972 Maltoni had discovered in a test rat exposed to vinyl chloride a liver angiosarcoma which was later found to be essentially identical with those that had killed the PVC workers. By January 1973 a team of three scientists representing the United States chemical industry had visited his laboratory to learn of his results. The data published in February 1974 are shown in Table 17.2. These results were supplemented at a May 1974 meeting of the New York Academy of Sciences and American Cancer Society where Maltoni re-

TABLE 17.2   Incidence of Liver Cancer in Test Rats

Exposure levels[a] (ppm)	Rats with liver angiosarcomas	No. of rats exposed
10 000	6	69
6 000	11	72
2 500	9	74
500	7	67
250	2	67
50	0	64

[a] Atmospheric exposure of Sprague-Dawley rats to vinyl chloride, 4 hours daily, 5 days a week, for 12 months. *Note:* no liver angiosarcomas were observed in controls or rats exposed similarly to 2500 ppm vinyl acetate. Source: Reprinted, with permission of the copyright owner, The American Chemical Society, from *Chem. Eng. News,* February 25, p. 16 (1974), from unpublished research by Professor Cesare Maltoni, Istituto di Oncologia, Bologna, Italy.

ported the incidence of angiosarcoma in rats at an exposure level of 50 ppm. Several other types of tumors were also found.

By May 1974 a cause and effect relationship between vinyl chloride and human angiosarcoma was generally accepted. Confirmed cases of cancer in workers associated with PVC plants had risen to 19, 17 of whom had already died. A recall of aerosol hair sprays using vinyl chloride as propellant was begun by the Food and Drug Administration, and the EPA asked makers of 23 insecticide sprays to reformulate their products in order not to contain vinyl chloride. However, the EPA order did not include recall of products already packaged, and the names of many manufacturers were withheld because the presence of vinyl chloride in their products was deemed to be proprietary information. A temporary standard of 50 ppm in air was also set for plants involving manufacture of vinyl chloride and PVC as well as PVC fabricating facilities. No standards were set for ambient air outside PVC plants because measurements done by EPA indicated that the lowest detectable level (1 ppm) of vinyl chloride was rarely exceeded in these locations.

In October 1974 the Occupational Safety and Health Administration (OSHA) of the U. S. Department of Labor set strict standards to reduce worker exposure to vinyl chloride in the workplace. Exposure to more than a 1-ppm 8-hour average or 5 ppm for any period of 15 min or more was prohibited, and it was required that workers be provided with respiratory equipment. Although appealed to the U. S. Supreme Court, these standards were upheld and went into effect April 1, 1975.

17.7   CONTROL OF TOXIC SUBSTANCES IN THE ENVIRONMENT

Several aspects of the vinyl chloride story are profoundly disturbing to anyone who feels that the release of toxic chemicals into the environment should be controlled. Aside from the controversy[11] over the possibility that industry withheld the results of Maltoni's study of carcinogenicity, which postponed regulation for no more than a year, a great deal of time elapsed between the introduction of vinyl chloride into commerce and the discovery of its toxicity. During this period production rose to the point that nearly 100×10^6 kg of vinyl chloride were lost to the environment each year and its importance to the economy grew so great that one industry spokesman could claim the OSHA limit of 1 ppm would throw "two million jobs down the drain."[12] Why did detection of carcinogenic effects take so long?

The principal difficulty is found in the characteristic behavior of chemical carcinogens. Usually there is a latency period between exposure and development of a cancerous tumor. The average length of exposure of affected B. F. Goodrich workers to vinyl chloride was 19 years, indicating a latency period within the normal range of 15–40 years in humans. Although there is evidence that increasing the dosage of a carcinogen increases the incidence of disease, there appears to be no requirement of chronic exposure. Once an individual has received enough of the compound, whether in a single dose or extended over a period of years, cancer may develop. Moreover there appears to be no threshold concentration below which carcinogenic effects do not occur, although the fraction of the population affected by disease decreases with reduced exposure.

As a result of the latency period, establishment of a cause and effect relationship between a chemical and cancer is difficult. Test animals such as rats have much shorter latency periods than humans, and the number of animals required for adequate statistics can be reduced if high dosages are used, but this introduces dual difficulties. Adequate models must be constructed for extrapolation from high to low doses, and differences in the metabolism of test animals and humans prevent complete certainty when results are extrapolated from one species to another. A substance carcinogenic in rats may be metabolized to a harmless form by humans, but innocuous substances may also be converted into carcinogens. It can be argued effectively that humans, because of the greater variety of their cell types, may be more susceptible than test animals. Further complicating the situation is the wide variation in susceptibility of humans to toxic agents. Thus levels that are safe for the majority may be injurous to a minority.

Because of these problems it is difficult and expensive to screen compounds for carcinogenicity. Of the nearly two million compounds registered by Chemical Abstracts Service, the National Cancer Institute estimates that no more than 6000 have been tested, perhaps only half of them

---

[11] *Chem. & Eng. News,* May 20, p. 16; May 27, p. 5; June 10, pp. 3 and 12; July 8, pp. 21 and 37 (1974).

[12] T. C. Walker, President of Firestone Plastics, quoted in *Chem. & Eng. News,* October 7, p. 5 (1974).

adequately. About 1000 of these have shown some signs of being carcinogenic, but no more than a few hundred have been clearly established as carcinogens. There are more than 9000 different chemicals synthesized commercially in quantities exceeding 500 kg/year in the United States today, and total production exceeds 55×10^9 kg/year. Clearly, only a small fraction of these have been adequately tested.

Indeed one might ask, in view of the difficulties of discovering carcinogens, is it not unusual that the effect of vinyl chloride on humans was detected at all? There appears to be a rather good chance that the answer is yes. Vinyl chloride induces a very unusual cancer, and when three cases out of a yearly average of twenty-five were found in one of its plants, B. F. Goodrich called the attention of health officials to them. The possibility of coincidence was too remote for such an indicator to be ignored. Two factors combined to reveal the effect. Workers in the PVC plant constituted as well-defined group with known exposure to vinyl chloride, and angiosarcoma of the liver is rare enough that even a single case is highly noticeable. It is generally agreed that had B. F. Goodrich not reported the deaths of its Louisville workers, several more years would have elapsed before the danger of vinyl chloride could have been established.

Other carcinogens may not produce such obvious telltale effects. Indeed, there is evidence that the greatest damage from vinyl chloride is not the result of liver angiosarcoma but of the other, more common cancers it induces. Studies done at the Harvard School of Public Health indicate a 60% increase in lung cancer and a 320% increase in brain cancer among deceased vinyl chloride workers.[13] Mutagenic effects have also been discovered now that angiosarcoma has focused attention on vinyl chloride. Another substance which "merely" increased the incidence of prevalent cancers such as those of the lung or brain by 10% and was not confined to a workplace environment would be almost impossible to detect. Lederberg[14] has suggested that a substance which reduced IQ by 10% but caused no overt symptoms would be completely unnoticed, despite its extremely deleterious effect on human society.

It is a well-known fact that the incidence of cancer has increased a good deal during the twentieth century. What is perhaps less well known is that all of this increase cannot be attributed to better methods of diagnosis, increased levels of cigarette consumption, and the fact that other killer diseases have been brought under control. Even within a particular age group and among the young cancer is on the rise. This is shown clearly by the data in Figure 17.8 for the 0–4-year age group in Sweden. During the decade 1958–1968 the incidence of cancer of the nervous system increased threefold.

No data are yet available to indicate whether this increase is associated

---

[13] *Chem. & Eng. News,* July 22, p. 9 (1974).
[14] S. S. Epstein and M. S. Legator, "The Mutagenicity of Pesticides," with a foreward by Joshua Lederberg, p. xiv. MIT Press, Cambridge, Massachusetts, 1971.

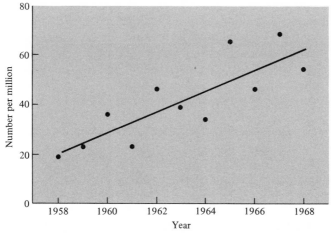

Figure 17.8 Incidence of cancer among Swedish children. The annual incidence of cancer of the nervous system in the 0–4-year-old age group Sweden, 1958–1968. The incidence is expressed as number of cases per million among children 0–4 years old. The absolute number of cases ranged from 10 in 1958 to 40 in 1967. About $7\frac{1}{2}$% or 600,000 of the total population in Sweden in 1968 falls in the 0–4-age group. Source: J. Schubert, *Ambio* **1**(3), 88 (1972).

with increased levels of industrialization, but it has been estimated that between 60 and 90% of all human malignancies are caused by environmental factors such as ionizing radiation or chemical carcinogens. In a recent study[15] the National Cancer Institute tabulated deaths resulting from cancer between 1950 and 1969 in all counties of the United States. Those counties having high rates of cancer deaths correlate rather well with regions which were heavily industrialized prior to 1940. In view of the 15–40-year latency period between exposure to a carcinogen and development of cancer, this does not contradict the hypothesis that increased levels of air and water pollution might lead to increased incidence of malignant tumors.

It has been pointed out by a number of observers of the vinyl chloride episode that the few cases of angiosarcoma of the liver observed so far may be only the tip of an iceberg. With a 20-year latency period and the greatly increased numbers of workers exposed (it is estimated that about 6500 were involved in contact with vinyl chloride in 1974), the number of angiosarcomas might well continue to rise for 20 years, even if all exposure were halted. This same type of effect may begin to show up for a variety of other substances in the next few decades. Vinyl chloride may well be an early warning signal of problems yet to come.

With testing of carcinogenic and mutagenic compounds so time-consuming and expensive, how can we check all of the compounds which have been and will be introduced into commerce? It seems likely that this will

---

[15] T. J. Mason and F. W. McKay, "U. S. Cancer Mortality by County: 1950–1969." National Cancer Institute, Department of Health, Education and Welfare, Washington, D. C., 1974.

prove impossible, but also unnecessary. Many of the new chemicals such as dyes used in food and paper or optical brighteners used in textiles serve purely cosmetic functions, and great hardship to society would not be caused by halting production entirely. In other cases, such as that of vinyl chloride, the substance serves a very useful function and deleterious effects must be balanced against this. Here testing must be carried out and at the same time considerable effort ought to be made to find viable substitutes in case the material is found to be extremely toxic.

The Toxic Substances Control Act of 1973 represents a start in the direction of adequate testing. It would require industry to test new chemicals and notify EPA before producing them in commercial quantities. Although versions of this bill were passed by the U. S. Senate and House of Representatives in 1973, they became stalled during 1974 in a House–Senate Conference Committee which was attempting to resolve the differences between the two legislative bodies. One such difference was whether industry would have to submit a notice of impending commercial production for all chemicals or just those included in the classes on the EPA danger list. Perhaps by the time this is read similar legislation will have been passed and signed into law.

Others[16] have suggested that better toxicological tests could be done with less economic disadvantage to small companies if government-supported and -operated regional testing laboratories, on the model of the Atomic Energy Commission National Laboratories, were established. These would provide greater expertise in this area than could be mustered by any but the largest corporations. Whatever approach is taken it will come none too soon. The lesson of vinyl chloride has only begun to unfold, but it is quite clear. The dispersion into the workplace and subsequently the environment of synthetic chemicals whose toxic, teratogenic, mutagenic, and/or carcinogenic properties have not been thoroughly evaluated may have very serious consequences. Moreover, these effects often cannot be readily detected and once found may not be immediately reversible.

EXERCISES

(The first group of exercises is based primarily on material found in this text.)

1. Distinguish between neutral fats and amphipathic lipids. Draw a schematic structure for each type.
2. Describe the structure of a biological membrane. Explain why such a structure is important in regulating the entrance of foreign compounds into the body.
3. What two body organs are chiefly responsible for removing toxic substances? Describe the role of each.

---

[16] J. Schubert, *Ambio* 1(3), 88 (1972).

4. How do enzyme activators and inhibitors work? Why are they important?
5. What is a mutation? Describe at least three molecular mechanisms by which mutation can occur.
6. What is angiosarcoma? Explain how it led to the discovery that vinyl chloride is a carcinogen.
7. Explain why the vinyl chloride episode may not be over, despite the fact that the Occupational Health and Safety Administration has set strict standards for exposure of workers.

(Subsequent exercises may require more extensive research or thought.)

8. Suppose you were the EPA administrator in charge of controlling release of toxic substances into the environment. From the list of compounds below choose those on which you would place a top priority for toxicity testing:

$$NaCl \qquad CH_3CH_2O{-}\overset{\overset{O}{\|}}{\underset{\underset{O}{\|}}{S}}{-}OCH_2CH_3$$

$$Ni(CO)_4 \qquad CH_3CH_2OH$$

$$CHCl_3$$

$$CH_3C\overset{\diagup O}{\underset{\diagdown O{-}O{-}NO_2}{}}$$

Explain the reasons for your choice in each case. With how many of these compounds have you come into contact in the laboratory?

9. How much monomeric vinyl chloride remains after polymerization to form polyvinyl chloride? Calculate roughly:
   a. The maximum concentration of vinyl chloride that might accumulate inside an automobile by vaporization from the seat covers.
   b. The maximum concentration of vinyl chloride that might accumulate in foods wrapped in PVC film.
10. *Chemical Engineering News* [May 5, p. 31 (1975)] has published a list of the "top 50" chemicals in the United States. Evaluate each with regard to potential toxicity, mutagenicity, and carcinogenicity. Is it possible to estimate the quantity of the material dispersed in the environment in each case? (The EPA, Public Information Office, 401 M Street, S. W., Washington, D. C., 20460, has available a report called "Activities of Federal Agencies Concerning Selected High-Volume Chemicals" which may be helpful.)

18

The Sustenance of Life

*If the earth must lose that great portion of its pleasantness
which it owes to things that the unlimited increase of wealth
and population would extirpate from it, for the mere purpose
of enabling it to support a larger, but not a happier or better
population, I sincerely hope, for the sake of posterity, that
they will be content to be stationary, long before necessity
compels them to it.*

John Stuart Mill

The role of population growth in increasing levels of pollution should
have been evident from the discussion in Section 13.1. The more people
there are the more pollution, assuming constant levels of affluence and
technology. Thus, control of rapidly growing human population plays a fundamental role in improving environmental quality. Modern medical, chemical, and other technologies have contributed to the population explosion by
drastically reducing death rates, and they have also made possible the
reduction of birth rates by means of contraceptives. However, accomplishing the *demographic transition,* in which birth rates drop to a corresponding level with death rates and a stable population is achieved,
depends on social as well as technological factors. In the "less-developed"
nations of the world children are far more economically valuable than in
those where adequate food supply and health care, industrialization, ur-
banization, and high standards of living and education are the rule. They
provide free labor, are relatively inexpensive to raise and educate, and
amount to a crude form of old-age insurance.

However, it is in just these less-developed countries, where the demographic transition to zero population growth has hardly showed signs of beginning, that Malthusian pressures of population against food supply

are greatest. Recently, the transfer of a number of practices from the highly technological United States agricultural system to the less-developed nations has provided additional time during which population controls may be instituted. However, intensive use of irrigation, pesticides, fertilizers, and free energy by such *Green Revolution* agriculture (and to much greater extent by United States "agribusiness") may have unwanted ecological consequences. Some observers have questioned whether even such intensive agricultural development will be able to keep up with growing numbers of individuals long enough for the demographic transition to take place.

18.1 FOOD QUALITY

Three components, vitamins and minerals, fats and carbohydrate, and protein, are required for optimal human nutrition. A deficiency in any of these categories can lead to reduced productivity, disease, or even death. *Minerals* correspond for the most part to the monatomic ions and trace elements found in the biological periodic table on back endpaper of this book. They are incorporated in the structures of numerous hormones and enzymes (Section 17.3) and some (calcium, phosphorus, and magnesium) are required in significant quantities in bones and teeth. *Vitamins* are complex organic or organometallic compounds for which the body has no adequate mechanism of synthesis. They serve as catalysts for certain metabolic processes, as enzyme activators, or as precursors from which such substances may be synthesized. Most vitamins and all but the structural minerals are required in quantities of the order of milligrams per day or less and therefore are readily supplied as supplements if the normal diet is deficient.

Carbohydrates and fats are storehouses for solar energy trapped by photosynthesis (Section 6.6). The standard measure for dietary consumption of these substances is the Calorie. (The capital "C" indicates that this is a nutritional "calorie" – actually, a kilocalorie or 4.186×10^3 J.) Fats and carbohydrates (and in certain cases proteins) combine with dioxygen in the process of *biological respiration*

$$C_6H_{12}O_6 + 6 O_2 \longrightarrow 6 CO_2 + 6 H_2O \qquad \Delta G^0_{298} = -2817 - T(0.1825) = -2871 \text{ kJ} \quad (18.1)$$
Glucose

releasing free energy which is then used to do the work necessary to maintain living cells. Chemical work is required to drive the nonspontaneous reactions by which lipids, proteins, nucleic acids, and other necessary biologically active molecules are synthesized from the monomers in Figure 3.2. Mechanical work is required for locomotion, other muscular activity, and cell division, and work is also necessary for active transport of solutes across membranes (Sections 12.5 and 17.2). Hence, a constant supply of carbohydrate and/or fat is essential for any organism which does not itself carry on photosynthesis.[1]

Depending on the amount of work done, the size of the organism, and the climate, Calorie requirements vary. For humans a range of from 11 700 kJ/day for an essentially sedentary person to about 18 400 kJ/day for someone engaged in heavy work seems reasonable according to the U. N. Food and Agricultural Organization. These numbers are not minimum survival values but rather quantities necessary to maintain good health. They may be compared with the first column of figures from Table 18.1, which indicates that average caloric intake in the developed countries of western Europe and North America runs approximately 4000 kJ/day above the 7500–9500 kJ/day found in the less-developed nations of Asia and Africa.

The third and perhaps most important aspect of nutrition is an adequate

TABLE 18.1 Estimated Daily Carbohydrate and Fat and Protein Content per Capita of National Average Food Supply (1966)[a]

Country	Carbohydrate and fat (kJ/day)	Total protein (g/day)	Animal protein (g/day)
Asia			
Taiwan	10 050	62.2	19.3
India	7 580	45.4	5.4
Japan	9 840	77.6	24.6
Malaysia	9 170	47.7	14.5
Africa			
Algeria	8 160	55.4	6.6
Ethiopia	8 580	72.3	10.8
Ghana	8 920	46.9	9.6
Kenya	9 380	67.9	13.3
Madagascar	9 880	54.0	9.9
Nigeria	9 080	59.5	5.1
Europe			
France	12 980	100.7	59.9
West Germany	12 010	79.9	51.5
Greece	12 181	98.2	41.7
Italy	11 970	85.4	35.5
Portugal	10 800	75.0	29.8
Sweden	12 140	79.8	53.6
United Kingdom	13 479	88.9	53.3
North America			
United States	13 310	95.9	64.2
Canada	13 230	93.8	65.1

[a] Source: United Nations Food and Agricultural Organization, "The State of Food and Agriculture." U. N. (FAO), New York, 1970.

---

[1] A reasonably detailed but quite readable description of the energy transfer process within living cells appears in A. L. Lehninger, "Bioenergetics." Benjamin, New York, 1965.

supply of *protein*. (The word *protein* is derived from Greek *proteios,* primary or fundamental.) During normal respiration and metabolism small amounts of protein will be consumed as fuel. Under conditions of stress such as inadequate supply of fats and carbohydrates, much larger quantities of proteins and their constituent amino acids may be destroyed. Because of the many important functions of protein and enzyme molecules in the human body, such losses must be replaced either by synthesis of the needed material within the body or its ingestion with food.

Polymeric protein molecules are readily synthesized in the human body under the direction of DNA molecules, provided that an adequate supply of their constituent amino acids is present. Of the twenty amino acids whose structures appear in Figure 3.2, ten can be synthesized in human children and twelve in adults. The eight or ten "essential" amino acids which cannot be synthesized must be constituents of the protein portion of human diet if proper growth and development is to occur. Once the ingested protein is hydrolyzed by digestive processes these individual amino acids become available for incorporation into different proteins or as starting materials for synthesis of nonessential amino acids. Table 18.2 indicates minimum daily requirements of amino acids in young men, young women, and infants.

Most protein sources will not be utilized at 100% efficiency because Liebig's law of the minimum (Section 14.3) applies to protein synthesis as well as to eutrophication. Construction of a polymeric protein molecule cannot proceed if even one of the constituent amino acids is unavailable. Thus protein *quality,* as determined by the ratios of amino acids present, is just as important as quantity. Table 18.3 indicates the quality of a number of protein sources. To summarize the quality of protein in different food-stuffs in a single number the concept of *net protein utilization* (NPU) has

TABLE 18.2 Minimum Requirements of Essential Amino Acids[a]

Essential amino acid	Young men (g/day)	Young women (g/day)	Infants (mg/kg day)
Leucine	1.10	0.62	150
Isoleucine	0.70	0.45	119
Lysine	0.80	0.50	103
Threonine	0.50	0.31	87
Tryptophan	0.25	0.16	22
Valine	0.80	0.65	105
Methionine	1.10	0.29	45
Phenylalanine	1.10	0.22	90
Histidine	—	—	34

[a] Data from President's Science Advisory Committee, "The World Food Problem," Vol. II, Report of the Panel on World Food Supply, p. 52. The White House, Washington, D. C., 1967.

TABLE 18.3 Percentage of Ideal Concentration of Essential Amino Acids Observed in Typical Proteins (using egg protein as 100%)[a]

Foodstuffs	Histidine	Threonine	Valine	Leucine	Isoleucine	Lysine	Methionine	Phenylalanine	Tryptophan
Beef	157	90	73	87	84	141	84	70	92
Fish muscle	124	96	86	106	105	148	100	79	109
Soybean meal, low fat	138	80	76	89	97	111	53	95	127
Whole rice	81	78	88	91	84	52	106	89	118
Whole wheat	100	67	62	78	64	44	78	91	109
Cottonseed meal	128	61	69	67	64	57	53	107	118
Whole corn	119	76	76	167	103	38	97	89	55
Peanut flour	100	57	66	79	66	57	25	88	72
Dried roast beans	104	79	78	78	89	106	62	89	73
Sesame meal	106	81	67	70	63	38	53	78	93

[a] Source: President's Science Advisory Committee, "The World Food Problem," Vol. II, Report of the Panel on World Food Supply, p. 315. The White House, Washington, D. C., 1967.

been developed. This measure takes account of the digestibility as well as the amino acid content of a protein source to determine how much is actually available to the human body. As an example of NPU consider the peanut flour listed in Table 18.3. Since it contains relatively small proportions of threonine, lysine, and especially methionine, its NPU is only about 40% as great as egg protein. (The latter has been chosen as a standard since its proportions of amino acids are nearly an exact match for human requirements.) The relative quantities and qualities of protein from various sources are indicated in Figure 18.1. Note that as a general rule protein derived from animal sources such as meat, poultry, and especially fish, milk, and eggs is of higher quality than that derived from grain. Most of these sources also contain larger percentages of protein as well. Although it is not entirely accurate (because of variability among different sources), the generalization that animal protein is high quality and plant protein low is often made. Hence, the distinction between the two in Table 18.1.

In the less-developed countries both quality and quantity of protein are sometimes deficient. Since requirements for protein are most crucial during periods of rapid growth and/or physiological stress, growing children and pregnant or lactating mothers are especially susceptible to protein deficiencies. Growth of the brain or other organs may be retarded and diseases such as kwashiorkor become prevalent. The latter affects newly weaned children whose protein-rich diet of mother's milk is replaced by starchy cereals or roots. Even though an adequate number of Calories is available,

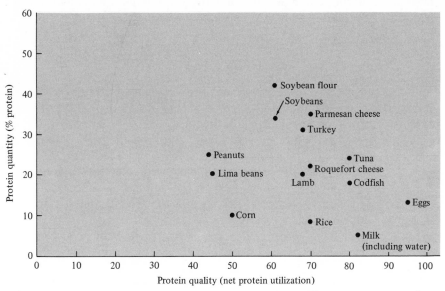

Figure 18.1 Relative quantity and quality of protein from different sources. Data from: F. M. Lappé, "Diet for a Small Planet," p. 48. Ballantine, New York, 1971.

protein deficiencies may exert very harmful effects both on the individual and on society as a whole.

Nearly the reverse of this situation is found in developed countries such as the United States which obtain the majority of their protein from animals. When more than the necessary quantities of essential amino acids are ingested the excess is treated as if it were carbohydrate or fat and burned as fuel, resulting in considerable waste of valuable protein. Moreover, despite the higher quality protein derived from animals, their large requirements for feed increase the acreage needed to provide the minimum daily requirements of amino acids for a human.

Overuse of animals as food also wastes large numbers of Calories, especially when the animals are fed grain (which could be consumed by humans) instead of being allowed to graze. The average animal converts less than 20% of its total feed to edible portions of the carcass. For beef cattle this percentage is less than five. Borgstrom[2] contends that the disparity in carbohydrate and fat consumption shown in Table 18.1 between countries such as India and the United States is a gross underestimate because it ignores the primary plant food consumed by animals which later will be consumed by man. When this is taken into account the daily Caloric consumption per capita in India increases by about 3350 kJ to 10 930 kJ, but United States consumption rises by 41 860 kJ/day. The total United States consumption, 55 170 kJ/day, is approximately five times that of India. Figures such as these have led a number of observers to the conclusion that a great many starving and undernourished citizens of the less-developed countries could be fed if only the United States and other developed countries reduced their overconsumption of meat protein.

18.2 THE GREEN REVOLUTION

Since a considerable fraction of the world's population is already under-nourished and population is expected to double in less than 50 years, the problem of providing for adequate nutrition is a difficult one. Most of the best agricultural land is already under cultivation and the most productive areas of the oceans are being fished at or near capacity. For the next 25–50 years the most satisfactory means of achieving adequate food supplies appears to be the improvement of crop yields on existing arable land. A number of highly developed countries such as the United States, Canada, Australia, and New Zealand have dramatically increased crop yields in recent decades. The "Green Revolution" refers to an attempt to transfer techniques learned in industrialized nations to the farms of the less-developed world.

---

[2] G. Borgstrom, "The Hungry Planet," pp. 27ff. Macmillan, New York, 1965.

The Green Revolution depends on the introduction of new plant strains with higher yield potential together with improved methods of husbandry. High yield varieties (HYV's) of rice and wheat have been bred for shorter straw and greater proportions of grain as well as increased adaptability over latitude, elevation, and other environmental factors. In order to obtain maximum benefit from these HYV's irrigation and more efficient use of water, more uniform and often mechanized methods of cultivation, increased use of fertilizers and pesticides, a more highly developed distribution system, and governmental and economic stimuli for greater harvests are necessary. The introduction of Green Revolution agriculture in India and Pakistan in 1967 was highly successful with yields of wheat increasing by as much as 270% and rice by 150% over yields of traditional varieties. As the revolution has spread to less fertile land in addition to the extremely high quality acreage chosen for the initial plantings, these percentages have dropped somewhat, but there is no doubt that large increases in yield are being achieved. Several other factors, however, have led to criticism of the Green Revolution and concern that too high a price may be paid for the larger quantities of food it produces.

The first of these is social. Because of economies of scale and the need for capital to purchase seed, tractors, pesticides, and fertilizers, rich farmers with larger land holdings are usually the first to adopt Green Revolution methods. In many cases governmental subsidies are offered to induce initial experimentation. Large profits accrue from increased crop yields as well as these subsidies, but by the time smaller farmers begin to employ Green Revolution techniques, the subsidies have usually ended and larger supplies of grain have reduced prices to the point that the poorer farmer achieves little increase in profit. Moreover, although the Green Revolution crops require considerably more care, much of this must be provided by mechanization. This increases rural unemployment and migration of farmers to the cities (as exemplified by the shift in United States population during the past half-century). The Green Revolution has the potential (and in general it seems it will be realized) for increasing the disparity in income between rich and poor.

A second latent difficulty is the vulnerability of Green Revolution crops to epidemics and pests. This largely results from genetic uniformity of the HYV's and from the recommended growing practices. Monoculture (growing a single variety of a single crop over a large area), luxuriant foliage, multiple cropping, close planting, and the transplantation of HYV's developed in one region to another all increase vulnerability to catastrophic attack by disease or insects.

The introduction of diverse new HYV's of wheat has reduced genetic uniformity, but if susceptibility to disease is linked to the dwarfing gene which is essential to HYV's they could all be infected simultaneously. The epidemic of southern corn blight which destroyed one-fifth of the United

States crop in 1970 was caused by a genetic linkage of this type, and an epidemic of tungro, which destroyed several hundred thousand hectares of rice in 1970 and 1971, has been attributed[3] to the Green Revolution practice of double cropping. In the case of the United States corn blight, disaster was only averted by intensive efforts of plant breeders to produce seed which was resistant to the blight prior to sowing for the 1971 growing season. While such breeding for resistance is possible it often leads to evolution of new strains of pests and may require the full-time effort of large numbers of scientists in well-equipped facilities. For example, the useful lifetime of a new wheat variety in the northwestern United States is about 5 years. There is some question as to the capability of less-developed countries to afford the facilities and trained personnel needed to keep up with such rapid changes.

Nutrition was not a major factor in the original breeding of HYV's, especially with regard to high quality protein. More recently, improvement of protein content has become an important goal, because HYV's tend to

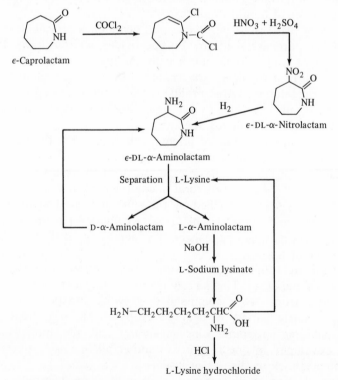

Figure 18.2   Synthesis of lysine from ε-caprolactam.

---

[3] National Academy of Sciences, "Genetic Vulnerability of Major Crops," pp. 183–187. Nat. Acad. Sci., Washington, D. C., 1972.

replace more nutritious crops such as beans, peas, and lentils as well as older varieties of wheat and rice. Thus, greater yields are offset by the fact that current HYV's contain half as much protein as some of the crops they replace. Poor quality may also be alleviated by direct chemical synthesis of the essential amino acids which are lacking. For example, lysine synthesized from ϵ-caprolactam (Figure 18.2) may be used as a supplement in those grains whose protein contains only a small percentage of the required amount (Table 18.3). (Caprolactam is a readily available industrial organic chemical—about 32×10^6 kg of Nylon-6 were synthesized from it in 1966.) Like breeding for resistance to pests, however, direct synthesis requires resources which may not be available to less-developed nations. The net effect to date, taking account of increases in population, seems to be very minor, if any, improvements in average diets as a result of the Green Revolution,[4] although just keeping up with population growth is no mean feat.

Another important general criticism of the Green Revolution is that it has not been tailored to the needs of less-developed countries but instead represents United States agricultural technology transferred abroad. This leads to greatly increased usage of fertilizer, pesticides, and fuel for tractors and other mechanized farm implements. Although peasant farmers tend to use less than the recommended quantities of fertilizer and pesticide, the same ecological consequences that have been observed in the United States may be expected. Moreover, it is highly questionable whether the energy intensiveness of United States agriculture is appropriate for countries which lack fossil fuel resources and cannot afford to import them. These aspects of United States "agribusiness" will be discussed in the next few sections.

18.3 PEST CONTROL

The production of fertilizers and some of the ecological consequences of their use have been discussed (Sections 11.11 and 11.12) and need not be considered further here. The effects of pesticides on the environment have not yet been elaborated, however, and we shall turn our attention to them now. The term *pesticide* refers to any substance which can poison or otherwise eliminate an organism (plant or animal) which is considered by man to be a pest. Pesticides may be classified in a variety of ways: (1) according to the target organism—insecticides, herbicides, fungicides, etc.; (2) according to chemical properties—inorganic compounds, chlorinated hydrocarbons, organophosphates, carbamates, etc.; and (3) according to their mode of action—for example, insecticides may be classified as stomach poisons, contact poisons, etc. The first set of categories is the one most

---

[4] N. Wade, *Science* **186,** 1187 (1974).

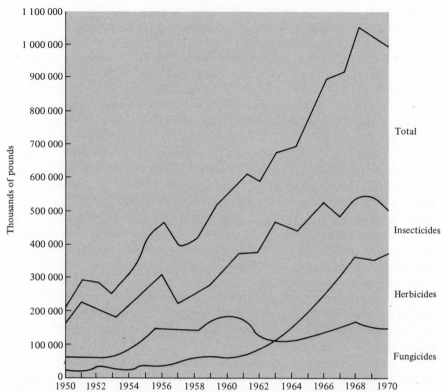

Figure 18.3   United States production of synthetic pesticides. Source: "Integrated Pest Management," p. 3. Council on Environmental Quality, Washington, D. C., 1972.

commonly used. Figure 18.3 indicates the rapid growth of production of a number of these classes during the past quarter century.

Prior to 1940 the majority of pesticides were inorganic materials, most of which were naturally occurring, and even the organic compounds (such as turpentine, nicotine, and pyrethrum) did not utilize the synthetic chemist's skills to any great degree. In the case of insecticides compounds such as lead and calcium arsenates, Paris green [copper acetoarsenite, $Cu(C_2H_3O_2)_2 \cdot 3\ Cu(AsO_2)_2$] and fluorides such as cryolite (Na_3AlF_6) and sodium fluoride were commonly used. All of these are highly poisonous to man and other animals as well as to insects. The discovery in 1939 of the insecticidal properties and relatively low human toxicity of DDT [the initials derive from *d*ichloro*d*iphenyl*t*richlorethane, an example of organic nomenclature which might better be replaced by the systematic 2,2-bis(*p*-chlorophenyl)-1,1,1-trichloroethane] began a period during which large numbers of chemists attempted to synthesize ever more nearly ideal pesticides.

Most of the new compounds were completely unknown to any ecological system on earth and many of them were quite persistent, decomposing very slowly under environmental conditions. By the late 1950's there were documented cases of decimation of bird populations as a result of broad-scale spraying programs[5] and in the 1960's contamination of entire watersheds (such as that of Lake Michigan) by DDT residues was not uncommon. Because of their persistence, affinity for nonaqueous, lipid tissues, consequent concentration along food chains (see Section 16.3), and the rapid growth in their rate of production, organic pesticides became a major environmental hazard. By 1972 they were ranked first on a semiquantitative list of environmental stresses.[6]

This potential for damage does not go completely unrecognized at the time that a new pesticide is introduced into use. In the United States all pesticides must be registered with the U. S. Department of Agriculture. An application for such registration must give evidence that the compound is useful for its intended purpose, and that it is not hazardous to humans or wildlife. In addition, its chemical composition and conditions of use must be clearly specified. If the pesticide is to be used on food products it also comes under the Federal Food, Drug and Cosmetic Acts, and the U. S. Department of Health, Education and Welfare is required to set a tolerance level for human consumption. The chemical identity of the substances in the pesticide formulation must be specified, and toxicity data showing its effects on laboratory animals must be submitted. Analytical techniques must be developed so that the amount of residue remaining after the compound has been applied according to the manufacturer's directions can be determined. The fact that such techniques are available has in many cases made possible the detection of pesticides in the biosphere and has aided in linking them to their unwanted side effects. A further important restriction is the so-called Delaney clause of the 1958 Federal Food and Drug Act, which specifies that no amount of any substance which causes cancer in test animals or humans may be included in any food product.

The structures of all pesticides whose production in the United States exceeded 10^7 lb/year in 1967 are shown in Figure 18.4. Some other compounds produced in smaller quantities or by decomposition of the primary pesticides are also included. The figure is organized by chemical properties with the target organisms indicated after each name. As a general rule the organochlorine compounds decompose more slowly than the others, with DDT being especially persistent. Typical half-lives range from 1 to 10 years for DDT, a few months for aldrin, less than a month for parathion, and a few days for malathion. The carbamates and organophosphorus com-

---

[5] For references to this and other effects of pesticides within the state of Michigan, see H. E. Johnson and R. C. Ball, *Advan. Chem. Ser.* **111**, 1–10 (1972).

[6] *Chem. & Eng. News,* January 10, p. 33. (1972).

a. Dienes:

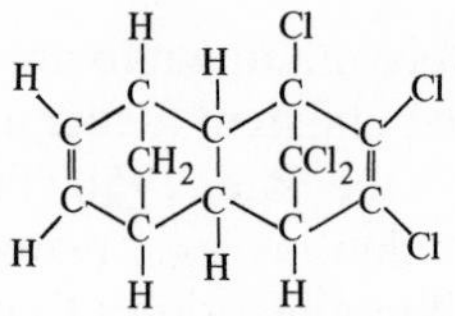

Aldrin (i)

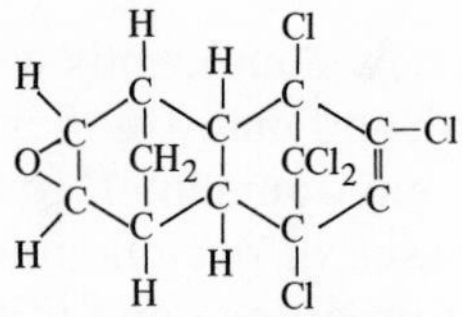

Dieldrin (i)   Endrin (i)
endo, exo   endo, endo

b. Chlorophenyls:

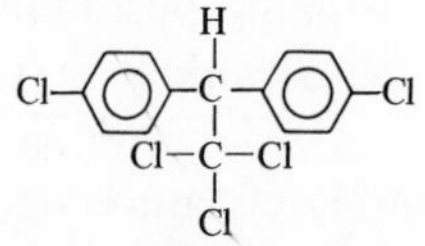

DDT (i)

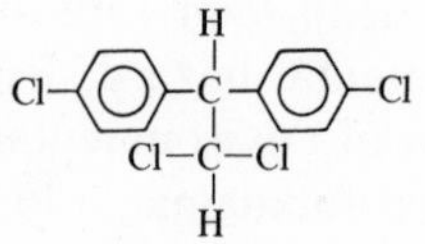

DDD (d)

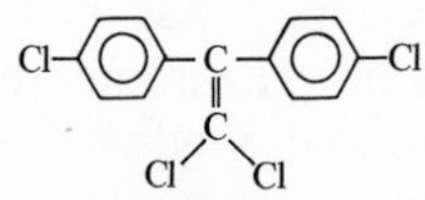

DDE (d)

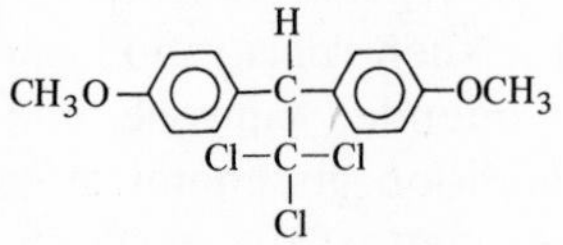

Methoxychlor (i)

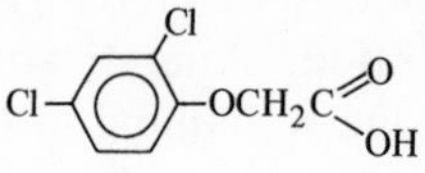

2,4-D (h)

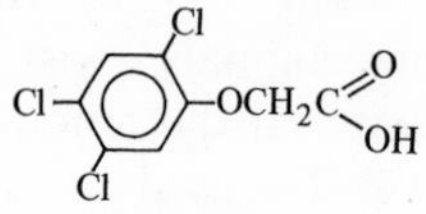

2,4,5-T (h)

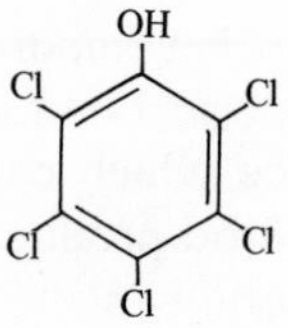

Pentachlorophenol (f)

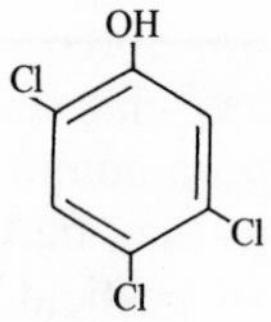

2,4,5-trichlorophenol (f)

c. Other:

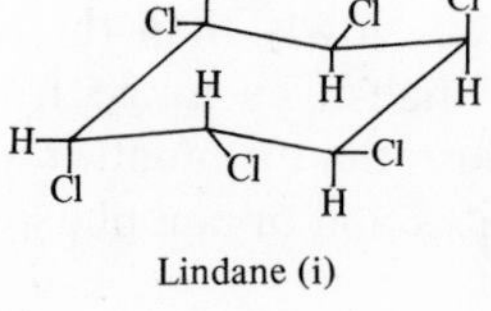

Lindane (i)

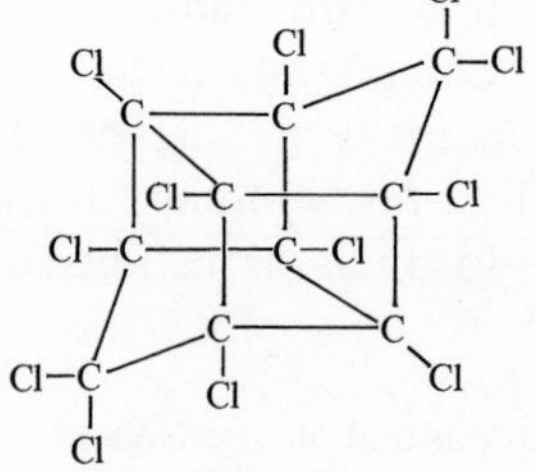

Mirex (i)

(A)

pounds are highly toxic to humans and generally require extreme care in handling. The difference between the effects on men and insects of the latter two classes are almost entirely accounted for by the difference in body weight, whereas for DDT a larger dose per kilogram of body weight is required to kill a human than an insect.

18.4 PERSISTENCE OF PESTICIDES

Persistence is important because the amount of pesticide residue in the environment depends on the rate at which the chemical compounds involved decompose to less harmful substances. This problem in applied chemical kinetics is complicated by the lack of control over the conditions

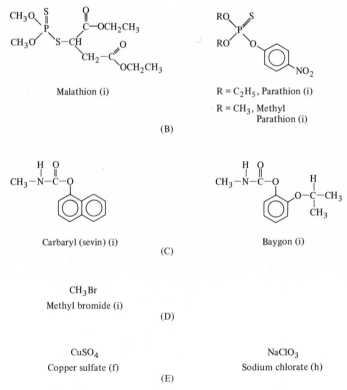

Malathion (i)

(B)

R = C$_2$H$_5$, Parathion (i)

R = CH$_3$, Methyl
Parathion (i)

Carbaryl (sevin) (i)

(C)

Baygon (i)

CH$_3$Br
Methyl bromide (i)

(D)

CuSO$_4$
Copper sulfate (f)

NaClO$_3$
Sodium chlorate (h)

(E)

Figure 18.4 Structures of common pesticides. (A) chlorinated hydrocarbons, (B) organophosphorus compounds, (C) carbamates, (D) other organics, and (E) inorganics. Compounds are identified according to the following code for target organism; i, insecticide; f, fungicide; and h, herbicide. If a structure listed is a decomposition product this is indicated by (d). Data from "Cleaning Our Environment: The Chemical Basis for Action," pp. 196–198. Amer. Chem. Soc., Washington, D. C., 1969.

TABLE 18.4   Accumulation of Pesticide Residues Assuming First-Order Decomposition

Half-life for residue disappearance (months)	Maximum accumulated residue (fraction of total annual dosage)	
	Single dose	Quarterly dose
1	1.00	0.29
2	1.02	0.40
3	1.07	0.50
6	1.33	0.85
12	2.00	1.57
24	3.40	3.04
36	4.84	4.44
48	6.29	5.90
120	14.9	13.7

(temperature, moisture, catalysts, etc.) under which the decomposition reactions must proceed. Often a first-order rate of decomposition is assumed, since this allows a relatively simple calculation of the amount of residue which may be expected to remain under steady-state conditions of repeated applications of the same quantity of pesticide. The length of time required for one-half of a given quantity of a substance to decompose is known as the half-life, and under first-order conditions it is independent of the quantity of substance present.

The figures in Table 18.4 give the ratio of the maximum accumulated residue to the total yearly dosage, assuming first-order kinetics, for a variety of half-lives. For nonpersistent compounds (half-lives of a few months or less) a considerable reduction in maximum soil residue may be effected by spraying four times per year at one-fourth the dosage, but when half-lives of 2–10 years are involved this technique merely slows down the rate at which nearly the same maximum soil accumulation is achieved. It should be clear from the table why considerable pressure has been exerted by environmentalists to insure the use of pesticides whose half-lives are relatively short.

Of course, using first-order kinetics to predict the accumulation of pesticides in the environment is a gross oversimplification. Many other factors must be taken into consideration. If there are large quantities of organic matter present in a soil, for instance, pesticides adsorbed on them will be kept from contact with air and will decompose more slowly. The pesticides are also less effective in such situations because they are less likely to contact the pests they are designed to eliminate. High levels of moisture often reduce the half-lives of pesticides in soil by helping to vaporize some compounds (by a process analogous to steam distillation) and causing increased rates of hydrolysis of others. Since chemical reactions usually occur more rapidly at higher temperatures, half-lives of pesticides are decreased in hot weather. In addition, the more volatile compounds are vaporized and exposed to air and sunlight which may accelerate other pathways for decomposition.

Even though a pesticide may decompose to another compound there is no guarantee that the product molecule will have less toxic properties than the parent. For instance, one of the principal decomposition products of DDT is DDE, which has been implicated in eggshell thinning in birds, and when aldrin decomposes it produces dieldrin, a slightly more toxic com-

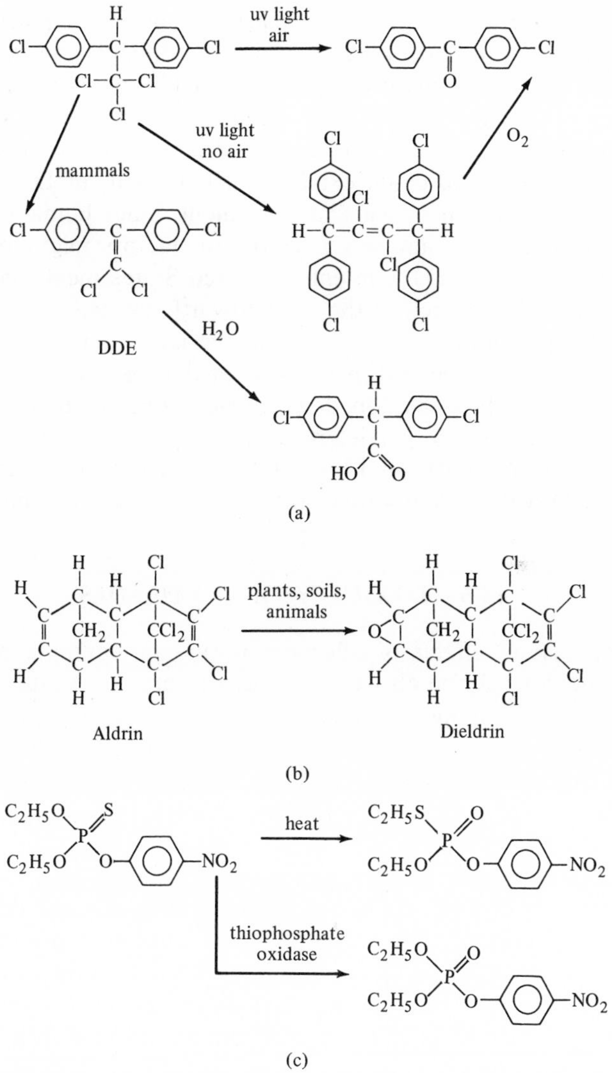

(a)

(b)

(c)

Figure 18.5 Pathways for decomposition of pesticides: (a) DDT, (b) aldrin, and (c) parathion.

pound. The reaction pathways by which pesticides decompose are also different under different conditions. It is essential in evaluating the environmental impact of any pesticide that the toxic effects of all possible decomposition products be taken into account in addition to the effects of the compound itself. The difficulty of identifying all such products and determining their effects makes the use of pesticides even riskier. Pathways for decomposition of DDT, aldrin, and parathion are indicated in Figure 18.5.

Residues of relatively persistent pesticides such as DDT and dieldrin in soil, air, and water are usually in the parts per billion or parts per trillion range—low enough to make accurate analyses difficult and too low to account for the quantities found in human diets. The phenomenon of bioamplification is quite efficient for these compounds, however, since they are quite lipid-soluble. Instances are known where birds of prey and other organisms high in a food chain exhibit concentrations of DDT more than 10^7 times those in the air or water environment. Since humans are often at the top of food chains, similar concentration factors might be expected. Actually, the average DDT content of United States meals has decreased steadily since 1954, reflecting the shift toward the use of less persistent pesticides in this country. At no time was the minimum level above which harmful effects were observed in test rats or dogs exceeded. On the other hand, much of the less-developed world has come to depend more and more on DDT and other persistent pesticides for control of agricultural and disease-carrying pests. In 1964, DDT concentrations found in Indian citizens were between 2 and 4 times those observed in the United States or Western Europe.

18.5 TOXICITY OF PESTICIDES

As in the case of effects of other environmental contaminants, the toxicity of pesticides may be classed as acute or chronic. Acute effects manifest themselves upon ingestion or other contact with a poison and are usually measured by an LD_{50} value. This is the dosage which is lethal to 50% of an exposed population. The LD_{50} values for a variety of pesticides are indicated in Table 18.5. Note that the smaller the LD_{50}, the greater the toxicity of the compound.

Chronic toxicity may involve much smaller doses than the LD_{50}, but over long periods of time. As in the vinyl chloride example discussed earlier, chronic effects may not become evident until after a lengthy latency period and are therefore much less readily detected. In the case of the less persistent pesticides, which do not remain in the environment for long periods, chronic toxicity is less of a problem, and there is little evidence so far demonstrating chronic effects in man even for persistent substances like DDT. On the other hand, aldrin and dieldrin have recently been banned[7] as

---

[7] *Chem. & Eng. News,* August 12, p. 4 (1974).

TABLE 18.5 Lethal Dosage (LD_{50} oral) for Pesticides in Test Rats[a]

Pesticides	LD_{50} oral (mg/kg)
Insecticides	
Aldrin	54–56
Arochlor (PCB)	250
Carbaryl	540
DDT	420–800 (60–75 dog)
Dichlorvos	56–80
Dieldrin	50–55
Endrin	5–43
Heptachlor	90
Lead arsenate	825 (192 sheep)
Lindane	125–200
Malathion	480–1500
Methoxychlor	5000–6000
Mirex	300–600
Nicotine	50–60
Parathion	4–30
Paris Green	22
Pyrethrins	820–1870
Herbicides	
2,4-D	666
Dioxin (impurity in 2,4,5-T)	0.03
Pentachlorophenol	27–80
Sodium arsenite	10–50
2,4,5-T	300 (100 dog)
Fungicides	
Copper sulfate	— (produces jaundice in sheep and chickens upon several months' exposure)
Mercurials (ethylmercury-*p*-toluenesulfonanilide)	100

[a] Note that the larger the LD_{50}, the less toxic the substance. Data from D. Pimentel, "Ecological Effects of Pesticides on Non-Target Species." Executive Office of the President, US Govt. Printing Office, Washington, D. C., 1971.

a result of demonstrated carcinogenicity in test animals. Since humans have a longer latency period, chronic effects may not yet have had time to appear. Certainly, it is prudent to use less persistent pesticides whenever possible, provided that their decomposition products are not harmful.

The purity of the substances involved in pesticide formulations also relates to their toxicity. This is perhaps best illustrated by the case of the herbicide 2,4,5-trichlorophenoxyacetic acid (2,4,5-T), which was found to provoke birth defects in rats as a result of a 1964 screening by the National Cancer Institute. Between this screening and governmental action to restrict its use in 1969 large quantities of 2,4,5-T as well as the related 2,4-dichlorophenoxyacetic acid (2,4-D) were sprayed as defoliants in

Vietnam. Some investigators attributed all teratogenic effects of 2,4,5-T to an impurity, dioxin (2,3,7,8-tetrachlorodibenzo-*p*-dioxin) TDD, whose structure is

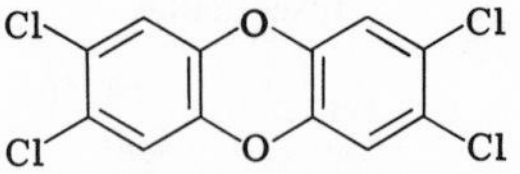

Dioxin (TDD)

Others claim that both the herbicide and the far more toxic dioxin are responsible for teratogenic effects in certain experimental animals. The data in Table 18.5 indicate that with regard to acute toxicity dioxin is an extremely hazardous substance.

The quantity of dioxin in 2,4,5-T is very dependent upon the conditions of manufacture. The synthesis of the herbicide is indicated in Figure 18.6, as is the side reaction which produces the dioxin impurity. Attempts to produce more 2,4,5-T from the same plant by increasing the temperature and pressure of the first reaction in Figure 18.6 also led to higher concen-

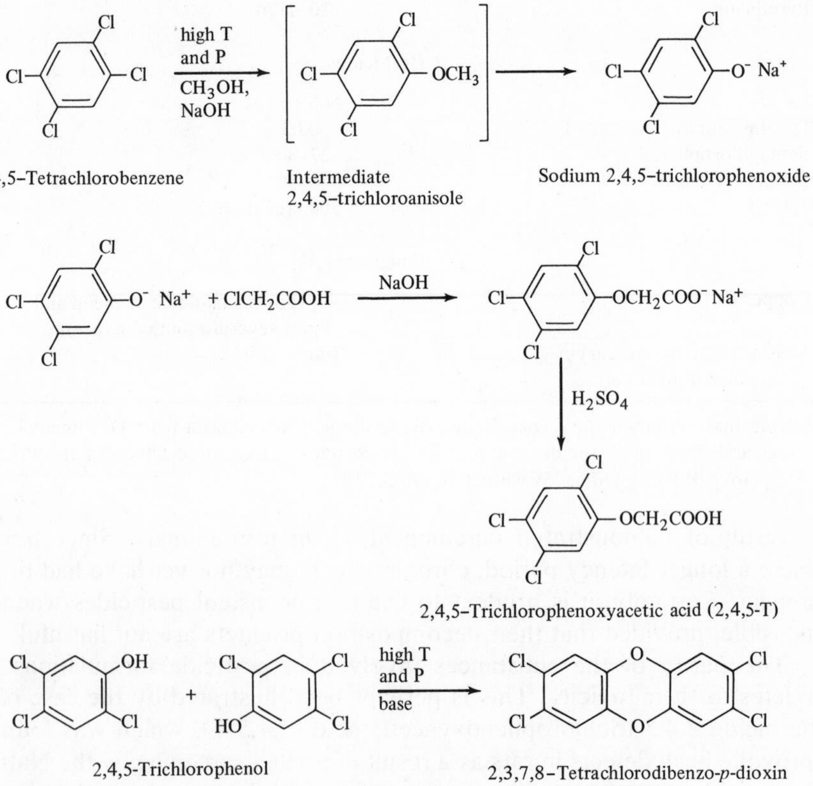

Figure 18.6 Synthesis of 2,4,5-T and dioxin.

trations of dioxin, because condensation of 2,4,5-trichlorophenol to form the latter requires similarly stringent conditions. Dow Chemical Company, which discovered the presence of dioxin as a result of an outbreak of chloracne among workers, closed its 2,4,5-T plant early in 1965 until methods were found to remove the impurity. Other companies, reacting to the demand of the military for defoliants to be used in Vietnam, did not adopt such procedures until 1969, by which time dioxin levels in all supplies of 2,4,5-T had been reduced below 1 ppm.

The point of this brief excursion into herbicide manufacture is to emphasize that all pesticide formulations containing the same active ingredient are not necessarily alike. Indeed this point can be extended to include synthetic chemicals in general use since their purities will depend on the conditions of synthesis. Most technical grade 2,4,5-T still contains 8–10% impurities, not counting various isomers of trichlorophenoxyacetic acid. Some samples contain as little as 80% 2,4,5-T. Moreover, the actual material applied in the field would also contain various "inert" ingredients—solvents, fillers, etc.—each of which might also contain impurities. All of these must be considered when evaluating the environmental impact of pesticide use.

18.6 MODE OF ACTION OF PESTICIDES

In many cases the precise mechanisms by which pesticides kill plants or animals are not known. The discovery of DDT's properties was somewhat fortuitous, and much pesticide development has consisted of synthesizing and testing structural analogs of substances known to be effective. Known or suspected modes of pesticide action are listed in Table 18.6. Although some other mechanisms do contribute, herbicides generally act by interrupting photosynthesis or, as in the case of 2,4-D and its analogs, promoting excessively rapid growth. Insecticides almost invariably act on the central nervous system, the part of the body of most animals which is least able to tolerate brief interruption of normal function.

On a molecular scale the action of 2,4-D apparently results from a complex series of reactions initiated by derepression of the gene regulating synthesis of the enzyme RNase. This permits excessive synthesis of RNA and protein which migrate to the stem of the plant, where cell division and proliferation occur on a massive scale. The plant's circulatory system is disrupted and the roots die of starvation.

The effects of insecticides on nerve transmission may be divided into two classes, depending on the point at which they occur. A *neuron* or nerve cell usually consists of a main body from which extend several short branches (dendrites) and one extremely long one (the *axon*). Ordinarily, the dendrites carry nerve impulses to the cell body and the axon carries them away. *Axonic transmission* refers to this propagation of the signal within

TABLE 18.6   Mode of Action of Pesticides[a]

Type of effect	Example compounds
Effects on Plants	
Inhibition of photosynthesis	Triazines; carbamates; substituted ureas
Inhibition of chlorophyll synthesis	3-Amino-1,2,4-triazole
Inhibition of oxidative phosphorylation (energy transfer)	Carbamates
Hormone analogs	2,4-D; 2,4,5-T
Inhibition of pantothenate (vitamin B) synthesis	Chlorinated aliphatic hydrocarbons
Unknown causes	Metals; sulfur
Effects on Animals	
Inhibition of acetylcholinesterase	Organophosphates; carbamates
Inhibition of neuromuscular junction	Nicotinoids
Neurotoxication with only partially known causes	Pyrethroids; chlorinated and brominated hydrocarbons

[a] Data from S. S. Epstein and M. S. Legator, "The Mutagenicity of Pesticides," p. 52. MIT Press, Cambridge, Massachusetts, 1971.

the neuron. It is accomplished by a reversal of electrical polarity, called an *action potential,* which moves rapidly along the axon. This is related to sudden changes in the permeability of the cell membrane to Na^+ and K^+ ions, but the causes of these changes remain obscure. When the nerve impulse reaches the end of the axon it is transmitted to an adjacent cell across the *synapse,* a distance of approximately 50 nm. Two substances, acetylcholine and norepinephrine, are known to be involved in this *synaptic transmission*:

$$CH_3\overset{\overset{\displaystyle O}{\|}}{C}OCH_2CH_2\overset{+}{N}(CH_3)_3$$

Acetylcholine

Norepinephrine

These neurotransmitters apparently diffuse across the synapse and trigger an appropriate response in the adjacent cell.

In order to restore the sensitivity of the synapse to additional impulses the neurotransmitters must be eliminated. Acetylcholine is promptly removed by the enzyme acetylcholinesterase (ACHE) which hydrolyzes it as shown in equation 18.2:

$$CH_3\overset{\overset{\displaystyle O}{\|}}{C}OCH_2CH_2\overset{+}{N}(CH_2)_3 \xrightarrow[H_2O]{ACHE} CH_3\overset{\overset{\displaystyle O}{\|}}{C}OH + H_2\overset{\overset{\displaystyle OH}{|}}{C}CH_2\overset{+}{N}(CH_3)_3 \qquad (18.2)$$

In the case of norepinephrine the enzyme monoamine oxidase carries out a similar function, but diffusion of the neurotransmitter away from the synapse is also quite important. Organophosphates and carbamates interfere with synaptic transmission by inhibiting ACHE, while nicotinoids interfere with synaptic transmission between nerve and muscle cells. On the other hand, DDT and its analogs apparently affect axonic transmission, producing prolonged volleys of impulses which result in tremors and finally paralysis.

The mechanism of ACHE action and inhibition is an interesting example of applied enzyme kinetics. It is diagrammed in Figure 18.7. The enzyme is indicated as EOH because of the presence of an —OH in a serine residue at the active site. Each substrate is indicated by AX where A represents a group (such as acetyl) from which the —OH can displace the X group to form a relatively stable molecule EOA. In both hydrolysis and inhibition the complex EOH · AX is formed rapidly and is fairly stable as indicated by relatively large values of the equilibrium constant $K_f = k_1/k_{-1}$. When acetylcholine is the substrate rate constants k_2 and k_3 are both large and the active enzyme EOH is rapidly regenerated. In the case of an organophosphate such as parathion k_2 is reasonably large but k_3 is extremely small and phosphorylated enzyme, EOA, accumulates. Practically no EOH or EOH · AX remains and acetylcholine cannot be hydrolyzed. With carbamates k_2 and k_3 are relatively small and K_f is exceptionally large, al-

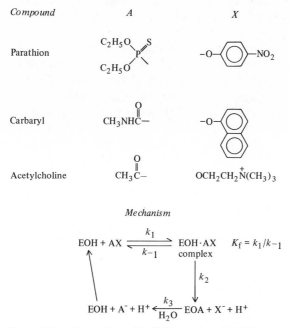

Figure 18.7   Mechanism of ACHE action and inhibition.

lowing most of the enzyme to accumulate as EOH · AX and EOA. In this case removal of the carbamate permits regeneration of active enzyme over a period of hours via dissociation of EOH · AX and slow decomposition of EOA. Thus, poisoning by carbamates is more readily treated than poisoning by organophosphates.

One final point needs to be made regarding the mode of action of insecticides. Much of the discussion above applies to the neurobiochemistry of vertebrates. Insects differ in that their neuromuscular junctions do not involve acetylcholine as transmitter, and the nature of the substance involved is unknown. Thus, although inhibition of ACHE is believed to be important in killing insects, it does not always correlate directly with toxicity. Metabolism of carbamates or organophosphates to less toxic forms before they can reach less accessible portions of the central nervous system may explain why certain excellent ACHE inhibitors are not as poisonous to insects as might be expected.

18.7 INTEGRATED PEST MANAGEMENT

The problem of insect resistance has plagued producers of insecticides almost from the beginning of their use. For instance, it was observed in Italy in 1948, only 3 years after the introduction of DDT, that strains of fruit flies were developing upon which the insecticide had no effect. Because of the short life-span of most insects reproduction rates are very rapid. When an effective insecticide is introduced only those insects which are slightly resistant survive to reproduce. A rapid natural selection produces large numbers of resistant individuals, requiring that more and more of a given pesticide be used, or in some cases that an entirely different substance be developed. Thus, whether or not an insecticide produces harmful environmental effects, its lifetime is usually limited by the rate at which insect resistance develops. In fact, most of the development of new insecticides has not been spurred by the threat of their being banned on environmental grounds, but rather because they had been rendered ineffective by the development of resistant strains of insects.

As a corollary to the above problem, it is disturbing to consider the slow rate at which humans reproduce — relative to insects, that is. People are much less rapidly adaptable to the introduction of environmental hazards than are insects, and the ironic possibility that we might poison ourselves with some persistent compound does exist. Combined with the insect resistance problem this has encouraged the development of a variety of alternatives to pesticides. The term *integrated pest management* refers to efforts to select the combination of such alternatives which is most appropriate for control of a given pest under specific local conditions. Economic gains, convenience, environmental effects, and long-term risks and benefits are all considered in arriving at such a choice. Although it can also be effective with respect to fungi or other plant pests, most of the development of integrated pest management has been in the area of insect control, to which further discussion in this section will be confined.

Most insects locate members of the opposite sex for mating by means of *pheromones* or "sex attractants," compounds which often are extremely potent and specific. For instance, only 20 mg of the gypsy moth sex attractant, whose structure is shown below, could be isolated from half a million females.

$$cis\text{-}d\text{-}CH_3(CH_2)_5CHCH_2CH{=}CH(CH_2)_5CH_2OH$$

$$\underset{O}{\overset{\displaystyle |}{\underset{\diagdown}{}}}$$

$$\underset{O}{\overset{\diagup}{}}CCH_3$$

Gypsy moth pheromone

Furthermore, in order to be effective the compound must have a cis structure. Numerous programs are underway in which such substances are isolated, their structures are determined, and methods are devised for their synthesis. If the compounds are available at low cost, large-scale spraying may be used to excite male insects to such a frenzy that they cannot locate females for mating. Smaller quantities are effective as bait for traps in which insects may be destroyed. Although traps are not as effective in the case of massive infestations (the male insect is more likely to find a female than the trap), they also serve to monitor the number of insects in a given area. This serves as a clear indicator by which the necessity for, and success of, pesticide spraying programs can be judged. Use of pheromones in traps often can serve double duty in reducing the introduction of general poisons into the environment.

A second control technique for insect populations which may be combined with the sex attractant approach is sterilization. Male insects bred in captivity or attracted to a trap may be sterilized by treatment with a chemosterilant or by means of radioactivity. When released to mate with females, sterile males often displace potent ones, greatly reducing the population. The problem of resistance does not occur since only insects which were *not* treated produce offspring, and there is no connection between the sterilant and future generations. In 1955 a sterile-male release program was successful in completely eradicating the screwworm from the Caribbean island of Curacao. Some sterilants have been found which affect both males and females. Application of such compounds in the way that we now use insecticides can be even more effective because sterile males may mate with normal females and vice versa. Thus, a fraction of the population even larger than that directly affected by the initial treatment is involved.

Of course, chemosterilants have some drawbacks, too. They work best on insect species where females mate only once. If this is not the case a small number of potent males may have a very significant effect. Moreover, the large-scale introduction of chemosterilants into the environment, if they were sprayed as is now done with insecticides, might well have harmful consequences. When feasible, the best approach would appear to be a combination of sex attractants with sterilization.

While there are not rules to predict which molecular structures may be useful as chemosterilants, it is often true that classes of related compounds have similar effects. One important group of such compounds are the derivatives of aziridine such as the one whose structure is

(You will recall that aziridine fell into one of the categories of toxic compounds in Section 17.5.) Sterilants are usually more general in their effects than are sex attractants, allowing the control of several species with the same compound. A wide variety of them has been found (more than forty are known for the housefly alone), and they are often amenable to relatively inexpensive synthetic production. It seems clear that their use will increase as insecticides lose potency or public favor.

Another group of chemicals which are useful in controlling insect damage to food crops are the antifeeding compounds. The mechanism by which they prevent insects from feeding is not clear, but in most cases their toxicity to humans is low. The pests die of starvation because they are prevented from eating the protected crop. Other insects and animals are not affected. The major disadvantage of antifeeding compounds is that they protect against surface feeders only. Species such as the worms which bore into apples would not be affected. Furthermore, new growth is not protected unless the compound is reapplied often. This means that large quantities would usually be distributed into the environment.

In addition to these chemical methods of pest control several biological means are available. It is possible, for example, to introduce genetic defects into a species. In some cases winter hibernation has been eliminated, insuring that the insects will perish during the cold months. As greater knowledge of the molecular basis of genetics becomes available the number of such techniques should increase. Other biological controls include the introduction of predators or diseases which may eliminate pests. These are often very specific with regard to individual insect species, as in the case of *Reesimermis nielseni,* a nematode which parasitizes mosquito larvae but not other insects or animals. Fifty to one hundred percent control of mosquitos has been recorded in U. S. Department of Agriculture studies. Predators and parasites are especially effective in cases, such as that of the

Japanese beetle in the United States, where the pest itself has been imported into an ecosystem and has few natural enemies. However, extensive testing is required to insure that they will not turn to other prey or hosts when the species to be controlled is gone.

Plant breeding programs to develop resistant crops were mentioned in conjunction with the green revolution. Agricultural practices such as crop rotation, strip planting (where different species are planted in alternating, adjacent areas), timing of plantings to avoid pests, and removal of crop residues and plants which host insects between growing seasons are also effective in controlling infestations. However, United States agriculture and the Green Revolution have tended to replace these by practices such as irrigation, fertilization, and multiple cropping. The assumption was that chemical pesticides were a panacea and required no help from older, more mundane methods of insect control. Insect resistance and environmental damage have provided strong challenges to such assumptions.

Modern pest control is coming to rely on persons who describe themselves as "applied insect ecologists" and are paid by farmers for producing maximum agricultural benefits regardless of the management techniques involved. Although usually educated as entomologists, these experts must have multidisciplinary backgrounds involving botany, chemistry, and agriculture to insure that they give fair consideration to all methods of insect control, not just chemicals. It seems certain that such persons will continue to recommend the use of chemical pesticides in a great many cases where other methods are ineffective. Indeed, on a worldwide scale it is difficult to argue that even the most maligned of chemicals, DDT, has done more harm than good. Nevertheless, the fact that harmful effects are put off to the future and may eventually exceed beneficial ones argues persuasively that the cycle of initial optimism followed by backlash and eventual banning of persistent pesticides will occur globally as it did in the United States. The era in which it was thought that pesticides were the final solution to the intense competition for food supplies between man and insects is rapidly coming to a close.

18.8 ENERGY AND AGRICULTURE

Most persons tend to think of agriculture as a means of converting solar energy into chemicals whose reactions can do the work necessary to maintain the structures and functions of the human body. It is only very recently that we have come to realize that this is no longer the entire story as a result of the rapid increase in both direct and indirect usage of fossil fuels in the United States food system. For example, Pimental *et al.*[8] have

---

[8] D. Pimentel, L. E. Hurd, A. C. Bellotti, M. J. Forster, I. N. Oka, O. D. Sholes, and R. J. Whitman, *Science* **182**, 443–449 (1973).

estimated that 29.9×10^9 J of fossil energy was used by farmers to raise the average hectare (10^4 m^2) of corn in 1970. This represents about 11% of the total energy (including solar) utilized by the plants. (The total solar energy input is 21.1×10^{12} J/hectare during the growing season, but the corn makes use of only 1.26%.)

Steinhart and Steinhart[9] have considered the entire United States food system from the standpoint of energy consumption. Their analysis, which includes energy expended in food processing, commercial distribution, and home preparation, is summarized in Table 18.7. Using these figures they concluded that nearly 13% of the total United States fossil fuel consumption was applied to the food system. About 3% was directly related to farming while the remaining 10% was used for processing, distribution, and preparation of food. Modern technologized agriculture requires about two units of fossil energy input for every unit of food energy produced on the farm and between six and nine units to deliver one unit of food energy to the consumer's table. Delivery of the average daily requirement (about 45 g) of protein consumes more than 39×10^6 J of fossil energy. Between 1960 and 1970 food-related energy use grew at about 3.3% a year in the United States—faster than population but more slowly than total energy use.

A number of interesting facts become evident upon study of Table 18.7. On the farm itself direct use of fuel for running tractors and other machinery, drying crops, etc., is the largest consumer of energy, followed by the manufacture of tractors and other machinery. Fertilizer production is also very energy-intensive. It should be noted that the often quoted figure of fifty persons who can be fed by one farmer ignores the large number of people involved in off-farm manufacture of things like machinery and fertilizer. Even more nonfarm workers are involved if the processing, transportation, and preparation portions of the food system are included.

The energy used for irrigation does not appear to be too large until one realizes that only 5% of United States agricultural land is irrigated. Thus, for an irrigated farm energy requirements for pumping and distribution of water often outstrip all others in the table. This is of special concern with regard to the Green Revolution countries, where crops which did not require irrigation are being replaced by high yield varieties which do.

When farming was a "cottage industry" food processing fell under the category of household energy expenditure. The benefits derived from economies of scale on the farm are partly canceled by the introduction of the food processing and transportation industries required to get food from where it is grown to where its consumers live. The preservation of food over the long time spans required increases the energy intensiveness in the commercial and household sectors by increasing the need for refrigeration.

---

[9] J. S. Steinhart and C. E. Steinhart, *Science* **184**, 307–316 (1974).

TABLE 18.7   Energy Use in the United States Food System (in units of 10^{15} J)[a]

Component	1940	1950	1960	1970
On Farm				
Fuel (direct use)	293.02	661.39	786.97	971.15
Electricity	2.93	137.72	192.97	267.07
Fertilizer	51.91	100.46	171.63	393.48
Agricultural steel	6.70	11.30	7.12	8.37
Farm machinery	37.67	125.58	217.67	334.88
Tractors	53.58	128.93	49.39	80.79
Irrigation	75.35	104.65	139.39	146.51
Subtotal	521.16	1270.03	1565.14	2202.25
Processing Industry				
Food processing industry	615.34	803.71	937.66	1289.29
Food processing machinery	2.93	20.93	20.93	25.12
Paper packaging	35.58	71.16	117.21	159.07
Glass containers	58.60	108.84	129.77	196.74
Steel cans and aluminum	159.07	259.53	360.00	510.69
Transport (fuel)	207.63	426.97	641.71	1033.52
Trucks and trailers (manufacturing)	117.21	207.21	185.02	309.76
Subtotal	1196.36	1898.35	2392.30	3524.19
Commercial and Home				
Commercial refrigeration and cooking	506.51	627.90	779.43	1100.92
Refrigeration machinery (home and commercial)	41.86	104.65	133.95	255.35
Home refrigeration and cooking	603.62	846.83	1157.85	2009.28
Subtotal	1151.99	1579.38	2071.23	3365.55
Grand total	2869.51	4747.76	6028.67	9091.99

[a] Data based on J. S. Steinhart and C. E. Steinhart, Energy use in the U. S. food system. *Science* **184**, 307, Table 1 (1974).

In many cases the convenience of rapid food preparation also requires greater energy expenditures.

A variety of changes could be made to reduce the energy intensiveness of the United States food system. Many of them correspond to suggestions in earlier sections of this book designed to reduce the vulnerability of Green Revolution crops or the use of chemical pesticides. For example, it has been estimated[8] that substitution of livestock manure (over 50% of which is produced on feedlots and often results in water pollution problems) for synthetic fertilizers in corn growing could result in savings of 11.3×10^9 J/hectare. Twenty percent of feedlot manure production would have been required to fertilize 30% of the United States crop in 1970. Even greater energy savings could be achieved by winter planting of nitrogen-fixing legumes such as vetch or clover. In the spring these could be plowed under as "green manure." The latter approach has the advantage of preventing winter erosion in addition to providing soil organic matter as ordinary manure would.

At the level of the food consumer even greater energy savings could be made. Eating habits might be directed toward less highly processed foods and those like fresh vegetables, eggs, dairy products, cereals, and soybean meat substitutes which provide maximum protein and carbohydrate per unit of fossil energy input. Reduction in packaging and reversal of the trend from rail to truck transport of foods would also produce significant savings, as would the replacement of individual automobile trips to the supermarket by well-planned truck delivery routes. Since a great deal of food system energy consumption is required for commercial and home refrigeration, smaller, better insulated, nonfrostless refrigerators and closable super-market freezers would produce significant savings.

A recent U. S. Government study of agricultural efficiency[10] has concluded that "Utopian dreams of returning to a relatively primitive agriculture destitute of mechanical power, fertilizers and pesticides are unrealistic because adequate food for the world's population cannot be produced under such systems." At the same time the committee reported that "the rate of increase of crop yields per pound of fertilizer added has been tapering since about 1965." They also added that legume crops, an increasingly important source of protein, had not participated in the agricultural revolution to the extent of cereals. Clearly, as energy prices increase and the agricultural return for a given energy input tapers off, many of the less energy intensive alternatives suggested in this section will become more popular.

EXERCISES

(The first group of exercises is based primarily on material found in this text.)

1. What is the importance of the demographic transition?
2. Explain the role of vitamins and minerals, fats and carbohydrates, and protein in the human diet.
3. Why is protein quality so important? What determines it? Name one disease caused by lack of quality protein.
4. Explain how overuse of animals as food wastes protein and Calories.
5. Describe the successes and problems of the Green Revolution.
6. What precautions must be taken before a pesticide can be produced and used commercially? Do you think they are adequate?

---

[10] Committee on Agricultural Production Efficiency, "Agricultural Production Efficiency." Board on Agriculture and Renewable Resources, Commission on Natural Resources, National Research Council–National Academy of Sciences, Washington, D. C., 1975.

7. Suppose that a pesticide whose half-life in the environment is 1 year is applied to crop land at a rate of 5.0 kg/hectare. If applications are made on a one-a-year basis, what is the maximum residue which can build up? Make a graph of residue concentration versus time to show how the buildup occurs.

8. What are some factors which make the assumption of first-order kinetics for pesticide decomposition (as in Exercise 7) an oversimplification?

9. From Table 18.5 determine:
   a. The most toxic insecticide.
   b. The least toxic herbicide.
   c. Which is more toxic, parathion or malathion?

10. Explain the importance of dioxin in the toxic effects of 2,4,5-T.

11. What is the difference between axonic and synaptic transmission of nerve impulses? Which of these types of nerve transmission is affected by DDT? By carbamates? By organophosphates?

12. Describe some of the techniques of integrated pest management. What are the strong and weak points of each?

(Subsequent exercises may require more extensive research or thought.)

13. Generalize the problem in Exercise 7 by deriving an equation which gives the maximum fraction of accumulated pesticide residue as a function of f_1, the fraction left after one unit of time [see J. W. Hamaker, Mathematical prediction of cumulative levels of pesticides in soil. *Advan. Chem. Ser.* **60**, 124 (1966)]. Hint: Derive an equation for $f = C/C_0$, where C is concentration at time t and C_0 is initial concentration, in terms of the half-life $t_{1/2}$. Then show that

$$\frac{r}{C_i} = 1 + f_1 + f_1^2 + f_1^3 + \cdots + f_1^{n-1}$$

where C_i is the incremental concentration added in one unit of time. The infinite series

$$1 + x + x^2 + x^3 + \cdots = 1/(1 - x)$$

14. Suppose you owned a 50-hectare farm. With which of the following types of crops could you supply the maximum amount of protein: rice, milk, soybeans, beef cattle, peas, wheat, hogs, poultry? (see Increasing high quality protein. *In* President's Science Advisory Committee, "The World Food Problem," Vol. II, pp. 295ff. The White House, Washington, D. C., 1967). Which of these crops would yield the most protein for the amount of energy needed to produce it? [see E. Hirst, Food-related energy requirements. *Science* **184**, 134 (1974)].

Suggested Readings

1. A. Karim Ahmed, Donald F. MacLeod, and James Carmody, Control of asbestos. *Environment* **14**(10), 16 (1972). A thorough yet readable treatment of the subject, ranging from early reports of occupational disease to effects on the community. There are extensive notes and references.

2. A. M. Altschul, Food: Protein for humans. *Chem. & Eng. News,* Nov. 24, pp. 68–81 (1969). This fairly complete discussion of the problem of adequate protein supply has an especially useful section on low-cost options for nutritional and esthetic improvements in diet.

3. A. Anderson, Jr., The hidden plague. *N. Y. Times Mag.* October 27, p. 20 (1974). A popular account of the problem of occupational disease associated with toxic chemicals.

4. F. M. Ashton and A. S. Crofts, "Mode of Action of Herbicides." Wiley (Interscience), New York, 1973. The title is descriptive of the material covered, but the reading is somewhat tough because of the number of specialized terms encountered.

5. A. H. Boerma, A world agricultural plan. *Sci. Amer.* **223**(2), 54–69 (1970). This article describes the Indicative World Plan for Agricultural Development sponsored by the Food and Agricultural Organization of the United Nations.

6. G. Borgstrom, "The Hungry Planet." Macmillan, New York, 1965. Much of Borgstrom's more recent writing is based on the figures in this extensive study.

7. M. S. Bretscher, Membrane structure: Some general principles. *Science* **181,** 622–629 (1973). A review of the contributions of lipid, protein, and carbohydrate to the structure and function of cell membranes.

8. H. Brown, J. Bonner, and J. Weir, "The Next Hundred Years." Viking Press, New York, 1957. Chapters 8, 9, and 10 on world food and agriculture are excellent.

9. L. R. Brown, Human food production as a process in the biosphere. *Sci. Amer.* **223**(3), 160–170 (1970). The question is raised: "How many people can the biosphere support without impairment of its overall operation?" A cooperative effort through a world environmental agency is suggested to monitor man's interventions in the environment, including those made in the quest for more food.

10. L. R. Brown, "Seeds of Change." Praeger, New York, 1970. This earliest of Brown's works to appear in this bibliography is a relatively optimistic view of the Green Revolution.

11. L. R. Brown, "In the Human Interest." Norton, New York, 1974. This most recent of Brown's writings develops the problem of overpopulation and presents a strategy to stabilize world population.

12. L. R. Brown and G. W. Finsterbusch, "Man and His Environment: Food." Harper, New York, 1972. This book considers not only the Malthusian question, "Can we produce enough food?" but also the more recently relevant, "What are the environmental consequences of trying to do so?"

13. R. A. Capaldi, A dynamic model of cell membranes. *Sci. Amer.* **230**(3), 26–33 (1974). An up-to-date account of the structure and mechanisms of operation of cell membranes.

14. L. J. Carter and R. Gillette, Cancer and the environment (I) and (II). *Science* **186,** 239–245 (1974). These two news reports give a good survey of the problem of chemical carcinogens. The regulation of aldrin, dieldrin, benzidine, and vinyl chloride is included.

15. *Chemical and Engineering News,* September 29, pp. 40–41 (1969); February 21, p. 19 (1972); August 6, pp. 14–15 (1973); August 19, p. 30 (1974); December 9, pp. 11–12 (1974). A series of reports on direct chemical synthesis of protein and carbohydrate.

16. *Chemical and Engineering News,* Cancer deaths among PVC workers cause concern. *Chem. & Eng. News* January 28, p. 6 (1974). This was the first public report of angiosarcoma of the liver among plastics workers.

17. W. Clark, U. S. agriculture is growing trouble as well as crops. *Smithsonian* January, pp. 59–65 (1975). The photographs by Tex Fuller depicting uniformity and mechanization of the United States monoculture system are striking.

18. Committee on Agricultural Production Efficiency, "Report on Agricultural Production Efficiency." Board on Agriculture and Renewable Resources, Commission on Natural Resources, National Research Council – National Academy of Sciences, Washington, D. C., 1975.

19. J. R. Corbett, "The Biochemical Mode of Action of Pesticides." Academic Press, New York, 1974. A thorough survey of the subject which probably requires some background in biochemistry for its understanding.

20. Council on Environmental Quality, "Toxic Substances." Superintendent of Documents, US Govt. Printing Office, Washington, D. C., 1971. An excellent summary of problems associated with release of toxic substances into the environment.

21. D. L. Dahlsten, R. Garcia, J. E. Laing, and R. van den Bosch, "Pesticides." Scientists' Institute for Public Information, New York, 1970. The environmental harm done by pesticides is summarized at an elementary level.

22. S. K. Dhar, ed., "Metal Ions in Biological Systems." Plenum, New York, 1973. The last four papers of this collection discuss transport and toxicity of heavy metals, especially lead, mercury, and cadmium, in the environment.

23. B. D. Dinman, The 'non-concept' of 'no threshold' chemicals in the environment. *Science* **175,** 495–497 (1972); also letters in *Science* **177,** 1152–1154 (1972). Dinman argues that for most chemicals in the environment the concept of "no threshold" below which no effects occur is counterproductive. The letters cited contain additional interesting views.

24. C. Djerassi, C. Shih-Coleman, and J. Diekman, Insect control of the future: Operational and policy aspects. *Science* **186,** 596 (1974). The difficulties faced in the development of new chemicals for the control of human or pest populations are similar and are summarized here.

25. D. E. Dumond, The limitation of human population: A natural history. *Science* **187,** 713–721 (1975). In this anthropologist's view human populations in hunting and gathering societies are governed by the perceived value of another child to the nuclear family – much the same mechanism which appears to produce the demographic transition in highly developed countries.

26. C. A. Edwards, "Persistent Pesticides in the Environment," 2nd ed., CRC Press, 1973. A comprehensive treatment; well written. The section that compares various models of DDT movement in the environment is especially good.

27. P. R. Ehrlich and J. P. Holdren, Impact of population growth. *Science* **171,** 1212–1217 (1971); see also correspondence, *ibid.* **173,** 278–280 (1971). The authors argue that population contributes a disproportionate negative environmental impact relative to that implicit in Commoner's multiplicative equation with affluence and technology.

28. S. S. Epstein and M. S. Legator, "The Mutagenicity of Pesticides." MIT Press, Cambridge, Massachusetts, 1971. The introduction by Joshua Lederberg is especially useful, as is the listing of all pesticides alphabetically by common names. The latter gives the formula, use, and manufacturer for each pesticide.

29. "Fate of Organic Pesticides in the Aquatic Environment," Advan. Chem. Ser. No. 111.

Amer. Chem. Soc., Washington, D. C., 1972. Chapter 1 in this collection of fairly advanced descriptions of decomposition pathways of pesticides summarizes the effects of DDT on the environment and economy of the state of Michigan.

30. L. Fishbein, W. G. Flamm, and H. L. Falk, "Chemical Mutagens." Academic Press, New York, 1970. This book gives a good introduction to the biochemistry and mode of action of mutagenic chemicals. Later chapters tabulate classes of compounds and specific substances known to be mutagenic.

31. "Genetic Vulnerability of Major Crops." Nat. Acad. Sci., Washington, D. C., 1972. The contents of this extensive report of a committee of the Agricultural Board of the Division of Biology and Agriculture, National Research Council, are obvious from its title.

32. F. A. Gunther and J. D. Gunther, eds., "Residue Reviews." Springer-Verlag, Berlin and New York, This multivolume series contains numerous useful articles on residues of pesticides and other foreign chemicals in foods and feeds.

33. C. Hansch. A quantitative approach to biochemical structure-activity relationships. *Accounts Chem. Res.* **2**, 232–239 (1969). An excellent account describing factors which control biological activity of chemicals.

34. R. Haque and V. H. Freed, eds., "Environmental Dynamics of Pesticides." Plenum, New York, 1975. A series of articles by experts on the transport and transformation of pesticides in air, soil, and water.

35. E. Hirst, "Energy Use for Food in the United States," ORNL-NSF-EP-57. Oak Ridge Nat. Lab., U. S. At. Energy Comm., Oak Ridge, Tennessee, 1973. This report, based on economic input–output data developed by the U. S. Department of Commerce and energy input–output data from Oak Ridge, concludes that 12% of 1963 energy consumption was required by the food system.

36. E. Hirst, Food-related energy requirements. *Science* **184**, 134–138 (1974). This article summarizes the results of Hirst's 1973 ORNL report.

37. A. Hollaender, ed., "Chemical Mutagens." Plenum, New York, 1971. This book describes the molecular mechanisms by which mutagens act as well as methods for detecting such effects.

38. "Integrated Pest Management." Council on Environmental Quality, Superintendent of Documents, Washington, D. C., 1972. This report surveys techniques other than broad-scale application of chemical pesticides which are available for pest control, and delineates the role of the federal government in their implementation.

39. P. R. Jennings, Rice breeding and world food production. *Science* **186**, 1085–1088 (1974). The role of international cooperation and revised research practices in continuing the Green Revolution is discussed.

40. A. Kappas and A. P. Alvares, How the liver metabolizes foreign substances. *Sci. Amer.* **232**(6), 22–31 (1975). An excellent description of how the liver enzymes convert nonpolar compounds into polar ones which can be excreted readily.

41. G. O. Kermode, Food additives. *Sci. Amer.* **226**(3), 15–21 (1972). This article considers the necessity and safety of the 2500 different substances currently being added to foods for flavoring, coloring, preservation, etc. A list of substances on the Food and Drug Administration's GRAS (generally recognized as safe) list is included.

42. D. E. Koshland, Jr., Protein shape and biological control. *Sci. Amer.* **229**(4), 52–64 (1973). An up-to-date account of theories of enzyme activity and its relation to the shape of protein molecules.

43. C. E. Knapp, Asbestos: Friend or foe? *Environ. Sci. Technol.* **4**(9), 727–728 (1970). Hazards, delayed effects, and regulations involving asbestos are described.

44. F. M. Lappé, "Diet for a Small Planet." Ballantine Books, New York, 1971. A how-to guide to the practice as well as the theory of protein conservation, this book is subtitled: "How to Enjoy a Rich Protein Harvest by Getting Off the Top of the Food Chain."

45. D. H. K. Lee and D. Minard, eds., "Physiology, Environment and Man." Academic Press, New York, 1970. This collection of papers presented at a 1966 National Academy of Sciences–National Research Council symposium includes penetration of environ-

mental agents into the body, metabolism, toxic effects, adaptation, and a good discussion of attempts to define an optimum environment.

46. A. L. Lehninger, "Biochemistry." Worth, New York, 1970. Chapters 1, 4, 5, 9, 10, 11, 21, 28, 29, 31, 32, and 34 of this general biochemistry text are especially useful with regard to the subjects included in this chapter.

47. L. R. Martin, "Agriculture as a Growth Sector – 1985 and Beyond," Staff Paper P74-4. Department of Agricultural and Applied Economics, University of Minnesota, St. Paul, 1974. This report examines the possibility that United States agriculture would be a candidate for rapid growth in exports. Among other things the conclusion is drawn that the farm sector of the food system is not extremely energy intensive.

48. J. L. Marx, Birth control: Current technology, future prospects, *Science* **179**, 1222–1224 (1973). A general survey of the technology of birth control.

49. R. L. Metcalf, DDT substitutes. *CRC Crit. Rev. Environ. Cont.* **3**(1), 25 (1972). An excellent survey of the uses and properties of DDT and substitute insecticides.

50. O. L. Miller, Jr., The visualization of genes in action. *Sci. Amer.* **228**(3), 34–42 (1973). An excellent account of the transcription and translation of the genetic code.

51. J. J. Mulvihill, Congenital and genetic disease in domestic animals. *Science* **176**, 132–137 (1972). Farm and household animals can warn of environmental hazards and provide models of human genetic disease.

52. The New Alchemy Institute in Falmouth, Massachusetts is a small group of people who believe that the deficiencies of modern agriculture are so great that it will collapse within a few decades. The Institute publishes a journal and is supported by associate memberships and foundation grants. One of its major projects is development of the "ark," a biologically diverse, compact food-producing system based on aquaculture and solar energy.

53. "New Approaches to Pest Control and Eradication," Advan. Chem. Ser. No. 41. Amer. Chem. Soc., Washington, D. C., 1963. Contains reports on nonpesticidal methods of insect control.

54. J. E. Newman and R. C. Pickett, World climates and food supply variations. *Science* **186**, 877–881 (1974). International cooperation in establishing food surplus storage centers in favorable regions near areas of expected shortage could do much to alleviate disasters brought on by short-term climatic variations.

55. R. D. O'Brien, "Insecticides: Action and Metabolism." Academic Press, New York, 1967. An excellent book, easily readable by anyone with a little background in organic chemistry, which describes the mode of action of insecticides without jargon.

56. E.-I. Ochiai, Environmental bioinorganic chemistry. *J. Chem. Educ.* **51**(4), 235–238 (1974). A good discussion of toxic effects resulting from inorganic substances, especially heavy metals.

57. "Organic Pesticides in the Environment," Advan. Chem. Ser. No. 60. Amer. Chem. Soc., Washington, D. C., 1966. Contains a number of useful chapters on toxicity, metabolism, and persistence of pesticides.

58. D. V. Parke, "The Biochemistry of Foreign Compounds." Pergamon, Oxford, 1968. This book has two sections. The first describes the mechanisms by which the human body handles foreign substances. The second applies these general mechanisms to specific examples involving food additives, pesticides, drugs, etc.

59. D. Pimentel, "Ecological Effects of Pesticides on Non-Target Species." Executive Office of the President, Office of Science and Technology, Washington, D. C., 1971. The properties and LD_{50} data of a large number of pesticides are included, as is a useful table relating trade names to chemical composition.

60. D. Pimentel, L. E. Hurd, A. C. Bellotti, M. J. Forster, I. N. Oka, O. D. Sholes, and R. J. Whitman, Food production and the energy crisis. *Science* **182**, 443–449 (1973); see also a letter, *ibid.* **187**, 560–561 (1975). An estimation of on-the-farm energy inputs to food production.

61. G. B. Pinchot, Marine farming. *Sci. Amer.* **223**(6), 15–21 (1970). Proposals are made for increasing food yields of the oceans by converting from hunting and gathering to farming.

62. N. W. Pirie, Orthodox and unorthodox methods of meeting world food needs. *Sci. Amer.* **216**(2), 27–35 (1967). Pirie's position is that orthodox methods of food production must be pressed to maximum capacity, but they cannot solve the problem without the aid of unorthodox ones. The latter will require basic changes in cultural attitudes for their adoption.

63. President's Science Advisory Committee, "The World Food Problem," Vols. I, II, and III. The White House, Washington, D. C., 1967. Volume II, the Report of the Panel on World Food Supply, contains much useful information and a myriad of figures and tables.

64. C. Ramel, ed., "Evaluation of Genetic Risks of Environmental Chemicals," Royal Swedish Academy of Sciences, Stockholm, 1973. This report of a symposium held in Sweden in 1972 gives a general survey of the problem of mutagenic chemicals in the environment.

65. "Report of the Secretary's Commission on Pesticides and Their Relationship to Environmental Health," Parts I and II. U. S. Department of Health, Education and Welfare, Washington, D. C., 1969. This report summarizes a great deal of information on the uses, toxicity, and interactions of pesticides as well as giving recommendations for controlling and monitoring their release into the environment.

66. "Report on 2,4,5-T," Panel on Herbicides of the President's Science Advisory Committee. Executive Office of the President, Washington, D. C., 1971. An excellent summary of the chemical and public policy aspects of commercial and military use of 2,4,5-T.

67. H. J. Sanders, Chemical mutagens. *Chem. & Eng. News* May 19, pp. 50–71; June 2, pp. 54–68 (1969). This two-part series gives a fairly complete treatment of possible mutagenicity of drugs. Less emphasis is placed on environmental hazards.

68. J. Schubert, A program to abolish harmful chemicals. *Ambio* **1**(3), 79–89 (1972). Schubert asserts that tests commonly used for detection of harmful chemicals are grossly inadequate. He lists a comprehensive series of tests which could protect the population and suggests a technical and administrative program by which such testing could be accomplished.

69. *Scientific American.* The human population. Special issue **231**(3) (1974). The eleven articles in this issue are invaluable in providing an understanding of the principles of population growth and the problems inherent in such growth.

70. R. A. Shakman, Nutritional influences on the toxicity of environmental pollutants. *Arch. Environ. Health* **28**(2), 105 (1974). Shakman describes the effects of nutritional factors such as protein, trace minerals, and vitamins A, C, D, and E on the toxicity of pollutants such as pesticides and oxidant air pollutants.

71. D. Spurgeon, The nutrition crunch: A world view. *Bull. Atom. Sci.* **29**(8), 50–54 (1973). A good general treatment of the worldwide nutrition problem.

72. C. E. Steinhart and J. S. Steinhart, "Energy." Duxbury Press, North Scituate, Massachusetts, 1974. Chapter 6 is devoted to the relationship between food production and energy consumption.

73. N. Wade, Green Revolution (I) and (II). *Science* **186**, 1093–1096 and 1186–1192 (1974). These two news articles delineate both social and ecological problems associated with increased food production.

74. G. L. Waldbott, "Health Effects of Environmental Pollutants." Mosby, St. Louis, Missouri, 1973. This readable book concentrates on air pollutants and toxic effects other than carcinogenicity and mutagenicity.

75. J. D. Watson, "Molecular Biology of the Gene," 2nd ed. Benjamin, New York, 1970. A complete, readable description of molecular genetics. Some speculations on the mechanism of carcinogenicity are given in the last chapter.

76. J. G. Wilson, "Environment and Birth Defects." Academic Press, New York, 1973. Mechanisms of teratogenesis, effects of environmental agents, and methods of testing for teratogens are described.

77. W. Worthy, Integrated insect control may alter pesticide use pattern. *Chem. & Eng. News* April 23, pp. 13–19 (1973). A good summary of the techniques of integrated pest management.

VII

Science, Ethics, and Ecology

NECESSARY ETHIC

Science and human values,
Is and Ought,
May be two sides of the same thing,
A Moebius-twist to time,
Merging, one into the other.

Science itself has a two-sided
  one-sidedness.

Science past,
'Ready made, public,
Is facts found.
Science-as-an-institution
Is about Is.

Science present,
Being made, private,
Is fact-finding.
Science-in-the-making
Is a bout with Ought:

One ought to behave
As Galileo did
To discover
What free-fall is.

Is is hen Ought's egg.
And the hen?
They say
One can't get Ought from Is.

But don't Is-finding Galileos
Behave as they do
Because free-fall is what it is?

Ought begets Is, Is begets Ought.
That is all we know,
    and all we need to know.
Darwin would understand.

Henry A. Bent

Science, Ethics, and Ecology

It is a fundamental assumption of all science that nature is lawful; that is, the results of observations and experiments may be correlated, understood, and summarized by means of descriptive natural laws. Furthermore, it is usually assumed (as in Part I) that such laws have been and will continue to be unchanging in time. The observations and laws of science—the "is" of Henry Bent's poem—are often separated from prescriptive, human laws or ethics which tell us what we "ought" to do, on the assumption that the latter cannot be verified by objective, value-free scientific procedures. Some scientists,[1] however, have disputed the absolute objectivity of what they do and of science itself. Others[2] have proposed a complementarity between subjective, intuitive knowledge (based on qualitative sensory perceptions) and quantitative, value-free knowledge (obtained by the objective techniques commonly ascribed to science). Thus, the segregation of "is" from "ought" is perhaps arbitrary, and even worse may reduce the benefits which scientific inquiry can provide.

This text has not attempted what in Polanyi's view might have been an impossible task of complete objectivity. Observation and study of the natural

---

[1] M. Polanyi, "Personal Knowledge." Harper, New York, 1959.
[2] T. R. Blackburn, *Science* **172**, 1003–1007 (1971).

world does yield certain characteristics or qualities which have a degree of universality, seem worthy of emulation, and thus may be classified as values. In the main these revolve about the perpetuation of the human race, and indeed the biosphere in all its diversity, over the geological time scale of millions of years. It has been argued that there is no rational, objective justification for such values, but most human consciences rebel at the thought that their action or inaction, individual or collective, might result in cutting short the tenure of humankind on earth.[3] Such a nonrational (but hardly irrational) value system has been at work behind the scenes, coloring the discussion of most of the problems already mentioned in this text.

Despite the previous inclusion of judgments relating to the efficacy of various procedures for alleviating environmental problems, it is worthwhile in these closing paragraphs to reflect on the "lawfulness" of environmental science and ask whether there are generalities which will serve well in approaching problems, old or new, which have not been discussed herein. Indeed we believe there are, and they have major implications for further study, research and action by you, the reader, as well as many others.

To begin with, many environmental problems have similar, if not identical, characteristics. They seem to appear suddenly and unexpectedly. Often, as in the case of photochemical smog or the freon–ozone interaction, they result from seemingly innocuous actions by the average person. They cannot be solved instantaneously, and the efforts of experts from a wide variety of disciplines are often required just to establish the existence and nature of the problem, much less indicate the approaches which may alleviate it. In most cases these characteristics are a direct consequence of the interdependence and interconnectedness of ecological systems, whose negative feedback loops are capable of handling the problem up to a point. This gives a false sense of security at first, and, when natural safeguards are eventually overcome, may result in a quite complicated pathway from effect to cause.

The fact that we have suddenly been faced with a great number of environmental crises reflects in part greater awareness of possible problems, but it is also related to the exponential increase in factors such as population, affluence, and technology. These are all rapidly approaching limits set by natural laws and thus the positive feedback which has led to exponential growth must be replaced by a negative feedback which slows growth and leads to a new steady state. Otherwise we may overshoot a limit; severely damage the capacity of the environment to maintain population, affluence, and technology; and therefore make a bad situation even worse. Bruce Catton, noted Civil War historian, has summed up the situation as follows: "What protected man in the old days was his awareness that there were things he just could not do. That awareness is gone. . . ." Recently, there

---

[3] See R. L. Heilbroner [*N. Y. Times Mag.* January 19, pp. 14–15 (1975)] for a discussion of opposing arguments.

have been attempts[4] to reinstill such awareness of impossibilities, but they have had little effect among *technological optimists,* who continue to believe that rapid scientific and technological growth will solve all our problems.

One thing is certain—already discovered scientific laws do place limits on what *Homo sapiens* (or any other organism in the biosphere) can do. The second law of thermodynamics comes most readily to the scientific mind, and a number of its important implications have been discussed throughout this book. It seems highly unlikely that, after a century of testing, the entropy law will be found to be incorrect, inadequate, or capable of circumvention. Human laws, projects, or proposals for cleaning up pollution which are incompatible with the second law will simply not work and therefore must be discarded. Other, similar limitations on human alternatives are required by insights from other branches of science.

SCIENCE AND TRANS-SCIENCE

Another type of limitation is inherent in the nature of science itself. Certain types of questions *appear* to be capable of scientific resolution but in fact are not, given the current state of scientific development. Weinberg[5] has dubbed such questions *trans-scientific* since they transcend the limitations of scientific inquiry. He divides trans-scientific questions into three types. A good example of the first is determining the effects of chronic exposure to low levels of a carcinogen (such as vinyl chloride) in the environment. The problems of extrapolating the results of experiments in which large doses are given to a small number of test animals to the situation where large numbers of humans are exposed to small doses have already been discussed. Weinberg argues that much of the furor attendant to discussion of such problems may be attributed to the fact that, because of the almost infinite expenditure of time and money necessary to resolve them, they are trans-scientific. His other two types of trans-scientific questions relate to the difficulty of using social scientific theories to predict the behavior of a single individual and judgments of "scientific value" which determine priorities for scientific research and government support.

The first and last of these three categories are obviously more apropos the subject of this book. With regard to the first, it must be emphasized that it often leaves a tremendous gray area between a proven hazard and a clean bill of health. One must be cautious of claims that "It cannot be proved that X is harmful," just as he must beware of demagogic use of "It

---

[4] B. Commoner, "The Closing Circle." Knopf, New York, 1971; P. R. Ehrlich, "The Population Bomb." Ballantine Books, New York, 1968; D. H. Meadows, D. L. Meadows, J. Randers, and W. W. Behrens, III, "The Limits to Growth," Universe Books, New York, 1972.

[5] A. M. Weinberg, *Minerva* **10**(2), 209–222 (1972); *Science* **177**, 211 (1972).

cannot be proved that X is safe." Both statements may be applied to exactly the same X, if the question of its safety is trans-scientific, and thus both should be avoided in rational discussions. On the other hand, one must resist the temptation to conclude that science cannot contribute to the resolution of trans-scientific questions. Weinberg suggests that the role of science and scientists is to make the facts known, insofar as there is scientific consensus, and then allow the question to be decided politically, in an arena where the scientist has no greater status or expertise than any other citizen.

There are two drawbacks inherent in this position. First, most legal systems require *proof* of damage to one party before decisions detrimental to the economic status of another can be made. Thus, trans-scientific questions will almost invariably be decided in favor of an already existing polluter. Yet, as in the case of environmental cancer, it may be clearly demonstrable that the absence of proof in no way indicates the absence of hazard. Under current legal practice *we are constrained to allow the experiment to proceed with all of society as the guinea pigs.*

The second problem with declaring a question to be trans-scientific is that science has not always waited for absolute proof of a hypothesis or theory before incorporating it as a major paradigm or way of thinking about nature. For example, the atomic theory is still treated as a *theory* in elementary textbooks, despite the fact that a century and a half of chemical progress has been based on it. Few scientists would argue against it now, but were it as controversial as Darwinian evolution contemporary chemists might be faced with tracts describing nonstoichiometric compounds and the ambiguities inherent in the "pictures" of atoms derived from x-ray or holographic electron diffraction experiments. Every chemist "knows" that atoms exist, but could it be proved in a court of law? A great deal of careful explanation is needed if the general public is to understand the paradigms upon which many scientific conclusions rest, and the problem is made more difficult when different disciplines, based on different paradigms, come to opposite conclusions.

THE TRAGEDY OF THE COMMONS

In addition to the limitations which natural laws place on the approaches to solving environmental problems and those which the nature of scientific proof impose on demonstration of the existence of such problems, rational behavior on the part of individuals does not invariably result in the common good. Noted economist Kenneth Boulding has summarized this social difficulty as follows: "The deep, crucial problem of social organization is how to prevent people from doing their best when the best in the particular, in the small, is not the best in the large."

Environmentalist Garrett Hardin[6] has treated the same topic in his

---

[6] G. Hardin, *Science* **162**, 1243–1248 (1968).

famous article, "The Tragedy of the Commons," where he uses the analogy of a number of shepherds who graze their animals in a common pasture. Eventually, of course, as the number of sheep increases, the pasture will begin to be overgrazed and its capacity to regenerate itself will be impaired by the removal of seed for next year's crop, erosion, etc.

Even in such a situation, however, it is still to the advantage of an individual shepherd to increase the size of his flock, because he obtains the entire benefit of a new animal, while its negative environmental impact is spread nearly evenly over all of society, since overgrazing affects every herdsman to nearly the same extent. The "tragedy" arises out of what Whitehead[7] has described as "the remorseless working of things" — despite the best efforts of individuals, the society as a whole fails.

Hardin makes the further point that the tragedy of the commons has *no technical solution,* although technology may be able to buy time during which other types of solutions may be attempted. Returning to the analogy of the common pasture, it is easy to conceive of ways in which science and technology might be applied to increase the number of animals which could be accommodated. The land might be fertilized to make grass grow faster, new more nutritious varieties of plants might be developed, techniques for erosion control might be developed, etc. But the second law of thermodynamics and other well-known laws of science place limitations on the extent of these improvements, while there is no limitation but social control or environmental catastrophe on the number of animals introduced into the commons. Furthermore, simulation of the tragedy of the commons by digital computer[8] indicates that the application of higher levels of technology exacerbates the rapidity of the collapse when commons is eventually exhausted.

Hardin has pointed out that the tragedy of the commons is inherent in population growth, national park usage, and air and water pollution. Its relationship to many of the problems described in this text should be obvious. Moreover, it can be shown to be a subset of an even larger class of problems having to do with the growth of cities and urban decay, ecological predator–prey relationships, arms races, and the activities of multinational corporations.[9] All can be cast in the same form when treated by the mathematical theory of simulations and games.

THE ROLE OF THE SCIENTIST

If tragedies of the commons have no technical solutions, what role can the scientist play in their attack? The answer is certainly not "none." The doubling of agricultural yields, for example, can provide valuable time during which human attitudes about population growth may change or be

---

[7] A. N. Whitehead. "Science and the Modern World," p. 17. Mentor, New York, 1948.
[8] J. M. Anderson, *IEEE Trans. Sys., Man & Cybernetics* **4**, No. 1, 103–105 (1974).
[9] P. Ray, University of Michigan, Ann Arbor (personal communication).

changed and social controls may be instituted. The fallacy lies in the all-too-common belief that advances in technology, *in and of themselves,* can provide complete solutions, when in fact they can only make contributions which must be augmented by those in other fields. Thus the scientist has a role to play, but must also interact with (or perhaps delve into) other disciplines.

Scientists, and especially chemists, are uniquely qualified to assess the objective "is" of environmental problems. Some time ago John Platt[10] argued that a large-scale mobilization of scientists was necessary in order to help solve the crisis problems facing the world. He suggested that some non-crisis areas (such as the space program and much basic research) were overstudied while pollution, ecological balance, population, and nuclear annihilation had not sufficiently captured the attention of scientists. More recently, Barry Commoner[11] has associated this imbalance with Weinberg's third type of trans-scientific question—the internal value system upon which are based the priorities for scientific research. Clearly, it is far easier to advance one's scientific reputation by choosing to do (and writing proposals for support of) research which has a high probability of rapid resolution and publication. Because they are long-term (and may even turn out to be trans-scientific), environmental problems often do not fall into such categories.

Nevertheless, many scientists have elected to concentrate on the environment, in many cases applying basic scientific principles in new, exciting, and significant ways. A number of examples of their work have been mentioned in this book, perhaps the most obvious being the applications of thermodynamics to the energy and food crises carried out by Berry, Hirst, Hannon, Pimentel, and Steinhart and Steinhart. There is no reason why such work need be any less "scientific" than other research, although the fact that it borders on trans-science may lead to such an accusation.

Quite recently another aspect of the scientific value system has garnered considerable attention. A group of prominent molecular biologists have recommended a voluntary moratorium on certain types of research.[12] Their reason was that artificially modified DNA molecules produced in such experiments might prove biologically hazardous, especially if they were to be dispersed from laboratories into the general environment. Their recommendation was made in full knowledge that there was only a *potential* (not a demonstrable) risk and that such a moratorium would entail postponement and possibly abandonment of scientifically valuable experiments. This situation contains all of the elements of a tragedy of the commons—valuable scientific information (and hence prestige for an experimenter) may be had by performing experiments which might be detrimen-

[10] J. Platt, *Science* **166**, 1115–1121 (1969).
[11] B. Commoner, *Chem. Technol.* **4**(5), 258 (1974).
[12] P. Berg *et al., Science* **185**, 303 and 332–334 (1974).

tal to society as a whole. It remains to be seen whether cooperative behavior on the part of the scientific community can avert the tragedy of the commons. If not, and if a real hazard is evident, regulation by outside agencies may be required.

Scientists, even those whose research is not directly connected to environmental problems, have yet another role to play in their discovery and solution. It arises out of Barry Commoner's first law of ecology: "Everything is connected to everything else," or as it has been paraphrased by Aleksandr Solzhenitsyn, "Mankind's sole salvation lies in everyone making everything his business." This can be accomplished in science by cultivating a more holistic view. Certainly one must specialize to do useful scientific work, but there will invariably be implications and consequences of such work which extend far beyond the specific reasons which led someone to undertake it. The wise scientist will attempt to explore these implications and consequences and if necessary, explain them to others, thus giving early warnings of aspects which might be beneficial or detrimental to society as a whole.

We hope that this text has not appeared to be pessimistic with regard to environmental problems. Certainly much work needs to be done, but informed action based on hope, by scientists and citizens alike, seems far preferable to the pessimism, technological optimism, or apathy which some have espoused. There will be no absolute solutions to many of the problems we have discussed—only approaches which show promise of success and which must constantly be reevaluated in light of more recent knowledge. Yesterday's solution becomes today's problem too often for vigilance to be relaxed. What is needed is wisdom as well as knowledge—the wisdom to see that humankind is a part of nature, not its master. Marston Bates has put it very well:

> In defying nature, in destroying nature, in building an arrogantly selfish, man-centered, artifical world, I do not see how man can gain peace or freedom or joy. I have faith in man's future, faith in the possibilities latent in the human experiment, but it is faith in man as a part of nature, working with the forces that govern the forests and the seas; faith in man sharing life, not destroying it.

It is hoped that the reading of this book has made some small contribution to the development of such wisdom.

Suggested Readings

1. T. R. Blackburn, Sensuous-intellectual complementarity in science. *Science* **172,** 1003–1007 (1971); see letters in *ibid.* **173,** 1191–1195 (1971). Blackburn argues that abstract, quantitative knowledge and direct sensuous information constitute complementary descriptions of nature in the same way that wave and particle descriptions of the electron are complementary.
2. K. E. Boulding, "Beyond Economics; Essays on Society, Religion and Ethics." Univ. of Michigan Press, Ann Arbor, 1968. Boulding is one of the few economists who transcends his discipline. This and his other writings are highly recommended.
3. B. Commoner, "The Closing Circle." Knopf, New York, 1971. An extremely important book on the nature and causes of environmental crises in the United States.
4. B. L. Crowe, The tragedy of the commons revisited. *Science* **166,** 1103–1107 (1969). Although in agreement with Hardin for the most part, Crowe argues that science can and must contribute to the saving of the commons. More interaction between natural and social scientists is essential.
5. R. J. Dubos, Humanizing the earth. *Science* **179,** 769–772 (1973). Dubos points out that many species other than man have had major environmental impacts. He believes that a combination of ecological wisdom and scientific knowledge can manage the earth to create environments favorable to the continued growth of civilization.
6. H. Gershinowitz, Applied research for the public good—a suggestion. *Science* **176,** 380–386 (1972). The author argues that "the application of research is a complex operation, involving continuing interaction and feedback, and is not a simple, orderly process of transmitting information from one place to another."
7. G. Hardin, The tragedy of the commons. *Science* **162,** 1243–1248 (1968). This article should be read by every scientist who is concerned about environmental problems.
8. R. L. Heilbroner, What has posterity ever done for me. *N. Y. Times Mag.* January 19, pp. 14–15 (1975). An essay on depriving the living to insure the survival of the unborn—the author concedes that no rational argument can be made for such a decision, but argues that most persons would value it.
9. M. H. Krieger, What's wrong with plastic trees? *Science* **179,** 446–455 (1973); see letters in *ibid.* **180,** 813–816 (1973). The author argues that the trade-off (which he assumes is necessary) between environmental preservation and aid to the poor in our society should be decided in favor of the latter.
10. H. J. Laski, Limitations of the expert. *Chem. Technol.* **4**(4), 198–202 (1974). This reprint of a 1930 *Harper's Magazine* article is amazingly up-to-date and well worth reading.
11. A. Leopold, "A Sand County Almanac." Sierra Club/Ballantine Books, New York, 1966. This environmental classic, first published in 1949, describes a style of life designed to merge with and protect the natural environment.
12. Konrad Lorenz, "Civilized Man's Eight Deadly Sins." Harcourt, New York, 1974. Lorenz catalogs and describes eight "pathological disorders of mankind" involving struc-

tures, institutions, and attitudes which serve no purpose in the survival of the species. Not all of them are related to environmental decay, but the analysis is worthy of a careful, critical audience.

13. D. H. Meadows, D. L. Meadows, J. Randers, and W. W. Behrens, III, "The Limits to Growth." Universe Books, New York, 1972. This report of the Club of Rome's Project on the Predicament of Mankind concludes on the basis of computer simulation models that limits should be placed on economic growth.

14. G. T. Miller, Jr., "Replenish the Earth." Wadsworth, Belmont, California, 1972; "Energy and Environment: Four Energy Crises." Wadsworth, Belmont, California, 1975. In these two books, both of which are easily readable by the nonscientist, Miller devotes considerable time to suggestions for future national policies in the area of energy and environment.

15. J. Platt, What we must do. *Science* **166,** 1115–1121 (1969). Platt's prescription is a large-scale reassignment of scientists to work on applied problems of national and international importance. Among these are ecological balance and pollution.

16. E. F. Schumacher, "Small is Beautiful." Harper Colophon Books, New York, 1973. The subtitle of this book is "Economics as if People Mattered." Schumacher argues that a more diversified system of production with much smaller individual units is more humane, more efficient, and ecologically sound.

17. C. P. Snow, "Public Affairs." Scribner's, New York, 1971. Snow, a scientist who participated in public affairs, argues that because the society of science is future-oriented, a great many more scientists should follow in his footsteps, bringing a dimension which is currently lacking in politics.

18. A. Spilhaus, Ecolibrium. *Science* **175,** 711–715 (1972). The author favors a social system in which diversity of choices (i.e., freedom) is maximized. He suggests that the challenge to science and technology is to continue providing for people's needs and wants while maintaining environmental quality.

19. C. Starr and R. Rudman, Parameters of technological growth. *Science* **182,** 358–364 (1973). Using the same computer models as the Limits to Growth study the conclusion is drawn that exponential growth of technology can save us. It is also argued that such growth is possible.

20. R. E. Train, The quality of growth. *Science* **184,** 1050–1053 (1974). Among other things this text of a lecture before the AAAS suggests that new types of political leaders, who take long-term views, will be necessary for solving environmental problems.

21. A. M. Weinberg, Science and trans-science. *Minerva* **10**(2), 209–222 (1972); see also *Science* **174,** 546–547 (1971); **177,** 211 (1972); **180,** 1123–1124 (1973). Weinberg's argument that certain questions which appear scientific have no objectively determinable answers is presented and commented upon.

Appendix

The Environmental Literature

Information of interest to chemists and other scientists concerned about environmental problems may be found across a broad spectrum of journals, documents, serials, and reference works. In this listing we have made no attempt to be comprehensive, but rather have selected those sources which appeared to us to be of greatest value. References covering the same subject are listed roughly in order of increasing degree of sophistication required of their readers. The Suggested Readings should also be consulted for more specific, annotated references.

I. General Sources (covering most of the topics in this book)
   A. Abstracts and Indexes
   Environment Abstracts (formerly *Environmental Information ACCESS*) and the accompanying annual *Index;* published by Environment Information Center, Inc.
   The Environment Film Review; an annual publication of the Environment Information Center, Inc.
   Pollution Abstracts and the accompanying annual *Index.*
   Selected U. S. Government Publications (free from Superintendent of Documents, U. S. Government Printing Office)
   Monthly Catalog, U. S. Government Publications
   Government Reports Index and Government Reports Announcements (companion volumes covering technical and sophisticated government and government-sponsored research)
   Chemical Abstracts
   Biological Abstracts
   B. Bibliographies and Directories
   "*Science for Society. A Bibliography,*" 5th ed. Amer. Ass. Advan. Sci., Washington, D. C., 1975.
   G. Paulson *et al.,* eds., "Environment USA. A Guide to Agencies, People and Resources." Bowker, New York, 1974.
   J. I. Shonle, Resource letter PE-1: Physics and the environment. *Amer. J. Phys.* **42**(4), 267 (1974).
   J. W. Moore and E. A. Moore, Resources in environmental chemistry I and II. An annotated bibliography of the chemistry of pollution and resources. *J. Chem. Educ.,* in press.
   G. S. Bonn, ed., "Information Resources in the Environmental Sciences." Univ. of Illinois Press, Urbana, 1973.
   H. N. M. Winton, ed., "Man and the Environment. A Bibliography of Selected Publications of the United Nations System, 1946–1971." Unipub, Inc./Bowker, New York, 1972.

"EPA Reports Bibliography," EPA-LIB-73-01. Environ. Protect. Ag., Washington, D. C., 1973; available from National Technical Information Service, Springfield, Virginia; to be up-dated.

C. Periodicals

Newsletters from regional EPA offices; contact the EPA office in your region; free.
Environmental News, national EPA newsletter free from EPA in Washington.
Environment
Chemistry in Britain
Chemical and Engineering News
Technology Review
Scientific American
Environmental Affairs
Environmental Science and Technology
Science
Nature
American Scientist
Environmental Letters
CRC Critical Reviews in Environmental Control
Advances in Environmental Science and Technology, serial, Wiley (Interscience)

II. Energy

A. Abstracts and Indexes

The Energy Index (annual compilation by the Environment Information Center, Inc.)
Energy Abstracts for Policy Analysis (formerly *NSF-RANN Energy Abstracts,* available from the Superintendent of Documents)
Energy Research & Technology: Abstracts of NSF Research Results (available from RANN, National Science Foundation)

B. Bibliographies and Directories

J. M. Fowler and K. E. Mervine, "Energy and the Environment," ERPP Project. Department of Physics and Astronomy, University of Maryland, College Park, 1973.
"Energy Facts." Compiled by W. Griffin for the Subcommittee on Energy of the Committee on Science and Astronautics, U. S. House of Representatives, by the Library of Congress, 1973.
J. W. Moore and E. A. Moore, Resources in environmental chemistry. An annotated bibliography of energy and energy-related topics. *J. Chem. Educ.* **52**(5), 288 (1975).
R. H. Romer, Resource letter ERPEE-1 on Energy: Resources, production, and environmental effects. *Amer. J. Phys.* **40**, 805 (1972).
"Energy Research & Technology: Interim Bibliography of Reports, with Abstracts," NSF 74-22. Nat. Sci. Found., Washington. D. C., 1974.
"Hydrogen Energy," Publ. No. PB 230-845. Available from National Technical Information Service, Springfield, Virginia, 1974.

C. Periodicals

Bituminous Coal Facts (published annually by the National Coal Association)
Oil & Gas Journal
American Gas Association Monthly
Physics Today
Bulletin of the Atomic Scientists (Science & Public Affairs)
Nuclear Safety
Solar Energy
Energy Conversion
Annual Reviews of Nuclear Science

III. Air

A. Abstracts and Indexes

Air Pollution Abstracts and its accompanying *Index*
KWIC Index of Air Pollution Titles

B. Bibliographies and Directories

U. S. Government, "Air Quality Criteria: Particulate Matter," AP-49; "Sulfur Oxides," AP-50; "Carbon Monoxide," AP-62; "Photochemical Oxidants," AP-63; "Hydrocarbons," AP-64; "Nitrogen Oxides," AP-84. Available from Superintendent of Documents, US Govt. Printing Office, Washington, D. C.

H. M. Englund, Literature of air pollution. *In* "Air Pollution Control" (W. Strauss, ed.), Part II, p. 255. Wiley (Interscience), New York, 1972.

J. S. Nader, Air pollution literature sources. *In* "Air Pollution" (A. C. Stern ed.), Vol. III, p. 813. Academic Press, New York, 1968.

The Environmental Protection Agency also has a series of annotated bibliographies available for many air pollutants; available from the Superintendent of Documents.

C. Periodicals

Journal of the Air Pollution Control Association

Atmospheric Environment

IV. Earth

A. Abstracts and Indexes

"Solid Waste Management: Available Information Materials," EPA Rep. SW-58.19 and SW-58.20. Available from Superintendent of Documents, US Govt. Printing Office, Washington, D. C., 1973. Contains information on publications, exhibits, and films.

B. Bibliographies and Directories

J. M. Fowler and K. E. Mervine, "No Deposit—No Return: The Management of Municipal Solid Wastes," ERPP Project. Department of Physics and Astronomy, University of Maryland, College Park, 1973.

Staff, "Mineral Facts and Problems." Bureau of Mines, U. S. Department of the Interior, Washington, D. C., 1970, or latest edition.

D. A. Brobst and W. P. Pratt, eds., "United States Mineral Resources," U. S. Geol. Surv. Prof. Pap. No. 820. Available from Superintendent of Documents, US Govt. Printing Office, Washington, D. C., 1973.

V. Water

A. Abstracts and Indexes

Selected Water Resources Abstracts

Water Quality Abstracts: A Special Compendium from Pollution Abstracts

Eutrophication: A Bimonthly Summary of Current Literature (University of Wisconsin)

Water Resources Abstracts and *HYDATA* (Water Resources Index Monthly)

B. Periodicals

Oceanus

American Water Works Association Journal

Journal of the Water Pollution Control Federation

Water Research: The Journal of the International Association on Water Pollution Research

VI. Life

A. Abstracts and Indexes

Selected References on Environmental Quality as It Relates to Health

Abstracts on Health Effects of Environmental Pollutants

Pesticides Abstracts (formerly *Health Aspects of Pesticides Bulletin*)

B. Bibliographies and Directories

"The State of Food and Agriculture." Published annually by the Food and Agriculture Organization of the United Nations.

S. S. Epstein and M. S. Legator, "The Mutagenicity of Pesticides." MIT Press, Cambridge, Massachusetts, 1971. The lengthy appendix contains a list of pesticides, their structural formulas, and major uses.

National Institute for Occupational Safety and Health, "The Toxic Substances List,"

Yearly ed. U. S. Pub. Health Serv., Washington, D. C. Available from the Superintendent of Documents.

C. Periodicals

Archives of Environmental Health

Ambio Journal of the Human Environment, Research and Management

"Environmental Quality and Safety; Chemistry, Toxicology and Technology." Thieme, Stuttgart; Academic Press, New York; serial publication.

Environmental Research. An International Journal of Environmental Medicine and the Environmental Sciences.

Essays in Toxicology, Academic Press (serial publication)

Pesticide Biochemistry and Physiology. An International Journal.

Residue Reviews, Springer-Verlag (serial publication)

Index

A

Abiotic synthesis experiments, 44, 45
Acetylcholinesterase, 426, 460–462
Acid mine drainage, 104, 105
Acid rain, 206
Activated carbon adsorption, 382
Activated sludge, 379, 390
Adiabatic lapse rate, 185
Adiabatic process, 80
Aerobic bacteria, 335, 390
  coal formation by, 93
Aerosol propellants, *see* Freons; Vinyl chloride
Affluence, 318, 478
Agricultural wastes, 314, 390
Agriculture, 418, *see also* Green Revolution
  and energy consumption, 465–468
Air, 181–255
  natural, 192
Air pollution, *see also* Asbestos; Carbon monoxide; Fluorides; Hydrocarbons; Metals; Nitrogen oxides; Particulate matter; Smog; Sulfur oxides
  abatement and control, 194
  automobile emission standards, 243
  from cement manufacture, 284
  from coal-fired power plants, 105
  control devices, 201, 212, 241
  cost of, 193, 222, 236
  criteria, 186, 208, 225, 232, 234, *see also* Automobile emissions, standards
  from energy production, 196–219
  from incinerators, 323–330
  industrial sources, 196–219, 284, 287, 290, 292; *see also* Air pollution, metals processing
  metals processing, 268, 269, 276, 279
  minor sources, 214–219
  from pulp and paper manufacturing, 407
  toxic effects, 432
  as tragedy of the commons, 481
  transportation related, 222–251
  U. S. emissions, 193
Algae, 151, 365, 376
Alkalinity, *see* Water
Alkylating agents, 431

Alkylation, in petroleum refining, 119
Alkylbenzenesulfonates (ABS), *see* Water pollution, detergents
Aluminosilicate, 34
Aluminum, *see* Metals
Amino acids, 443–449
  synthesis on primitive earth, 42, 43
Ammonia stripping, 389
Anaerobic bacteria, 57, 322, 377
  coal formation by, 93
Anesthetics, 432
Anion exchange capacity, *see* Soils, ion exchange
Antifeeding compounds, 464
Aromatization, 94, 119
Asbestos, as air pollutant, 218
Asbestosis, *see* Environmental illness
Atmophiles, 25, 26
Atmosphere, 61, 183–194
  formation of, 41, 42, 56–60, 63
  photochemistry of, 187–189
  structure of, 183–185
  temperature–altitude profile, 184, 185
Atomic absorption spectroscopy, 402
Atomic Energy Commission (AEC), *see* Nuclear Regulatory Commission; Energy Research and Development Administration
Automobile emissions, 241
  control of, 245–247
  effects of, 243–245
  standards, 243
Automobile engines, *see* Engines; Electric powered automobiles
Automobile industry, energy analysis, 342
Automobiles, fuel economy of, 121
Autotrophic organisms, 57, 376
Azotobacteraceae, 290, 378

B

Bacteria, 153, 290, 310, 376, 378, 387; *see also* Aerobic bacteria; Anaerobic bacteria
Bag houses, *see* Filters
Batteries, rechargeable, 164

491

Heat Content of Fuels*

Fuel	Gigajoules	Btu	Kilocalories
1 tonne bituminous coal	31	29×10^6	7.4×10^6
1 tonne lignite	16	15×10^6	3.8×10^6
1 barrel crude petroleum	6.1	5.8×10^6	1.46×10^6
1 barrel gasoline	5.5	5.2×10^6	1.31×10^6
1 MCF natural gas	1.1	1.0×10^6	0.26×10^6
(1000 ft³ at 1 atm, 288.7 K)			
1 gram uranium-235	78	74×10^6	17×10^6
1 gram natural uranium	0.56	0.53×10^6	0.12×10^6
1 cord wood	22	21×10^6	5.2×10^6

* In 1971 the United States consumed 66.5×10^{18} J of energy resources. Using the above figures this would be equivalent to a *daily* consumption of

5.9×10^6 tonnes bituminous coal

or

29.9×10^6 barrels crude petroleum

or

166×10^6 MCF natural gas

or

330×10^6 gram natural uranium

or

8.3×10^6 cords wood

Figure 7.1 shows the actual distribution of 1971 energy consumption among the resources listed above.